MVT: A Most Valuable Theorem

Craig Smoryński

MVT: A Most Valuable Theorem

Springer

Craig Smoryński
Westmont, IL
USA

ISBN 978-3-319-52955-4 ISBN 978-3-319-52956-1 (eBook)
DOI 10.1007/978-3-319-52956-1

Library of Congress Control Number: 2017931055

Printed on acid-free paper

This Springer imprint is published by Springer Nature
The registered company is Springer International Publishing AG
The registered company address is: Gewerbestrasse 11, 6330 Cham, Switzerland

Further Reading: Books by the Same Author

University of Illinois at Chicago Circle:

Notes on Hilbert's Tenth Problem; An Introduction to Unsolvability 1972

Springer-Verlag:

Self-Reference and Modal Logic 1985

Logical Number Theory I; An Introduction 1991

History of Mathematics; A Supplement 2008

College Publications:

Adventures in Formalism 2012

Chapters in Mathematics 2012

Chapters in Probability 2012

A Treatise on the Binomial Theorem 2012

Acknowledgements

For various reasons I have relied more heavily on the help and advice of others in writing this book than I have generally done in the past. Those I am indebted to, in alphabetical order, are Kirsti Andersen, Mark van Atten, John Baldwin, Ádám Besenyei, Reinhard Bölling, Michel Bourdeau, Robert Burckel, Pete L. Clark, Dirk van Dalen, P.P. Divakaran, Enrico Giusti, Prakash Gorroochurn, Eckart Menzler, David Mumford, K. Ramasubramanian, Fred Thulin, and Peter Ullrich. Of these, special mention should be made of Ádám Besenyei and Robert Burckel. Besenyei has himself delved into the history of the Mean Value Theorem and has shared the results of his research with me, sent me materials, and made helpful comments. Burckel brought to my attention the fascinating results on the Mean Value Theorem in Complex Analysis, sent me materials, and made many suggestion improving my exposition on this and other subjects as well.

Additionally I wish to thank John J. O'Connor and Edmond F. Robertson, Jason Ross, and Lou Talman for permission to quote from their works published online, and the following corporations for permission to use copyrighted material from the indicated sources:

American Mathematical Society:
Notices of the AMS
A Century of Mathematics III
Cengage:
Dictionary of Scientific Biography
Dover Publications Inc.:
La Geometrie
Mathematical Association of America:
American Mathematical Monthly
College Mathematics Journal
Mathematics Magazine
Selected Papers on Calculus
Oxford University Press:
Mathematics Emerging: A Sourcebook 1540–1900

The Oxford Handbook of the History of Mathematics

Princeton University Press:

Proclus, A Commentary on the First Book of Euclid's Elements

Crest of the Peacock, 3rd edition

Springer Science+Business LLC:

Archive for History of Exact Sciences

Cauchy's Cours d'analyse: An Annotated Translation

Conflicts between Generalization, Rigor, and Intuition: Number Concepts Underlying the Development of Analysis in 17–19th Century France and Germany

Denkweise grosser Mathematiker

Giuseppe Peano between Mathematics and Logic

Mathematical Conversations

Peano, Making a Career

The Historical Development of the Calculus

University of Wisconsin Press:

The Science of Mechanics in the Middle Ages

Contents

Chapter 1
Introduction

An odd candidate for controversy in mathematics education is the Mean Value Theorem. It is easily, if imprecisely, described to the man in the street: Draw a smooth curve and choose two points A and B on the curve. Somewhere on the arc of the curve between these two points is another point C at which the line tangent to the curve will be parallel to the line connecting A and B. Figures 1.1 and 1.2 illustrate this nicely.

With such pictures in mind, the method of finding such a point C and the reason for the truth of the Theorem suggest themselves. C is the point (or any of the points) on the curve farthest from the line AB as possible and is found by sliding the line AB upward (in Fig. 1.1) or leftward (in Fig. 1.2) or in whatever appropriate direction — without rotating AB at all — until it exits the curve. The last point (or any of the last points if there are more than one) at which AB had contact with the curve is this farthest point C. Why is the tangent at C parallel to AB? Referring to Fig. 1.3, we see that as we move along the curve from A to C, the tangents are always steeper than AB but they are becoming less steep. And as we move from C to B, the tangents continue to grow less steep and none are as steep as AB. The level of steepness, or *slope*, of the tangent at C must lie between these two trends and thus bears the same level of steepness as AB itself, i.e., the tangent is parallel to AB.

The argument given is not mathematically rigorous, but it can be made so first by offering precise definitions of the terms involved ("curve", "tangent") and isolating the conditions under which the Theorem is to hold, and then appealing to more advanced theory to justify the individual steps. This will be done in the course of this book. Here let me just indicate that some conditions must be met. The hyperbola of Fig. 1.4 demonstrates an obvious condition: the Mean Value Theorem presupposes A and B to lie on the same branch of the curve in question. If they don't, there are no points lying on the arc of the curve connecting them simply because there is no such arc. One could try to find a point maximally distant from AB elsewhere on the curve, but again no such point exists. Indeed, on the hyperbola, if A and B lie on different branches, there is no point C at which the tangent is parallel to AB. There

C. Smoryński, *MVT: A Most Valuable Theorem*,
DOI 10.1007/978-3-319-52956-1_1

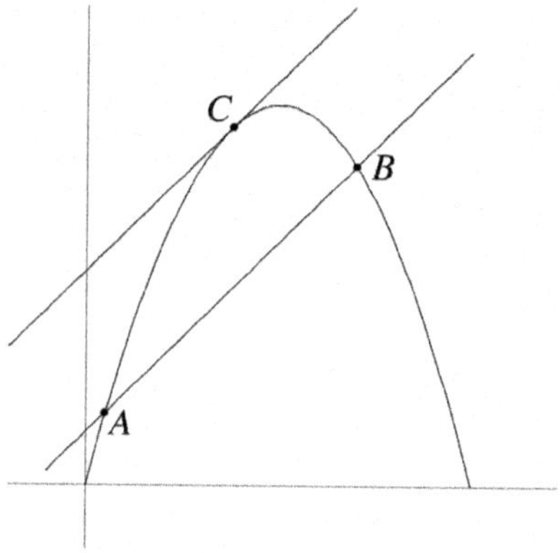

Fig. 1.1 Mean Value Theorem; Example 1

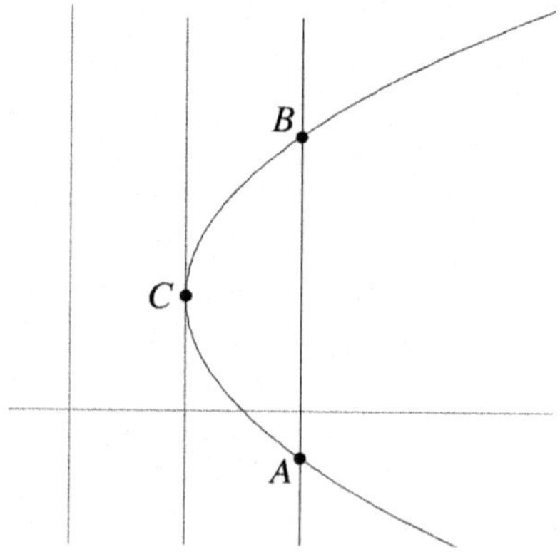

Fig. 1.2 Mean Value Theorem; Example 2

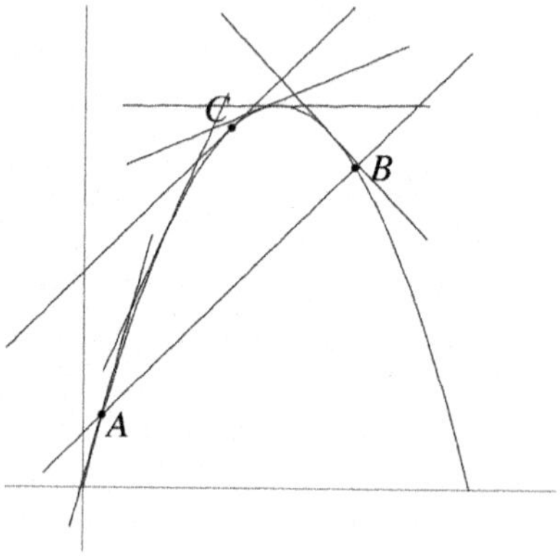

Fig. 1.3 Changing slopes

is a simple reason for this: the slope of AB lies strictly between the slopes of the two asymptotes, while the slopes of the tangent lines are all strictly outside this range.

Also, the Theorem can fail if the curve fails to have tangents everywhere, as in Fig. 1.5. Here, depending on how one defines "tangent", the curve possesses infinitely

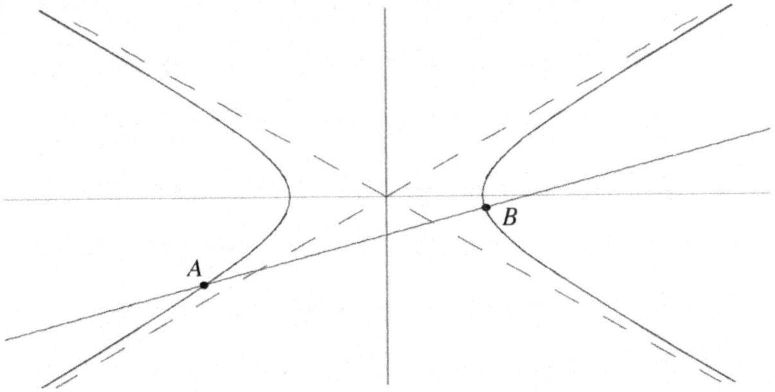

Fig. 1.4 Hyperbola

Fig. 1.5 Corner

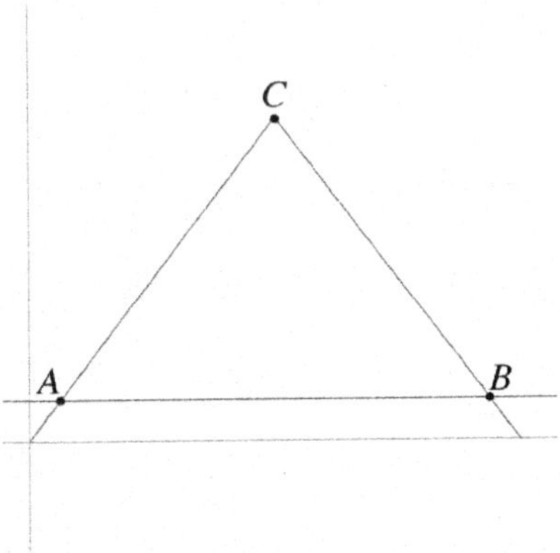

many, two, or no tangents at the point C farthest from the line AB. The grammar school definition of a tangent to a curve being a line meeting the curve at a single point without crossing the curve allows for infinitely many tangents through C, one of which is indeed parallel to AB. The kinematic definition of the tangent at C as being the line a particle travelling along the curve would follow if at C, instead of following the curve, the particle would continue in the direction it was going when it arrived at C, yields two tangents — the lines extending past C the segments AC and BC, respectively. And the definition of tangent used in the Calculus as the line passing through C with slope given by the derivative of the curve at that point yields no tangent because the curve is not differentiable at the point C.

The curve of Fig. 1.6 has a cusp rather than a corner. The Mean Value Theorem fails for the same reason. C has infinitely many tangents according to the grammar school definition, one according to the kinematic definition (namely, the vertical line), and none according to the definition used in the Calculus.

These counterexamples dictate the conditions a curve must satisfy for the Mean Value Theorem to apply. It must be "connected" or "continuous", i.e., it must consist

Fig. 1.6 Cusp

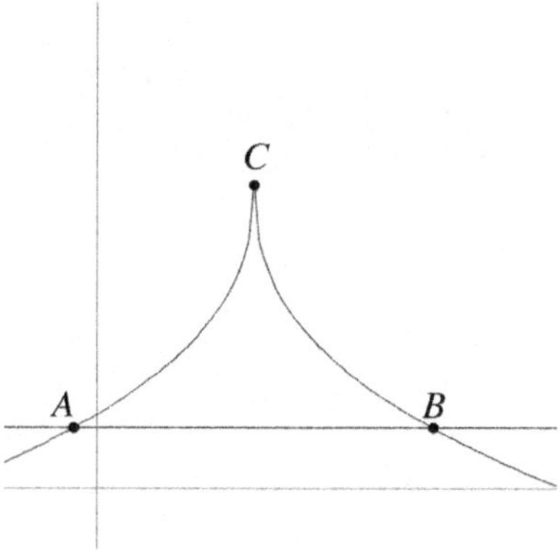

of a single branch. And it must be "smooth", i.e., it must have no sudden changes of direction as given by the corner and cusp of the last two figures.

The sequel divides into three chapters. The first (Chap. 2) provides precise formalisations of the concepts of curve, tangent, continuity, and smoothness, and finishes with a rigorous proof of the Mean Value Theorem as geometrically described above. It is a leisurely presentation that seems to take its time getting to the point. However, much of it is referred to later in the book, occasionally explicitly and occasionally implicitly as absorbed background for the discussion of issues that are later raised. I beg the reader's indulgence in what may at first appear excessive background information.

The Chap. 3 is devoted to the Mean Value Theorem itself, in its more familiar analytic form. This includes the Theorem and related results, some applications, and the history of attempts to prove it. In this I stick to the Theorem as it is usually presented in a single-variable Calculus course and do not attempt to discuss generalisations except where such directly impinge on the discussion of the standard version of the Theorem.

Finally, the last chapter discusses the educational question: Is the Mean Value Theorem itself or one of several proposed replacements more appropriate for the introductory course in the Calculus? I present and criticise some of the points that have been raised in the debate, but offer no final solution. My purpose is not to dictate the contents of the Calculus course, but to provide enough information for a more informed discussion than has thus far taken place.

Ostensibly the issues of the present book will only interest those members of departmental curriculum committees deciding on what must be covered in the basic Calculus course, or, more broadly, those who teach Calculus. I think, however, the issues might be of broader interest, say to the student who is taking such a course. And I suppose such a book could have been written for him or her, particularly as the topic comes from such a course. The temptations, however, to pepper the exposition with examples assuming some computational facility and to approach the subject from a higher point of view have played havoc with this good intention. The student who attempts to read the present work while first learning the Calculus might find some of the material a bit heavy going: the present book discusses the *theory* of the Calculus, not the Calculus *per se*, and although the main prerequisite is the sort of mathematical maturity that can be acquired by a reasonably, but not overly, rigorous course in the Calculus, it does presuppose some knowledge of the Calculus as well.

In this book I include many citations from the literature, so many that the book might be viewed as an absurdly heavily annotated source book. I have included excerpts from original mathematics, as well as comments from various historians and other expositors. In doing this I have tried to be as true to the originals as possible, short of including facsimile reproductions of the pages. Where I have taken liberties is mostly footnoted. Exceptions are that: lacking the old "long s", I have used only the standard "s" in reproducing older material; I omit references to the literature of the form "[ABC]" or "[1]" as these are not relevant; and, in place of indentation, paragraphs in quoted material are separated by extra space between them. I am fairly confident in the overall correctness of those passages I've translated. Occasionally I have used an English idiom in place of the literal translation of idioms, and I have been a bit loose when it comes to the tenses of verbs.

The decision to include the mathematical sources needs no justification. Citing historians and other expositors I justify on several counts. Aside from acknowledging the sources of some of my information, I do this in part to introduce whatever students may be reading this book to some of the experts. In part I do this when I have not checked the primary sources myself because I don't read the language in question (Sanskrit, Latin), the original was unavailable to me, or the point was too tangential to check myself. Sometimes I find the historians or expositors to express matters so well it would be impossible to improve upon them and I feel I must quote them; and sometimes I cannot resist quoting someone so that I can spout my disagreement.

There is one by-product of quoting original sources; the notation shifts from author to author. When discussing a particular paper, after quoting a passage, I may stick to the author's notation if it is clear enough, or I may translate the material into more (currently) standard notation. My hope is that this will not confuse the reader.

Chapter 2
Curves and Tangents

2.1 Curves

In her sourcebook of mathematics,[1] Jacqueline Stedall represents Euclid via facsimile reproductions of selections from Isaac Barrow's 1660 edition of the *Elements* written for undergraduates. The definition given of a curve reads

> A Line is a longitude without latitude.[2]

The most familiar English translation, available in multiple editions, is due to Thomas Little Heath.[3] In this translation the definition reads

> *A line is breadthless length.*[4]

The word "line" in these quotes is used to mean a curve, our line being called a "straight line". The annotated edition of Heath's translation follows this statement with several pages of discussion,[5] beginning with the attribution of this definition to Plato's school. He cites Aristotle's quibble about the negative form of the definition and also offers a couple of alternative definitions from Aristotle (384–322 B.C.) already cited in the commentary of Proclus, and then goes on to discuss classifications of curves and offering a number of examples of such. I quote Proclus (412–485):

---

[1] Jacqueline Stedall (ed.), *Mathematics Emerging: A Sourcebook 1540–1900*, Oxford University Press, Oxford, 2008.

[2] *Ibid.*, p. 10.

[3] Thomas Little Heath, *The Thirteen Books of Euclid's Elements*, 3 volumes, Cambridge University Press, Cambridge, 1908. This translation has been reprinted a number of times. The edition put out by Dover Publications includes all the annotations. Two other editions currently in print but lacking the annotations are that in the series *Great Books of the Western World* and an attractively typeset single volume published by Green Lion Press.

[4] *Ibid.*, vol. 1, p. 158.

[5] *Ibid.*, pp. 158–165.

C. Smoryński, *MVT: A Most Valuable Theorem*,
DOI 10.1007/978-3-319-52956-1_2

> II. *A line is length without breadth.*
> The line is second in order[6] as the first and simplest extension, what our geometer calls "length," adding "without breadth" because the line also has the relation of a principle to the surface. He taught us what the point is through negations only, since it is the principle of all magnitudes; but the line he explains partly by affirmation and partly by negation. The line is length, and in this respect it goes beyond the undividedness of the point; yet it is without breadth, since it is devoid of the other dimensions. For everything that is without breadth is also without depth, but the converse is not true. Thus in denying breadth of it he has also taken away depth, and this is why he does not add "without depth," since this is implied in the absence of breadth.
>
> The line has also been defined in other ways. Some define it as the "flowing of a point," others as "magnitude extended in one direction." The latter definition indicates perfectly the nature of the line, but that which calls it the flowing of a point appears to explain it in terms of its generative cause and sets before us not the line in general, but the material line. This line owes its being to the point, which, though without parts, is the cause of the existence of all divisible things; and the "flowing" indicates the forthgoing of the point and its generative power that extends to every dimension without diminution and, remaining itself the same, provides existence to all divisible things.[7]

As the quotation from Proclus makes clear, the Euclidean definition takes its one-dimensionality as the defining property of a curve, thus distinguishing it from the zero-dimensionality of a point, the two dimensionality of a surface, and the three-dimensionality of a solid. There is, of course, a certain intuitiveness to this definition, but as a basis on which to establish theorems it leaves a lot to be desired. It is not something for the Calculus course, but a matter of the more advanced field of Topology. The definition of the "flowing of a point", i.e., a kinematic approach, which can be found in Aristotle, is more accessible and is abstractly what the standard definitions in the Calculus are based on. The "magnitude extended in one direction" also harks back to Aristotle. Heath explains

> A line is, according to Aristotle, a magnitude "*divisible* in one way only", in contrast to a magnitude divisible in *two* ways, or a surface, and a magnitude divisible "in all or in three ways", or a body; or it is a magnitude "*continuous* one way (or in one direction)," as compared with magnitudes continuous in *two* ways or *three* ways, which curiously enough he describes as "breadth" and "depth" respectively, though he immediately adds that "length" means a line, "breadth" a surface, and "depth" a body.[8]

Without some more abstract topological concepts, one would be hard put to make these definitions mathematically precise. About the only thing that seems clear is that a line, i.e., curve, is to have only one dimension, whatever that might mean.

---

[6]Euclid first defined the point in Definition I as that which has no part.

[7]Proclus (Glenn R. Morrow, ed.), *A Commentary on the First Book of Euclid's Elements*, Princeton University Press, Princeton, 1970, pp. 79–80.

[8]Heath, *Elements,op. cit.*, vol. 1, pp. 158–159. I have omitted his parenthetical insertions of Greek terms and page references in Aristotle.

Our intuition, however, is often misleading. In an oft-cited paper published in 1933,[9] Hans Hahn (1879–1934) discusses this point with reference to the notion of a curve. Our two intuitions of a curve as the "flowing of a point" and as a one-dimensional entity, though properties common to most things we consider to be curves, are not equivalent and, when given rigorous formal definitions, both can be shown not to agree completely with our intuition of what constitutes a curve. The best that one can usually hope for in mathematics is to replace a vague intuitive notion by a precise, formally defined one that agrees with intuition in all familiar cases and all future cases not too dissimilar to these. It can happen, after giving such a definition, that at the fringes of our experience there are objects our intuition might accept but our formal definition rejects or *vice versa*. When this happens we may amend our formal definition to accommodate or to exclude the new objects, or we may define a new class of objects. In the case of curves, Hahn offers two definitions based on the "flowing of a point" and one-dimensionality, respectively, and shows by example that each concept accepts some questionable curves and rejects some things we might intuitively accept as curves. The question is not one of giving a definition that precisely captures the intuitive notion — after all, different people have different intuitions — but to offer, as Hahn puts it, a "serviceable definition":

> Since the time-honoured definition of a curve fails to cover the fundamental concept, what other more serviceable definition can be substituted for it?[10]

The word "serviceable" is relative, not absolute. A definition of "curve" is serviceable in a given context if it applies to those curves we are likely to come across in that context and does not apply to the non-curves we are likely to meet. Our context is the first-year Calculus course, not an advanced Topology course, and we don't need Hahn's abstract definitions that we would be hard pressed to apply usefully in the Calculus course.

So how are we to define a curve here?

The first step toward isolating a formal notion of curve, one precise enough on which to base proofs, is to catalogue some curves and look for a commonality in their modes of definition. Traditional Geometry doesn't have a lot of curves. If you look into Euclid's *Elements*, you will find straight lines and circles. Other curves were known. However, Boyer, in his *History of Analytic Geometry*, says that the Greeks "did not discover more than half a dozen new curves in all of their enormous mathematical activity, and these were not systematically classified".[11] The first of

---

[9]Hans Hahn, "Die Krise der Anschauung", in: *Krise und Neuaufbau in den exakten Wissenschaften*, F. Deuticke, Leipzig and Vienna, 1933. An English translation, "The crisis of intuition", appears in: Hans Hahn (Brian McGuinness, ed.), *Empiricism, Logic, and Mathematics: Philosophical Papers*, D. Reidel Publishing Company, Dordrecht, 1980.

[10]Hahn, "The crisis of intuition", *op. cit.*, p. 88.

[11]Carl B. Boyer, *History of Analytic Geometry*, The Scholars Bookshelf, Princeton Junction (NJ), 1988, p. 20. This work was originally published in 1956 as numbers 6 and 7 of *The Scripta Mathematica Studies*. Incidentally, the numerical estimate given here is figurative, not literal: Boyer cites at least half a dozen curves known to the Greeks and on page 35 announces, "yet scarcely a dozen curves were familiar to the ancients".

these was the *quadratrix* of Hippias of Elis (*fl.* 400 B.C.). The most important were the conic sections (ellipse, parabola, and hyperbola) discovered by Menæchmus in the mid-fourth century B.C. Following these were a scattering of curves including the *conchoid* of Nicomedes (*fl.* 250 B.C.), the spiral of Archimedes (*c.* 287–212 B.C.), and the *cissoid* of Diocles (*c.* 150 B.C.). The true flowering of curves would not come about until the invention of Analytic Geometry at the hands of René Descartes (1596–1650) and Pierre de Fermat (1601–1665) in the 17th century, but this development was a direct descendent of the work of the Greeks.

The quadratrix is historically important for a variety of reasons. It was the first curve after the line and circle to make an appearance in mathematics. It was the first curve not constructible even weakly by ruler and compass. Indeed, it cannot be drawn by any mechanical device but must be graphed by tabulating and plotting points. And, a matter of only minor importance here, it could be used to solve two of the three problems of classical geometry: using the quadratrix, the angle is easily trisected and the circle is squared.

The definition of the quadratrix is kinematic. One considers two line segments AB and BC of equal length perpendicular to each other at B. Segment BC moves in the direction of A without rotation, while segment AB rotates around A. These movements are of uniform speeds and are such that at the end of a unit of time the two segments coincide. The quadratrix is the path traced out by the points of intersection of the two line segments, as illustrated in Fig. 2.1.

As mentioned, the only way to graph the quadratrix is to plot a lot of points. There is no simple mechanical linkage that will perform the drawing for us. To this end, consider Fig. 2.1 where we assume B and C'' to lie one unit away from A and we assume BC to be descending at 1 unit per second, AB to be rotating clockwise at a quarter revolution per second. After t seconds (t in the interval $[0, 1]$) BC will coincide with $B'C'$ and AB with AB''. The coordinates $\langle x, y\rangle$ of the point D of intersection are fairly easy to determine. (See Fig. 2.2.) That $y = 1 - t$ is obvious. And $x/(1 - t)$ is the tangent of $\angle BAD$, which angle is $t\frac{\pi}{2}$. Thus

$$\frac{x}{1-t} = \tan\frac{t\pi}{2},$$

Fig. 2.1 Quadratrix

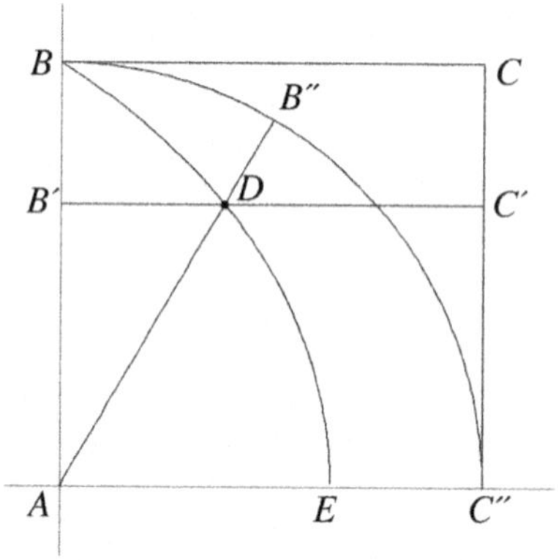

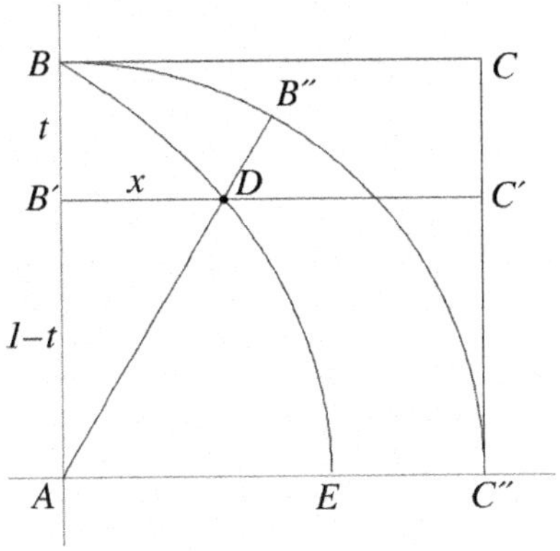

Fig. 2.2 Quadratrix, a Second View

i.e.,

$$x = (1 - t)\tan\frac{t\pi}{2}, \quad y = 1 - t, \quad \text{for } 0 \le t < 1. \tag{2.1}$$

Formula (2.1) provides a *parametric definition* of the curve, defining x and y as functions $x(t)$ and $y(t)$ of the parameter t. Using this we can readily solve for x as a function of y:

$$\begin{aligned} x &= (1-t)\tan\frac{t\pi}{2} = y\tan\frac{(1-y)\pi}{2} = y\tan\left(\frac{\pi}{2} - \frac{y\pi}{2}\right) \\ &= y\cot\frac{y\pi}{2}, \text{ as } \tan\left(\frac{\pi}{2} - \theta\right) = \cot\theta. \end{aligned} \tag{2.2}$$

Here, $0 < y \le 1$.

A glance at the graph informs us that y is also a function of x for some appropriate values of x. However, it is unlikely one will be able to find a closed form for expressing y in terms of x. Moreover, specifying the domain of such a function $y(x)$ is not trivial. The variable ranges over the interval $[0, a)$, where a is the x-coordinate of E. a should be $x(1)$ if we express x as a function of t *à la* (2.1) or $x(0)$ if we express x as a function of y. Unfortunately, for $t = 1$, (2.1) expresses x in the form $0 \cdot \infty$ and (2.2) expresses x in the form $0/0$. Either way it yields no value. Indeed, if we look to the definition of the quadratrix to see where the point E should be, namely at "the" point of intersection of $B'C'$ and AB'' at $t = 1$, we see the problem: the line segments intersect at all points of AC''. We must determine E by some other means, as we shall do in the next chapter when we discuss L'Hôpital's Rule. Before doing this, however, we can see why the quadratrix as plotted in Figs. 2.1 and 2.2 does not quite reach the point E. The coordinate x not being calculable at $t = 1$ where $y = 0$, the curve stops short at the last calculated value of t. Depending on the resolution of one's display and the density of the values of t in one's table, one may or may not notice the gap. It is not visible on my graphing calculator, but is clearly visible with the computer software I used to generate the graphs.

1. Bisect AE at F.
2. Draw the segment BF.
3. Through C'' draw a line parallel to BF.
4. Extend this line and AB until they intersect at a point G.
5. The length of AG is π.
6. The rectangle $AGHC''$ has area π.

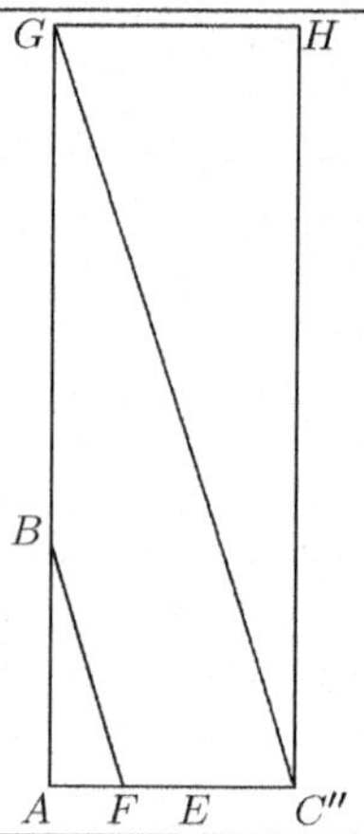

Fig. 2.3 Squaring the circle

Anticipating this calculation, I can report that the coordinates of the point E are $\langle 2/\pi, 0\rangle$, i.e., if AB is taken to be the unit, the length of AE is $2/\pi$. Standard ruler and compass constructions readily yield a segment of length π, whence a rectangle of area π from the segment AE. (See Fig. 2.3.) And Euclid's *Elements* shows how to find a square of area equal to that of any given rectangle. Thus we have used the quadratrix to square the circle of radius 1.

It was this application of the quadratrix to the problem of quadrature that gave the curve its name. Boyer,[12] admittedly not the most up-to-date reference, but a convenient one, tells us this was a later development due to Dinostratus (*fl. c.* 350 B.C.); Hippias himself seems to have invented it to solve the trisection and other multisection problems. This is an easy matter: Given an angle $\angle ABC$ of less than 90°, copy it to one's diagram of the quadratrix as in Fig. 2.4. Let D be the point of intersection of the elevated ray of the angle with the quadratrix. Draw a horizontal line through D and let it intersect the vertical line through the vertex B of the angle at a point E. Let BF equal a third of BE and let G be the point of intersection of the horizontal line through F with the quadratrix. The angle $\angle GBC$ is one third the angle $\angle ABC$, this last because of the uniform motions of the lines generating the quadratrix.

In the larger scheme of things, the applications of the quadratrix to the solution of two of the three classical problems are no more than amusing asides. However, these problems seem to have been the inspiration behind the next family of curves to arrive on the scene — the conic sections. Menæchmus, often cited as the mathematical tutor of Alexander the Great, was also the brother of Dinostratus, a connexion of greater immediate relevance here. One of Plato's contemporaries, Archytas (*fl. c.* 395 B.C.) had solved the problem of duplicating the cube through application of a cone, a cylinder, and a torus. Menæchmus realised that conic sections alone could do the trick.

[12]Boyer, *op. cit.*, p. 11.

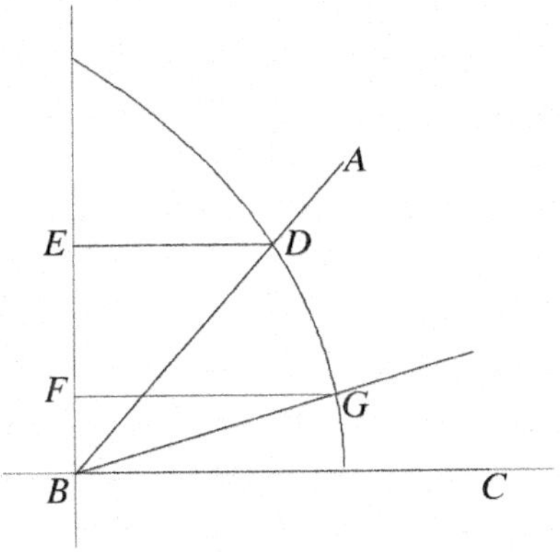

Fig. 2.4 Trisection

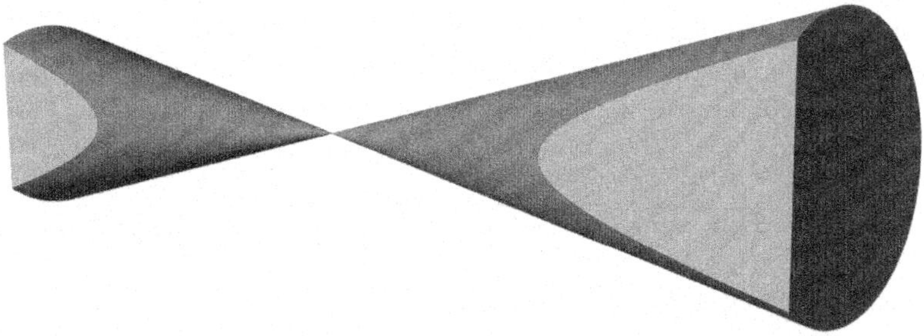

Fig. 2.5 Hyperbola as a conic section

The conic sections were quickly taken up by Greek mathematicians. Menæchmus was followed by Aristæus (*fl. c.* 350–330 B.C.) and then by Euclid, whose four books on the conics are believed to be based on the work of Aristæus. None of these works survives. However, we do have some work by Archimedes, most notably his quadrature of the parabola; the *Conics* of Apollonius of Perga (*c.* 262–*c.* 190 B.C.), the great work on the subject; and the commentaries of Pappus of Alexandria (*fl. c.* 300–350), which give information on some of the lost works of Greek mathematics.

Conic sections are, by definition, the curves that result from the intersection of the surface of a right circular cone with a plane. Working directly with this definition is not the easiest thing in the world, so textbooks often merely pay lip service to it by calling the curves conic sections, perhaps even including some illustrations like Fig. 2.5 before giving an alternative definition. The Greeks had several characterisations of these curves as loci of various sorts that can be used to introduce them in an Analytic Geometry course. The definition in terms of focus, directrix, and eccentricity, first found in the later Greek work of Pappus, may not be the best motivated definition, but it does yield an almost unified approach to the three types of curves.

2.1.1 Definition Let a line L, a point F not lying on L, and a positive real number e be given. Consider the *locus* (i.e., set) of all points P such that the distance from P to F equals e times the distance from P to L:

$$\gamma = \{P | \text{dist}(P, F) = e \cdot \text{dist}(P, L)\}.$$

γ is called an *ellipse* if $e < 1$, a *parabola* if $e = 1$, and a *hyperbola* if $e > 1$. F, L, and e are called the *focus*, *directrix*, and *eccentricity*, respectively, of γ.

The advent of symbolic algebra in 16th century Europe and the growing shift in emphasis from geometry to algebra made inevitable the invention of Analytic Geometry, as evidenced by the fact that the invention occurred twice at the hands of Fermat and Descartes around 1630. Both men studied conic sections and their analytic expression.

Equational representations for the conic sections are readily derivable. Suppose F, L, and e are given. We can take L to be the x-axis, the normal connecting F to L to be the y-axis, and the distance from F to L to be the unit. With respect to these axes, F has coordinates $\langle 0, 1\rangle$. A point P with coordinates $\langle x, y\rangle$ lies on γ just in case

$$\text{dist}(P, F) = \sqrt{(x-0)^2 + (y-1)^2} = e \cdot |y| = e \cdot \text{dist}(P, L).$$

Squaring the terms of the central equation yields

$$x^2 + (y-1)^2 = e^2 y^2, \tag{2.3}$$

i.e.,

$$x^2 + (1-e^2)y^2 - 2y + 1 = 0. \tag{2.4}$$

For $e = 1$, (2.4) becomes

$$y = \frac{1}{2}x^2 + \frac{1}{2},$$

which we can quickly enough graph to obtain a recognisably parabolic shape.

For $e \neq 1$, we can divide by $1 - e^2$ to get

$$\frac{x^2}{1-e^2} + y^2 - \frac{2}{1-e^2}y = -\frac{1}{1-e^2}$$

$$\frac{x^2}{1-e^2} + \left(y - \frac{1}{1-e^2}\right)^2 = -\frac{1}{1-e^2} + \frac{1}{(1-e^2)^2} = \frac{e^2}{(1-e^2)^2}. \tag{2.5}$$

If $e < 1$, one has $0 < 1 - e^2 < 1$, and Eq. (2.5) assumes the form

$$ax^2 + b(y-\beta)^2 = c$$

with a, b, c all positive. And, if $e > 1$, $1 - e^2 < 0$ and (2.5) assumes the form

$$ax^2 + b(y - \beta)^2 = c$$

with a negative and b, c positive. In either case one can solve for y to get

$$y = \beta \pm \sqrt{\frac{c - ax^2}{b}}, \tag{2.6}$$

graph the two resulting functions, and recognise the familiar elliptical shape when $e < 1$ and hyperbolic one when $e > 1$.

2.1.2 Exercise Carry out the above derivation for the following values of e:

i. $e = 1/2$
ii. $e = 1$
iii. $e = 2$.

Graph the resulting curves to verify they are of the appropriate forms.

Definition 2.1.1 does not capture all conic sections. Missing are the degenerate conics — points, lines, and circles. There are two exceptions for $e = 0$ and $e = \infty$: For fixed F and L, if one graphs (2.6) for successively smaller values of e, one gets ellipses that become more and more circular. But they also become progressively smaller and at $e = 0$, the graph consists solely of the focus F. Indeed, plugging 0 in for e the Eq. (2.3) results in

$$x^2 + (y - 1)^2 = 0,$$

the equation of the circle of radius 0 centred at $\langle 0, 1\rangle$. At the other extreme, larger and larger values of e give graphs of hyperbolas hugging more and more closely to their asymptotes, which themselves are closing scissors-like towards the x-axis. And, indeed, plugging[13] ∞ in for e in (2.5) results in the equation,

[13] One can go a long way calculating with ∞ taking ∞ as an ideal element and applying rules like

$$a \pm \infty = \pm\infty, \quad a \cdot \infty = \infty, \quad a/\infty = 0$$

for real a. Terms like $0 \cdot \infty$, $\infty - \infty$, and ∞/∞ are indeterminate and simple algebra doesn't apply. In fact, I have cheated in writing $\infty^2/\infty^4 = 1/\infty^2 = 0$. One should first manipulate (2.5) to express

$$\frac{e^2}{(1 - e^2)^2} = \frac{1}{(1 - e^2)^2/e^2} = \frac{1}{(1/e - e)^2}$$

and only then plugging ∞ in for e:

$$\frac{1}{(1/\infty - \infty)^2} = \frac{1}{(0 - \infty)^2} = \frac{1}{\infty^2} = 0.$$

$$
\begin{aligned}
\frac{1}{1-\infty^2}x^2 + \left(y - \frac{1}{1-\infty^2}\right)^2 &= \frac{\infty^2}{(1-\infty^2)^2} \\
\frac{1}{-\infty^2}x^2 + \left(y - \frac{1}{-\infty^2}\right)^2 &= \frac{\infty^2}{\infty^4} \\
0x^2 + (y-0)^2 &= \frac{1}{\infty^2} \\
0x^2 + y^2 &= 0,
\end{aligned}
$$

i.e., $y = 0$. Thus $e = \infty$ yields the directrix itself as the resulting conic section.

Relative to a pre-existing pair of coordinate axes, the focus will not necessarily have as simple a pair of coordinates and the equation of the directrix will be more complicated than $y = 0$. The computation becomes more involved, but it follows roughly the same lines.

2.1.3 Example Let F be $\langle 1, 2\rangle$, L be given by $x + 2y = 3$, and $e = 2$. The first step is to determine the distance from a point $\langle \alpha, \beta\rangle$ to L. To this end note that a line perpendicular to L has an equation $2x - y = c$ for some constant c. For $\langle \alpha, \beta\rangle$ to lie on the perpendicular in question one must have $c = 2\alpha - \beta$. The point on L closest to $\langle \alpha, \beta\rangle$ is the point $\langle x, y\rangle$ of intersection of the lines:

$$
\begin{aligned}
x + 2y &= 3 \\
2x - y &= 2\alpha - \beta.
\end{aligned}
$$

Doubling the first of these and subtracting the second from the result yields

$$
\begin{aligned}
5y &= 6 - 2\alpha + \beta \\
y &= \frac{6 - 2\alpha + \beta}{5} \\
x &= 3 - 2\frac{6 - 2\alpha + \beta}{5} = \frac{15 - 12 + 4\alpha - 2\beta}{5} \\
&= \frac{3 + 4\alpha - 2\beta}{5}.
\end{aligned}
$$

The distance from $\langle \alpha, \beta\rangle$ to L is thus the square root of

$$
\begin{aligned}
&\left(\alpha - \frac{3 + 4\alpha - 2\beta}{5}\right)^2 + \left(\beta - \frac{6 - 2\alpha + \beta}{5}\right)^2 \\
&\qquad = \left(\frac{-3 + \alpha + 2\beta}{5}\right)^2 + \left(\frac{-6 + 2\alpha + 4\beta}{5}\right)^2 \\
&\qquad = \frac{\alpha^2 + 4\alpha\beta - 6\alpha + 4\beta^2 - 12\beta + 9}{5}. \qquad (2.7)
\end{aligned}
$$

And the square of the distance from $\langle \alpha, \beta\rangle$ to $\langle 1, 2\rangle$ is

$$(\alpha-1)^2+(\beta-2)^2=\alpha^2-2\alpha+1+\beta^2-4\beta+4$$
$$=\alpha^2+\beta^2-2\alpha-4\beta+5. \tag{2.8}$$

Combining (2.7) and (2.8) we see that the equation of the hyperbola in question is thus

$$x^2+y^2-2x-4y+5=2^2\left(\frac{x^2+4xy+4y^2-6x-12y+9}{5}\right),$$

i.e.,

$$5x^2+5y^2-10x-20y+25=4x^2+16xy+16y^2-24x-48y+36,$$

i.e.,

$$x^2-16xy-11y^2+14x+28y-11=0.$$

In general every conic section will have an equation of the form

$$ax^2+bxy+cy^2+dx+ey+f=0, \tag{2.9}$$

and, conversely, every Eq. (2.9) will define a (possibly degenerate) conic section. In my student days a goodly portion of an Analytic Geometry course was devoted to graphing conic sections and determining the type and basic parameters of the curve defined by (2.9) from the coefficients. The first step was to transform the equation into one with no mixed term,

$$Au^2+Cv^2+Du+Ev+F=0, \tag{2.10}$$

by performing a substitution,

$$x=u\cos\theta-v\sin\theta$$
$$y=u\sin\theta+v\cos\theta,$$

where $\theta=45°$ if $a=c$ and $\tan 2\theta=\dfrac{2b}{a-c}$ otherwise. This represented a simple rotation of the xy-axes into a new pair of uv-axes. The type was then easily determined: if A and C had the same sign one had an ellipse; opposite signs meant the curve was a hyperbola; one of the two coefficients being 0 indicated a parabola; and both being 0 made for a straight line. The exact details are easily forgotten and books of mathematical tables and formulæ would include a table outlining the classification.

But we need not stop here.

If $A=C=0$, (2.10) is the equation of a straight line and is not very interesting. If $A\neq 0$, but $C=0$, we have the parabola

$$Au^2+Du+Ev+F=0,$$

which, if not degenerate (i.e., a line or pair of lines when $E = 0$), can be solved for v in terms of u,

$$v = \frac{-Au^2 - Du - F}{E},$$

thus yielding the parametrisation,

$$\begin{aligned} u(t) &= t \\ v(t) &= \frac{-At^2 - Dt - F}{E}, \quad t \in (-\infty, \infty). \end{aligned}$$

And this yields the following parametric equations for the original curve,

$$\begin{aligned} x(t) &= u(t)\cos\theta - v(t)\sin\theta \\ y(t) &= u(t)\sin\theta + v(t)\cos\theta, \end{aligned}$$

for θ as before and $t \in (-\infty, \infty)$.

The case $A = 0$ and $C \neq 0$ is treated similarly.

In the elliptic and hyperbolic cases, when $A \neq 0$ and $C \neq 0$, one first makes the substitution,

$$\begin{aligned} u &= U + \frac{D}{2A} \\ v &= V + \frac{E}{2C} \end{aligned}$$

to complete the squares and transform (2.10) into

$$AU^2 + CV^2 = \frac{D^2}{4A^2} + \frac{E^2}{4C^2} - F,$$

i.e.,

$$AU^2 + CV^2 = G,$$

for some G. Taking A positive, another substitution,

$$\begin{aligned} U &= \sqrt{\left|\frac{G}{A}\right|}X \\ V &= \sqrt{\left|\frac{G}{C}\right|}Y, \end{aligned}$$

results in an equation in one of the forms,

$$X^2 + Y^2 = 1, \quad X^2 - Y^2 = 1, \quad X^2 - Y^2 = -1,$$

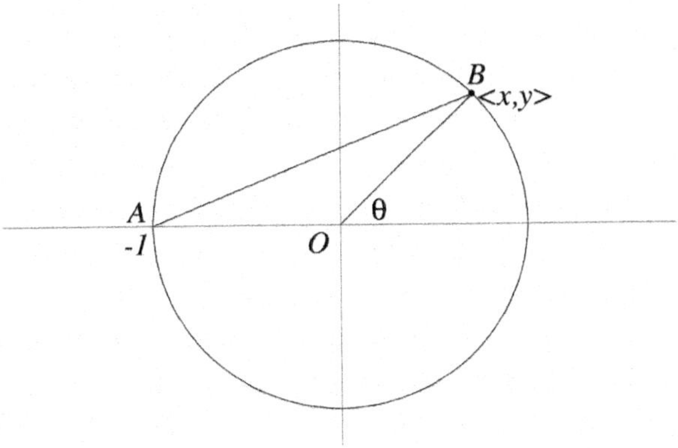

Fig. 2.6 Parametrisation of the circle

which define the unit circle, a standard left-right opening hyperbola, and a standard up-down opening hyperbola, respectively. These are easily parametrised.

The parametrisation of the unit circle most familiar in the literature is trigonometric:

$$\begin{matrix} x(\theta) = \cos\theta \\ y(\theta) = \sin\theta \end{matrix}, \quad \theta \in [0, 2\pi].$$

Some Calculus texts express $\cos\theta$ and $\sin\theta$ in terms of $t = \tan(\theta/2)$:

$$\begin{matrix} x(t) = \dfrac{1-t^2}{1+t^2} \\ y(t) = \dfrac{2t}{1+t^2} \end{matrix}, \quad t \in (-\infty, \infty] \text{ or } [-\infty, \infty).$$

The second of these parametrisations is readily established. Consider Fig. 2.6. The angle $\angle BAO$ is half the angle θ and its tangent t is

$$t = \frac{y}{1+x}.$$

This makes

$$y = t(1+x). \tag{2.11}$$

Combining this with the equation $x^2 + y^2 = 1$ of the unit circle successively yields

$$\begin{gathered} x^2 + t^2(1+x)^2 = 1 \\ (1+t^2)x^2 + 2t^2x + t^2 - 1 = 0. \end{gathered}$$

The solution to the quadratic equation yields

$$\begin{aligned} x &= \frac{-2t^2 \pm \sqrt{4t^4 - 4(t^2+1)(t^2-1)}}{2(1+t^2)} \\ &= \frac{-2t^2 \pm \sqrt{4t^4 - 4(t^4-1)}}{2(1+t^2)} \\ &= \frac{-2t^2 \pm \sqrt{4}}{2(1+t^2)} \\ &= \frac{-t^2 \pm 1}{1+t^2} = -1, \frac{1-t^2}{1+t^2}. \end{aligned}$$

Now, $x = -1$ occurs only when $\theta = \pi$ and $t = \tan\theta/2$ is undefined. For other x we have

$$x = \frac{1-t^2}{1+t^2}.$$

If we now plug this value back into (2.11), we get

$$y = t(1+x) = t\left(\frac{1+t^2}{1+t^2} + \frac{1-t^2}{1+t^2}\right) = t\frac{2}{1+t^2} = \frac{2t}{1+t^2},$$

as promised.

Thus, every point on the unit circle other than $\langle -1, 0\rangle$ is given by

$$x(t) = \frac{1-t^2}{1+t^2}, \quad y(t) = \frac{2t}{1+t^2} \tag{2.12}$$

for some $t \in (-\infty, \infty)$. Writing

$$x(t) = \frac{\frac{1}{t^2} - 1}{\frac{1}{t^2} + 1}, \quad y(t) = \frac{\frac{2}{t}}{\frac{1}{t^2} + 1},$$

and plugging $\pm\infty$ in for t yields $x(\pm\infty) = -1$, $y(\pm\infty) = 0$.

2.1.4 Exercise We have seen that every point $\langle x, y\rangle$ on the unit circle other than $\langle -1, 0\rangle$ is of the form $\langle x(t), y(t)\rangle$ for $x(t), y(t)$ defined by (2.12). Complete the proof that these functions parametrise the circle for $t \in [-\infty, \infty)$ by showing the point $\langle x(t), y(t)\rangle$ to lie on the circle, i.e., show that

$$x(t)^2 + y(t)^2 = 1$$

for all $t \in (-\infty, \infty)$.

The hyperbola $x^2 - y^2 = 1$ is similarly parametrised. One starts with the analogous Fig. 2.7. Letting t be the tangent of the angle $\angle BAO$ which is no longer half the

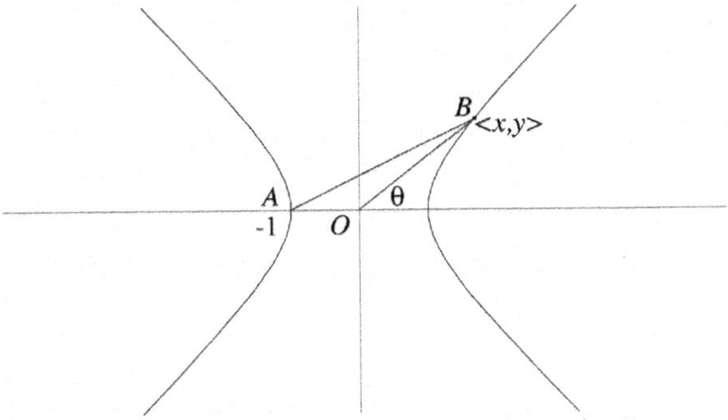

Fig. 2.7 Parametrisation of the hyperbola

angle θ but is the slope of AB, we again have $t = y/(1+x)$ and Eq. (2.11). If one now plugs $t(1+x)$ into the equation $x^2 - y^2 = 1$ for the hyperbola, and performs the algebra, i.e., the simplification, one obtains

$$x(t) = \frac{1+t^2}{1-t^2}, \quad y = \frac{2t}{1-t^2}. \tag{2.13}$$

2.1.5 Exercise Perform the algebraic derivation just described and show, for $t \neq \pm 1$, that $x(t)$, $y(t)$ defined by (2.13) do indeed satisfy

$$x(t)^2 - y(t)^2 = 1.$$

The domain of t is more complicated in the hyperbolic case than in the circular one. Consider first the right branch. The points on this branch are given by allowing t to range over the open interval $(-1, 1)$. Geometrically this is obvious because the slope of AB must lie between the slopes ± 1 of the asymptotes of the hyperbola. Algebraically we note that x is undefined, or infinite, for $t = \pm 1$, negative for $|t| > 1$, and positive for $t \in (-1, 1)$. For points B on the left branch, one must have $|t| > 1$, t negative for B on the upper half of the branch and positive for B on the lower portion. Once again, A corresponds to the choice $t = \pm\infty$.

2.1.6 Exercise One can explore this nicely on a graphing calculator, which, unlike a computer, is slow enough that one can see the curve as it is being drawn. On the TI-83 or TI-84 from Texas Instruments, I suggest setting the MODE to Par, entering

```
X1T= (1 + T²)/(1 − T²)
Y1T= 2T/(1 − T²)
```

in the equation editor, entering

```
Tmin = ⁻5
Tmax = 5
Tstep = .1
```

in the WINDOW menu, and then choosing ZDecimal from the ZOOM menu. First the upper portion of the left branch will be drawn, starting just above the point $\langle -1, 0\rangle$ and continuing in an upward-leftward direction for $t \in [-5, -1)$. Then the right branch, from lower right towards the centre and thence to the upper right, will be traced out as t covers the interval $(-1, 1)$. Finally, for $t \in (1, 5]$, the lower portion of the left branch will be drawn, proceeding from lower left toward the centre. (The drawing of the right branch being rather quick, one might prefer choosing Tstep=.05. One can then speed up the drawing of the left branch by choosing Tmin=$^-$4 and Tmax=4. However, this does leave an even larger gap in the graph near $\langle -1, 0\rangle$.)

I mentioned earlier that Menæchmus applied conic sections to duplicate the cube. This is actually quite simple. Let $\langle \alpha, \beta\rangle$ be the point of intersection of the two parabolas,

$$y = x^2 \tag{2.14}$$
$$2x = y^2. \tag{2.15}$$

Then

$$\begin{aligned} \beta^3 &= \beta \cdot \beta^2 \\ &= \alpha^2 \cdot \beta^2, \text{ by (14)} \\ &= \alpha^2 \cdot 2\alpha, \text{ by (15)} \\ &= 2\alpha^3. \end{aligned}$$

The line connecting the origin with $\langle \alpha, \beta\rangle$ thus has slope $\beta/\alpha = \sqrt[3]{2}$ and choosing for x the length of the edge of any cube, the corresponding y will have a cube of twice the volume. See Fig. 2.8.

So all three problems — squaring the circle, trisecting the angle, and doubling the cube — were solved by the Greeks through the addition of new curves. Unlike the Chinese or Indian mathematicians who excelled in numerical methods, Greek

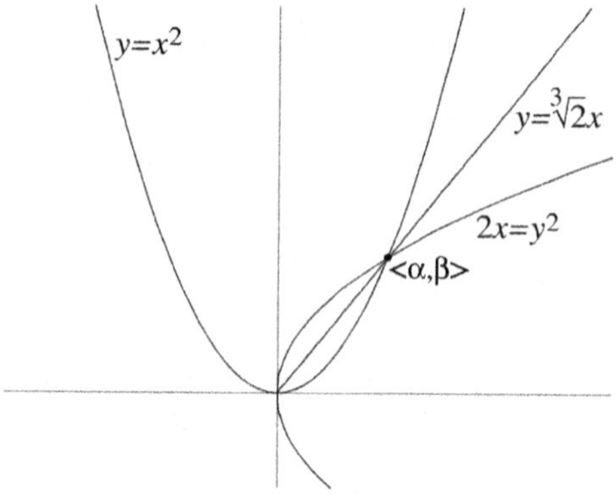

Fig. 2.8 Duplication of the cube

mathematicians were geometrically oriented. Curvilinear solutions were only good if one could graph the curves. None of the new curves could be graphed by ruler and compass alone, a fact that would only first be proven in the 19th century. The conic sections can be graphed mechanically using other tools, most crudely by slicing a cone. The string construction of the ellipse is known to every school child. And mechanical linkages can be constructed for all the conic sections. The same does not hold, however, for the quadratrix and we can say that, from a draughtsman's perspective, only the duplication of the cube had thus far been achieved.

The angle can similarly be trisected by appeal to conic sections, as would be done centuries later in Persia by 'Umar al-Khayyāmī (1048–1122). The reason for this is algebraic: the relation between the cosine of an angle and that of its tripled angle can be expressed algebraically as a cubic equation and al-Khayyāmī could solve cubic equations by intersecting conic sections. The Greeks were unfamiliar with this but they successfully solved the trisection problem by various other means. One solution can be had by means of the conchoid of Nicomedes, the next major curve to come along after the conic sections.

The conchoid, so named because of its resemblance to the curve of a conch shell, is defined kinematically. One starts with a line L and a point O not on the line. One takes another line L' anchored at O and rotates it around O. The locus of all points P at a fixed distance d from L as measured along L' gives the conchoid. It has two branches, one on either side of L. There are three types of conchoids determined by the relationship between d and the distance a of O from L. When $d < a$, the branch on the same side of L as O has a dip near O, but is smooth; when $d = a$, the dip reaches O in a cusp; and when $d > a$, the curve not only dips toward O, but passes through it as a loop. See Fig. 2.9.

Let us analyse the branch of the conchoid lying on the opposite side of L from O. To this end, draw two additional lines through O, one parallel to L to serve as the x-axis and one perpendicular to L to serve as the y-axis. The line L' is then completely determined by the angle θ it makes with the x-axis at the origin O. It intersects L in some point $P = \langle x, y \rangle$ for $0 < \theta < \pi$. (See Fig. 2.10, where I have rotated the

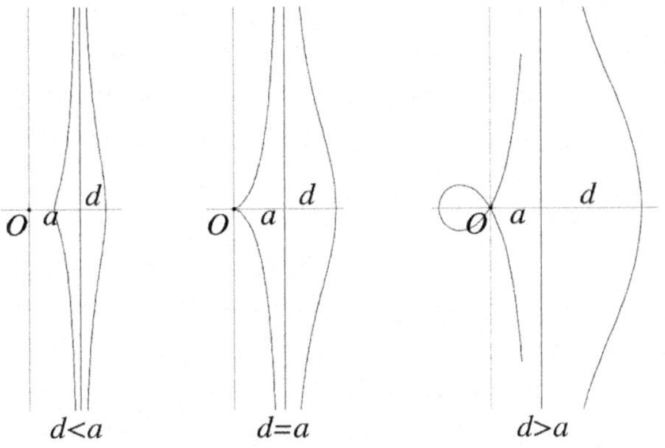

Fig. 2.9 Conchoids

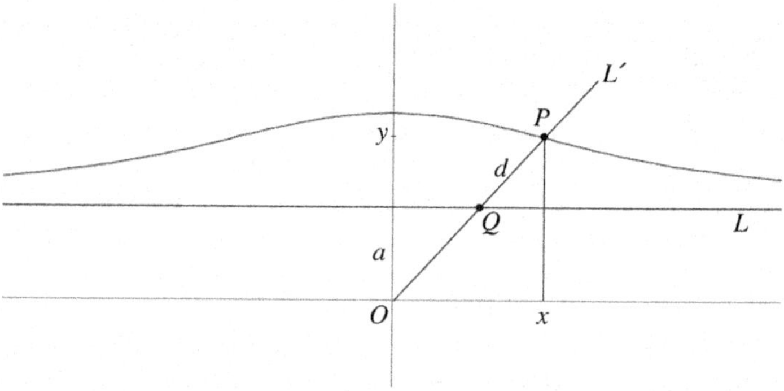

Fig. 2.10 Parametrisation of the conchoid

graph to better fit the allotted space.) From the figure we see that $y = a + d\sin\theta$. But $\tan\theta = y/x$, whence $x = (a + d\sin\theta)/\tan\theta$. Thus we have the parametric equation:

$$\begin{aligned} x(\theta) &= \frac{a + d\sin\theta}{\tan\theta} \\ y(\theta) &= a + d\sin\theta \end{aligned}, \quad 0 < \theta < \pi.$$

We can also find an algebraic parametrisation in terms of $t \in (-\infty, \infty)$ where $t = \cot\theta$. First rewrite

$$x = (a + d\sin\theta)\cot\theta = a\cot\theta + d\cos\theta = at + d\cos\theta. \tag{2.16}$$

Again, from $t = \cot\theta$, we have $t\sin\theta = \cos\theta$, and

$$1 = \sin^2\theta + \cos^2\theta = \sin^2\theta + t^2\sin^2\theta = (1 + t^2)\sin^2\theta,$$

i.e.,

$$\sin^2\theta = \frac{1}{1 + t^2},$$

and

$$\sin\theta = \frac{1}{\sqrt{1 + t^2}}.$$

And

$$\begin{aligned} \cos\theta &= \sqrt{1 - \sin^2\theta} = \sqrt{1 - \frac{1}{1 + t^2}} \\ &= \sqrt{\frac{1 + t^2 - 1}{1 + t^2}} = \frac{\pm t}{\sqrt{1 + t^2}}. \end{aligned}$$

Each choice of the plus or minus sign will yield a parametrisation of a branch of the conchoid. Recalling (2.16) and choosing the positive sign results in the parametrisation[14]

$$\begin{aligned} x(t) &= at + \frac{dt}{\sqrt{1+t^2}} \\ y(t) &= a + \frac{d}{\sqrt{1+t^2}} \end{aligned}, \quad -\infty < t < \infty,$$

of the upper branch. The negative sign yields the corresponding parametrisation for the other branch:

$$\begin{aligned} x(t) &= at - \frac{dt}{\sqrt{1+t^2}} \\ y(t) &= a - \frac{d}{\sqrt{1+t^2}} \end{aligned}, \quad -\infty < t < \infty.$$

2.1.7 Exercise As with the trigonometry-free parametrisation of the hyperbola, we can verify this on a graphing calculator. On the TI-83 or TI-84, in parametric graphing mode enter the functions

X_{1T}= AT + {1, −1}DT/$\sqrt{}$(1 + T^2)
Y_{1T}= A + {1, −1}D/$\sqrt{}$(1 + T^2),

with parameters A, D to be chosen later. Choose the window

Tmin = ⁻6.5
Tmax = 6.5
Tstep = .1
Xmin = ⁻10
Xmax = 10
Ymin = ⁻3
Ymax = 5.

Then graph the curves for the following A, D pairs:

i. A = 2, D = 1
ii. A = 2, D = 2
iii. A = 2, D = 3.

In each case, the upper branch will be graphed first from left to right. Two brief periods of inactivity will occur before and after graphing this branch of the curve. This is because the points on this branch corresponding to values of T less than −3.9 and greater than 3.9 are offscreen. Following this, the lower branch of the curve will be drawn, again from left to right. [A final small warning: The axes are drawn to different scales and the usual calculator distortion will occur.]

The construction problems of antiquity are only remotely relevant to any discussion of the Mean Value Theorem, our central concern in this book; however, they have been a running thread throughout this section and I suppose I should comment on

[14] Apologies to the reader: dt here denotes multiplication by d, not the differential.

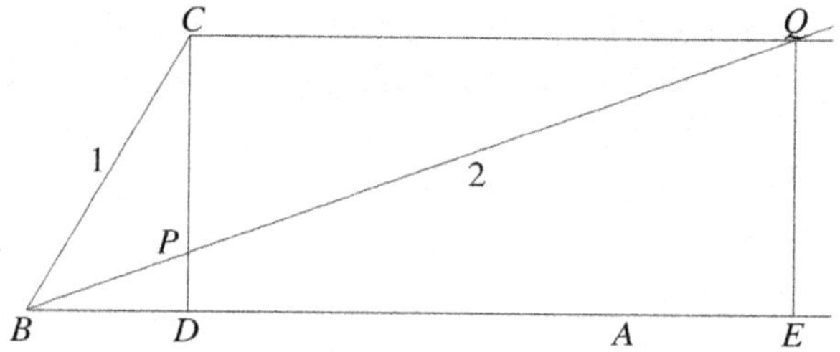

Fig. 2.11 Nicomedes's trisection

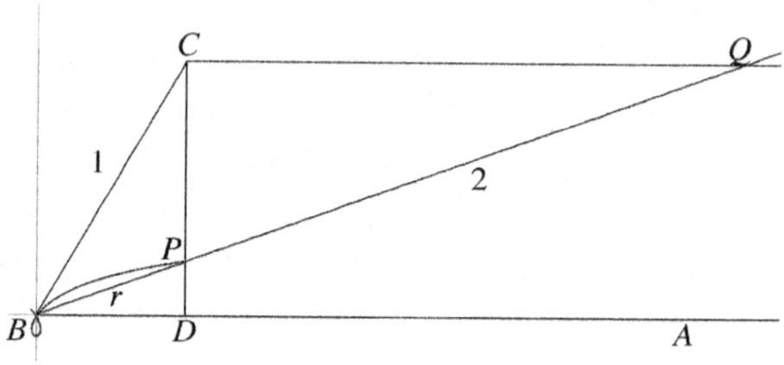

Fig. 2.12 Trisection via the conchoid

the use of the conchoid in trisecting the angle. The reader with no particular interest in the matter is invited to skip ahead to page 32. There is, however, a small pædagogical point illustrated by the construction that mirrors a criticism of the classroom presentation of the proof of the Mean Value Theorem. This is that the construction is given with no explanation for the choice of a crucial parameter.

Bunt, Jones, and Bedient[15] begin their explanation with a diagram like Fig. 2.11. Their labelling is different and the perpendicular QE to AB isn't drawn, but overall their diagram agrees with this one. They explain that $\angle ABC$ is a given angle θ that is to be trisected. One performs the trisection by dropping a perpendicular CD to AB and drawing a line perpendicular to CD at C, i.e., a parallel to AB passing through C. If we choose BC as the unit, they suggest finding P as the intersection of the line CD and the conchoid given by choosing B as the pivot point O, CQ as the line L, $CD = \sin\theta$ as a (where we here use the geometric convention of writing XY for $\mathrm{dist}(X, Y)$ for points X, Y), and finally 2 as d. Then $\angle ABP$ trisects $\angle ABC$. (See Fig. 2.12.)

Figure 2.12 is not the prettiest picture in the world, especially when graphed on the small screen of one's calculator, and some prefer to use the conchoid based on $O = B$, $L = CD$, $a = BD = \cos\theta$, and $d = 2$. Then Q is the point of intersection of the conchoid with the horizontal line passing through C.

2.1.8 Exercise I should include an illustration of the second conchoid for comparison with Fig. 2.12, but the image is so clear on the calculator and the graph can be redrawn for various choices of θ by using the variable θ on the calculator keyboard,

[15] Lucas N.H. Bunt, Phillip S. Jones, and Jack D. Bedient, *The Historical Roots of Elementary Mathematics*, Prentice-Hall, Inc., Englewood Cliffs (NJ), 1976, pp. 105–106.

so I choose instead to instruct the reader to do the diagram himself. Set the graphing mode to Par and enter the equation for the conchoid based on $a = \cos\theta$, $d = 2$ for a generic angle θ (noting that the rôles of x, y are reversed from those in our determination of the parametric equations for the conchoid):

```
X1T= cos(θ) + 2sin(T)
Y1T= (cos(θ) + 2sin(T))/tan(T).
```

Then enter the equations of the lines BC determining the angle, CQ determining Q, and that of the angle trisector:

```
X2T= T
Y2T= tan(θ)T
X3T= T
Y3T= sin(θ)
X4T= T
Y4T= tan(θ/3)T.
```

In the WINDOW menu set

```
Tmin = 0
Tmax = π
```

to avoid drawing any of the other branch of the conchoid and the intrusive near vertical lines connecting the two branches. Then use ZDecimal to draw the graphs for a variety of angles including $\pi/2, \pi/3, \pi/4, 0$. For the angles $\pi/2$ and 0 you might want to enter the FORMAT menu and choose AxesOff. Note that the choice $\pi/3$ is the one of Fig. 2.12.

As I say, the graph is very nice, but one might like to zoom in on it a bit. The default zoom factor is 4, which is too large a zoom. One can use the MEMORY submenu accessed by the ZOOM button to access SetFactors... and set XFact and YFact equal to 2. Or one can enter parameters in the WINDOW menu. I found the following values worked well:

```
Xmin = ⁻1.3
Xmax = 3
Ymin = ⁻1.55
Ymax = 1.55.
```

The reader who has faithfully carried out this latest exercise has seen that the construction works, at least for acute angles.[16] But he hasn't seen why the construction works or how the value 2 was chosen. We have seen that this choice works, but not how we knew to use 2 rather than, say, 3.

My favourite explanation is given by elaborating on the trigonometric proof cited by Coolidge,[17] which yields the sought-after rationale. See Fig. 2.13.

---

[16]To the Greeks, angles were between 0° and 180°. As every obtuse angle is the sum of a right angle and an acute angle, and as the right angle is easily trisected, we need only concern ourselves here with acute angles.

[17]Julian Lowell Coolidge, *A History of Geometrical Methods*, Dover Publications, Inc., New York, 1963, pp. 46–47. This is a reprint of a volume originally published by Oxford University Press in 1940.

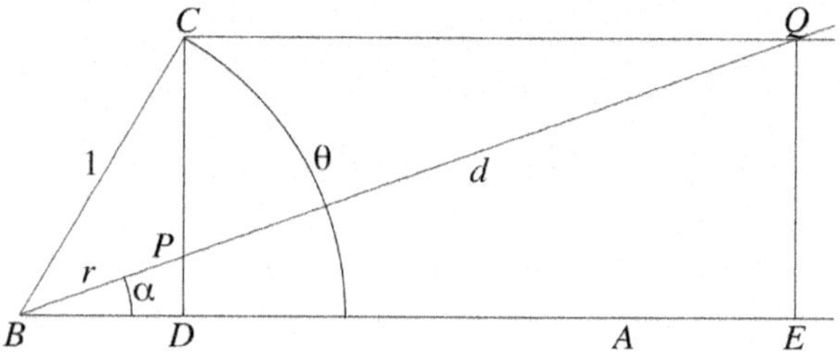

Fig. 2.13 Showing $d = 2$

First, note that if we are given $\theta = \angle ABC$ and another angle α we hope to be $\theta/3$, we can find the points P and Q where the line with angle α intersects CD and the horizontal line passing through C, respectively. If we let BC be the unit, we can ask for the value of $PQ = d$. To this end, let $r = BP$ and note that $BD = \cos\theta$, whence

$$\cos\alpha = \frac{BD}{BP} = \frac{\cos\theta}{r},$$

i.e.,

$$r = \frac{\cos\theta}{\cos\alpha}. \tag{2.17}$$

But, looking at triangle QBE,

$$\sin\alpha = \frac{QE}{BQ} = \frac{CD}{d+r} = \frac{\sin\theta}{d+r},$$

whence

$$d + r = \frac{\sin\theta}{\sin\alpha} \tag{2.18}$$

Combining (2.17) and (2.18) we have

$$\begin{aligned} d = d + r - r &= \tfrac{\sin\theta}{\sin\alpha} - \tfrac{\cos\theta}{\cos\alpha} \\ d\sin\alpha\cos\alpha &= \sin\theta\cos\alpha - \sin\alpha\cos\theta \\ d\sin\alpha\cos\alpha &= \sin(\theta - \alpha). \end{aligned} \tag{2.19}$$

But $\theta = 3\alpha$ iff $\theta - \alpha = 2\alpha$ and, by (2.19),

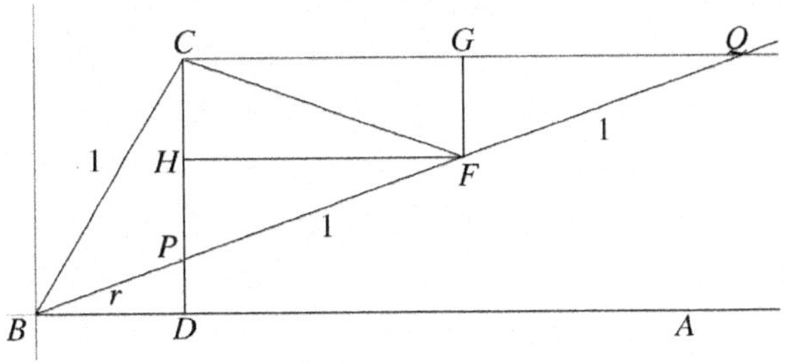

Fig. 2.14 Geometric proof of trisection

$$\sin 2\alpha = \sin(\theta - \alpha) = d \sin\alpha\cos\alpha \text{ iff } d = 2.$$

Thus, to trisect the angle we must have $d = 2$. And, conversely, if we choose $d = 2$, the construction trisects the angle.

The trigonometric solution would not have been directly available to Nicomedes. The Greeks used chords rather than sines and cosines, which were later introduced by Indian mathematicians, and the trigonometric addition formulæ were, I believe, first proven some centuries after Nicomedes by Claudius Ptolemy (*c*. 85–*c*. 165). A more traditionally geometric proof that $\angle ABQ$ trisects $\angle ABC$, given the assumption that $PQ = 2$, proceeds as follows: Bisect PQ at F. From F drop perpendiculars to CD and CQ as in Fig. 2.14. Triangles PHF and FGQ are similar and $PF = FQ = 1$, whence they are congruent. Thus, $CG = HF = GQ$ and triangles CGF and QGF are congruent, sharing as they do two pairs of equal sides and equal right angles between them. Thus $\angle GCF = \angle GQF$, and $\angle GQF$ equals $\alpha = \angle ABQ$. We have

$$\angle CFP = \pi - \angle CFQ = \pi - (\pi - 2\alpha) = 2\alpha.$$

But $BC = 1 = CF$, whence BCF is an isosceles triangle and we have $\theta - \alpha = \angle CBF = \angle CFP = 2\alpha$, i.e., $\theta = 3\alpha$. Thus, $\angle ABQ$ is indeed the trisector of $\angle ABC$.

2.1.9 Remark I find adding the lines HF and GF to clutter up the diagram unnecessarily. To conclude that $CF = QF$, note that

$$\cos\alpha = \frac{CQ}{PQ} = \frac{CQ}{d},$$

i.e., $CQ = d\cos\alpha$. If we now apply the *Law of Cosines* to the triangle CQF and $\angle CQF = \alpha$, we get

$$
\begin{aligned}
CF^2 &= CQ^2 + QF^2 - 2 \cdot CQ \cdot QF \cdot \cos\alpha \\
&= d^2\cos^2\alpha + \frac{d^2}{4} - 2 \cdot d\cos\alpha \cdot \frac{d}{2} \cdot \cos\alpha \\
&= d^2\cos^2\alpha + \frac{d^2}{4} - d^2\cos^2\alpha = \frac{d^2}{4},
\end{aligned}
$$

whence $CF = \frac{d}{2} = QF$.

The next step of showing $\angle CFB = 2\alpha$ can be accomplished as before, assuming $d/2 = 1 = BC$, i.e., $d = 2$, or we can appeal to another result in Euclid[18]: If the same chord in a circle is subtended both by an angle with vertex at the centre and a vertex on the same side of the chord as the centre and lying on the circumference, then the former angle is twice the latter. In this case, one takes the circle centred at F of radius $d/2$, lets PC be the chord and Q the second vertex. One automatically has

$$\angle CFP = 2\angle CQF = 2\alpha.$$

One completes the proof by noting

$$\angle CBF = \angle CFB = \angle CFP \text{ iff } BC = CF \text{ iff } 1 = \frac{d}{2},$$

i.e., iff $d = 2$. Thus again we see that the choice of $d = 2$ does lead to the conclusion that $\angle ABQ = \alpha$ trisects $\angle ABC = \theta$. And we see again where the choice of $d = 2$ came from.

2.1.10 Remark The Law of Cosines is an important identity and pops up in Vector Analysis. Somewhat less generally important, but useful here, is the *Law of Sines*. Applied to triangle BCQ, not assuming a specific value for $d = PC$, it yields

$$\frac{\sin\alpha}{BC} = \frac{\sin(\theta - \alpha)}{CQ},$$

i.e.,

$$\frac{\sin\alpha}{1} = \frac{\sin(\theta - \alpha)}{d\cos\alpha}.$$

Thus $\sin(\theta - \alpha) = d \sin\alpha\cos\alpha$ and we conclude $\theta - \alpha = 2\alpha$ iff $d = 2$, i.e., the construction trisects $\angle ABC$ iff $d = 2$.

I've overindulged myself in presenting all these alternatives. But I think it is important here to stress that the choice of $d = 2$ is forced upon us by a simple

[18]The Law of Cosines, in what we might call a disguised form, appears as Propositions 12 and 13 in Book II of the *Elements*. To make this proof non-trigonometric and purely geometric requires merely a change in terminology.

consideration and is not an act of inspiration or omniscience on the part of the presenter. The construction, which appears to come from nowhere, is actually quite natural. Some things like the drawing of Fig. 2.11 are fairly automatic. One would begin an analysis of the problem by drawing an angle $\angle ABC$ and its trisector $\angle ABQ$ as presented there. Dropping the perpendicular CD is a fairly natural thing to do in one's exploration, as is choosing $BC = 1$: we are dealing with angles and it is natural to place the vertex B in the centre of a circle of radius 1. Once this is done, the problem is to find P. P is determined by any of the distances BP, CP, and DP. At some stage one may draw the parallel CQ, perhaps to have a right triangle of altitude CD opposite $\angle ABP$. One then realises that PQ can also be used to determine P. If one knows about the conchoid, one now merely has to choose the right d and verify that it works. The most mysterious part of the presentation is the often unexplained choice of $d = 2$.

An analogous situation arises in the classroom proof of the Mean Value Theorem in which an auxiliary function is used. One of the criticisms levelled against this proof, which we will encounter in the next chapter, is the lack of motivating explanation behind the choice of this function. As with the choice here of $d = 2$, the choice there can be explained. The lack of explanation in a textbook or in a lecture speaks only of the laziness of the expositor, and not of the mysteriousness of the proof. This is not to say that such laziness is always bad: the expositor may choose not to explain such things if his intended readers or classroom students have sufficient background and ability to work out the details for themselves, or if the point is too minor to justify the necessary page count or classroom time. Is this the case here regarding the choice of d? It seems not to be the case with the auxiliary function used in the proof of the Mean Value Theorem in the standard Calculus course.

But this is a matter for consideration later. For now we have to finish up with the conchoid, discuss two additional curves, and then give a tentative formal definition of a curve.

I don't have too much more to say about the conchoid. In his introductory essay on the work of Archimedes and his predecessors, Heath[19] informs us that, according to Pappus, Nicomedes introduced the conchoid for the purpose of duplicating the cube. He also says that Pappus and another commentator Eutocius tell us that Nicomedes also constructed a mechanical instrument for use in drawing the conchoid. Such a device is illustrated in figure above, below, on the next page.

[19]T.L. Heath, *The Works of Archimedes Edited in Modern Notation with Introductory Chapters by T.L. Heath with a Supplement The Method of Archimedes Recently Discovered by Heiberg*, Dover Publications, Inc., New York, no date given. Heath's original edition was published in 1897 by Cambridge University Press, the supplement appearing subsequently in 1912. Cf. pp. cvi–cvii for his remarks on the conchoid.

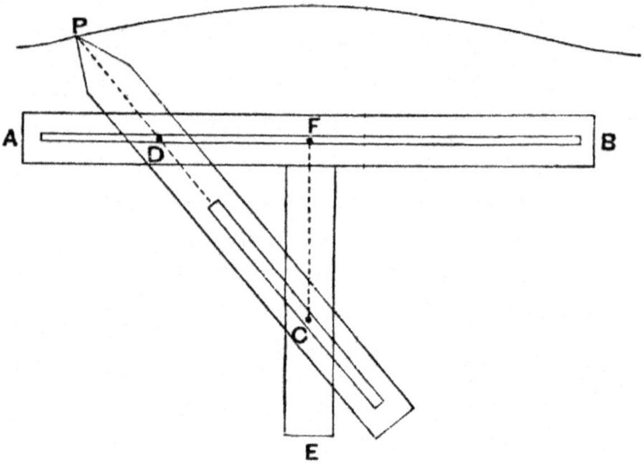

Conchoidograph

Pictured above is Heath's drawing of a conchoidograph and below is a working model based on it that I made from scrap material found around the house. The point O is fixed and represents the pivot point of the conchoid (i.e., the point O of Fig. 2.10). The slider labelled L represents the line L of that figure. The pencil pointing to P represents the point P on the conchoid and Q the point on the line L. The screw at Q can be temporarily loosened to allow adjustment of the distance from P to Q; when tightened PQ is fixed and Q is only allowed to move along L. With respect to my skill in constructing such, I must admit that my pencil holder is a bit wobbly and that the line L being fastened at only one end has a tendency to change its orientation while one is drawing the curve. But, when operated with three hands, it works quite well ... Presumably a trip to the local hardware store will remedy these defects.

The conic sections can also be drawn by means of simple mechanical devices called linkages. Such devices show that these solutions to the duplication and trisection problems were genuine solutions, albeit solutions involving more than ruler and compass. The same cannot be said of the purely theoretical solutions to the trisection and quadrature problems afforded by the quadratrix. This solution shows that these problems can be solved *if* one can draw the quadratrix. It provides a reformulation of the problem rather than a solution. And, indeed, it can be shown that no similar linkage exists for drawing the quadratrix.

The same is true of another famous curve that can be used to trisect the angle and square the circle. This is the *spiral of Archimedes*. Exactly why Archimedes was drawn to the spiral is unclear. His work, *On Spirals*, is extant[20] and no explanation of his interest in spirals is given. The work is prefaced by a letter to a colleague named Dositheus, more than half of which summarises work not discussed in the book he is sending:

> After these came the following propositions about the *spiral*, which are as it were another sort of problem having nothing in common with the foregoing; and I have written out the proofs of them for you in this book. They are as follows. If a straight line of which one extremity remains fixed be made to revolve at a uniform rate in a plane until it returns to the position from which it started, and if, at the same time as the straight line revolves, a point move at a uniform rate along the straight line, starting from the fixed extremity, the point will describe a spiral in the plane. I say then that the area bounded by the spiral and the straight line which has returned to the position from which it started is a third part of the circle described with the fixed point as centre and with radius the length traversed by the point along the straight line during the one revolution. And, if a straight line touch the spiral at the extreme end of the spiral, and another straight line be drawn at right angles to the line which has revolved and resumed its position from the fixed extremity of it, so as to meet the tangent, I say that the straight line so drawn to meet it is equal to the circumference of the circle.[21]

He cites a few more results before beginning the actual work of the book, but the two just cited are impressive enough.

Today, with Analytic Geometry and Calculus, these results are actually quite easy. First, referring to Fig. 2.15, one expresses the curve parametrically in terms of time t. Let P be the moving point, and assume the line segment L to be rotated is on the x-axis, with the fixed extremity at O and P initially coinciding with O. As P moves away from O along L, L is rotating around O, crossing the x-axis at D. Thus the motion of P is a composite of two motions, both assumed to be uniform. Thus there are constants a, b representing these uniform rates so that, at time t, the position of P in polar coordinates is given by

$$\theta = at, \quad r = bt.$$

Solving for t in terms of θ, we have $r = b\theta/a$. Thus, the spiral of Archimedes can be expressed in polar coordinates as

---

[20] An English translation can be found in Heath's book cited in the preceding footnote. The work *On Spirals* occupies pp. 151–188.

[21] *Ibid.*, pp. 153–154.

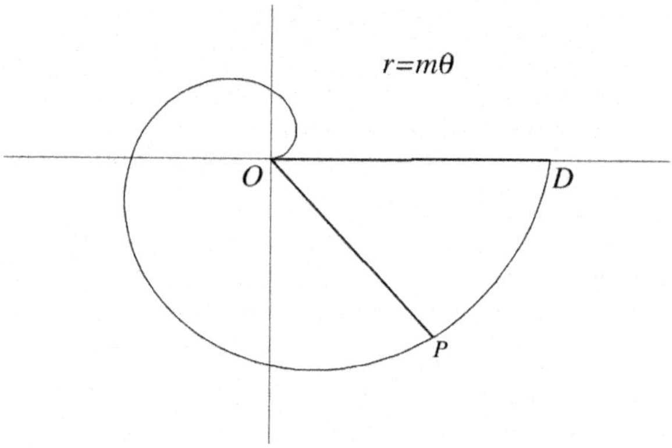

Fig. 2.15 Spiral of Archimedes

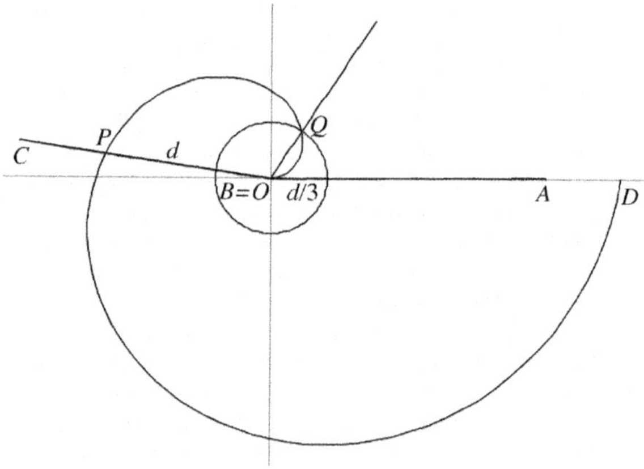

Fig. 2.16 Trisection via the spiral

$$r = \rho(\theta) = m\theta, \text{ for some } m.$$

Figure 2.15 shows the graph of this spiral for $0 \leq \theta \leq 2\pi$.

One can also express x and y parametrically in terms of θ:

$$\begin{matrix} x = r\cos\theta = m\theta\cos\theta \\ y = r\sin\theta = m\theta\sin\theta \end{matrix}, \quad 0 \leq \theta \leq 2\pi. \tag{2.20}$$

From the polar equation, for example, we can see immediately how to trisect an angle once the spiral is known. Let $\alpha = \angle ABC$ be given. Lay AB along OD, with B at O and find the point P where BC intersects the spiral. This has a certain length d, which is easily trisected using ruler and compass. Draw the circle of radius $d/3$ centred at $O = B$ and find the point Q where this circle intersects the spiral. Since $OQ = \frac{1}{3}OP$, we also have $\angle AOQ = \frac{1}{3}\angle AOP$. See Fig. 2.16.

By the same method any angle can be divided into any number of equal parts: the spiral solves the general multisection problem.

The first result cited by Archimedes in the preface to his book concerns the area of the region with boundary given by the spiral as θ ranges from 0 to 2π and the line OD as in Fig. 2.15. His claim is that this area is one third the area of the circle of radius OD. Today this is an easy calculation accessible to any student of the Calculus who has got as far as finding areas of polar curves by integration: For $r = \rho(\theta) = m\theta$, the area of the given region is

$$\begin{aligned}\text{Area} &= \int_0^{2\pi} \pi \cdot \rho(\theta)^2 \frac{d\theta}{2\pi} = \int_0^{2\pi} \frac{(m\theta)^2}{2} d\theta \\ &= \frac{m^2}{2} \cdot \frac{\theta^3}{3}\bigg|_0^{2\pi} = \frac{m^2(2\pi)^3}{2 \cdot 3} = \frac{4m^2\pi^3}{3},\end{aligned}$$

while the area of the circle with radius $OD = \rho(2\pi) = 2m\pi$ is $\pi \cdot (2m\pi)^2 = 4m^2\pi^3$.

Of greater interest here is the second property of the spiral cited by Archimedes in his letter to Dositheus. According to it, if one draws the tangent to the spiral at D in Fig. 2.15, it will meet the y-axis at a point E of distance $2\pi \cdot OD = 2\pi \cdot m2\pi = 4m\pi^2$ from the origin. Again, this is easy with modern Calculus. The slope of the tangent is

$$\frac{dy}{dx} = \frac{dy/d\theta}{dx/d\theta} = \frac{m \sin\theta + m\theta\cos\theta}{m\cos\theta - m\theta\sin\theta},$$

where we use the parametric representation (2.20). At $\theta = 2\pi$ this equals

$$\frac{m \cdot 0 + m \cdot 2\pi \cdot 1}{m \cdot 1 - m \cdot 2\pi \cdot 0} = \frac{2m\pi}{m} = 2\pi.$$

The equation of the tangent is thus

$$\frac{y - 0}{x - 2m\pi} = 2\pi,$$

i.e., $y = 2\pi x - 2\pi \cdot 2m\pi$, and the y-intercept E is given by $y = -2\pi \cdot 2m\pi = -2\pi \cdot m \cdot 2\pi$, whence $OE = 2\pi \cdot m2\pi = 2\pi \cdot OD$, the circumference of the circle of radius OD. From this the quadrature of the circle is an easy exercise.

The spiral of Archimedes can be continued by allowing $\theta > 2\pi$. If one does this, the curve spirals outward ever more, but at a constant rate of movement away from the origin. The radial distance between successive passes of the curve remains a constant $2m\pi$. Allowing θ to be negative produces a mirror image of the original curve reflected across the y-axis. The full curve thus intersects itself infinitely often in a rather attractive pattern, as the reader can see in Fig. 2.17.

Another famous spiral discovered some centuries later, the *logarithmic spiral* has completely different properties. Its definition in polar coordinates is

$$r = \rho(\theta) = ae^{b\theta}, \quad \theta \in (-\infty, \infty),$$

where a, b are positive constants and e is the base of the natural logarithms. As θ assumes larger positive values, its radial growth is unbounded, the radial distance between passes increasing without bound. At $\theta = 0$, the curve does not begin at the origin; in fact, as θ assumes larger and larger negative values, the curve spirals in towards the origin at an ever decreasing rate. See Fig. 2.18.

Our final curve for consideration is the *cycloid*. Boyer introduces the cycloid as follows:

> However, it is reported that the imaginative Nicholas of Cusa (1401–1464) had noted the curve traced out by a point on the rim of a cart wheel as the wheel rolled along the road. Although he seems to have been unable to determine its nature or properties, this observation constituted a significant step in the study of curves, for it seems to represent the first modern instance in which a new curve was suggested by natural phenomena. The ancients had invented new curves *ad hoc* to solve specific geometrical problems: they had not discovered

Fig. 2.17 Full archimedean spiral

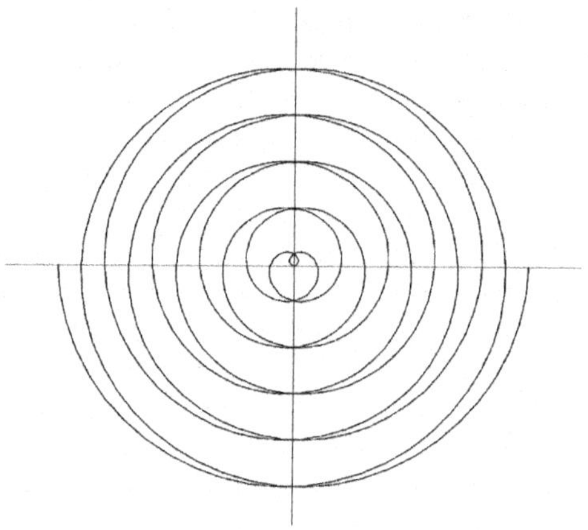

Fig. 2.18 Logarithmic spiral

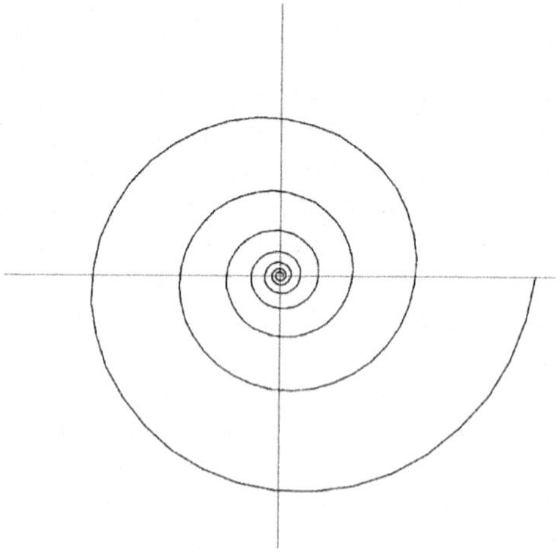

> these, except for the line and the circle, in the world of nature. The new curve of Cusanus[22] was followed two centuries later by other curves which were disclosed by, and useful in the study of physical science.[23]

The cycloid has its own mini-history, having been studied by a number of excellent mathematicians over the ensuing centuries. Apparently, Nikolaus von Kues played no rôle in it aside from a later misinterpretation on the part of John Wallis (1616–1703) of one of the diagrams in the Oxford manuscript of the work *De mathematicis complementis* of Kues.[24] A more accurate introduction to the cycloid and its history reads

> The seventeenth century is one of the most exciting periods in the history of mathematics. The first half of the century saw the invention of analytic geometry and the discovery of new methods for finding tangents, areas, and volumes. These results set the stage for the development of the calculus during the second half. One curve played a central role in this drama and was used by nearly every mathematician of the time as an example for demonstrating new techniques. That curve was the cycloid.
>
> The cycloid is the curve traced out by a point on the circumference of a circle, called the *generating circle*, which rolls along a straight line without slipping. It has been called the "Helen of Geometry," not just because of its many beautiful properties but also for the conflicts it engendered...
>
> The earliest mention of a curve generated by a point on a moving circle appears in 1501, when Charles de Bouvelles used such a curve in his mechanical solution to the problem of squaring the circle.[25]

The history of the cycloid is a matter of some interest, but this interest is tangential to our purpose here. The reader curious about its central rôle in the development of the Calculus, the controversies it engendered (criticisms of proofs, and priority conflicts), and the fascinating physical properties of the curve is referred to the paper of John Martin from which the above quote has been taken.[26]

What is relevant here, aside from its introduction of the physical world as a source of curves, is that it is, with our modern algebraic notation, easy to obtain a parametric representation of the curve and therewith a means to divine its properties.

The cycloid is depicted in Fig. 2.19. Here, the point P is assumed to coincide with the origin at the time $t = 0$, but remains fixed relative to the circle as it rolls along at

---

[22]Nikolaus von Kues is often cited under variants of his name. The Latin form is Nicolaus Cusanus, though Cusanus often suffices. Other variants are Nikolaus von Cusa, Nicholas of Cusa, or simply Nicholas Cusa.

[23]Boyer, *op. cit.*, p. 72.

[24]Nikolaus von Kues, *Die mathematischen Schriften*, 2nd. edition, Verlag von Felix Meiner, Hamburg, 1979, p. 220. The volume contains translations of Kues's manuscripts from the Latin by Josepha Hofmann and an introduction and notes by Joseph Ehrenfried Hofmann. Footnote 37 on page 217 includes the remark, "The figure contained in the *Oxford* manuscript has led WALLIS to the rash claim that CUSANUS had aleady arrived at the construction of the cycloid".

[25]John Martin, "The Helen of Geometry", *The College Mathematics Journal* 41, no. 1 (2010), pp. 17–28; here: p. 17.

[26]I also suggest V. Frederick Rickey, "Build a brachistochrone and captivate your class" in: Amy Shell Gellasch (ed.), *Hands on History. A Resource for Teaching Mathematics*, Mathematical Association of America, 2007.

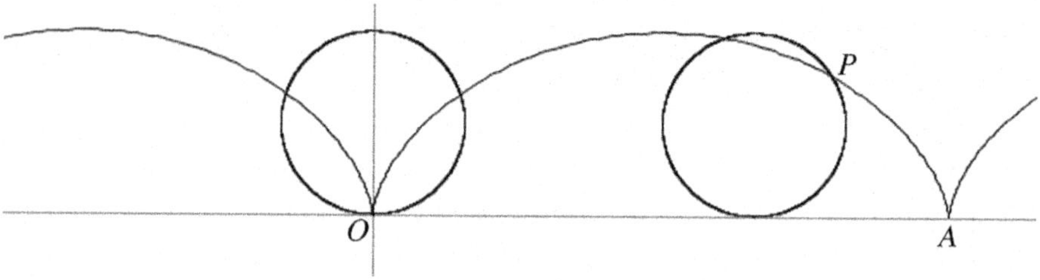

Fig. 2.19 Simple cycloid

a constant rate. If we affixed a movie camera to a dolly and filmed the wheel as the camera was moved along with the wheel, the film would show the point P moving around the circumference of a circle in a clockwise direction. Thus, if we imagine the axes moving too, the position of P would be described by

$$x = r\cos(\theta), \quad y = r + r\sin\theta,$$

where r is the radius of the circle and θ is the angle at which P sits on the circle as measured from its centre. For simplicity's sake we can take $r = 1$ and assume the rate of rotation is 1 radian/second. Thus in terms of time t, we have[27] $\theta = -\frac{\pi}{2} - t$ and

$$x = \cos\left(-\frac{\pi}{2} - t\right) = -\sin t, \quad y = 1 + \sin\left(-\frac{\pi}{2} - t\right) = 1 - \cos t.$$

Now, viewed from a stationary position, the vertical position is unchanged; thus we still have $y = 1 - \cos t$. But, horizontally, the circle itself has moved t radians: $x = t - \sin t$. The parametric equations of the cycloid are thus

$$\begin{matrix} x = t - \sin t \\ y = 1 - \cos t \end{matrix}, \quad t \in (-\infty, \infty).$$

It is not relevant to our purposes here, but the cycloid, like the conchoid, has a couple of variants if P is not on the rim but is elsewhere on the radius of the circle. If P lies inside the circle there is a dip in place of the cusp, and if P is outside the circle there is a loop. There are also *epicycloids* where we imagine a wheel rolling not in a straight line but around the exterior of the circle, and *hypocycloids* obtained when the wheel rolls around inside a circle. I suppose one could roll it along spirals as well and see what develops. It is a good subject for experimentation with one's graphing software or graphing calculator.

2.1.11 Exercise Had the Greeks been aware of the cycloid, would they have accepted it as a mechanical curve that effectively squares the circle? Presumably the distance OA between the places where P touches the x-axis equals the circumference of

---

[27] t has a minus sign because the clockwise rotation is the reverse of the usual rotation.

the circle. However, they might have had some difficulty with this because of *Aristotle's Wheel*, a geometric paradox dubiously ascribed to Aristotle. One imagines two wheels, a larger and a smaller, rigidly fixed to each other at the hub. After a complete revolution, points on their rims have traced out lines of equal distance. (See Fig. 2.20.) If the radii of the wheels are r and R, respectively, have the wheels travelled a distance of $2\pi r$ or $2\pi R$? The ratio $P'P'$ to PP should be R/r, but the distances are clearly equal. How is one to explain this?

This paradox puzzled scholars for centuries before Galileo (1564–1642) accounted for the discrepancy. Not everyone would accept Galileo's explanation today[28] and one refers blithely to "slippage". What is happening is clearest if one imagines the small wheel rolling on a rail and carrying the large wheel with it as it turns. Assume for convenience that $r = 1$, $R = 2$ as in Fig. 2.20 and graph the paths of P and P' on your calculator as t goes from $-\pi$ to π. In the equation editor enter

```
X1T= T − sin(T)
Y1T= 1 − cos(T)
X2T= T − 2sin(T)
Y2T= 1 − 2cos(T) .
```

In the WINDOW screen enter

```
Tmin=⁻π
Tmax=π
```

and graph the functions using ZDecimal. You will see the paths of P and P' between the times when they are at the high points of the wheel. Notice that P travels only from left-to-right in the x-direction, while P' does some backtracking. If one does the calculation of the total horizontal movement of P' without regard for direction, one will find P' has actually travelled $10\pi/3$, still $2\pi/3$ short of the expected 4π, but at least one sees some of the discrepancy simply explained.

It is time we reconsider the problem of defining what a curve is. With our modern knowledge of algebraic notation, we spot immediately that all the curves cited have one thing in common: they all have parametric definitions with the parameter ranging over some interval or intervals. The Greeks had no such symbolism and resorted either to vague descriptions of what their curves had in common, or classified them according to their obvious differences.

The Euclidean definition of a curve as "breadthless length" and the Aristotelian definition as "magnitude extended in one direction", both nods to the one-dimensionality expected of a curve, are too vague to serve as actual definitions. To do anything with these, one needs to define the term "dimension", which would only be done adequately in the 20th century by L.E.J. Brouwer, P.S. Urysohn (1898–1924), and eventually Karl Menger (1902–1985). Aristotle's definition of a curve as the "flowing of a point" is more promising. It suggests to the modern mind

---

[28]Cf., e.g., my exposition: Craig Smoryński, *Adventures in Formalism*, College Publications, London, 2012, pp. 99–104.

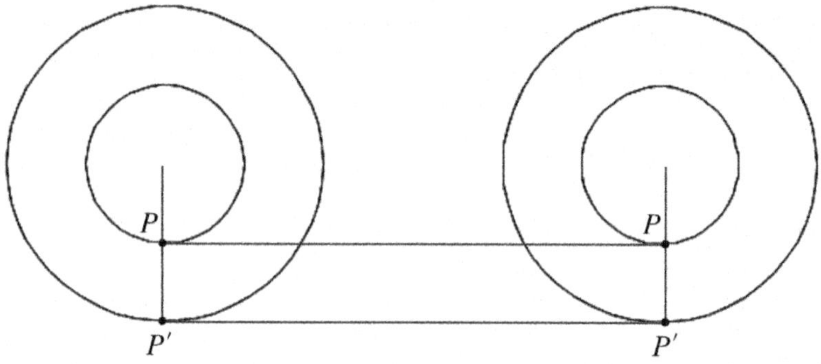

Fig. 2.20 Aristotle's wheel

a parametrisation of the position of the point P as a function of the time t from some interval during which the point is doing its flowing.

Heath's discussion of these definitions in his annotated edition of the *Elements* was followed by a discussion of two classifications of curves by Geminus (*c*. 10 B.C.–*c*. 60 A.D.). The first of these is very crude: Lines (i.e., curves) can be composite (broken lines forming angles) or incomposite; the latter can then form a figure (circle, ellipse) or not form a figure (straight line, parabola, hyperbola, conchoid). His second classification was more elaborate, but still crude.[29]

Boyer cites a better classification:

> The classical Greek geometers divided curves into three ranks or orders: the highest place was reserved only for the perfect curves, the line and the circle. These were called plane loci. Second place was granted to the Menaechmian conics which, probably on account of their original mode of definition, were known as solid loci. All other curves, whether algebraic or transcendental, were grouped together under the heading linear loci. Pappus described this last category as made up of those curves "the origin of which is more complicated and less natural [than that of the plane and solid loci], as they are generated from more irregular surfaces and intricate movements." In this description we see the two types of curve definition which the Greeks recognized — the kinematic and the stereometric.[30]

Since the Greeks individual curves could be defined in any of three ways:

(1) kinematically, as the "flowing of a point" — as we've defined the conchoid, the spiral of Archimedes, and the cycloid;
(2) as loci — as we've defined the individual conic sections;
(3) stereometrically, i.e., in terms of solids — as the conic sections were originally defined.

It is not clear what the curves given rise to by these three modes of definition have in common. The kinematic approach is nowadays naturally formulated in terms of parametrisation. The various loci we've encountered were naturally supplied with

[29] Heath, *Elements*, *op. cit.*, pp. 160–165.

[30] Boyer, *op. cit.*, p. 32. The bracketed insertion is Boyer's.

parametrisations in part because they were described as loci of points satisfying certain distance requirements easily expressed equationally. But can one easily isolate those modes of definition of loci the solutions to which yield curves? And, although once given a curve in the plane one can readily concoct a solid, one of the edges of which happens to be that curve, using stereometry to define the notion of a curve presupposes a definition of a solid and its surface or of a solid and the boundary of a region obtained by intersecting the solid with a plane. And defining these concepts entails the same difficulties as defining a curve, but in higher dimensions.

So one's best bet for a usable formal definition of a curve to replace the vague informal conception seems to be the kinematic one encapsulated by a parametrisation of the position of a point P in terms of time t taken over an interval or intervals. Before making this our official choice, however, note that we have seen another way of defining a curve, namely another algebraic method that became available in the 17th century: With Eq. (2.9) we noted that every conic section was the locus of points satisfying a quadratic equation

$$f(x, y) = 0.$$

The ability to define curves as solution sets to equations was ushered in by the near simultaneous invention of Analytic Geometry by Fermat and Descartes. Fermat applied the algebraic symbolism of François Viète (1540–1603) to classical locus problems, deriving equations and declaring that such equations described curves. Boyer waxes eloquently on Fermat's statement that the equations described curves:

> This brief sentence represents one of the most significant statements in the history of mathematics. It introduces not only analytic geometry, but also the immensely useful idea of an algebraic variable. The vowels in Viète's terminology previously had represented unknown, but nevertheless fixed or determinate, magnitudes. Fermat's point of view gave meaning to indeterminate equations in two unknowns — which previously had been rejected in geometry — by permitting one of the vowels to take on successive line-values, measured along a given axis from an initial point, the corresponding lines representing the other vowel, as determined by the given equation, being erected as ordinates at a given angle to the axis.[31] In ancient Greek works, certain lines associated with a given curve had played a role equivalent to that of a coordinate system, and the properties of the curve had been expressed in terms of these lines by means of rhetorical algebra.[32] The curve came first, the lines were then superimposed upon it, and finally the verbal description (or algebraic equation) was derived from the geometrical properties of the curve. Fermat's genius made it possible to reverse this situation. *Beginning* with an algebraic equation, he showed how this equation could be regarded as defining a locus of points — a curve — with respect to a given coordinate

[31] A brief word of explanation: For some time mathematicians viewed curves as the paths traced out by the intersection of two lines, eventually a vertical line moving along the x-axis and a horizontal one moving up and down the y-axis. With Fermat, however, the axes were not necessarily perpendicular but met at a given angle. The variables thus stood for the positions of the lines parallel to these axes. Viète had begun a short-lived practice of using vowels to denote variables and consonants to denote unspecified constants and Fermat adhered to this tradition.

[32] Mathematical historians distinguish 3 phases in the development of algebraic symbolism: *rhetorical*, in which everything is expressed in words; *syncopated*, in which some abbreviations are introduced; and *symbolic*, in which everything is expressed in abstract symbols and calculations follow strict term rewriting rules.

> system. Fermat did not invent coordinates and he was not the first one to use graphical representation. Analytic reasoning had long been used in mathematics, and the application of algebra to geometry had become a commonplace. However, there appears to have been no appreciation before the times of Fermat and Descartes of the fact that, in general, a given algebraic equation in two unknown quantities determines, *per se*, a unique geometric curve. The recognition of this principle, together with its use as a formalized algorithmic procedure, constituted the decisive contribution of these two men.[33]

Fermat went so far as to demonstrate that quadratic equations yielded exactly the plane and solid loci, i.e., the conic sections, but did not consider general algebraic equations that arise when f is a polynomial in two variables of arbitrary degree. Descartes did. Boyer summarises Descartes's approach as follows:

> Whereas Viète had been interested in the constructibility of determinate problems, Descartes went further and applied the criteria to loci as well. It was here that he found it necessary to use a coordinate system. One may say that, in a general sense, the invention of analytic geometry by Descartes consisted in the extension of the analytic art of Viète[34] to the construction of *indeterminate* equations, just as in the case of Fermat it was the study of loci, by the analytic art, which led to the same result. But Descartes continued to regard the construction of *determinate* equations as his ultimate purpose.
>
> The plotting of curves in the now customary manner was not a part of Cartesian analytic geometry. Even the Pappus loci are not sketched. Descartes knew that an equation in two unknowns determines a curve, but oddly enough, he seems not to have regarded such an equation as an adequate definition of the curve, and felt constrained to exhibit an actual mechanical construction in each case. It has been conjectured that the ancient Greeks stressed constructions because these served as existence theorems. One is tempted to apply this idea to Descartes and say that he doubted the existence of a curve corresponding to an equation unless he could supply a kinematic construction for it. Like the ancient Greeks, he felt that a locus had to be legitimized by associating it geometrically or kinematically with another known curve. Perhaps it was the traditionally axiomatic form of geometry that led him in this direction... This represents, of course, a clear-cut break with the Platonic limitation of instruments to compasses and straight-edge, and Descartes makes free use of various linkages and mechanical devices. The concept of movement plays a far more prominent role in his work than in that of Fermat.[35]

The success of analytic (i.e., algebraic) techniques in solving locus problems dictates that we want an algebraic definition of what a curve is. We have two candidates at our disposal — kinematic, defining them via parametric equations, and algebraic, defining them as the solution set of an equation. Neither definition is perfect, but the latter appears more immediately recognisably imperfect than the former.[36] Descartes's elevation of the kinematic over the equational as the definitive hallmark

---

[33]Boyer, *op. cit.*, pp. 75–76.

[34]This "analytic art" was the beginning of symbolical algebra. The adjective "analytic" here referred to the algebraic analysis of a problem — its expression in algebraic terms and the solution of the resulting equations. Except for "Analytic Geometry", the adjective "analytic" today refers more generally to those areas of mathematics that the Calculus evolved into, Calculus itself having evolved from Analytic Geometry.

[35]Boyer, *op. cit.*, pp. 88–89.

[36]Consider, e.g., the "curve" defined by the constant function $f(x, y) = 0$.

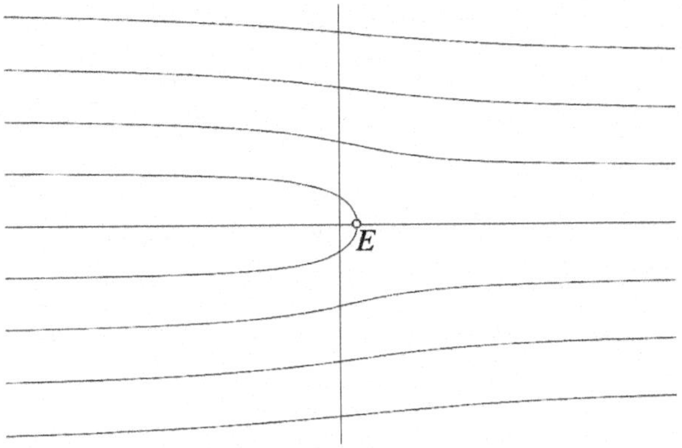

Fig. 2.21 Full quadratrix

of *curveness* may have indicated lingering doubts about an equation's always defining a curve. Moreover, singling out those functions f which define curves is merely an algebraic reformulation of the basic problem of defining what a curve is in the first place. So we will tentatively define a curve as one that is parametrically definable.

2.1.12 Definition Let I be an interval and $\gamma : I \to \mathbb{R} \times \mathbb{R}$ a function from I to the real plane. The *curve* defined by γ is the range $\gamma(I) = \{\gamma(t) \mid t \in I\}$ of γ; γ itself is called a *parametrisation* of the curve $\gamma(I)$.

The interval I can be open, closed, half-open, bounded, or unbounded. Moreover, we can relax the requirement that I be an interval to I being a union of disjoint intervals so as to accommodate curves with multiple branches.

Note that the *graph* of a function $y = f(x)$ or $x = f(y)$ falls under the scope of this definition: one simply defines $\gamma(t) = \langle t, f(t)\rangle$ or $\gamma(t) = \langle f(t), t\rangle$, respectively.

Definition 2.1.12 is tentative as it is still a bit too inclusive. Not every parametrically defined function $\gamma(t) = \langle x(t), y(t)\rangle$ has as its range something we would call a curve. We have already noted the two branches of the hyperbola. And if one has used (2.1) to graph the quadratrix on one's graphing calculator, one will have noticed that it consists of lots (in fact, infinitely many) branches. (Cf. Fig. 2.21. Note that the point E is not on the curve.) And one will have seen the graphs of the trigonometric functions $\sec x$, $\csc x$, $\tan x$, and $\cot x$ with their infinite collections of branches.

Even if the range of γ has only a single branch, it might not be something we want to call a curve. Consider the graph of the function

$$y = \begin{cases} \sin 1/x, & x \neq 0 \\ 0, & x = 0 \end{cases}$$

given in Fig. 2.22. In topological terms, it is *connected*, but it is not nicely connected in that in going from the left of the y-axis to the right one does not pass smoothly through the origin at $x = 0$. As x moves closer and closer to 0, y infinitely often

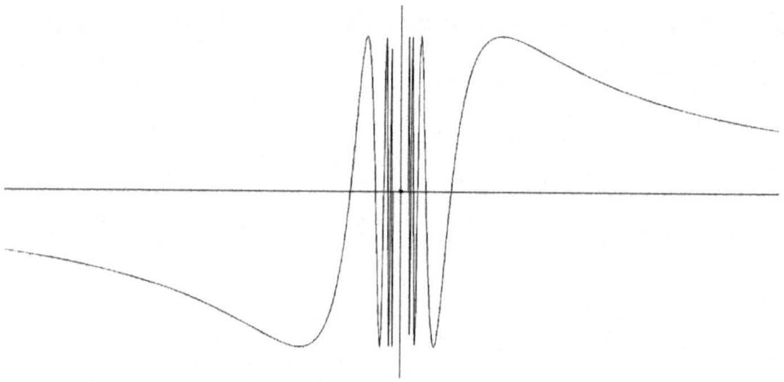

Fig. 2.22 Graph of $\sin(1/x)$

assumes the value 0 only to veer away from it. It does not have 0 as its unique limiting value as x goes to 0. Indeed, we could have defined $y(0)$ to be any value in the interval $[-1, 1]$ with the same result.

More seriously, it can happen that the range of γ can be two dimensional. Every real number in the interval $[0, 1]$ has a decimal expansion $.r_0r_1r_2\ldots$ which does not end in an infinite repeating sequence of 9's (with the exception of the value $1 = 1.\bar{0}$). If we define γ by

$$\gamma(1) = \langle 1, 1\rangle$$
$$\gamma(.r_0r_1r_2\ldots) = \langle .r_0r_2r_4\ldots, .r_1r_3r_5\ldots\rangle,$$

the range of γ is the entire unit square $[0, 1] \times [0, 1]$.[37]

In the next section we will narrow the definition further and define the class of continuous curves, which more closely fit our intuitive geometric conceptions of curve and kinematic motion, the latter at least as conceived before the advent of the quantum leap.

[37]The only tricky part is recognising that

$$\gamma(.r_09r_19r_29\ldots) = \langle .r_0r_1r_2\ldots, .999\ldots\rangle = \langle .r_0r_1r_2\ldots, 1\rangle$$
$$\gamma(.9r_09r_19r_2\ldots) = \langle .999\ldots, .r_0r_1r_2\ldots\rangle = \langle 1, .r_0r_1r_2\ldots\rangle.$$

2.2 Continuous Curves

2.2.1 Defining Continuity

The notion of a curve as the "flowing of a point" would seem to entail a bit more than a succession of positions through time described parametrically by some listing function γ defined on a time interval as tentatively specified in Definition 2.1.12. The word "flowing" also promises some smoothness to the motion, with no gaps or sudden jumps. It may also suggest no sudden changes in direction, as given by corners and cusps. To accommodate these additional expectations of curveness, we have two refinements of our definition of a curve — definitions of *continuous curves* and *smooth curves*. Continuous curves have no gaps or strange jumps, but may have corners and cusps; smooth curves are allowed none of this sort of bad behaviour. As our central interest in this book is the Mean Value Theorem, we will eventually want to consider smooth curves. But in the Calculus in general, one wants to consider the broader class of continuous curves. Not all motions, after all, are simple smooth "flows". We have seen cusps, for example, in the motion-defined cycloid; and corners will appear when moving objects are reflected (i.e., bounced) or refracted. In the present section we will consider continuous curves and in the next section we will consider smooth curves.

Modern definitions of "curve" were given in the early decades of the 20th century. Continuity was adequately defined in the 19th, yet it too required some preparation. Where today we would first define a continuous function and then declare a curve to be continuous if it possessed a continuous parametrisation, continuity was a property of curves long before one spoke of functions. A slight familiarity with the history of continuity is not necessary here for our understanding of the Mean Value Theorem, but it will bear on the later history of this Theorem.

Philosophical discussions of continuity generally begin with Aristotle and his belief in the *potentially* infinite divisibility of the line. Any line segment can be divided into two properly smaller segments, each of which can again be divided, and so on — where "so on" means the process can be repeated any finite number of times without bound, not that one can actually do it infinitely often and still have a line rather than a single point at the end. A relatively modern version of this is discussed by Bertrand Russell (1872–1970):

> It is generally held by philosophers that numbers are essentially discrete, while magnitudes are essentially continuous. This we shall find to be not the case. Real numbers possess the most complete continuity known, while many kinds of magnitude possess no continuity at all.[38]

[38]Bertrand Russell, *Principles of Mathematics*, 2nd ed., W.W. Norton & Company, Inc., New York, no date given, p. 193. The first edition was published in 1903, the second originally in 1938. The printing I quote from is a paperback that I acquired new in the late 1960 s or early 1970 s and is thus a reprint of the second edition.

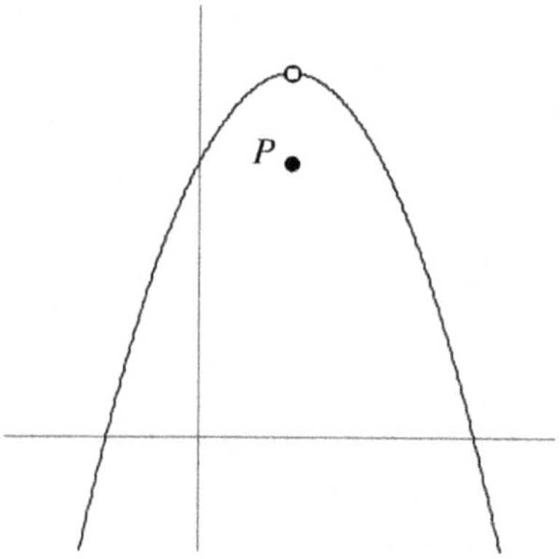

$$x(t) = t$$

$$y(t) = \begin{cases} 4 - (t-1)^2, & t \neq 1 \\ 3, & t = 1. \end{cases}$$

Fig. 2.23 An isolated point

Russell proceeds to define an ordered set to be continuous if, between any two elements of the set, an intermediate one exists. This makes the rational numbers, as well as the reals, continuous. Later in his book he admits that this is insufficient and calls such sets *compact*.[39] His new definition of continuity is due to Georg Cantor and Russell spreads the definition over two chapters. This requires the introduction of two additional notions of *perfect* and *cohesive* sets.

Russell reverses the order of the terms and defines the second one first. A set T is *cohesive* iff for any $t, t' \in T$ and any $\epsilon > 0$, a chain $t_0 = t, t_1, t_2, \ldots, t_n = t'$ can be found such that each of the distances $d(t_i, t_{i+1})$ is less than ϵ.

As Russell states, cohesiveness is a sort of connectedness condition, by which the rational numbers would be considered connected. There are gaps, but they all have length 0. If we consider Fig. 2.21 we can see that that part of the full quadratrix consisting of the two branches that approach each other at E is cohesive, but the whole graph is not because the distances between points on any other pair of branches are all at least $\pi > \epsilon$ for small ϵ.

2.2.1 Exercise Is the graph of Fig. 2.22 cohesive? Does your answer depend on whether or not you measure the distances taken along the curve or "as the crow flies", i.e., the distances between these successive points in the plane?

The second condition, that the set T be *perfect* refers to limits and has two subconditions, namely, i. every point on T is the limit of a sequence of elements of T, and ii. T contains all of its limits. The first subcondition rules out *isolated* points, examples of which have not yet appeared in our illustrations, but are readily given as in Fig. 2.23. The second subcondition requiring that the set contain all of its limits is, even without a formal definition of limit, clearly not satisfied by the curves of Figs. 2.1, 2.21, 2.22, or 2.23.

---

[39]The modern term for this is "dense"; "compact" has an altogether different meaning in mathematics.

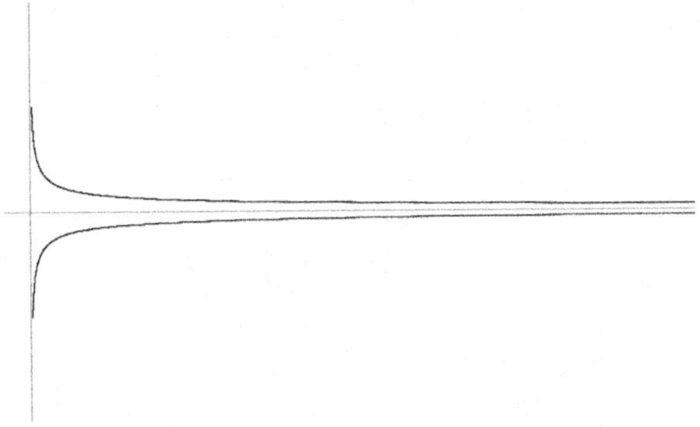

Fig. 2.24 Second order hyperbola

Cantor's definition is not the best possible. According to it, an open interval, say (0, 1), is not continuous because it fails to include its endpoints. And the simple quadratrix of Fig. 2.1 is not continuous because it lacks the limit point E.

On the other hand, some discontinuous curves are cohesive in Cantor's sense. A particularly simple example would be given by simply removing the point P from Fig. 2.23. A slightly more subtle example is given by the second order hyperbola

$$\begin{aligned} x(t) &= \frac{1}{t^2}, \quad t \in [-4, 0) \cup (0, 4]. \\ y(t) &= t \end{aligned}$$

(See Fig. 2.24.)

Inadequacy aside, Cantor's definition of continuity won't do here because it is too technically advanced for the first year Calculus course. With the post-Cantor advent of Topology, more adequate definitions of the continuity of a set emerged. Roughly, one defines a set T to be *connected* if it contains no gaps, where one recognises a gap by its ability to separate two portions of T. One does not necessarily determine the gap or separation by distance. In Fig. 2.23, the distance from P to the rest of the curve is 3/4 (*Exercise.*); the distance between the two branches of the quadratrix converging to each other at E in Fig. 2.21 is 0. A disconnecting gap can have measure 0.

Topology offers a formal definition of connectedness in terms of open sets:

2.2.2 Definitions A set U is said to be *open* if, for every point $u \in U$, all points of the space sufficiently close to u also lie in U:

$$\forall u \in U \exists \epsilon > 0 \forall x \big(\text{dist}(x, u) < \epsilon \Rightarrow x \in U\big).$$

If U is a subset of the plane, this means that if $P \in U$, then some *open disc* centred at P (i.e., the interior of a circle with centre P) is a subset of U. A set T is *disconnected* by a pair of disjoint open sets U, V if there are nonempty sets $X, Y \subset T$ such that

$$X \subseteq U,\ Y \subseteq V,\ T = X \cup Y,\ U \cap V = \emptyset.$$

A set T that is not disconnected is called *connected*.

2.2.3 Examples i. The graph of the hyperbola $x^2 - y^2 = 1$ of Fig. 2.7 is disconnected. Here we can take

$$U = \{\langle x, y\rangle \mid x < 0\}, \quad V = \{\langle x, y\rangle \mid x > 0\};$$

ii. The two branches of the conchoids of Fig. 2.9 are disconnected by the sets

$$U = \{\langle x, y\rangle \mid x < a\}, \quad V = \{\langle x, y\rangle \mid x > a\};$$

iii. The full quadratrix of Fig. 2.21 is disconnected by, among others,

$$U = \{\langle x, y\rangle \mid y > 0\}, \quad V = \{\langle x, y\rangle \mid y < 0\};$$

iv. The graph of $\sin(1/x)$ as given by Fig. 2.22 is connected, but if one drops the point $\langle 0, 0\rangle$ what remains is disconnected by

$$U = \{\langle x, y\rangle \mid x < 0\}, \quad V = \{\langle x, y\rangle \mid x > 0\}.$$

In each of these examples, the openness of the sets U, V is easy to establish and we see that it can be quite easy to show that a curve is disconnected when it is. It can be a lot harder to prove connectivity, as Example 2.2.3.iv illustrates.

Russell and Georg Cantor (1845–1918) were late arrivals on the continuous scene and I mention them first because their divisibility criterion for continuity traces back to Aristotle. Additionally there was always the unconscious assumption that curves which crossed each other actually met in a point. However, until Fermat and Descartes opened up the field by introducing numerous new curves algebraically, curves could be discussed on a case-by-case basis and the need for general definitions never arose. In his classic *La Géométrie*, Descartes used the word continuous, but offered no attempt to analyse the notion or to explain what he meant. It was the explanation, the informal intuition behind the scenes:

> ...if we think of geometry as the science which furnishes a general knowledge of the measurement of all bodies, then we have no more right to exclude the more complex curves than the simpler ones, provided they can be conceived of as described by a continuous motion or by several successive motions, each motion being completely determined by those which precede; for in this way an exact knowledge of the magnitude of each is always obtainable.[40]

The importance of continuity is again noted later:

---

[40]René Descartes (David Eugene Smith and Marcia L. Latham, trans.), *The Geometry of René Descartes*, Dover Publications, Inc., New York, 1954, pp. 42 (original French version) and 43 (English translation). The French original was published in 1637 as an appendix to Descartes's philosophical work *Discours de la Methode*. The English translation was first published in 1925 by the Open Court Publishing Company.

But the fact that this method of tracing a curve by determining a number of its points taken at random applies only to curves that can be generated by a regular and continuous motion does not justify its exclusion from geometry.[41]

La Géométrie divides into three "books". The first introduces the work, describing a locus problem of Pappus that he solved using his methods. In the second book he solves various locus problems by deriving equations describing the curves defined, and uses this algebraic formulation to solve various problems involving the curves. This included a computationally intense method of finding tangents and normals to such curves by determining circles that "touch" the curves and then finding the tangents and normals to these circles. The third book deals mainly with the Theory of Equations and the problem of finding roots of polynomials. Descartes did not introduce functions in this work.

Fermat considered functions given by expressions and even came close to the notion of the derivative with a technique for finding the maximum of a curve $y = f(x)$ by manipulating the difference quotient[42]

$$\frac{f(A+E)-f(A)}{E}.$$

This did not yet bring the notion of function to centre stage. With Descartes and Fermat algebraic expressions entered the stage but there were still other perspectives. Isaac Newton (1642–1727) envisioned a curve as the path traced out by the points of intersection of two lines, a vertical one moving horizontally and a horizontal one moving vertically. Their respective positions x and y were dependent on time, thus, in our modern terminology parametric functions $x(t)$ and $y(t)$ of time. His younger contemporary Gottfried Wilhelm Leibniz (1646–1716) introduced the term "function", but two mathematical generations later, Leonhard Euler (1707–1783) considered a function to be continuous if the same expression was used throughout an interval — "continuity" meant "continuity of definition". And curves were still on one's mind when one began to speak of continuous functions as opposed to continuous curves.

As seen from Euler's standard textbook *Introductio in Analysin infinitorum* of 1748, continuity was at first understood as a geometrical quality: as a quality of curves. Continuous curves were characterized by the fact that they could be expressed by an analytic expression. In contrast, discontinuous curves consisted of several segments that belonged to different functions and hence did not correspond to just one analytic expression, but to several. This explains why Euler called the non-continuous curves "discontinuous" or "mixed" curves...

In his later treatise of 1763 *De usu functionum discontinuarum in analysis*..., Euler, specifying the concept of continuity, stressed that it is necessary for continuous curves to obey a single analytic law. A hyperbola's two branches thus form a continuous curve.

The historical literature always refers to Arbogast's treatise of 1791 as to that which offered new conceptual proposals. This is said firstly because he explicitly formulated the interme-

---

[41] *Ibid.*, pp. 90 (French) and 91 (English).

[42] Fermat followed Viéte in using vowels A, E, I, O, U for variables $x, y, \ldots$ Cf. p. $83f$, below, for a more precise description of Fermat's technique.

> diate value property for continuous functions and secondly because he introduced a new term: "discontigue." While curves, according to Euler's specification, had been considered to be discontinuous as well if their various parts were attached to one another, provided that these were defined by different "laws," Arbogast now called curves *discontigue* if their various parts were unconnected. In all these works, this continual conceptual differentiation is emphasised as an important achievement, because with it, and with the novel term, the discipline had come closer to the meaning of discontinuity as it is understood today.
>
> It must be pointed out, however, that Arbogast's reflections on the meaning of *continue*, *discontinue* and *discontigue* still refer to curves, and that functions, for him, were only of secondary importance for representing particular parts of a curve. Arbogast assumed functions as basic concepts only when reflecting on intermediate values.[43]

The distinction between continuity and contiguity, something of a non-issue today, actually lies at the heart of the matter. It was the climax of a four-and-a-half decade long controversy over the vibrating string problem. In 1746 Jean le Rond d'Alembert (1717–1783) wrote his first paper on the vibrating string. This was a geometric problem, but the method of attack was analytic. D'Alembert insisted on functions continuous in the Eulerian, algebraic sense of continuity of expression. Euler, who also considered the problem, allowed mixed functions, both in determining the initial shape of the string and in specifying the solution. These solutions involved the solution of what are called partial differential equations and in 1787 the St. Petersburg Academy proposed a prize competition concerning these solutions:

> "Do they belong to any curves or surfaces either algebraic, transcendental,[44] or mechanical, either discontinuous or produced by a simple movement of the hand? Or shouldn't they legitimately be applied only to continuous curves susceptible of being expressed by algebraic or transcendental equations?"[45]

The prize was awarded to Louis Arbogast (1759–1803), whose treatise on the matter was published in 1791.

Historian Judith Grabiner sums up Arbogast's contribution succinctly:

> To get a feeling for the climate of mathematical opinion about continuous functions in which Bolzano and Cauchy worked, it will suffice to quote from the work of L.F.A. Arbogast, who won the St. Petersburg Academy's prize competition in 1787 by giving the best characterization of those functions that would be allowable solutions to the vibrating-string equation. Arbogast described these functions in several ways: The functions make "no jumps"; they have the intermediate-value property; and they increase in increments whose sizes correspond to the sizes of the increments in the variable. For instance, if the function is represented by

[43]Gert Schubring, *Conflicts between Generalization, Rigor, and Intuition: Number Concepts Underlying the Development of Analysis in 17–19th Century France and Germany*, Springer Science+Business Media, Inc., New York, 2005, pp. 26–27. In quoting this I have omitted Schubring's citations to the literature.

[44]Descartes and Fermat had introduced algebraic descriptions $f(x, y) = 0$ for curves, where f was a polynomial; very quickly transcendental functions like sines, cosines, logarithms, etc., were introduced into the composition of f.

[45]Citation from Jean Itard, "Arbogast, Louis François Antoine", in: Charles Coulston Gillispie (ed.), *Dictionary of Scientific Biography*, vol. 1, Charles Scribner's Sons, New York, 1970, p. 207. Itard adds, "The Academy was thus requesting a drastic settlement of the dispute between Jean d'Alembert, who adopted the second point of view, and Leonhard Euler, partisan of the first".

> two different formulas on adjacent intervals, "the last ordinate of the old form, and the first of the new, are equal to each other, or differ only by an infinitely small quantity." Again, "The ordinate y cannot pass brusquely from one value to another; there cannot be a jump from one ordinate to another which differs from it by an assignable quantity." This "no-jumps" characterization, though it helps call attention to the crucial property of continuous function as defined by Cauchy and Bolzano, is not in itself an anticipation of that definition; it deals with functions that are piecewise continuous, and discusses the behavior of the function in detail only at the break. Thus a definition of the continuity of the "piece" is still lacking. But Arbogast was concerned about this question. He linked his no-jumps property to the intermediate-value property, saying that the functions had to obey what he called the "law of continuity": "A quantity cannot pass from one state to another without passing through all the intermediate states subject to the same law." The closest Arbogast came to the Cauchy-Bolzano definition was to say "assuming that the variable increases continually, the function receives corresponding variations," though the language is not sufficiently precise to be a real anticipation of that definition.[46]

The property cited is sufficiently important to be singled out and given a formal definition.

2.2.4 Definition A real-valued function defined on an interval I has the *intermediate value property* if, whenever $a, b \in I$ and d are such that $f(a) < d < f(b)$, there is some c between a and b for which $f(c) = d$.

In the standard course in the Calculus one learns that every continuous function $f : I \to \mathbb{R}$ has the intermediate value property. The intermediate value property is clearly a version of the no gaps requirement for the continuity of the curve $y = f(x)$ and is the requirement most explicitly stated by Arbogast for the continuity of such a function. The question arose: does the intermediate value property characterise continuity of real-valued functions of reals?

2.2.5 Example Consider the function graphed in Fig. 2.22,

$$f(x) = \begin{cases} \sin\left(\frac{1}{x}\right), & x \neq 0 \\ 0, & x = 0. \end{cases}$$

Intuitively it is clear that f has the intermediate value property. But it fails to be continuous in two senses. First, the graph, considered as a set, is not continuous in Cantor's sense: The upper points of the oscillations of the curve, i.e., the points

[46] Judith Grabiner, "Cauchy and Bolzano: tradition and transformation in the history of mathematics", in: Everett Mendelsohn (ed.), *Transformation and Tradition in the Sciences: Essays in Honor of I. Bernard Cohen*, Cambridge University Press, Cambridge, 1984, p. 112. A similar, earlier, discussion of the matter was given by Grabiner in: Judith V. Grabiner, *The Origins of Cauchy's Rigorous Calculus*, The MIT Press, Cambridge (Mass.), 1981, pp. 91–92. This book was reprinted by Dover Publications, Inc., in 2005. Accessible fuller quotations from Arbogast can be found in: C.H. Edwards, Jr., *The Historical Development of the Calculus*, Springer-Verlag, New York, 1979, pp. 303–304; Umberto Bottazzini (Warren van Egmond (trans.)), *The* Higher Calculus*: A History of Real and Complex Analysis from Euler to Weierstrass*, Springer-Verlag, New York, 1986, pp. 34–35; and (in German) Klaus Volkert, *Geschichte der Analysis*, Bibliographisches Institut & F.A. Brockhaus AG, Zürich, 1988, pp. 170–171.

$$\left\langle \frac{2}{(4n+1)\pi}, \sin\frac{(4n+1)\pi}{2} \right\rangle = \left\langle \frac{2}{(4n+1)\pi}, 1 \right\rangle$$

for $n = 0, 1, 2, \ldots$ converge to the point $\langle 0, 1 \rangle$, which is not on the curve. Viewed as a set, the curve is not closed under taking limits and is thus not perfect in Cantor's sense, i.e., he wouldn't accept it as continuous.

And it does not satisfy Arbogast's less emphasised condition mentioned by Grabiner that "functions...increase in increments whose sizes correspond to the sizes of the increments in the variable". This definitely fails at $x = 0$ where the tiniest increment Δx can take one from $f(0) = 0$ to $f(\Delta x) = \pm 1$.

2.2.6 Remark I confess to ignorance of the origin of this Example. Augustin Louis Cauchy (1789–1857) cites[47] the function $\sin(\frac{1}{x})$ as one which, as x tends to 0, "admits an infinity of limits between the limits -1 and $+1$"[48] but doesn't use the function as an explicit counterexample to anything. In an unfinished manuscript written in the 1830s or so, Bernard Bolzano (1781–1848) addresses the problem, but his explanation is a bit vague and only seems to yield a weak counterexample.[49] However, a bit later in the same work,[50] he cites the function

$$f(x) = \sin \ln(1 - x)$$

as one which oscillates infinitely often on the interval $[0, 1)$. This function readily yields an example analogous to that of Example 2.2.5, namely

$$f(x) = \begin{cases} \sin \ln |1 - x|, & x \neq 1 \\ 0, & x = 1. \end{cases}$$

Klaus Volkert credits Gaston Darboux (1842–1917) with being in 1875 the first to definitively answer in the negative the question of whether or not the intermediate value property implies continuity.[51] The oscillating sine function has become a popular example. Multiplication by x,

$$g(x) = \begin{cases} x \sin\left(\frac{1}{x}\right), & x \neq 0 \\ 0, & x = 0, \end{cases}$$

---

[47] Augustin Louis Cauchy, *Cours d'analyse de l'École Royale Polytechnique; I.*^{re} *Partie. Analyse algébrique* [*Course in Analysis of the Royal Polytechnical School; Part I. Algebraic Analysis*], de Bure, Paris, 1821. English translation: Robert E. Bradley and C. Edward Sandifer (eds. and trans.), *Cauchy's* Cours d'analyse*; An Annotated Translation*, Springer Science+Business Media, LLC, New York, 2009. The function $\sin(\frac{1}{x})$ is cited on p. 12 of the Bradley/Sandifer edition.

[48] The word "limit" is used in two senses here. The first occurrence refers to what we now call *limit points*; the second refers to the endpoints of the interval $[-1, 1]$ on the y-axis.

[49] The manuscript is called "Functionenlehre" ["Theory of functions"] and can be found in: Steve Russ (ed.), *The Mathematical Works of Bernard Bolzano*, Oxford University Press, Oxford, 2004. For Bolzano's counterexample, cf. pp. 471–472 (§§83–84, but see also §46, pp. 453–454).

[50] *Ibid.*, p. 481, §102.

[51] Volkert, *op. cit.*, p. 187. Cf. Lemma 3.1.5 on page 187, below.

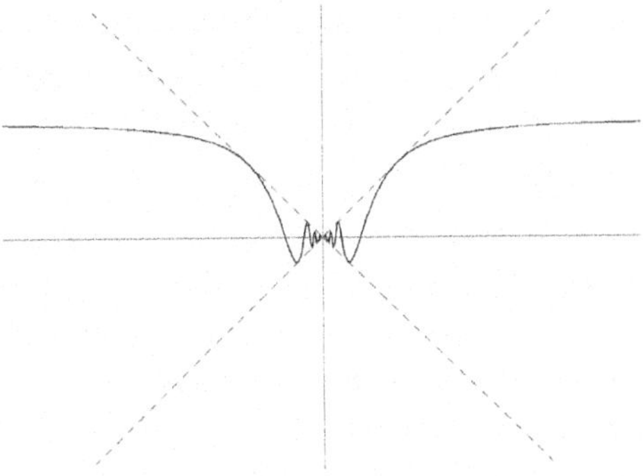

Fig. 2.25 Graph of $x\sin(1/x)$

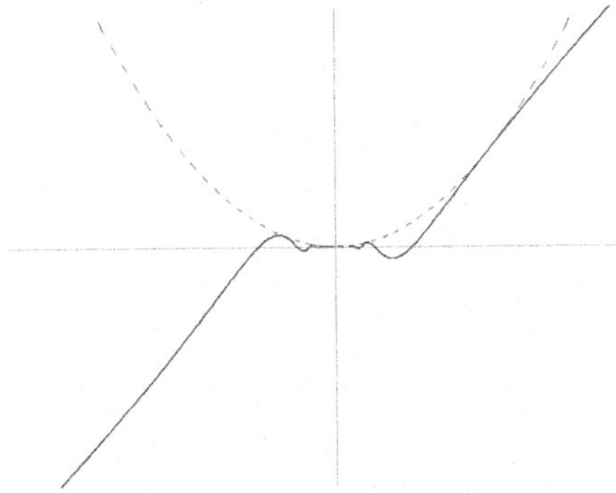

Fig. 2.26 Graph of $x^2\sin(1/x)$

results in a continuous function (see Fig. 2.25.) with infinite arc length over any interval containing 0. And Darboux uses the variant,

$$h(x) = \begin{cases} x^2 \sin\left(\frac{1}{x}\right), & x \neq 0 \\ 0, & x = 0, \end{cases}$$

(See Fig. 2.26.), which we will refer to in the sequel as *Darboux's function*, as an example of an infinitely oscillating differentiable function with a discontinuous derivative at $x = 0$.[52] Functions of the form $f(x) = x^\alpha \sin(\frac{1}{x})$ for $\alpha > 0$ form a rich class of counterexamples in the Calculus.[53]

Returning to the intermediate value property, we see in it a manifestation of continuity — the "flowing of a point" —, but not a sufficient condition for an adequate definition thereof. Moreover, if we reflect on general curves $\gamma : I \to \mathbb{R}\times\mathbb{R}$, we must ask: if we were to try to define continuity for $f : I \to \mathbb{R}$ by means of the intermediate value property, how could we extend this definition to γ, i.e., what is the proper ana-

[52]Gaston Darboux, "Mémoire sur les fonctions discontinues", *Annales scientifiques de l'École Normale Supérieure*, 2nd series, vol. 4 (1875), pp. 57–112; here: p. 109.

[53]H. Turgay Kaptanoğlu, "In praise of $y = x^\alpha \sin(\frac{1}{x})$", *American Mathematical Monthly* 108 (2001), pp. 144–150.

logue of the intermediate value property in two-dimensional space? The intermediate value property emerges as more of a deflection from our goal than a path towards it. Nevertheless, it is a useful property for a function to have and numerous mathematicians attempted to prove that those functions defined by "analytic expressions" possessed the property. Bolzano, who gave the first correct proof for continuous functions, cites attempts by Abraham Gotthelf Kästner, Alexis Claude Clairaut, Sylvestre François Lacroix, Mathias Metternich, Georg Simon Klügel, Joseph Louis Lagrange, and Christian Lebrecht Rösling, and others.[54] To carry out his proof, Bolzano gave the first "correct" definition of continuity.

I have put cautionary quotation marks around the word "correct" to emphasise that correctness here is not absolute. The definition is correct in that it more-or-less agrees with our modern definition. And it is correct in that it works — those functions we think of as being continuous are continuous under his definition, and theorems we would expect to hold for continuous functions are indeed provable. Without further ado, I quote Bolzano's definition.

> Following the *correct definition* one understands by the expression that *a function* $f(x)$ *varies according to the law of continuity for all values of* x*, which lie inside or outside certain bounds, only so far as, if* x *is any such value, the difference* $f(x+\omega) - f(x)$ *can be made to be smaller than any given value, if* x *is so small as one wishes to make it.*[55]

In a footnote, Bolzano explains the conditions on the domain:

> There are functions which vary continuously for all values of their arguments, e.g. $\alpha x + \beta x$. But there are others which vary according to the law of continuity only within or without certain limits of their roots. So $x + \sqrt{(1-x)(2-x)}$ varies continuously only for all values of x which are $< +1$ or $> +2$; but not for the values which lie between $+1$ and $+2$.[56]

And in the text itself he remarks on the intermediate value property as not defining continuity:

> But...the continuous function never reaches a higher value without first going through all lower values, i.e., that $f(x+n\Delta x)$ can take on every value lying between $f(x)$ and $f(x+\Delta x)$, if one takes n arbitrarily between 0 and $+1$: this is indeed a very true conjecture, but it cannot be seen as the *definition* of the concept of continuity but rather is a *theorem* about the same.[57]

---

[54] Bernard Bolzano, *Rein analytischer Beweis des Lehrsatzes, daß zwischen je zwey Werthen, die ein entgegengesetztes Resultat gewählen, wenigstens eine reelle Wurzel der Gleichung liege*, Gottlieb Haase, Prague, 1817. The work also appeared the following year in volume 5 of the *Abhandlungen der königlichen böhmischen Gesellschaft der Wissenschaften*, and was edited and reprinted by Philip E.B. Jourdain in 1905 as half of number 153 of *Ostwalds Klassiker der exakten Wissenschaften*. English translations by Steve Russ and William Ewald appeared first in 1980 and 1996, respectively. The most recent English version appears in Russ's edition of *The Mathematical Works*, *op. cit.* Below, I shall refer to the *Ostwald Klassiker* reprint as "Bolzano, *Klassiker*" in what follows, but will also give references to *The Mathematical Works* for English translations. Thus, the above list of names can be found in: Bolzano, *Klassiker*, p. 6; Russ, *op .cit.*, p. 253.

[55] Bolzano, *Klassiker*, pp. 3–4, Russ*op. cit.*, p. 256.

[56] *Ibid.* Presumably Bolzano intends "<" here to read "less in absolute value than".

[57] Bolzano, *Klassiker*, p. 6; Russ, *op. cit.*, pp. 256–257.

It was in 1817 that Bolzano published his definition of continuity in a short pamphlet, the long title of which translates to *Purely Analytic Proof of the Theorem that between any two Values, which give Results of Opposite Sign, there lies at least one real Root of the Equation*. This is one of the major works in the history of the foundations of the Calculus, including the definition of continuity, the introduction of Cauchy sequences and a proof of their convergence,[58] a proof of the least upper bound property as a corollary, yielding the Bolzano-Weierstrass Theorem[59] in the process, and, finally, the statement and proof of the Intermediate Value Theorem:

> *Theorem.* If two functions of x, $f(x)$ and $\varphi(x)$, vary *according to the law of continuity* either for *all* values of x or for all which lie between α and β, if further $f(\alpha) < \varphi(\alpha)$ and $f(\beta) > \varphi(\beta)$, then between α and β there is always a value of x for which $f(x) = \varphi(x)$.[60, 61]

Four years after Bolzano's paper was published, Cauchy's lectures, the famous *Cours d'analyse*, offering an independent treatment, was published. This work is the first part of what was intended to be a two-part textbook on the Calculus. The second part, which would have included the Differential and Integral Calculus was delayed a couple of years and published under a different title. The first part, which runs several hundred pages, does not get as far as differentiation or integration, but lays the foundations of the Calculus, treating the real numbers, continuity, infinite series, complex numbers, and related topics. In the midst of all of this is Cauchy's definition of continuity:

---

[58] A *Cauchy sequence* is a sequence $a_0, a_1, a_2, \ldots$ of numbers satisfying: for any $\epsilon > 0$ a number n_0 can be found such that for all $m, n > n_0$ one has $|a_m - a_n| < \epsilon$. The convergence of such sequences had been used without note by Euler. Bolzano drew attention to them and proved their convergence relative to his notion of real number as incompletely treated in a later work not published in his lifetime. Jacqueline Stedall finds the proof "incorrectly argued" (Stedall, *op. cit.*, p. 496):

> It turned out to be more difficult than it might seem, and Bolzano was forced to introduce [as] a fresh assumption the existence of a quantity X to which the terms of the series approach as closely as we please. Such a hypothesis, Bolzano claimed 'contains nothing impossible'..., but it was precisely what he was trying to prove in the first place. The problem was deeper than Bolzano realized. Convergence of Cauchy sequences requires *completeness* of the real numbers or, simply speaking, that the number line is an unbroken continuum with no gaps. Convergence of Cauchy sequences is in fact mathematically *equivalent* to completeness: either must be assumed in order to prove the other. Without some such assumption, Bolzano was forced to introduce his hypothetical quantity X.

This is a fair criticism, but I give Bolzano full credit nonetheless as he later offered some justification for his variant of completeness on which his proof of the convergence of Cauchy sequences was based. I discuss this sort of thing in some detail in Smoryński, *Formalism*, *op. cit.*, pp. 232–265.

[59] The Bolzano-Weierstrass Theorem asserts that any bounded sequence $a_0, a_1, a_2, \ldots$ of numbers, i.e., any such sequence for which there is a bound $B > |a_n|$ for all n, has a convergent subsequence. It is a fundamental result of Analysis.

[60] Bolzano, *Klassiker*, p. 31; Russ, *op. cit.*, p. 273.

[61] Bolzano is a little sloppy here: In his example cited above, the law of continuity is two-sided and does not apply to the endpoints α, β of an interval, but his proof of the Theorem assumes the one-sided continuity of f and ϕ at the endpoints of the interval.

> Let $f(x)$ be a function of the variable x, and suppose that for each value of x between two given limits, the function always takes a unique finite value. If, beginning with a value of x contained between these limits, we add to the variable x an infinitely small increment α, the function itself is incremented by the difference
>
> $$f(x+\alpha) - f(x),$$
>
> which depends both on the new variable α and on the value of x. Given this, the function $f(x)$ is a *continuous* function of x between the assigned limits if, for each value of x between these limits, the numerical value of the difference
>
> $$f(x+\alpha) - f(x)$$
>
> decreases indefinitely with the numerical value of α. In other words, *the function $f(x)$ is continuous with respect to x between the given limits if, between these limits, an infinitely small increment in the variable always produces an infinitely small increment in the function itself.*
>
> We also say that the function $f(x)$ is a continuous function of the variable x in a neighborhood of a particular value of the variable x whenever it is continuous between two limits of x that enclose that particular value, even if they are very close together.
>
> Finally, whenever the function $f(x)$ ceases to be continuous in the neighborhood of a particular value of x, we say that it becomes discontinuous, and that there is a *solution of continuity* for this particular value.[62]

In 1817 Bolzano had been careful to avoid using the words "infinitely small". And half a century later, in completing the work of Bolzano and Cauchy on the "arithmetisation of analysis" as their programmes of bringing rigour to the Calculus came to be called, Karl Weierstrass (1815–1897) treated these words as a mere figure of speech. Cauchy, however, used infinitesimals in an essential way. To him, a function continuous in an interval was continuous at all numbers in the interval,[63] not just at the real numbers in the interval. This has powerful consequences and Cauchy's notion of continuity is strictly stronger when the domain is an open interval than is ordinary continuity.

Speaking of ordinary continuity, Weierstrass gives an equally prosaic definition in his lectures of 1861:

> If $f(x)$ is a function of x and x is a definite value, then the function will change into $F(x+h)$[64] if x passes from x to $x+h$; the difference $f(x+h) - f(x)$ is called the change which the function experiences through the passage of the argument from x to $x+h$. If it is now possible to determine a bound δ for h so that for *all* values of h of absolute value smaller than δ, $f(x+h) - f(x)$ will be smaller than any ε however small, one says to infinitely small changes in the argument correspond infinitely small changes of the function. Because one says, if the absolute value of a quantity can be made smaller than any arbitrarily chosen value, however small, it can be chosen infinitely small. If now a function is so obtained that [to] infinitely small changes in the argument correspond infinitely small changes in

---

[62] Bradley and Sandifer, *op. cit.*, p. 26. The editors explain that "solution of continuity" is to be read as "dissolving of continuity", i.e., the breakdown of continuity is meant. Note again, as in footnote 48, the use of the word "limits" to mean "endpoints".

[63] I.e., at every number $r+\eta$ in the interval, where r is real and η is infinitesimal.

[64] *Sic*. This should read $f(x+h)$.

> the function, then one says this function is a *continuous function* of its argument, or that it changes continuously with this argument. — For individual values of the arguments of functions, which are in general continuous, the continuity can become interrupted. For such values the function will be discontinuous.[65]

The truth be told, this is not at first sight any clearer than Bolzano's or Cauchy's definitions. All three are at least partly ambiguous. Bolzano and Cauchy depart from our modern practice of defining what it means for a function to be continuous at a point, and define what it means for a function to be continuous on an open interval or intervals. Both authors can produce δ for given ϵ when necessary, but it is not clear from the given definitions whether δ depends only on ϵ and the interval in question, or if it is allowed to depend on x as well. Reading their proofs suggests Bolzano allows δ to depend on ϵ and x (ordinary continuity) while Cauchy insists δ depends only on ϵ (uniform continuity). Weierstrass starts out defining continuity at a point, but his continued clarification makes this less clear.

Bolzano was writing a major work on the foundations of the Calculus when he died and he never completed the task, his impressive partial work only first published in the 20th century. Cauchy prepared lectures on the subject and published them. They were widely read in France and Germany. Weierstrass lectured regularly on *Functionenlehre*, the Theory of Functions, covering real and complex number systems, and the foundations of the Calculus and that of the theory of functions of a complex variable. He generally did not publish the results of these lectures, but copies were deposited in Mathematical Institute libraries around Germany and, additionally, his students were not shy about publishing expositions of the work of their master. One of these was Eduard Heine (1821–1881), whose "Die Elemente der Functionenlehre"[66] ["The elements of the theory of functions"] is occasionally cited as the first published modern definition of continuity.

In this paper, Heine begins by singing the praises of Weierstrass:

> The advance of the Theory of Functions is actually limited by the circumstance, that certain of its elementary propositions, although proven by a penetrating researcher, will still be doubted, so that the results of an investigation are not held universally as correct, if they rest on these indispensable fundamental assertions. The explanation I find therein, that indeed the principles of Mr. Weierstrass, directly through his lectures and other oral communications, indirectly through copies of exercise books, which would have been worked out following these lectures, themselves having been disseminated in wider circles, that however have not been published by him himself in authentic versions, so that there is no place at which one can find these propositions *developed in context*.[67]

Heine's paper divides into a Part A on numbers and a Part B on functions. Part A gives an infinitistic construction of the real numbers from the rational numbers

---

[65] Karl Weierstrass and Hermann Amandus Schwarz, *Differential Rechnung, nach einer Vorlesung des Herrn Weierstrass im Sommersemester 1861*, Hdschr. Koll. N 37 (Humboldt-Universität zu Berlin), pp. 2–3.

[66] Eduard Heine, "Die Elemente der Functionenlehre", *Journal für die reine und angewandte Mathematik* 74 (1872), pp. 172–188.

[67] *Ibid.*, p. 172.

using Cauchy sequences and depends more on Cantor than on Weierstrass.[68] Part B on functions followed Weierstrass more closely.

In Part B, §1 Heine gave a modern definition of function and in §2 he led off with a definition of what is now called *pointwise continuity*:

> 1. *Definition.* A function $f(x)$ is called *continuous at a given individual value* $x = X$ if, for every given magnitude ε, however small, there exists another positive number η_0 of such a nature, that for no positive magnitude η, which is smaller than η_0, does the absolute value of $f(X \pm \eta) - f(X)$ exceed ε.[69]

There may be some awkwardness in the phrasing, but there is no ambiguity. It agrees substantially with the modern definition. The only major difference is that we are more explicit in assuming f to be defined on an interval around the given point at which f is to be continuous:

2.2.7 Definition Let I be an interval, $f : I \to \mathbb{R}$ a function defined for all elements of I, and $x \in I$ a point in the interval. f is *continuous at* x if for every $\epsilon > 0$ there is a $\delta > 0$ such that for all $y \in I$ whenever $|y - x| < \delta$ we have $|f(y) - f(x)| < \epsilon$:

$$\forall \epsilon > 0 \exists \delta > 0 \forall y \in I \big(|y - x| < \delta \Rightarrow |f(y) - f(x)| < \epsilon \big).$$

Defining $|\langle x, y\rangle| = \sqrt{x^2 + y^2} = \text{dist}(\langle 0, 0\rangle, \langle x, y\rangle)$ for $\langle x, y\rangle \in \mathbb{R} \times \mathbb{R}$, we can readily adapt this definition to functions $\gamma : I \to \mathbb{R} \times \mathbb{R}$:

2.2.8 Definition Let I be an interval, $\gamma : I \to \mathbb{R} \times \mathbb{R}$ a function defined for all elements of I, and $t \in I$ a point in the interval. γ is *continuous at* t if

$$\forall \epsilon > 0 \exists \delta > 0 \forall t' \in I \big(|t' - t| < \delta \Rightarrow |\gamma(t') - \gamma(t)| < \epsilon \big).$$

We need one more definition before we can define what it is for a curve to be continuous.

2.2.9 Definition Let I be an interval and f a function mapping I either to $\mathbb{R}$ or $\mathbb{R} \times \mathbb{R}$. f is said to be *continuous on* I if f is continuous at all points in I.

With this, we have the following.

2.2.10 Definition Let $C \subseteq \mathbb{R} \times \mathbb{R}$ be a curve. C is a *continuous curve* if there is an interval I and a continuous function $\gamma : I \to \mathbb{R} \times \mathbb{R}$ such that $C = \gamma(I)$, i.e., C is a continuous curve if it has a continuous parametrisation.

---

[68] In the 1830s Bolzano offered a description of real numbers that nowadays one would treat as such a construction, but this went unpublished until the 20th century. At some later, undetermined, date (cf. pp. 334–335, below), Weierstrass offered such a construction treating real numbers as abstract sums of rationals. And in 1858 Richard Dedekind independently constructed the reals using sets of rationals. None of this was published until 1872 when several such constructions, new and old, simultaneously made it into print. Cantor's, Charles Méray's, and Heine's constructions used Cauchy sequences.

[69] Heine, *op. cit.*, p. 182.

2.2.2 Properties of Continuity

We have finally fulfilled our promise that we would define what a continuous curve is in this section, and presumably we could now turn to our next task, namely, that of defining what we mean by a smooth curve. The reader who has had his fill of the discussion of continuity will doubtless want to skip ahead to the next section for this topic. I beg the reader's indulgence, however, as there is a bit more to be said about continuity, both generalities and results that will be needed later.

The first thing to note is that the definition of continuity at a point is an unnatural and non-intuitive notion. For centuries continuity referred to a flow, a smoothness of motion, a blending (as of shades of colour), or the non-existence of gaps — it had nothing whatsoever to do with a single stationary point. The natural notion refers to continuity on an interval: the point flows over an interval of time, the motion is smooth for a while, etc. What we have with Definition 2.2.7 is a technically useful generalisation of the expected formalisation of the concept of a function's being continuous on an interval. Bolzano's and Cauchy's definitions were of the continuity of a function on an interval. It is not clear which of the two notions Weierstrass is referring to in the translated quotation from his 1861 lecture; he begins by referring to a "definite value" x, but finishes with the sentence,

> If now a function is such that to infinitely small changes of the argument...

Does "the argument" still refer to a fixed x or is he referring to arbitrary elements in the function's domain? In the intervening years, as he lectured repeatedly on the Theory of Functions, Weierstrass undoubtedly cleared up the ambiguity, but in print we have Heine to thank for this clarification.

Heine leads off §3 of his paper defining *two* notions of continuity on a closed interval:

> 1. *Definition.* A function $f(x)$ is called *continuous* from $x = a$ to $x = b$, if it is continuous (B, §2, Def. 1)[70] for each individual value $x = X$ between $x = a$ and $x = b$, with the inclusion of the values a and b; it is called *uniformly continuous* from $x = a$ to $x = b$, if for every magnitude ε, however small, there is a positive magnitude η_0 such that for all positive values η, which are smaller than η_0, $f(x \pm \eta) - f(x)$ remains below ε. Whatever value one may give to x, assuming only that x and $x \pm \eta$ belong to the region from a to b, *the same* η_0 must yield the demanded [inequality].[71, 72]

Once again, the use of natural language makes for awkward phrasing and mathematicians might prefer using more precise formal language:
f is continuous on I if

$$\forall x \in I\, \forall \epsilon > 0\, \exists \delta > 0\, \forall y \in I \big(|y - x| < \delta \Rightarrow |f(y) - f(x)| < \epsilon \big);$$

---

[70] The reference is to his earlier definition cited on page 183, above.

[71] *Ibid.*, p. 184.

[72] One way of visualising this is to imagine a rectangle $[a-\delta, a+\delta] \times [f(a)-\epsilon, f(a)+\epsilon]$ of fixed size $2\delta \times 2\epsilon$. As one moves $\langle a, f(a)\rangle$ along the curve, the graph over the interval $[a-\delta, a+\delta]$ always remains inside the rectangle.

and f is uniformly continuous on I if

$$\forall \epsilon > 0 \exists \delta > 0 \forall x \in I \forall y \in I \big(|y - x| < \delta \Rightarrow |f(y) - f(x)| < \epsilon \big).$$

One would then point out that in the former definition δ depends on x as well as on ϵ, while in the latter case, given ϵ, the choice of δ is uniform for all x.

But I betray here my background as a mathematical logician. Mathematicians are not generally so happy with alternations of quantifiers and prefer instead to introduce the *modulus of continuity*, by which is meant a function yielding δ — a function $\delta(x, \epsilon)$ of two arguments in the former case and a function $\delta(\epsilon)$ in the latter.

Without great care in its formulation, the definition of continuity on an interval can be read either as continuity on the interval or as uniform continuity on the interval. It is generally agreed that Bolzano had in mind the former concept and it has been put forward without yet achieving universal agreement that Cauchy meant the latter. In work that lay unpublished until the 20th century, in the 1830s Bolzano recognised the distinction and proved[73] that a function continuous on a closed bounded interval is uniformly continuous there. Cauchy apparently never noted the distinction, always dealing with the uniform notion.

In the United States uniform continuity is not mentioned in the introductory Calculus course, being deemed a topic for an advanced course in the subject. It is not deeper or more difficult a concept than ordinary continuity on an interval. Indeed, direct proofs that various functions are continuous usually yield uniform continuity and I imagine the failure to mention the notion is the fact that one would feel obliged to discuss the relation between the two notions of continuity on an interval, a relation easy enough to state but requiring a proof taking one to a higher level of abstraction. There are, however, several results that are asserted without proof in the first year Calculus course — the Intermediate Value Theorem, the Extreme Value Theorem, the existence of the integral of a continuous function on a closed bounded interval, etc.

In a footnote in his paper "Die Elemente der Functionenlehre",[74] Heine remarks that the results of §3 of his paper generally follow the principles of Weierstrass, Heine himself contributing only to the details of execution. The important results in fact predate Weierstrass: the Intermediate Value Theorem (Bolzano 1817, Cauchy

[73]His exposition is muddled and not everyone accepts it, but a correct proof was certainly within his grasp. Cf. pages 301–302, below, for details.

[74]Heine, *op. cit.*, p. 182.

1821), the Extreme Value Theorem (Bolzano 1830s),[75] and the Uniform Continuity Theorem (Bolzano 1830s).

Both the Intermediate Value Theorem and the Extreme Value Theorem are intimately connected with the Mean Value Theorem and I ought to say something about their proofs. The Uniform Continuity Theorem is not obviously[76] as central to our present purpose, but its discussion is not a great digression, and, in any event, the result will be used repeatedly in the sequel.

2.2.11 Theorem (Intermediate Value Theorem) *Let* $a < b$ *and let* $f : [a, b] \to \mathbb{R}$ *be continuous on* $[a, b]$. *Suppose* $f(a) < 0 < f(b)$. *Then there is some* $c \in (a, b)$ *such that* $f(c) = 0$.

Proof sketch. Probably the simplest proof uses infinite integers and infinitesimals, *à la* Euler and Cauchy[77] Let N be an infinite integer and consider the *finite[78] set of values

$$f(a), f\left(a + \frac{b-a}{N}\right), f\left(a + 2\frac{b-a}{N}\right), \ldots, f\left(a + N\frac{b-a}{N}\right).$$

Let K be the smallest integer such that $f\left(a + K\frac{b-a}{N}\right) \geq 0$. Then $a + K\frac{b-a}{N}$ differs from a standard real c by an infinitesimal amount. $f(c)$ must of necessity be 0 as, by continuity it is infinitesimally close to $f\left(a + (K-1)\frac{b-a}{N}\right) < 0$ and to $f\left(a + K\frac{b-a}{N}\right) \geq 0$. □

For one not familiar with Nonstandard Analysis, this may make no sense at all, but it is quite rigorous. The nice thing about it, in addition to its simplicity, is that it adapts quickly to a proof of the Extreme Value Theorem.

2.2.12 Theorem (Extreme Value Theorem) *Let* $a < b$ *and let* $f : [a, b] \to \mathbb{R}$ *be continuous on* $[a, b]$. *There are* $c, d \in [a, b]$ *such that for all* $x \in [a, b]$,

---

[75]In Craig Smoryński, *A Treatise on the Binomial Theorem*, College Publications, London, 2012, p. 138, I also credit Cauchy with a proof of this Theorem. In glancing over his two main textbooks I have not found the result proven although it is appealed to in the *Résumé des leçons données a l'École Royale Polytechnique sur le calcul infinitesimal*, de Bure, Paris, 1823. The nonstandard proofs of the Intermediate Value Theorem, which Theorem is proven in the *Cours d'analyse*, and the Extreme Value Theorem being virtually identical, I must have simply assumed Cauchy had proven the latter. It would naturally have fit into the projected second volume of the *Cours*. As this volume was intended as a textbook and policies at the École had changed, Cauchy did not include as much foundational material in the *Résumé* when he came to write this later. So he might have proven the result and simply neglected to include the proof in any of his textbooks.

[76]But see Sect. 6, below.

[77]Such a proof, long discredited, is nowadays acceptable thanks to the rigourous foundation and development of Nonstandard Analysis.. The reader unfamiliar with these modern developments may consider the proof merely heuristic. The curious reader who would like to know more is referred to Chapter II, Sect. 6, of Smoryński, *Formalism*, for an introduction to and some references on the subject.

[78]In Nonstandard Analysis, a set of nonstandard numbers is called *finite if it can be put into one-to-one correspondence with an integer, finite or infinite, by an "internal" function. In simple terms, a *finite set is a possibly infinite set that behaves like a finite set.

$$f(c) \leq f(x) \leq f(d),$$

i.e., f assumes minimum and maximum values on $[a, b]$.

Proof sketch. Again, let N be an infinite integer and again consider the values

$$f(a), f\left(a + \frac{b-a}{N}\right), f\left(a + 2\frac{b-a}{N}\right), \ldots, f\left(a + N\frac{b-a}{N}\right).$$

This set has a least element $f\left(a + K_1\frac{b-a}{N}\right)$ and a greatest element $f\left(a + K_2\frac{b-a}{N}\right)$. Then c and d are the "standard parts" of $a + K_1\frac{b-a}{N}$ and $a + K_2\frac{b-a}{N}$, respectively, i.e., the real numbers infinitesimally close to these. □

Those not schooled in Nonstandard Analysis might find these proofs too simple to be believable. The usual proofs are a bit deeper, requiring an iterated partitioning, a sequence of approximations, an appeal to the convergence of Cauchy sequences, and, in the latter proof, an invocation of the Bolzano-Weierstrass Theorem. Against this I point out that much of the difficulty in the standard proofs has been transferred in Nonstandard Analysis to the construction of the nonstandard reals. There is thus no reason to distrust the nonstandard proofs on the grounds of their simplicity.

Note that the Extreme Value Theorem of necessity is valid for closed intervals $[a, b]$ and not open ones (a, b). The open interval (a, b) being without extreme values itself maps trivially to an interval without extreme values via any strictly increasing function, e.g., $f(x) = x$.

2.2.13 Exercise Graph the following functions and find open intervals over which they fail to satisfy the Extreme Value Theorem:

i. $y = f(x) = \dfrac{x}{(x+1)(x-1)}$.

ii. $y = \dfrac{x^2}{(x+1)(x-1)}$.

Bolzano originally proved Theorem 2.2.11 by appeal to the Least Upper Bound Principle, which he proved by appeal to the convergence of Cauchy sequences. In standard Analysis, some completeness axiom must be assumed. My own preference is to choose the Least Upper Bound Principle itself.

2.2.14 Definitions Let $X \subseteq \mathbb{R}$ be a nonempty set of real numbers. A number B is an *upper bound* on X if no element of X exceeds B:

$$\forall x \in X (x \leq B).$$

A number B_0 is the *least upper bound* of X if B_0 is an upper bound on X and if $B_0 \leq B$ for any other upper bound B on X.

Note that B_0 need not be an element of X itself. For example, 1 is the least upper bound of the interval $(0, 1)$ yet does not belong to the interval. It does, however, belong to the set of upper bounds on the interval, being itself the least element of this set.

The Least Upper Bound Principle is simply the assertion that bounded nonempty sets possess least upper bounds and optionally follows from or is taken as the completeness axiom of the reals.

2.2.15 Axiom (*Completeness Axiom; Least Upper Bound Principle*) Let $X \subseteq \mathbb{R}$ be a bounded nonempty set of real numbers. There is a least upper bound of X.

2.2.16 Remark One can also define the Greatest Lower Bound of a set bounded below and postulate a Greatest Lower Bound Principle. But this new Principle is redundant: if B_0 is a least upper bound for $\{-x \mid x \in X\}$, then $-B_0$ is the greatest lower bound of X.

From this axiom Theorems 2.2.11 and 2.2.12 are easily derived.

Proof of Theorem 2.2.11. Let $f : [a, b] \to \mathbb{R}$ be continuous with $f(a) < 0 < f(b)$. Define

$$X = \big\{x \in [a, b] \,\big|\, \forall y \in [a, b]\big(y \le x \Rightarrow f(y) < 0\big)\big\}.$$

X is nonempty since $a \in X$ and it has b as an upper bound as $X \subseteq [a, b]$. By the Least Upper Bound Principle, X has a least upper bound c. The claim is that $f(c) = 0$. To prove this we need a simple lemma.

2.2.17 Lemma *Let $f : [a, b] \to \mathbb{R}$ be continuous and let $x \in [a, b]$ be such that $f(x) \neq 0$. There is a $\delta > 0$ such that, for all $y \in [a, b]$,*

$$|y - x| < \delta \Rightarrow f(y) \neq 0 \text{ and } f(y) \text{ has the same sign as } f(x).$$

Proof Assume for the sake of definiteness that $f(x) > 0$. Let $\epsilon = f(x)/2$ and choose $\delta > 0$ such that, for all $y \in [a, b]$,

$$|y - x| < \delta \Rightarrow |f(y) - f(x)| < \epsilon,$$

i.e.,

$$\begin{aligned}
|y - x| < \delta \Rightarrow |f(y) - f(x)| &< \frac{f(x)}{2} \\
\Rightarrow -\frac{f(x)}{2} &< f(y) - f(x) < \frac{f(x)}{2} \\
\Rightarrow f(x) - \frac{f(x)}{2} &< f(y) - f(x) + f(x) \\
\Rightarrow f(y) > \frac{f(x)}{2} &> 0.
\end{aligned}$$

□

Returning to the proof of Theorem 2.2.11, assume $f(c) \neq 0$. Choose δ according to the Lemma such that for $x \in [a, b]$ $f(x)$ has the same sign as $f(c)$ whenever $x \in (c-\delta, c+\delta)$. If $f(c) < 0$, then $c < b$, and for $\eta = \min\{\delta, b-c\}$,

$$\begin{aligned} c < x \leq c + \frac{\eta}{2} &\Rightarrow x \in (c-\delta, c+\delta) \;\&\; x \in [a, b] \\ &\Rightarrow f(x) < 0 \\ &\Rightarrow c + \frac{\eta}{2} \in X \\ &\Rightarrow c \text{ is not an upper bound of } X. \end{aligned}$$

Similarly, $f(c)$ cannot be positive.

It follows that $f(c) = 0$ and, since neither $f(a)$ nor $f(b)$ is 0, c lies in the interior of the interval. □

Proof of Theorem 2.2.12. I outline the proof for the existence of a maximum.

This is slightly more complicated than the proof of Theorem 2.2.11. First we show that $\{f(x) \mid x \in [a, b]\}$ is bounded by considering the set

$$X = \{x \in [a, b] \mid \exists B \, \forall y \in [a, b] (y \leq x \Rightarrow f(y) \leq B)\}.$$

X is again nonempty because $a \in X$ and it is again bounded by b because $X \subseteq [a, b]$. Let c be its least upper bound. The claim is that $c = b$. To prove this we need another simple lemma.

2.2.18 Lemma *Let* $f : [a, b] \to \mathbb{R}$ *be continuous and let* $x \in [a, b]$. *There is a* $\delta > 0$ *such that* f *is bounded on* $(x-\delta, x+\delta)$.

Proof Let $\epsilon > 0$ and choose δ such that, for all $y \in [a, b]$,

$$|y - x| < \delta \Rightarrow |f(y) - f(x)| < \epsilon.$$

For such y one has $f(y) < f(x) + \epsilon$, whence $f(x) + \epsilon$ is the bound sought. □

Continuing the proof of the Theorem, note that $c \neq a$ since the Lemma yields a bound on $\{f(y) \mid a \leq y < a + \delta\}$ for some δ. In particular $a + \frac{\delta}{2} \in X$ and $a < a + \frac{\delta}{2} < c$.

If $c < b$, apply the Lemma to c: For any choice of ϵ, e.g., $\epsilon = 1$, there is a $\delta > 0$ such that, for all $y \in [a, b]$,

$$|y - c| < \delta \Rightarrow f(y) < B_1,$$

for some bound B_1. Let $\eta = \min\{\delta, c-a, b-c\}$. Now $c - \frac{\eta}{2} < c$ and is an element of $[a, b]$, whence $c - \frac{\eta}{2} \in X$ and there is a bound B_0 such that

$$\forall x \in [a, b] \left(x \leq c - \frac{\eta}{2} \Rightarrow f(x) < B_0\right).$$

But $\left[c - \frac{\eta}{2}, c + \frac{\eta}{2}\right] \subseteq (c - \delta, c + \delta)$, so

$$\forall x \in [a, b] \left(x \in \left[c - \frac{\eta}{2}, c + \frac{\eta}{2}\right] \Rightarrow f(x) < B_1\right).$$

Hence, if $x \in [a, b]$ is less than or equal to $c + \frac{\eta}{2}$, $f(x) < \max\{B_0, B_1\}$. Thus $c + \frac{\eta}{2} \in X$, contrary to the assumption that c is an upper bound of X. Hence $c = b$ and f is bounded on $[a, b]$.

Now define a new function on $[a, b]$ by

$$g(x) = \text{least upper bound of } \left\{f(y) \,\middle|\, a \leq y \leq x\right\}.$$

g is defined because $\left\{f(y) \,\middle|\, a \leq y \leq x\right\} \subseteq f([a, b])$ which we have just shown is bounded.

2.2.19 Exercise Show that g is continuous on $[a, b]$.

Finally, let c be the least upper bound of

$$Y = \left\{x \in [a, b] \,\middle|\, g(x) < g(b)\right\}.$$

2.2.20 Exercise Show that $f(c) = g(b) =$ maximum value of f on $[a, b]$.

With this Exercise the reader has finished the alternative proof of the Extreme Value Theorem. □

2.2.21 Remark I am beginning to think it might have been a mistake not to have given the usual proof using Cauchy sequences and the Bolzano-Weierstrass Theorem. These proofs of Theorems 2.2.11 and 2.2.12 can be motivated by two words: *continuous induction*. The Least Upper Bound Principle is a continuous analogue to the *Least Number Principle* in arithmetic whereby every non-empty set of natural numbers contains a least element. Contrapositive to the Least Number Principle is the *Strong Form of Mathematical Induction*, also called the *Principle of Complete Induction*, which is equivalent to the usual *Principle of Mathematical Induction* one learns in pre-Calculus courses and applies quite often in the Calculus course. There is an analogous principle based on the Least Upper Bound Principle called continuous induction. According to it, to prove a property P holds for all $x \in [a, b]$ one has only to show that it holds in some interval $[a, b_0)$ and that, if it holds in any interval $[a, b_0)$, it must hold in some interval extending $[a, b_0)$, either $[a, b]$ if $b = b_0$ or $[a, b_1)$ for some $b_1 > b_0$ if $b \neq b_0$. I leave the proof of the principle of continuous induction by appeal to the Least Number Principle as a nice exercise and invite the reader to apply it to either replace or explain the proof of the Uniform Continuity Theorem, which is coming up, in terms of such an induction.

Continuous induction has not been traditionally presented in Analysis courses and I confess not to have recognised it initially on presenting these proofs. When writing the next section, on similarly proving a theorem called the Strictly Increasing

Function Theorem, I added a comment on continuous induction as a heuristic to take the edge off what appears to be an overly complicated proof. It did not occur to me to carefully formulate and apply such a principle. It was only after completing the book when Robert B. Burckel mentioned it with respect to the proof given in Chap. 3, Sect. 3.10.2 of the Heine-Borel Theorem that the principle fully entered my consciousness. It is a venerable principle, going back at least to Lebesgue who outlined its use in proving the Heine-Borel Theorem in 1904[79] and more explicitly to Y.R. Chao who in 1919[80] seems first to have explicitly formulated and named a variant of the principle. Since then it has been repeatedly forgotten and rediscovered. I am of two minds on the use of continuous induction in proving these results. On the one hand, it seems to be more elegant than the approach I have taken. On the other, the details of the inductive proof are pretty much the same as those I've given. The difference is that in the induction step of the inductive proof, one goes from $[a, b_0)$ to $[a, b_1)$ while I go from $[a, c)$ to $[a, b_1)$, where c is the demarcation between constant validity of P and occasional non-validity of P. In the inductive proof one concludes P holds in $[a, b]$ by induction, while I conclude c must be b and P holds in $[a, b]$. As I like arguing from first principles, I haven't replaced my proofs by the more elegant approach. I thus leave the conversion of my proofs to applications of continuous induction to the more ambitious reader. For the curious, but less ambitious, reader I note that Pete L. Clark has written a very nice survey[81] of continuous induction with several applications and bibliographic references.

2.2.22 Theorem (Uniform Continuity Theorem) *Let* $f : [a, b] \to \mathbb{R}$ *be continuous on* $[a, b]$. *Then:* f *is uniformly continuous on* $[a, b]$.

Proof Let $\epsilon > 0$ be given. Define

$$X = \{x \in [a, b] \mid \exists \delta > 0 \forall y \in [a, x] \forall z \in [a, x] \big(|y - z| < \delta \Rightarrow |f(y) - f(z)| < \epsilon \big)\}.$$

Trivially $a \in X$ since $y \in [a, a]$ and $z \in [a, a]$ imply $y = z = a$, whence $|f(y) - f(z)| = |f(a) - f(a)| = 0 < \epsilon$. As usual, we have a lemma extending the possible boundary of X:

2.2.23 Lemma *Let* $f : [a, b] \to \mathbb{R}$ *be continuous on* $[a, b]$ *and let* $x \in [a, b]$. *For any* $\epsilon > 0$ *there is a* $\delta > 0$ *such that, for all* $y, z \in [a, b]$,

$$|y - x| < \delta \ \&\ |z - x| < \delta \Rightarrow |f(y) - f(z)| < \epsilon.$$

Proof Using the continuity of f at x, choose $\delta > 0$ so that for all $y \in [a, b]$

$$|y - x| < \delta \Rightarrow |f(y) - f(x)| < \frac{\epsilon}{2}.$$

---

[79] H. Lebesgue, *Leçons sur l'intégration et la recherche des fonctions primitives*, Gauthier-Villars, Paris, 1904, p. 105.

[80] Y.R. Chao, "A note on 'Continuous mathematical induction' ", *Bulletin of the American Mathematical Society* 26 (1919), pp. 17–18.

[81] Pete L. Clark, "The instructor's guide to real induction", online at http://arxiv.org/abs/1208.0973.

If both $|y - x| < \delta$ and $|z - x| < \delta$, then

$$\begin{aligned} |f(y) - f(z)| &= |f(y) - f(x) + f(x) - f(z)| \\ &\leq |f(y) - f(x)| + |f(x) - f(z)| \\ &< \frac{\epsilon}{2} + \frac{\epsilon}{2} = \epsilon. \end{aligned}$$

□

Let c be the least upper bound of X and suppose by way of contradiction that $c < b$. Choose δ_1 according to the Lemma so that, for all $y, z \in [a, b]$,

$$|y - c| < \delta_1 \ \& \ |z - c| < \delta_1 \Rightarrow |f(y) - f(z)| < \epsilon.$$

By the Lemma we know that $c > a$, so choose $\eta = \min\{\delta_1, c - a, b - c\}$ so that

$$(c - \eta, c + \eta) \subseteq [a, b],$$

and choose δ_0 so small that, for all $y, z \in [a, b]$,

$$y \leq c - \frac{\eta}{2} \ \& \ z \leq c - \frac{\eta}{2} \Rightarrow |f(y) - f(z)| < \epsilon.$$

Finally, let $\delta = \min\left\{\delta_0, \delta_1, \frac{\eta}{2}\right\}$ so that for $y, z \in [a, b]$, if $y, z \leq c + \frac{\eta}{2}$,

$$|y - z| < \delta \ \Rightarrow \ y, z < c - \frac{\eta}{2} \text{ or } y, z \in (c - \eta, c + \eta).$$

Either possibility yields $|f(y) - f(z)| < \epsilon$. Thus $c + \frac{\eta}{2} \in X$, and the usual contradiction occurs. □

As with the Extreme Value Theorem, the Uniform Continuity Theorem fails for open intervals.

2.2.24 Exercise Let $f : (a, b) \to \mathbb{R}$ be uniformly continuous.

i. Show that f is bounded on (a, b): there is a B such that, for all $x \in (a, b)$, $|f(x)| < B$.
ii. Show that f can be extended to a continuous function $\overline{f} : [a, b] \to \mathbb{R}$.
iii. Show that $g(x) = \sin(1/x)$ is not uniformly continuous on $(0, 1)$.
[Part i is fairly straightforward. A nonstandard proof of part ii is also straightforward. The easiest standard proof of part ii uses the convergence of Cauchy sequences as the Completeness Axiom. A proof based directly on the Least Upper Bound Principle, like that of the Extreme Value Theorem, is a little trickier.]

2.2.25 Remark The nonstandard proofs given earlier for the Intermediate Value Theorem and the Extreme Value Theorem are fairly intuitive and have a heuristic value even for the mathematician unfamiliar with Nonstandard Analysis. There is also a nonstandard proof of the Uniform Continuity Theorem. Indeed, it is even more trivial than the proofs given for Theorems 2.2.11 and 2.2.12, but it is not as intuitive in one particular and hence is of limited heuristic value. This proof rests on three facts:

(1) A function $f[a, b] \to \mathbb{R}$ is continuous on $[a, b]$ iff, for every real $x \in [a, b]$ and every infinitesimal η for which $x + \eta \in [a, b]$, the difference $f(x + \eta) - f(x)$ is infinitesimal;
(2) f is uniformly continuous on $[a, b]$ iff, for every number $x \in [a, b]$, real or nonstandard, and every infinitesimal η for which $x + \eta \in [a, b]$, the difference $f(x + \eta) - f(x)$ is infinitesimal; and
(3) every nonstandard $\alpha \in [a, b]$ is infinitesimally close to a standard real $r \in [a, b]$.

Now, (1) is the usual interpretation of continuity in terms of infinitesimals and most mathematicians would accept it without hesitation. Point (3) has not entered universal consciousness, but when one considers that α defines a Dedekind cut, it becomes quite plausible. Point (2), though easy enough to prove to one familiar with the logical setting in which the existence of nonstandard numbers is established, is not intuitively obvious. It is essentially Cauchy's definition of continuity and only becomes clear when considering non-uniformly continuous functions like

$$f(x) = \frac{1}{x} \text{ on } (0, 1)$$

or

$$g(x) = x^2 \text{ on } [0, \infty).$$

[In the latter case, if η is a positive infinitesimal, $1/\eta \in [0, \infty)$ is infinite and

$$g\left(\frac{1}{\eta} + \eta\right) = \left(\frac{1}{\eta} + \eta\right)^2 = \frac{1}{\eta^2} + 2\eta \cdot \frac{1}{\eta} + \eta^2 = g\left(\frac{1}{\eta}\right) + 2 + \eta^2,$$

and the difference $2 + \eta^2 > 2$ is not infinitesimal.] On a closed, bounded interval, however, by virtue of (3), the nonstandard conditions for continuity and uniform continuity given by (1) and (2) are equivalent.

As mentioned earlier, the Intermediate Value Theorem and the Extreme Value Theorem will have a direct bearing on our later discussion of the Mean Value Theorem, the Uniform Continuity Theorem less so. Thus our digression to discuss these important theorems is not a digression from our main path, but is a small diversion from our immediate goal of discussing continuous curves.

Together, the Intermediate Value Theorem and the Extreme Value Theorem tell us that the range of a continuous function $f : [a, b] \to \mathbb{R}$ is a closed and bounded interval $[m, M]$, where m, M are the minimum and maximum values, respectively, attained by f on $[a, b]$. But this tells us nothing yet about the graph of f, i.e., the curve itself. Nor does it apply directly to more general curves with continuous parametrisations γ. It remains to see how well our formally defined curves match our intuition. If I is an interval and $\gamma : I \to \mathbb{R} \times \mathbb{R}$ is a continuous function, what can be said about the curve $C = \gamma(I)$? Is C perfect and cohesive in Cantor's sense? Is it connected? Is it one-dimensional?

Definitive answers to some of these questions cannot be given without formal definitions of the concepts involved. Cantor's definition of a perfect set as one that is closed under taking limits and which has no isolated points requires a formal definition of limit, which we haven't given yet. And we never did answer the question here of what constituted one-dimensionality. Fortunately, some of the questions can be answered using only intuitive notions of the properties involved.

2.2.26 Examples Not every continuous curve is perfect.

i. The quadratrix (Fig. 2.1), defined by

$$\gamma(t) = \left\langle (1-t)\tan\frac{t\pi}{2}, 1-t \right\rangle, \quad t \in [0, 1),$$

is not perfect because the point $E = \langle 2/\pi, 0\rangle$ is the limit of $\gamma(t)$ as $t \to 1$, but E is not on the curve.

ii. The logarithmic spiral (Fig. 2.18), defined by

$$\gamma(t) = \langle ae^{bt}\cos t, ae^{bt}\sin t\rangle, \quad t \in (-\infty, \infty),$$

for fixed $a, b > 0$ is not perfect because it does not contain the point $\langle 0, 0\rangle$ which is the limit of $\gamma(t)$ as $t \to -\infty$.

iii. The point, defined by $\gamma(t) = \langle a, b\rangle$ for t on any interval, is not perfect because it consists of an isolated point.

On the other hand, it can be shown that for a closed, bounded interval $I = [a, b]$ and any nonconstant $\gamma : I \to \mathbb{R} \times \mathbb{R}$, the curve $\gamma(I)$ is perfect.

It can also be shown that every continuous curve is cohesive in Cantor's sense. This is a fairly easy consequence of the Uniform Continuity Theorem generalised to functions $\gamma : [a, b] \to \mathbb{R} \times \mathbb{R}$.

2.2.27 Theorem *Let I be an interval and $\gamma : I \to \mathbb{R} \times \mathbb{R}$ be continuous. The curve $C = \gamma(I)$ is cohesive.*

Proof Let $P, Q \in C$ and $\epsilon > 0$ be given. We have to show that there exist $P_0, P_1, \ldots, P_n$ such that $P_0 = P$, $P_n = Q$, and, for $i = 0, 1, \ldots, n-1$, $\text{dist}(P_i, P_{i+1}) < \epsilon$.

Choose $a, b \in I$ such that $\gamma(a) = P$ and $\gamma(b) = Q$. Relabelling P and Q if necessary we can assume $a < b$. Restricted to $[a, b]$, γ is continuous and hence uniformly continuous (as the reader will show in the next exercise). Thus there is a $\delta > 0$ such that for all $x, y \in [a, b]$, if $|x - y| < \delta$ then $|\gamma(x) - \gamma(y)| < \epsilon$. Let n be so large that $n > (b-a)/\delta$, i.e., $\delta > (b-a)/n$. For $i = 0, 1, \ldots, n$, let $P_i = \gamma(a + i(b-a)/n)$. Note that

$$a + (i+1)\frac{b-a}{n} - \left(a + i\frac{b-a}{n}\right) = \frac{b-a}{n} < \delta,$$

whence $\text{dist}(P_i, P_{i+1}) < \epsilon$. □

2.2.28 Exercise Prove the version of the Uniform Continuity Theorem appealed to in the proof of Theorem 2.2.27: If $\gamma : [a, b] \to \mathbb{R} \times \mathbb{R}$ is continuous on $[a, b]$, then γ is uniformly continuous on $[a, b]$.

By the cohesiveness of continuous curves we have ruled out their having large gaps, but not isolated gaps like that of Fig. 2.23. For this — at least for curves that look like curves — we have to show that $C = \gamma(I)$ is connected.

2.2.29 Theorem *Let I be an interval and $\gamma : I \to \mathbb{R} \times \mathbb{R}$ be continuous. The curve $C = \gamma(I)$ is connected.*

Proof Suppose C were not connected, i.e., suppose $C \subseteq U \cup V$, where U, V are disjoint open sets and there are points P, Q of C with $P \in U$ and $Q \in V$. Let a, b be such that $P = \gamma(a)$, $Q = \gamma(b)$ and assume without loss of generality that $a < b$. Define a function g on $[a, b]$ by

$$g(t) = \begin{cases} -1, & \gamma(t) \in U \\ 1, & \gamma(t) \in V. \end{cases}$$

The claim is that g is continuous on $[a, b]$. To see this, let $\epsilon > 0$ be given.

For any $t \in [a, b]$, $\gamma(t) \in U$ or $\gamma(t) \in V$. Consider the case $\gamma(t) \in U$. Because U is open, there is some $\epsilon_0 > 0$ such that, for all points R in the plane, if $\text{dist}(\gamma(t), R) < \epsilon_0$, then $R \in U$. Now, γ is continuous, so there is a $\delta > 0$ such that, for all $t' \in [a, b]$,

$$\begin{aligned} |t - t'| < \delta &\Rightarrow |\gamma(t) - \gamma(t')| < \epsilon_0 \\ &\Rightarrow \gamma(t') \in U \\ &\Rightarrow g(t') = -1 \\ &\Rightarrow |g(t) - g(t')| = |-1 - (-1)| = 0 < \epsilon. \end{aligned}$$

Similarly, if $\gamma(t) \in V$, we can find δ such that

$$|t - t'| < \delta \Rightarrow |g(t) - g(t')| < \epsilon.$$

Thus g is continuous on $[a, b]$ with $g(a) = -1 < 0 < 1 = g(b)$. By the Intermediate Value Theorem there is some c between a and b at which $g(c) = 0$, which is not the case. □

The establishment of a precise definition of the continuity of a curve allows us to make rigorous the first step in the informal argument outlined in the Preface for the truth of the Mean Value Theorem. This is the proof that, if $\gamma : [a, b] \to \mathbb{R} \times \mathbb{R}$ is a continuous parametrisation of a curve, there is a number $c \in (a, b)$ at which the distance from $\gamma(c)$ to the line passing through $\gamma(a)$ and $\gamma(b)$ is maximum.

2.2.30 Lemma *Let L be a line in $\mathbb{R} \times \mathbb{R}$ and define $d_L(x, y)$ to be the distance from the point $P = \langle x, y \rangle$ to L. Then: d_L is a continuous function.*

Proof We have only defined continuity for functions of one variable, but the definition for functions of two variables is the same: f is continuous at a point $P_0 \in \mathbb{R} \times \mathbb{R}$ if for any $\epsilon > 0$ there is a $\delta > 0$ such that, for all P in the domain of f,

$$\text{dist}(P, P_0) < \delta \Rightarrow \left|f(P) - f(P_0)\right| < \epsilon.$$

The function d_L is in fact uniformly continuous. To see this, let $\epsilon > 0$ be given and consider two points $P, Q \in \mathbb{R} \times \mathbb{R}$ as in Fig. 2.27. Let d_p, d_q denote the respective distances of P, Q from L, $a = \left|d_L(P) - d_L(Q)\right| = \left|d_p - d_q\right|$, and, for the sake of definiteness, let $d_p \geq d_q$ as in the figure. To get $\left|d_p - d_q\right| = a < \epsilon$, note that $a^2 + b^2 = c^2$, whence $a^2 \leq c^2$ and $a \leq c = \text{dist}(P, Q)$. Thus, for $\delta = \epsilon$, we have

$$\text{dist}(P, Q) < \delta \Rightarrow \left|d_L(P) - d_L(Q)\right| = a \leq \text{dist}(P, Q) < \delta = \epsilon.$$

□

2.2.31 Lemma *Let $\gamma : [a, b] \to \mathbb{R} \times \mathbb{R}$ be a continuous parametrisation of a curve and let L be the line passing through $\gamma(a)$ and $\gamma(b)$. Define, for $t \in [a, b]$,*

$$d_\gamma(t) = d_L(\gamma(t)) = d_L(x(t), y(t)).$$

Then: d_γ attains a maximum on $[a, b]$.

Proof d_γ is continuous: Let $\epsilon > 0$ be given, $t_0 \in [a, b]$. Choose δ_1 so that, for all $t \in [a, b]$,

$$\text{dist}(\gamma(t), \gamma(t_0)) < \delta_1 \Rightarrow \left|d_L(\gamma(t)) - d_L(\gamma(t_0))\right| < \epsilon,$$

by the continuity of d_L; and choose $\delta > 0$ so that, for all $t \in [a, b]$,

$$\left|t - t_0\right| < \delta \Rightarrow \text{dist}(\gamma(t), \gamma(t_0)) < \delta_1,$$

by the continuity of γ. The Extreme Value Theorem applies: For some $c \in [a, b]$ $d_\gamma(c)$ is maximum. Since $d_\gamma(a) = d_\gamma(b) = 0$, we can take c to be in the interior of the interval. □

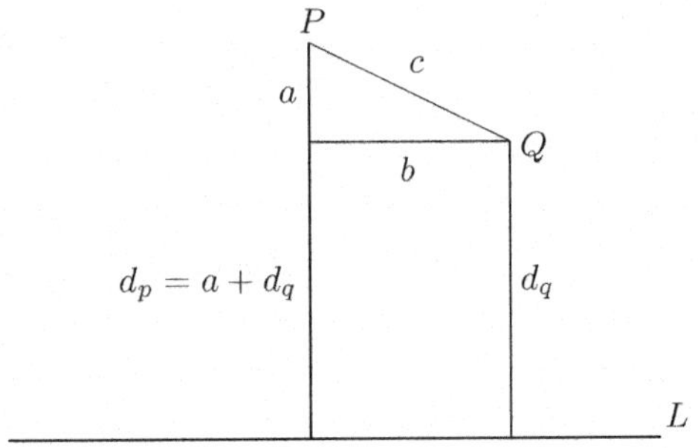

Fig. 2.27 Continuity of the distance function

To continue from here to prove the Mean Value Theorem, we need to guarantee that the curve $C = \gamma([a, b])$ has a tangent at all interior points $c \in (a, b)$, and that the tangent at a point of maximum distance on the curve from the line L passing through $\gamma(a)$ and $\gamma(b)$ is necessarily parallel to L. The first step in this is to define what a tangent line is. But before we do that there is one last property of curves to consider.

2.2.3 *Peano's Space-Filling Curve*

So far the only failure of the formal definition of a continuous curve has been the failure of curves defined on open or semi-open intervals to have endpoints as required by Cantor, but not so required by our intuition. It begins to look as if the formal definition of a continuous curve has captured the intuitive notion of curve. Alas, this is not the case. Whatever we mean by one-dimensionality, there are continuous curves that definitely are not one-dimensional. For, in 1890 Giuseppe Peano (1858–1932) proved the existence of continuous *space-filling curves*.

2.2.32 Theorem *The unit square* $[0, 1] \times [0, 1]$ *is a continuous curve, i.e., there is a continuous function* $\gamma : [0, 1] \to \mathbb{R} \times \mathbb{R}$ *such that* $\gamma([0, 1]) = [0, 1] \times [0, 1]$.

This came as a shock to mathematicians. It was known since Cantor that $[0, 1]$ and $[0, 1] \times [0, 1]$ could be put into one-to-one correspondence with each other, and thus there were "curves" γ for which $\gamma([0, 1])$ was $[0, 1] \times [0, 1]$. But, as shown by Eugen Netto, no such one-to-one correspondence could be continuous. The question of the nature and meaning of dimension arose. Peano's curve γ was not one-to-one and hence did not prove that $[0, 1]$ and $[0, 1] \times [0, 1]$ had the same dimension, and, indeed, it is not too difficult to prove by a simple appeal to connectivity that there cannot be a one-to-one continuous function γ such that $\gamma([0, 1]) = [0, 1] \times [0, 1]$: removing an interior point disconnects $[0, 1]$, but the removal of a single point will not disconnect $[0, 1] \times [0, 1]$.[82] Still, the general problem of invariance of dimension had been raised and the result was only first proven in 1910 by Brouwer.

---

[82]The one-to-one correspondence γ given at the end of the first section (p. 70, above) can be shown directly not to be continuous. The point $t_0 = .01$ is mapped by γ to the pair $\langle 0, .1\rangle$. The points

$$t_n = .00\underbrace{9999\ldots 99}_{2n}$$

can be chosen as close to t_0 as one wishes by choosing n large enough, yet

$$\begin{aligned} |\gamma(t_n) - \gamma(t_0)| &= |\langle .0\underbrace{9\ldots 9}_{n}, .0\underbrace{9\ldots 9}_{n}\rangle - \langle 0, .1\rangle| \\ &= \sqrt{(.09\ldots 9 - 0)^2 + (.09\ldots 9 - .1)^2} \\ &> .09\ldots 9 > .09, \end{aligned}$$

i.e., the points $\gamma(t_n)$ are bounded away from $\gamma(t_0)$.

Peano's curve was an important milestone in the history of mathematics, and, even though it will not be needed in our discussion of the Mean Value Theorem, it clearly belongs in any discussion of curves such as that we have been involving ourselves in. Thus, like many a less experienced writer, who doesn't know how to prune his creation, I have succumbed to the temptation to include a proof. The reader with no interest in this proof may safely skip ahead to the next section on page 79.

The construction of a space-filling curve is usually described geometrically[83] and the presentation of a rigorous proof can be a bit challenging combinatorially as one must translate the intuitive description into a sufficiently sharp analytic one. Moreover, such a proof relies on another deeper concept of analysis we haven't discussed yet in the present book, namely *uniform convergence*. Fortunately, Peano's original proof is much more elementary. And his paper is so simply written, with little extraneous material, that I have decided to present it in its entirety (after translation[84]) here.

Peano's construction is very similar to the one-to-one correspondence given, but, instead of mapping $.r_0r_1r_2\ldots$ to $\langle .r_0r_2\ldots, .r_1r_3\ldots\rangle$, he uses a clever device to occasionally replace r_n by its *complement*, which is like $9-r_n$, but he uses the base 3 representation of numbers instead of the usual base 10 version for a reason explained at the end of his paper.

Without further ado, I present Peano's paper:

On a curve which fills an entire plane area.

by

G. PEANO in Turin.

In this note we determine two well-defined[85] and continuous functions x and y, of a (real) variable t, which, when t varies over the interval (0, 1), takes each pair of values such that $0 \le x \le 1, 0 \le y \le 1$. If, as is customary, one calls the set of points at which the coordinates are continuous functions of the variable, a *continuous curve* and one has thus an arc of a curve which passes through all the points of the square. Thus, being given an arc of a continuous curve, without making any other hypothesis, it is not always possible to contain it in an arbitrarily small area.

---

[83]Cf., e.g., Hahn, *op. cit.*, pp. 85–87, or Bernard R. Gelbaum and John M.H. Olmsted, *Counterexamples in Analysis*, Holden-Day, Inc., San Francisco, 1964, pp. 133–134. The publication of Gelbaum and Olmsted has been taken over by Dover Publications and the book is still in print. The authors also cite a couple of variant constructions.

[84]After making this translation, I was reminded by Ádám Besenyei that an excellent English translation can be found in: Hubert C. Kennedy (ed. and trans.), *Selected Works of Giuseppe Peano*, George Allen & Unwin Ltd, London, 1973. I bought a copy of this book decades ago and, being a logician, read some of the logical papers, storing the book on my general logic shelf. In my memory, the book was a selection of the logical papers of Peano and I thus neglected to consult it until receiving the reminder. Kennedy accompanies his translation with an excerpt from a later (1908) work of Peano in which a geometric construction is discussed.

[85]Peano writes "uniformes", which I take to mean "well-defined". Kennedy translates this as "single-valued", which is perhaps a more felicitous choice.

Let us adopt the number 3 as the number base; we call each of the numbers 0, 1, 2 a *cipher*[86]; and consider an infinite sequence of ciphers $a_1, a_2, a_3, \ldots$ which we write[87]

$$T = 0, a_1 a_2 a_3 \ldots.$$

(For the moment, T is only a sequence of ciphers).

If a is a cipher, designate by $\mathbf{k}a$ the cipher $2 - a$, *complementary* to a; that is to say, put

$$\mathbf{k}0 = 2, \quad \mathbf{k}1 = 1, \quad \mathbf{k}2 = 0.$$

If $b = \mathbf{k}a$, we conclude $a = \mathbf{k}b$; we also have $\mathbf{k}a \equiv a$ (mod. 2).

Let $\mathbf{k}^n a$ denote the result of repeating the operation $\mathbf{k}$ n times on a. If n is even, we have $\mathbf{k}^n a = a$; if n is odd, $\mathbf{k}^n a = \mathbf{k}a$. If $m \equiv n$ (mod. 2), we have $\mathbf{k}^m a = \mathbf{k}^n a$.

There correspond to the sequence T the two sequences

$$X = 0, b_1 b_2 b_3 \ldots, \quad Y = 0, c_1 c_2 c_3 \ldots,$$

where the ciphers b and c are given by the relations

$$\begin{gathered} b_1 = a_1,\ c_1 = \mathbf{k}^{a_1} a_2,\ b_2 = \mathbf{k}^{a_2} a_3,\ c_2 = \mathbf{k}^{a_1 + a_3} a_4,\ b_3 = \mathbf{k}^{a_2 + a_4} a_5, \ldots \\ b_n = \mathbf{k}^{a_2 + a_4 + \ldots + a_{2n-2}} a_{2n-1},\ c_n = \mathbf{k}^{a_1 + a_3 + \ldots + a_{2n-1}} a_{2n}. \end{gathered}$$

Thus b_n, the n-th cipher of X, is equal to a_{2n-1}, the n-th cipher of odd rank of T, or to its complement, according as the sum $a_2 + \ldots + a_{2n-2}$ of the ciphers of even rank which precede it is even or odd. Analogously for Y. We can thus write these relations in the form:

$$\begin{gathered} a_1 = b_1,\ a_2 = \mathbf{k}^{b_1} c_1,\ a_3 = \mathbf{k}^{c_1} b_2,\ a_4 = \mathbf{k}^{b_1 + b_2} c_2, \ldots, \\ a_{2n-1} = \mathbf{k}^{c_1 + c_2 + \ldots + c_{n-1}} b_n,\ a_{2n} = \mathbf{k}^{b_1 + b_2 + \ldots + b_n} c_n. \end{gathered}$$

If we are given the sequence T, then X and Y are consequently determined, and if we are given X and Y, then T is determined.

We call the *value* of the sequence T the quantity (analogous to a decimal number having the same notation)

$$t = \text{val. } T = \frac{a_1}{3} + \frac{a_2}{3^2} + \ldots + \frac{a_n}{3^n} + \ldots.$$

To each sequence T corresponds a number t, and we have $0 \leq t \leq 1$. Conversely, the numbers t in the interval (0, 1) are divided into two classes:

(α) The numbers, different from 0 and 1, which multiplied by a power of 3 yield an integer; they can be represented by two sequences, one

$$T = 0, a_1 a_2 \ldots a_{n-1} a_n 222 \ldots$$

where a_n is equal to 0 or to 1; the other

[86] Kennedy uses the word "digit", more in line with standard English usage. I tend to think of "digit" as referring to base 10 unless some modifier is added. In the present case this would result in "ternary digit", which I didn't like. So I stuck with the more literal "cipher".

[87] The European fashion is to use commas and periods in decimal representations where Americans use periods and commas, respectively. I have followed Peano more closely in these small details than Kennedy, for better or for worse.

$$T' = 0, a_1 a_2 \ldots a_{n-1} a'_n 000 \ldots$$

where $a'_n = a_n + 1$.

(β) The other numbers; they are represented by a unique sequence T.

Now the correspondence established between T and (X, Y) is such that if T and T' are two sequences of different form, but val. $T =$ val. T', and if X, Y are the sequences corresponding to T, and X', Y' those corresponding to T', we have

$$\text{val. } X = \text{val. } X', \quad \text{val. } Y = \text{val. } Y'.$$

Indeed, consider the sequence

$$T = 0, a_1 a_2 \ldots a_{2n-3} a_{2n-2} a_{2n-1} a_{2n} 2222 \ldots$$

where a_{2n-1} and a_{2n} are not both equal to 2. This sequence represents a number of the class α. Letting

$$X = 0, b_1 b_2 \ldots b_{n-1} b_n b_{n+1} \ldots,$$

we have

$$b_n = \mathbf{k}^{a_2+\ldots+a_{2n-2}} a_{2n-1}, \; b_{n+1} = b_{n+2} = \ldots = \mathbf{k}^{a_2+\ldots a_{2n-2}+a_{2n}} 2.$$

Letting T' be the other sequence coinciding with val. T,

$$T' = 0, a_1 a_2 \ldots a_{2n-3} a_{2n-2} a'_{2n-1} a'_{2n} 0000 \ldots$$

and

$$X' = 0, b_1 \ldots b_{n-1} b'_n b'_{n+1} \ldots .$$

The first $2n - 2$ ciphers of T' coincide with those of T; thus the first $n - 1$ ciphers of X' also coincide with those of X; the others are determined by the relations

$$b'_n = \mathbf{k}^{a_2+\ldots+a_{2n-2}} a'_{2n-1}, \; b'_{n+1} = b'_{n+2} = \ldots = \mathbf{k}^{a_2+\ldots+a_{2n-2}+a'_{2n}} 0.$$

We distinguish two cases, according to whether $a_{2n} < 2$ or $a_{2n} = 2$.

If a_{2n} has the value 0 or 1, we have $a'_{2n} = a_{2n} + 1$, $a'_{2n-1} = a_{2n-1}$, $b'_n = b_n$,

$$a_2 + a_4 + \ldots + a_{2n-2} + a'_{2n} = a_2 + \ldots + a_{2n-2} + a_{2n} + 1,$$

whence

$$b'_{n+1} = b'_{n+2} = \ldots = b_{n+1} = b_{n+2} = \ldots = \mathbf{k}^{a_2+\ldots+a_{2n}} 2.$$

In this case the two sequences X and X' coincide in form and in value.

If $a_{2n} = 2$, we have $a_{2n-1} = 0$ or 1, $a'_{2n} = 0$, $a'_{2n-1} = a_{2n-1} + 1$, and on putting

$$s = a_2 + a_4 + \ldots + a_{2n-2}$$

we have

$$b_n = \mathbf{k}^s a_{2n-1}, \quad b_{n+1} = b_{n+2} = \ldots = \mathbf{k}^s 2$$
$$b'_n = \mathbf{k}^s a'_{2n-1}, \quad b'_{n+1} = b'_{n+2} = \ldots = \mathbf{k}^s 0.$$

Now, seeing that $a'_{2n-1} = a_{2n-1} + 1$, the two fractions $0, a_{2n-2} 222 \ldots$ and $0, a'_{2n-1} 000 \ldots$ have the same value; and applying the same operation $\mathbf{k}^s$ to the ciphers we obtain the two fractions $0, b_n b_{n+1} b_{n+2} \ldots$ and $0, b'_n b'_{n+1} b'_{n+2} \ldots$, which have likewise, as one can easily see, the same value; thus the fractions X and X', although of different forms, have the same value.

Analogously one proves that val. $Y =$ val. Y'.

Thus if we set $x =$ val. X, and $y =$ val. Y, we conclude that x and y are two well-defined functions of the variable t over the interval $(0, 1)$. They are continuous; indeed if t tends to t_0, terminating the first $2n$ ciphers in the development of t coincide with those of the development of t_0, if t_0 is a β, or with those of one of the two developments of t_0, if t_0 is an α; and then the first n ciphers of x and y corresponding to t coincide with those of x and y corresponding to t_0.

Finally to each pair (x, y) such that $0 \leq x \leq 1, 0 \leq y \leq 1$ corresponds at least a pair of sequences (X, Y), which express the value; to (X, Y) corresponds a T, and to that its t; thus we can always determine t in such a manner that the two functions x and y take arbitrarily given values in the interval $(0, 1)$.

One comes to the same conclusion if, in place of 3, one takes any odd number whatsoever for a number base. One can also take an even number for the base, but then it is necessary to establish a more complicated correspondence between T and (X, Y).

One can form an arc of a continuous curve which entirely fills a cube. Make correspond to the fraction (in base 3)

$$T = 0, a_1 a_2 a_3 a_4 \dots$$

the fractions

$$X = 0, b_1 b_2 \dots, \quad Y = 0, c_1 c_2 \dots, \quad Z = 0, d_1 d_2 \dots$$

where

$$b_1 = a_1, \; c_1 = \mathbf{k}^{b_1} a_2, \; d_1 = \mathbf{k}^{b_1 + c_1} a_3, \; b_2 = \mathbf{k}^{c_1 + d_1} a_4, \; \dots$$

$$b_n = \mathbf{k}^{c_1 + \dots + c_{n-1} + d_1 + \dots + d_{n-1}} a_{3n-2},$$

$$c_n = \mathbf{k}^{d_1 + \dots + d_{n-1} + b_1 + \dots + b_n} a_{3n-1},$$

$$d_n = \mathbf{k}^{b_1 + \dots + b_n + c_1 + \dots + c_n} a_{3n}.$$

One proves that $x =$ val. X, $y =$ val. Y, $z =$ val. Z are well-defined and continuous functions of the variable $t =$ val. T; and if t varies between 0 and 1, x, y, z take on all the triples of values which satisfy $0 \leq x \leq 1, 0 \leq y \leq 1, 0 \leq z \leq 1$.

Mr. Cantor, (Crelle's Journal[88] 84, p. 242.) has demonstrated that one can establish a one-to-one correspondence between the points of a line and those of a surface. But Mr. Netto (Crelle's Journal 86, p. 263), and others have demonstrated that such a correspondence is necessarily discontinuous (see also G. Loria, *La definizione dello spazio ad n dimensioni ... secondo le ricerche di G. Cantor*, Giornale di Matematiche, 1877). In my note we demonstrate that one can establish well-definedness and continuity in one direction, that is to say, the points of the line can be corresponded to the points of a surface, in such a fashion that the points of the surface are a continuous function of the points of the line. But this correspondence is not at all one-to-one and onto, for to the points (x, y) of the square, if x and y are of [class] β, there rightly corresponds only one value of t, but if x, or y, or both of the two are of [class] α, the corresponding values of t are 2 or 4 in number.

[88] More formally, *Journal für reine und angewandte Mathematik*. This journal was founded by August Crelle and is often called *Crelle's Journal* in his honour.

One has demonstrated that one can enclose an arc of a plane continuous curve in an arbitrarily small area:

(1) If one of the functions, e.g., x coincides with the independent variable t; one has then the theorem of integrability of continuous functions.
(2) If the two functions are of bounded variation (Jordan, Cours d'Analyse, III, p. 599). But, like the demonstration of the preceding example, this is not true if we suppose only the continuity of the functions x and y.

These x and y, continuous functions of the variable t, everywhere fail to have a derivative.

Turin, January 1890.[89]

Following Peano's announcement of the existence of a space-filling curve, David Hilbert (1862–1943), a rising young German mathematician who would become the leading mathematician of the early decades of the 20th century, gave a geometric presentation of such a curve (see Fig. 2.28) at a meeting of the Society of German Natural Scientists and Physicians in Bremen. Felix Klein (1849–1925), the editor of the journal in which Peano's paper appeared, wrote to Hilbert on 23 November 1890:

Two additional wishes concerning the Annalen:

(1) Could you not give us a note furnished with figures on the curve which you treated in Bremen. That you have *returned this matter to geometric intuition* is to me the essential thing. Indeed: I and probably many other mathematicians with me have not read the abstract presentation of Peano at all; however, with the figure, it becomes to me immediately accessible and I feel the whole importance of the matter.[90]

Hilbert's note[91] duly appeared in the 1891 volume of the same journal in which Peano's paper had appeared. Unlike Peano, Hilbert did not give a rigorous proof, just the geometric intuition behind his construction and a graphical display of the first few curves in his sequence. (See Fig. 2.28, for these curves.)

On the matter of the geometric approach, Peano's biographer tells us

In 1891 Hilbert published the first intuitive geometrical example of such a curve. His curve results as a limit of a sequence of curves. It is probable that Peano was led to the construction of his curve by such considerations. This is shown by his publication in the last edition (1908) of the *Formulario* of such a sequence of curves. He also had one of the curves in this sequence constructed on the terrace of the villa he purchased in the summer of 1891, where the curve showed up as black tiles on white. His 1890 publication, however, is purely analytic. Ugo Cassina has suggested that this is probably because he wished no doubt about the validity of his result and because he typically suppressed everything unnecessary to the goal set. "Besides," Cassina added, "the difficulty does not lie in becoming aware intuitively of the fact that a planar region can be conceived as the limit of a variable polygon, but in

[89] G. Peano, "Sur une courbe, qui remplit toute une aire plane", *Mathematische Annalen* 36 (1890), pp. 157–160.

[90] Günther Frei (ed.), *Der Briefwechsel David Hilbert – Felix Klein (1886–1918)*, Vandenhoeck & Ruprecht, Göttingen, 1985, pp. 70–71.

[91] David Hilbert, "Ueber die stetige Abbildung einer Linie auf ein Flächenstück", *Mathematische Annalen* 38 (1891), pp. 459–460.

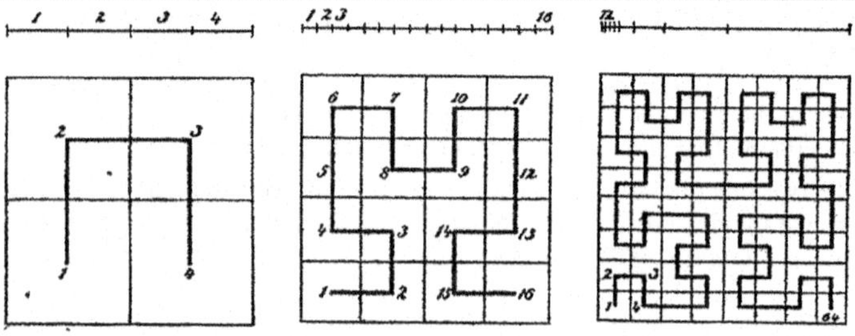

Beginning with Bolzano's nowhere differentiable function, pathological functions were often produced as limits of sequences of continuous functions and were not as explicitly described as Peano's space-filling curve. As remarked by his biographer, Peano probably first envisioned his curve geometrically as the limit of a sequence of continuous curves. The first published version of such an iteration was made by Hilbert in 1891 and his representation from David Hilbert, "Ueber die stetige Abbildung einer Linie auf ein Flächenstück", *Mathematische Annalen* 38, (1891), pp. 459 – 460, is reproduced above.

Fig. 2.28 Iterates of Hilbert's space filling curve

> giving the explicit expression of the coordinates of a point of a planar region as *continuous* functions of a variable parameter in the interval."
>
> Peano closes his note with the observation that the parametric functions are nowhere differentiable. We may add that this curve also has the property that, given any two points on the curve, the arc length between the two points is infinite. A curve often cited as having this property was invented by Helge von Koch in 1904.[92]

Cassina, as quoted here, makes a good point. Geometrically, one presents the ranges of the curves $\gamma_0, \gamma_1, \gamma_2, \ldots$, but not the parametrisations. It is not these ranges, but the functions $\gamma_0, \gamma_1, \gamma_2, \ldots$ that tend to a limit. The same curves, under different parametrisations can have wildly different limits. If γ is the limit of Hilbert's parametrised curves and we define, for each γ_n a new parametrisation γ_n^* by

$$\gamma_n^* = \begin{cases} \gamma_n(nt), & 0 \leq t \leq \dfrac{1}{n+1} \\ \gamma_n\left(\dfrac{n}{n+1} + \dfrac{1}{n}\left(t - \dfrac{1}{n+1}\right)\right), & \dfrac{1}{n+1} < t \leq 1, \end{cases}$$

then the limit $\gamma^*(t) = \lim_{n\to\infty} \gamma_n^*(t)$ is discontinuous; indeed, one has

$$\gamma^*(t) = \begin{cases} \gamma(0), & t = 0 \\ \gamma(1), & t \neq 0 \end{cases}.$$

[92] Hubert C. Kennedy, *Peano, Life and Works of Giuseppe Peano*, D. Reidel Publishing Company, Dordrecht, 1980, p. 32.

The range of γ^* thus consists, not of the whole unit square, but of the two points $\gamma(0) = \langle 0, 0\rangle$ and $\gamma(1) = \langle 1, 0\rangle$.

An actual proof that Hilbert's curve, like Peano's, actually covers the unit square requires a careful presentation of the parametrisations, proof that the functions $\gamma_0, \gamma_1, \gamma_2, \ldots$ converge uniformly to a continuous function γ, and then a proof that γ actually does map $[0, 1]$ onto $[0, 1] \times [0, 1]$. All of this is fairly routine and expositors do not always feel the need to present the details.[93, 94]

Space-filling curves do not match our intuition of what a curve is or should be. Nonetheless, continuous curves like Peano's are accepted as curves in general mathematics. In Topology, one can work a bit harder and define one-dimensionality to refine the formal concept of curve even further, but in less specialised areas of mathematics, such as our discussion, a simpler refinement is often more useful — this is the notion of a *smooth curve*, which will be the topic of the next section.

2.3 Smooth Curves

2.3.1 Traditional Views of Tangents

We have already remarked in the Preface that it is not at all obvious what one should mean by the tangent to a curve. Euclid defines the tangent to a circle in Book III of *The Elements* as follows:

> 2. A straight line is said to **touch a circle** which, meeting the circle and being produced, does not cut the circle.[95]

Heath offers little elucidation other than that there is a distinction between "meeting" the circle and "touching" it, and that the distinction was used by later geometers. The most important of these are Apollonius who determined the tangents to all the conic sections and Archimedes who found a single tangent to his spiral. A cursory

---

[93] Whence, of course, follows Klein's preference for Hilbert's geometric presentation.

[94] Two examples are the paper of Hahn and the book of Gelbaum and Olmsted cited in footnote 83 a few pages back. Hahn accompanies the pictures of some of the curves in Hilbert's sequence with the announcement, "It is now possible to give a rigorous proof that the successive motions considered here approach without limit a definite course, or curve, that takes the moving point through all the points of the large square in unit time". Gelbaum and Olmsted give the parametrisation, but leave the details that the limit is a continuous function and that it fills the square as an exercise to the reader. E. Hairer and G. Wanner, *Analysis by Its History*, Springer-Verlag New York, Inc., New York, 1996, pp. 289–290 repeat Hilbert's graphical presentation and give the parametric representation for a more general construction, proving the continuity of the limit, but leaving unproven the more intuitive fact that the range of the function is the entire square. They also present Peano's construction geometrically as an exercise on page 298. A cursory check of my personal library found no fuller proof for the geometrical construction. Indeed, most of my textbooks on Analysis do not even mention the result.

[95] Heath, *Elements*, *op. cit.*, vol. 2, p. 2.

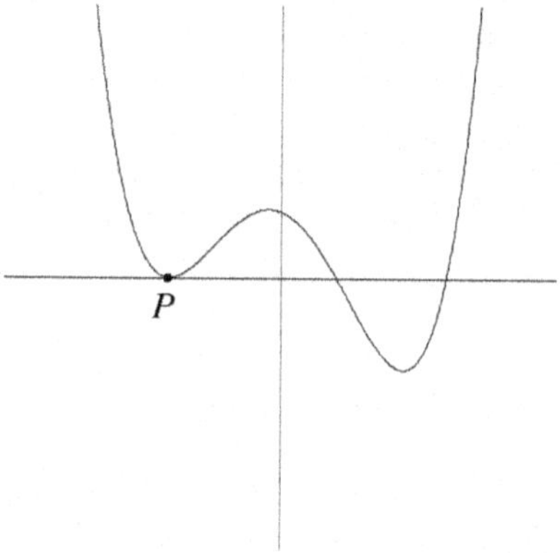

Fig. 2.29 A quartic

inspection of these later works reveals no new definition of "touching", but Heath comments on the next definition in Euclid:

> 3. **Circles** are said to **touch one another** which, meeting one another, do not cut one another.[96]

About this Heath says

> Todhunter remarks that different opinions have been held as to what is, or should be, included in this definition, one opinion being that it only means that the circles do not cut in the neighbourhood of the point of contact, and that it must be shown that they do not cut elsewhere, while another opinion is that the definition means that the circles do not cut at all. Todhunter thinks the latter opinion correct. I do not think this is proved; and I prefer to read the definition as meaning simply that the circles meet at a point but do not cut *at that point*.[97]

We have already seen a failing (see Figs. 1.5 and 1.6 of the Preface) of the definition of the tangent line as one which intersects the curve at only one point without crossing it, namely curves with points through which infinitely many tangents can pass. Figs. 2.29 and 2.30, give two more examples. In Fig. 2.29 the x-axis is clearly a tangent at P and yet clearly "cuts" the curve — but does not cut the curve *at* P. In Fig. 2.30, however, the x-axis does cross the curve at P, and yet one would like to consider this axis a tangent there for kinematic reasons: a particle travelling along the curve and allowed to go "off on a tangent" at P would follow the axis.

Another problematic curve is given in Fig. 2.31. Here I have simply taken a parabola, split it in two at the vertex, moved half of it over, and connected the two halves with a straight line. It is obvious that the straight line is the tangent at all points of the curve lying on it, yet it does not satisfy the usual condition of "touching" the curve. It, in fact, *coincides* with the curve for an entire interval, without "cutting" the curve.

---

[96] *Ibid.*.

[97] *Ibid.*, p. 3.

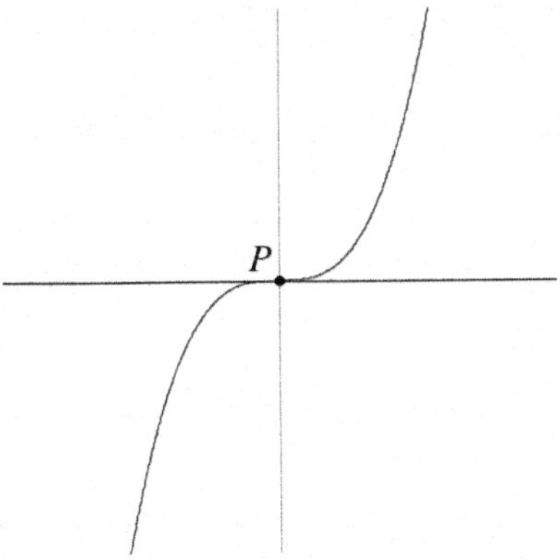

Fig. 2.30 A cubic

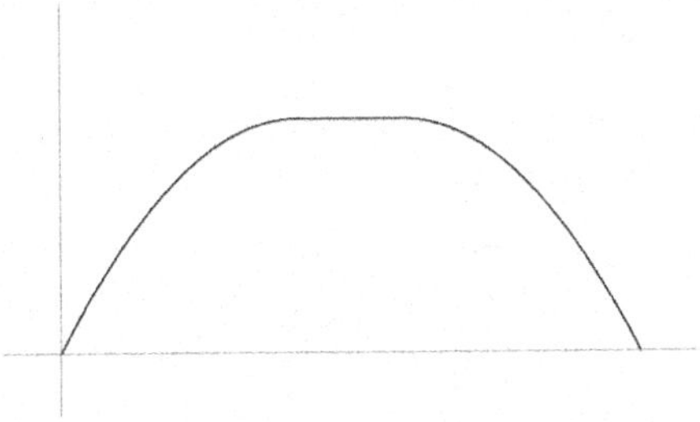

$$y = \begin{cases} 1-(x-1)^2, & 0 \le x < 2 \\ 1, & 2 \le x < 3 \\ 1-(x-3)^2, & 3 \le x \le 5. \end{cases}$$

Fig. 2.31 A curve partially coinciding with a tangent

An even more problematic curve is Darboux's function graphed in Fig. 2.26. Again one would like to think of the x-axis as a horizontal tangent to the curve at the origin. It crosses the curve infinitely many times in any neighbourhood of the origin, but does it "cut" the curve at the origin? Put differently, does Darboux's curve *cross* the x-axis at the origin or just touch it there?

The intuitive geometric description of the tangent is too vague to determine definitively the tangency or non-tangency of lines in these questionable cases. In the more clear-cut ones, however, they are not totally useless. We can tentatively define the *Greek tangent* to a curve C at a point P on C to be the unique line, if it exists, which passes through P without crossing the curve. Further defining a continuous curve to be *smooth in the Greek sense* if every point other than an end point has a Greek tangent, we can almost get an easy proof of the Mean Value Theorem for curves that are smooth in the Greek sense. For, let $\gamma : [a, b] \to \mathbb{R} \times \mathbb{R}$ be a continuous curve smooth in the Greek sense, and let $c \in (a, b)$ be a point on C, as given by Lemma 2.2.31, of maximum distance from the line connecting $\gamma(a)$ and $\gamma(b)$. The line through $\gamma(c)$ parallel to the line connecting $\gamma(a)$ and $\gamma(b)$ cannot cross the curve at $\gamma(c)$ as those points of the curve on one side of the parallel are farther from the line connecting $\gamma(a)$ and $\gamma(b)$ than is $\gamma(c)$.[98, 99]

In *La Géométrie*, Descartes introduced another definition of tangent and used it to determine the tangents to certain curves using geometric intuition and algebraic calculation. The geometric intuition is that if a circle and a line touch a curve at a given point P, the line is then simultaneously tangent to the circle and the curve at P. Hence the tangent to the curve at P is perpendicular to the normal to the circle. Descartes was rather pleased with himself over this method:

> Finally, all other properties of curves depend only on the angles which these curves make with other lines. But the angle formed by two intersecting curves can be as easily measured as the angle formed by two straight lines, provided that a straight line can be drawn making right angles with one of these curves at its point of intersection with the other. This is my reason for believing that I shall have given here a sufficient introduction to the study of curves when I have given a general method of drawing a straight line making right angles with a curve at an arbitrarily chosen point upon it. And I dare say that this is not only the most useful and most general problem in geometry that I know, but even that I have ever desired to know.[100]

The last sentence would no doubt be considered an exaggeration no matter what it referred to, but his excitement was understandable.

2.3.1 Remark Before leaving Descartes, note that the definition of the tangent as the line perpendicular to the normal of a circle touching the curve at a given point has its uses. Consider the cubic of Fig. 2.30: The circles of radius 1/2 centred at the points $\langle 0, 1/2\rangle$ and $\langle 0, -1/2\rangle$ each touch the curve at P and have the x-axis as their tangents. It is easy to see that the cubic $y = x^3$ and, for example, the circle $x^2 + (y - 1/2)^2 = (1/2)^2$ have $P = \langle x, y\rangle = \langle 0, 0\rangle$ in common. Verifying that they have no other common root requires a little work, but is not too hard:

$$\begin{aligned} x^2 + \left(y - \frac{1}{2}\right)^2 = \frac{1}{4} &\Rightarrow x^2 + y^2 - y = 0 \\ &\Rightarrow x^2 + (x^3)^2 - x^3 = 0, \text{ for } y = x^3 \\ &\Rightarrow x^6 - x^3 + x^2 = 0. \end{aligned} \tag{2.21}$$

(2.21) points to a double root at the origin. Eliminating this root results in the equation,

$$x^4 - x + 1 = 0. \tag{2.22}$$

[98] I say this is "almost" a proof because we have not defined precisely what is meant by "crossing". In algebraic terms we note that a line $Ax + By = C$ partitions the plane into three disjoint sets according as $Ax + By$ is $< C$, $= C$, or $> C$. A line may be said to cross the curve C at $P = \langle\alpha, \beta\rangle$ if $A\alpha + B\beta = C$ and in any neighbourhood of P there are points of the curve in each of the sets $\{\langle x, y\rangle \mid Ax + By < C\}$ and $\{\langle x, y\rangle \mid Ax + By > C\}$. Can we give a precise, purely geometric definition of the notion? How about the notion of two curves crossing each other?

[99] Another problem is: how can we tell algebraically or analytically that a curve given by a continuous parametrisation γ is smooth in the Greek sense?.

[100] Descartes, *op. cit.*, p. 95.

Three ways of verifying this to have no real roots come to mind. First, one can graph it on one's pocket calculator and notice that the graph does not appear to touch the x-axis.

Today we would probably use the Calculus to verify that our calculator isn't lying and differentiate: if $f(x) = x^4 - x + 1$ as in (2.22), then $f'(x) = 4x^3 - 1$ has the unique root $x = \sqrt[3]{1/4}$. $f''(x) = 12x^2$ is positive there, whence $f\left(\sqrt[3]{1/4}\right)$ is a minimum and it happens to be positive.

Descartes was pre-Calculus, so he would have had to devise other methods. This he did in the third part of *La Géométrie*, wherein he enunciated *Descartes's Rule of Signs*: A polynomial $P(x) = a_n x^n + a_{n-1}x^{n-1} + \ldots + a_1 x + a_0$ cannot have more positive roots than the number of sign changes in the sequence $a_n, a_{n-1}, \ldots, a_1, a_0$. In the present case, this sequence is $1, 0, -1, 0, 1$ and has only two sign changes, hence has at most two positive real roots. And $f(x)$ cannot have any negative roots since $f(-x) = x^4 + x + 1 > 1$ for $x > 0$. Moreover, $f(y+1) = y^4 + 4y^3 + 6y^2 + 3y + 1$ has no sign changes, whence no roots for $y > 0$, i.e., f(x) has no roots for $x = y + 1 > 0 + 1 = 1$. Thus we have restricted any possible real root to the interval $(0, 1)$. And now the simplest thing to do is to make an *ad hoc* observation:

$$\begin{aligned} x \in (0,1) &\Rightarrow 0 < x < 1 \;\&\; 0 < 1 - x^3 < 1 \\ &\Rightarrow 0 < x(1 - x^3) < 1 \\ &\Rightarrow x - x^4 = x(1 - x^3) < 1 \\ &\Rightarrow 0 < 1 - (x - x^4) = x^4 - x + 1. \end{aligned}$$

2.3.2 Exercise Show that the circle of radius $1/2$ centred at $\langle 0, 1/2\rangle$ intersects the parabola $y = x^2$ in only one point. Draw the same conclusion for this circle and Darboux's curve from Fig. 2.26. Conclude that Descartes should have accepted the x-axis as the tangent to Darboux's curve at $\langle 0, 0\rangle$ had he been aware of the curve.

2.3.3 Exercise Show that the following circles all meet the curve $y = \sqrt[3]{|x|}$ at $\langle 0, 0\rangle$ and nowhere else:

i. $\left(x - \frac{1}{2}\right)^2 + y^2 = \frac{1}{4}$
ii. $x^2 + (y+1)^2 = 1$
iii. $(x+1)^2 + (y+1)^2 = 2$.

What tangent lines do they suggest? Should the multiplicity of candidates for being the tangent line tell us that this curve has no tangent? Can you make a case for claiming that ii and iii cross the curve at $\langle 0, 0\rangle$ and thus that the y-axis constitutes a Cartesian tangent to the curve? (Consider the curve $y^6 - x^2 = 0$.)

The implicit Cartesian definition of the tangent as the line perpendicular to the normal of a circle meeting the curve in only one point thus has something to offer. If we don't share his excitement today it is because the method can be computationally horrendous and we now have much simpler methods. The more adventurous reader is referred elsewhere[101] for the details of his method.

In January of 1638 Descartes received a letter from Fermat in which the latter laid out his method for finding maxima and minima, and showed how to use his own method for finding tangent lines. Fermat explained the procedure, but not the rationale:

> We will express the maximum or minimum quantity in terms of a, by means of terms of any degree. We will then substitute $a+e$ for the primitive unknown a, and express the maximum or minimum quantity in terms containing a and e to any degree. We will *ad-equate*,[102] to speak like Diophantus, the two expressions of the maximum or minimum quantity, and we will remove from them the terms common to both sides. Having done this, it will be found that on both sides, all the terms will involve e or a power of e. We will divide all the terms by e, or by a higher power of e, so that on at least one of the sides, e will disappear entirely. We will then eliminate all the terms where e (or one of its powers) still exists, and we will consider the others equal, or if nothing remains on one of the sides, we will equate the added terms with the subtracted terms, which comes to be the same. Solving this last equation will give the value of a, which will lead to the maximum or the minimum, in the original expression.[103]

Translated into mathematical terms, given an expression $f(x)$, Fermat assumes f has a maximum at $x = a$ and $x = a + e$ and writes $f(a+e) \sim f(a)$.[104] For example, to find the maximum of $f(x) = x(b-x)$ he writes

$$\begin{gathered}(a+e)\big(b-(a+e)\big) \sim a(b-a)\\ ab - a^2 - ae + be - ae - e^2 \sim ab - a^2.\end{gathered}$$

He then deletes the common terms

$$be - 2ae - e^2 \sim 0.$$

---

[101]*Ibid.*, pp. 95 *ff.* But see also Edwards, *op. cit.*, pp. 125–127.

[102]The Latin original is "adæquentur", later rendered into the French as "adégalera". I suppose the most direct English translation would be "equate to", but it is not clear that he really means "equate". Thus historians of mathematics agree to keep the "ad". The rest, i.e., what the term means, is hotly debated among the historians. Cf. Mikhail G. Katz, David M. Schaps, and Steven Shnider, "Almost equal: the method of adequality from Diophantus to Fermat and beyond", arXiv:1210.7750v1.

[103]Pierre de Fermat, "Methodus ad disquirendum maximum & minimam". Fermat did not publish the contents of this letter during his life, and it first appeared, in 1679 in Latin, in the *Varia opera mathematica* edited by his son Samuel de Fermat. A couple of centuries later, when it was translated into French for inclusion in the third volume (1896) of his collected works, *Œuvres de Fermat*, his antiquated notation was updated, the result being much more readable. In both these works, the letter to Descartes was accompanied by a number of later items on the method of maxima and minima. A translation into English of the modernised French translations of the letter to Descartes and its immediately following letter to Gilles Personne de Roberval (1602–1675) appeared in: Dirk Struik (ed.), *A Source Book in Mathematics, 1200–1800*, Harvard University Press, Cambridge (Mass.), 1969, pp. 222–227. The quotation reproduced above is from a more recent translation from the French edition by Jason Ross, which I found online. Ross translates all seven parts of Fermat's method of maxima and minima.

[104]$\sim$ is the symbol used in the French translation to stand for ad-equality.

The common terms are essentially $f(a)$ and he has basically formed $f(a+e) - f(a) \sim 0$. He now divides by e (or a higher power — whatever power is common to all the terms):

$$\frac{f(a+e) - f(a)}{e} = \frac{be - 2ae - e^2}{e} = b - 2a - e \sim 0. \tag{2.23}$$

And finally, he deletes all the remaining terms containing e, resulting in an equation

$$b - 2a = 0, \tag{2.24}$$

i.e., $a = b/2$ maximises $f(x) = x(b - x)$.

It is very hard not to recognise the difference quotient in (2.23) and the derivative $f'(a) = b - 2a$ in (2.24). That Fermat was not thinking in terms of the difference quotient (2.23) and its limit as $e \rightarrow 0$ becomes apparent when one continues to read his application to the construction of tangent lines, which strikes us today as roundabout. In later notes Fermat reveals that he is using a property of maxima and minima of continuous curves pointed out by Pappus:

> ...if one poses a question regarding given magnitudes which is satisfied in general by two points, then for the maximum or minimum values there would only be one point. It is for this reason that Pappus calls the smallest possible ratio for the question *minimum* and *singular* (that is, unique).[105]
>
> The maxima and minima are indeed unique and singular, as Pappus has said and as the ancients already knew... It follows that on both sides of the limit point, one could find an ambiguous equation; that the two ambiguous equations are then correlative, equal and alike.[106]

The point is that if the maximum or minimum occurs at a, for the typical curve, $f(x)$ will not equal $f(a)$ nearby, but we will always have $f(x) < f(a)$ in the case of a maximum or $f(x) > f(a)$ for a minimum, provided $x \neq a$ is sufficiently close to a. Moreover, $f(x)$ will be paired with an $f(x')$ for x' on the opposite side of a. So he ad-equates the two expressions $f(a+e)$ and $f(a)$, removes common terms, and divides by e. Then

> To find the maximum, we must equate the roots of the two equations...
>
> Thus we must equate $a + e$ with a, whence $e = 0$...[107]

Thus, Fermat is pursuing an algebraic solution justified by geometric considerations; he is not finding the slopes of secant lines and letting them rotate into the tangent's position. This would come later. When it did come, however, Fermat's method of finding maxima and minima would prove useful in proving the Mean Value Theorem.

---

[105] "III. On the same method", p. 5 of Ross, *op. cit.*.

[106] "IV. The method of maximum and minimum", Ross, *op.cit.*, p. 7.

[107] *Ibid.*, p. 9.

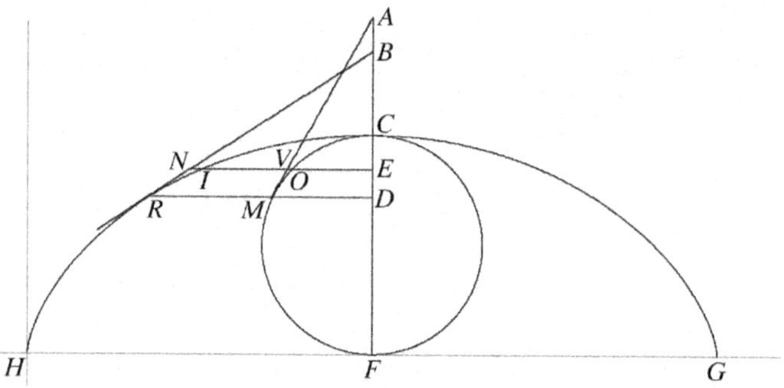

Fig. 2.32 The tangent of a cycloid

The impression to be had from Fermat's verbal description of his method of finding maxima and minima, as repeated above, is that it would apply only to polynomial expressions. This is not the case and, in a second note to bear the heading "On the same method", he explains how to find the tangents to the cissoid, the conchoid, the cycloid, and the quadratrix, giving the most detailed treatments for the cissoid and the cycloid. As we are already familiar with the cycloid, we shall consider how he determines the tangent to this curve.

Fermat begins with a cluttered diagram (Fig. 2.32). On this diagram, R represents the point at which we desire to draw the tangent line. RB is this tangent line. RD is the horizontal line through R. It intersects the circle inscribed in the centre of the cycloid at M and MA is the tangent to this circle at M. Finally, one draws a horizontal line from E to the tangent at a point E of distance e from D on the diameter CF of the circle. NE intersects the tangent RB at N, the cycloid at I, the circle's tangent MA at V, and the circle at O.

The tangent RB is determined by its slope $BD/RD = a/z$, which we hope to express in terms of known quantities. Given the cycloid, the circle, and the point R, these would be b, d, r, and z.

There are two pairs of similar triangles we can use to derive equations. From the pair RBD and NBE we have

$$\frac{NE}{RD} = \frac{BE}{BD}, \text{ i.e., } \frac{NE}{z} = \frac{a-e}{a}.$$

Thus

$$NE = \frac{za - ze}{a}. \tag{2.25}$$

And from VAE and MAD we have

$$\frac{EV}{MD} = \frac{AE}{AD}, \text{ i.e., } \frac{EV}{r} = \frac{b-e}{b}.$$

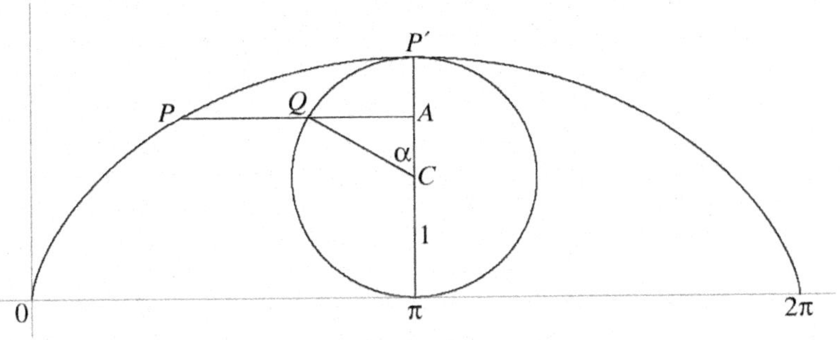

Fig. 2.33 A specific property

Thus

$$EV = \frac{rb - re}{b}. \tag{2.26}$$

At some point one must use some information about the cycloid itself. Fermat uses a "specific property of the curve", one sufficiently non-obvious as to require isolation here as a lemma — or, better yet, as an exercise, which I state without loss of generality for a circle of radius 1:

2.3.4 Exercise Consider Fig. 2.33. Show that the segment PQ equals arc $QP' = \alpha$. [Hint: Use the parametrisation $P = \langle t - \sin t, 1 - \cos t\rangle$ to express PQ, QA, the height $1 + AC$ of A above the base line, and α in terms of t.]

In the present situation this means RM = arc CM and IO = arc OC. Fermat now ad-equates

$$\begin{aligned} NE = NO + OE &\sim IO + OE \\ &\sim \text{arc } OC + OE \\ &\sim \text{arc } OC + EV. \end{aligned} \tag{2.27}$$

Now

$$\begin{aligned} \text{arc } OC &= \text{arc } CM - \text{arc } OM \\ &= RM - \text{arc } OM \\ &= RD - MD - \text{arc } OM \\ &= z - r - \text{arc } OM \\ &\sim z - r - VM \end{aligned} \tag{2.28}$$

as VM is very close to arc OM for e small, though Fermat omits this explanation.

Combining (2.25)–(2.28) we have

$$\frac{za-ze}{a} \sim z - r - VM + \frac{rb-re}{b}$$
$$-\frac{z}{a}e \sim -VM - \frac{r}{b}e. \tag{2.29}$$

To determine VM, we use the similarity of VAE to MAD again:

$$\frac{AV}{AM} = \frac{AE}{AD}, \quad \text{i.e.,} \quad \frac{AV}{d} = \frac{b-e}{b},$$

whence

$$AV = \frac{db-de}{b}$$

and

$$VM = AM - AV = d - \frac{db-de}{b} = \frac{de}{b}. \tag{2.30}$$

Combining (2.29) and (2.30) we have

$$-\frac{z}{a}e \sim -\frac{d}{b}e - \frac{r}{b}e,$$

and dividing by $-e$ yields

$$\frac{z}{a} = \frac{d+r}{b}.$$

Fermat continues from here to construct the tangent line. Today we would do this by inverting the fractions to get

$$\text{slope of the tangent} = \frac{a}{z} = \frac{b}{d+r} = \frac{AD}{MA+MD}.$$

I think I have organised the details a little more clearly than Fermat, and it is definitely not as bad as the tangent determinations of Descartes, but it still isn't pretty. It is not the familiar use of the difference quotient in slightly disguised form as reading his introductory remark on his method of maxima and minima would have suggested to the modern reader. It is computational, but far from algorithmic. And:

2.3.5 Exercise What happens if the point D in Fig. 2.32 lies at or below the centre of the circle?

In a nice exposition of the history of the derivative, historian Judith Grabiner summarises Fermat's contribution:

> …he did not explain why he could first divide by E (treating it as nonzero) and then throw it out (treating it as zero). Furthermore, he did not explain what he was doing as a special case of a more general concept, be it derivative, rate of change, or even slope of tangent. He did not even understand the relationship between his maximum-minimum method and the way one found tangents; in fact he followed his treatment of maxima and minima by saying

> that the same method — that is, adding E, doing the algebra, then suppressing E — could be used to find tangents.
>
> Though the considerations that led Fermat to his method may seem surprising to us, he did devise a method of finding extrema that worked, and it gave results that were far from trivial. For instance, Fermat applied his method to optics...
>
> Though Fermat did not publish his method of maxima and minima, it became well known through correspondence and was widely used. After mathematicians had become familiar with a variety of examples, a pattern emerged from the solutions by Fermat's method to maximum-minimum problems.[108]

By the 1650 s there was any number of methods for finding tangents of and areas under curves. As regards tangents, two people in particular deserve mention — Johann Hudde (1628–1704) and René François de Sluse (1622–1685). Their actual methods need not be discussed here[109] as they are not directly related to the Mean Value Theorem, but they deserve to be noted because

> ...the principal significance of the rules of Sluse and Hudde lay in the fact that they provided general algorithms by which tangents to algebraic curves could be constructed in a routine manner. It was no longer necessary to resort to special devices adapted to particular curves, nor to give in every case a complete demonstration of the process. For these reasons, the rules of Sluse and Hudde were perhaps the first methods to exhibit fully the algorithmic approach that is a distinctive feature of the calculus....
>
> The introduction in the 1650 s of the algebraic rules of Hudde and Sluse was soon followed by infinitesimal derivations of these and similar methods. These newer derivations and methods owed more to the ideas of Fermat than those of Descartes, and involved the concept of a tangent line at the point P of a curve as the limiting position of a secant line PQ as Q approaches P along the curve.[110]

Judith Grabiner reports on the next stage in the development:

> By the year 1660, both the computational and the geometric relationships between the problem of extrema and the problem of tangents were clearly understood; that is, a maximum was found by computing the slope of the tangent, according to the rule, and asking when it was zero. While in 1660 there was not yet a general concept of derivative, there was a general method for solving one type of geometric problem.[111]

The two names to reckon with in the 1660 s are Isaac Barrow (1630–1677) and Isaac Newton.

---

[108] Judith V. Grabiner, "The changing concept of change: the derivative from Fermat to Weierstrass", *Mathematics Magazine* 56 (1983), pp. 195–206; here: p. 197. Grabiner is, of course, using "E" where Fermat used "e".

[109] Readable accounts of their contributions can be found in: Margaret E. Baron, *The Origins of the Infinitesimal Calculus*, Pergamon Press, Oxford, 1969, pp. 214–220; and Edwards, *op. cit.*, pp. 127–132.

[110] Edwards, *op. cit.*, pp. 131–132.

[111] Grabiner, "Changing concept...", *op. cit.*, p. 198.

2.3.2 *Isaac Barrow*

In the mid-1660s Barrow lectured on Geometry at Cambridge, his lecture notes, the *Lectiones geometricæ*, first being published in Latin in 1670. His treatment of tangent and area problems is in the Greek geometric style, defining tangents as lines which touch the curve at single points. At the end of Lecture X, however, Barrow writes

> Thus I have in some sort accomplished the chief Part of my proposed Design. As a Supplement to which, I shall annex our Method of determining Tangents by Calculation. Tho' I scarcely perceive the Use of so doing, considering the several Methods of this Nature now become common and published. I do this at least by the Advice of a Friend[112]; and indeed so much more willingly as it seems to be compendious and general with respect to what else I have handled. The Thing is thus.
>
> Let AP, PM be right Lines given in Position (whereof PM cuts the proposed Curve in M,) and let MT touch the Curve in M, and cut the right Line AP in the Point T. Now to determine the length of the right Line PT, I suppose the Arch MN of the Curve to be indefinitely small, and draw the right Lines NQ, NR parallel to MP, AP; I call MP, m; PT, t; MR, a; NR, e; and give Names to other Lines useful to our purpose [See Barrow's Fig. 115 in Fig. 2.34.[113]], determin'd from the particular Nature of the Curve; and then compare MR, NR expressed by Calculation in an Equation, and by their means MP, PT themselves; observing the following Rules at the same Time.
>
> 1. I reject all the Terms in the Calculation, affected with[114] any Power of a or e, or with the product of them; for these Terms will be equal to nothing.
> 2. After the Equation is formed, I reject all the Terms wherein are Letters expressing constant or known Quantities; or which are not affected with a, or e; for these Terms brought over to one side of the Equation will be always equivalent to nothing.
> 3. I substitute a for m (MP), and t (PT) for e; by which means the Quantity of PT will be found.
>   When any indefinitely small Particle of the Curve enters the Calculation, I substitute in its stead a Particle of the Curve properly taken; or any right Line equal to it, because of the indefinitely Smallness of the Part of the Curve.
>   All of this will appear more evident by the following Examples.
>
> ## EXAMPLE I.[115]
>
> Let ABH be a right Angle [As in Barrow's Fig. 116 in Fig. 2.34.], and let the Curve AMO be such, that drawing any right Line AK thro' A, cutting the right Line BH in K, and the Curve AMO in M, the Subtense AM may be equal to the Absciss BK; it is required to draw the Tangent (at M) of this Curve, or find the Value of the right Line PT.
>
> Proceed according to the Directions above, and (drawing ANL) call AB, r, and AP, q. Then[116] AG $= q - e$; also QN $= m - a$. Therefore it is $qq + ee - 2qe + mm + aa - 2ma = (\text{AQ}q + \text{QN}q) = \text{AN}q) = \text{BL}q$; that is, (rejecting according to the Rule above)

---

[112]Scholars have identified this friend as Newton, who helped prepare the work for publication.

[113]Note that the character that looks like an ℓ is the Q of the text.

[114]I.e., multiplied by.

[115]The following is a bit opaque and the reader may wish to skip ahead to the modern explanation following this quotation.

[116]The G here is clearly a misprint for Q.

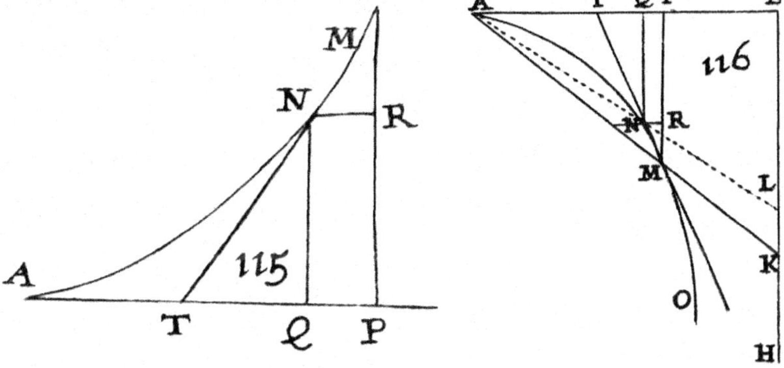

Fig. 2.34 Illustrations from Barrow

$qq - 2qe + mm - 2ma = \mathrm{BL}q$. Again it is AQ:QN :: AB:BL; that is $q - e : m - a ::$ $r : \mathrm{BL} = \dfrac{rm - ra}{q - e}$. Wherefore $\dfrac{rrmm + rraa - 2rrma}{qq + ee - 2qe} = \mathrm{BL}q$. Or (casting away what is superfluous)[117] $\dfrac{rrmm - 2rrmm}{qq - 2qe} = \mathrm{BL}q = qq - 2qe + mm - 2ma$. Or $rrmm - 2rrma =$ $q4 - 2q3e + qqmm - 2qqma - 2q3e + 4qqee - 2qmme + 4qmae$, that is (rejecting as *per* Rule) $-2rrma = -4q3e - 2qqmma - 2qmme$, or $rrma - qqma = 2q3e + qmme$. Or at length substituting m for a, and t for e, it is[118] $rrmm - qqmm = 2q3t - qmmt$, or $\dfrac{rrmm - qqmm}{2q3 - qmm} = t = \mathrm{PT}$.[119]

This excerpt can be understood with a great deal of patience and some guesswork about the notation, or with a little explanation. There are several obstacles for the modern reader. Barrow's book is geometrical in style and his analytic treatment here is not quite the modern presentation. He does not yet have the concept of function,

[117]There is a typo here: $-2rrmm$ should be $-2rrma$.

[118]There is a double typo here as the + between the two terms containing e accidentally changes to a −.

[119]Isaac Barrow (Edmund Stone trans.), *Geometrical Lectures: Explaining the Generation, Nature and Properties of Curve Lines*, London, 1735, pp. 171–175. This edition is available in facsimile online. The copy I downloaded, however, was very imperfectly done, some pages being repeated, and the fold-out plates scanned without being unfolded — whence not all the illustrations are available. One can, however, find all the illustrations online at ECHO (European Cultural Heritage Online) by searching, not for the *Lectiones geometricæ* of 1670, but for the larger work *Lectiones opticæ & geometricæ* of 1674 in which the former is incorporated. Figure 2.34, combines screen captures of pieces of one of the plates (indexed by thumbnail 361 at ECHO) cleaned up with photo-retouching software.

A more recent annotated, but abridged, translation by J.M. Child, *The Geometrical Lectures of Isaac Barrow*, was published in 1916 by the Open Court Publishing Company (Chicago and London). This translation is available in several reprinted editions and can also be found online.

Struik, *op. cit.*, excerpts a couple of important passages from Barrow, including that portion of the above quotation omitting the Example.

Barrow illustrates his technique with five examples, of which I have cited the first, Child the fifth.

he does not emphasise the slope of the tangent, and his notation is archaic and not quite consistent. Finally, printing standards were not what they are today: the run-on structure of the final paragraph does nothing to help the reader along, but it is a positive advance of what had gone before, some lectures printed page after page in single unbroken paragraphs.

Historians of the Calculus explain Barrow's general remarks as follows. One has a curve $f(x, y) = 0$. To find the tangent at a point $\langle x, y \rangle$, one moves infinitesimally to a nearby point $\langle x - e, y - a \rangle$ on the curve:

$$f(x - e, y - a) = 0 = f(x, y).$$

One expands both sides and removes those terms containing no a or e (i.e., one subtracts $f(x, y)$ from both sides of the equation) (Barrow's Rule 2). Since a and e are infinitesimal, a^2, e^2, ae and all higher powers are infinitesimally small compared to a, e and can be deleted (Rule 1). This leaves an equation linear in a, e and one can solve for a/e or e/a. Referring to Barrow's Fig. 2.34, think of AP and MP as the axes, AP the y-axis (as NR = e), A on the positive side, and MP the x-axis, with M on the positive side. From the equation we determine e/a, but from the picture[120] we know the triangles TMP and NMR are adequately[121] similar, whence

$$\frac{e}{a} \sim \frac{\mathrm{NR}}{\mathrm{MR}} \sim \frac{\mathrm{TP}}{\mathrm{MP}} \sim \frac{\mathrm{TP}}{m}.$$

We know the ratio e/a and m, whence we know PT = TP and can draw the tangent line. Barrow does not explicitly take the ratio here, but equivalently replaces e and a by t = PT and m, respectively, in the linear equation. This is an improvement on Fermat in that he explicitly appeals to the infinitesimal nature of a and e in applying Rule 1 and implicitly appeals to the ad-equality of the ratios e/a and t/m in the application of Rule 3.

The example illustrates his method nicely if not completely clearly. First, there is no function f. The curve is defined as a locus and he derives an equation, but not the equation of the curve. The notation needs to be explained, and the equations laid out more readably.

The locus is described as follows. One takes a line BH and a point A not on BH but on the perpendicular to BH passing through B. Think of BH as the x-axis and AB as the y-axis, with B denoting the origin. (See Fig. 2.35, for a more familiar orientation.) The curve in question is the locus of points M = $\langle x, y \rangle$ such that when one extends the line AM to meet BH in a point K, then AM = BK. Where we would

[120] His picture is imperfect here. The line TN is supposed to be tangent to the curve at M, not at N. The disposition of T, N, and M in his Fig. 2.35 is slightly better in that the tangent passes through M. Whether N lies on the curve or the tangent, however, is not discernible from these pictures. Fermat's Fig. 2.32, separating N from I and V from O is clearer in this respect. Barrow himself did better in his later Fig. 2.36 — cf. p. 96. Indeed, Struik reproduces Fig. 2.36 in place of Fig. 2.34 in his excerpt cited in the preceding footnote.

[121] See the preceding footnote.

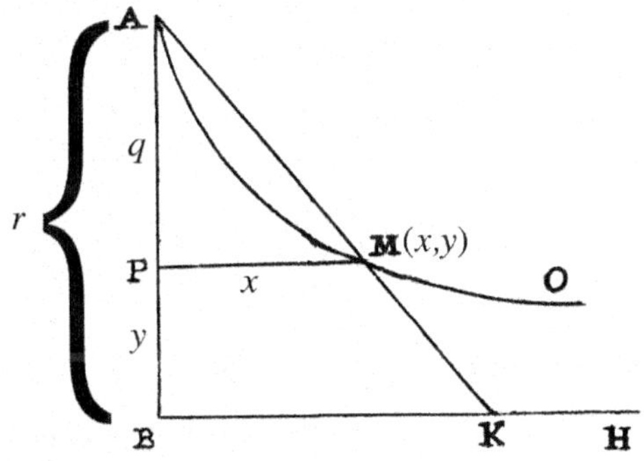

Fig. 2.35 Simplified Fig. 116 of Barrow

use x and y to determine the equation of the curve AMO, Barrow sets AB $= r$, PM $= m$, and AP $= q$. r is a constant, $q = r - y$ is a variable, as is $m = x$. Thus his equation is in the variables q, m with unspecified constant r.

To obtain an equation $f(q, m) = 0$ for the curve, note first that

$$\mathrm{AM} = \sqrt{q^2 + x^2} = \sqrt{q^2 + m^2}. \tag{2.31}$$

Also note that, by the similarity of the triangles PAM and BAK,

$$\frac{\mathrm{BK}}{\mathrm{PM}} = \frac{\mathrm{AB}}{\mathrm{AP}},$$

i.e.,

$$\frac{\mathrm{BK}}{m} = \frac{r}{q}, \quad \text{whence BK} = \frac{r}{q}m. \tag{2.32}$$

Plugging the values (2.31) and (2.32) into the locus equation AM = BK, we have

$$\frac{r}{q}m = \sqrt{q^2 + m^2}.$$

Thus

$$r^2m^2 = q^2(q^2 + m^2) = q^4 + q^2m^2,$$

and we have

$$f(q, m) = q^4 + q^2m^2 - r^2m^2 = 0$$

as the equation of the curve. If $\langle q - e, m - a\rangle$ is a nearby point on the curve, then

$$\begin{aligned} 0 = f(q - e, m - a) &= (q - e)^4 + (q - e)^2(m - a)^2 - r^2(m - a)^2 \\ &= q^4 - 4q^3e + 6q^2e^2 - 4qe^3 + e^4 + \\ &\quad (q^2 - 2qe + e^2)(m^2 - 2ma + a^2) - r^2(m^2 - 2ma + a^2). \end{aligned}$$

Applying Rule 1,[122]

$$\begin{aligned} f(q-e, m-a) &= q^4 - 4q^3e + (q^2 - 2qe)(m^2 - 2ma) - r^2m^2 + 2r^2ma \\ &= q^4 - 4q^3e + q^2m^2 - 2q^2ma - 2qm^2e + 4qmea - \\ &\quad r^2m^2 + 2r^2ma \\ &= q^4 - 4q^3e + q^2m^2 - 2q^2ma - 2qm^2e - r^2m^2 + 2r^2ma, \end{aligned}$$

applying Rule 1 again. Subtracting $f(q, m)$ (i.e., applying Rule 2) yields

$$f(q-e, m-a) - f(q, m) = -4q^3e - 2q^2ma - 2qm^2e + 2r^2ma = 0,$$

whence

$$(2q^3 + qm^2)e = (r^2m - q^2m)a$$

and (in essence, Rule 3)

$$\frac{t}{m} = \frac{e}{a} = \frac{r^2m - q^2m}{2q^3 + qm^2}.$$

Thus,

$$t = \frac{r^2m^2 - q^2m^2}{2q^3 + qm^2},$$

which would agree with Barrow had not the plus sign in $2q3e + qmme$ suddenly changed to a minus sign in the last sentence of the quotation.

Before discussing Barrow's derivation, the reader might want to check the result using ordinary Calculus:

2.3.6 Exercise Recalling that $m = x$, $q = r - y$, $f(q, m) = q^4 + q^2m^2 - r^2m^2$, verify that

$$\frac{dx}{dy} = \frac{dm}{-dq} = \frac{r^2m - q^2m}{2q^3 + qm^2}$$

by implicit differentiation.

Let us now consider Barrow's presentation of this example. He has the curve AMO and a point N on the curve infinitesimally close to M. Dropping the perpendicular NQ to AB, he has a right triangle AQN for which $\mathrm{AQ}^2 + \mathrm{QN}^2 = \mathrm{AN}^2$. Moreover, AN = BL by the defining property of the locus and

$$\mathrm{AQ}^2 + \mathrm{QN}^2 = \mathrm{AN}^2 = \mathrm{BL}^2,$$

telling us to read the q following AQ, QN, etc., in the equation[123]

[122]Perhaps we should use Fermat's adequality $\sim$ here.

[123]The parenthesis following ANq is merely a typographical error.

$$(\text{AQ}q + \text{QN}q) = \text{AN}q) = \text{BL}q$$

as squaring. This was a well-established practice at the time, and would give way to our modern exponential notation. A bit later Barrow approximates our modern notation in writing $q4e$ and $q3e$ for q^4e and q^3e, respectively. He nowhere uses 2 as an exponent, preferring qq, mm, etc. or AQq, etc. to represent the taking of squares. Exponential notation was still in its beginning stage of being used as an abbreviation, its functionality (indeed, functionality itself) not yet recognised.

Using the similarity of the triangles AQN and ABL he notes that

$$\frac{\text{AQ}}{\text{QN}} = \frac{\text{AB}}{\text{BL}},$$

in the quaint colonic notation AQ:QN :: AB:BL which survives today in standardised tests of verbal skills:

dog : puppy :: cat : ? .

The point here is to express BL as a ratio as we did earlier for BK. In functional terms, he is deriving an algebraic expression for $f(q-e, m-a)$ directly without first determining $f(q, m)$ and then making the substitution as we did. The determination of $f(q, m)$ is implicit, however, in the application of Rule 2.

Barrow's second use of colons is the sort of thing that makes modern maths teachers cringe when they see it in their students' papers.

$$q - e : m - a :: r : \text{BL} = \frac{rm - ra}{q - e}$$

is to be read as

$$\frac{q-e}{m-a} = \frac{r}{\text{BL}} \text{ and therefore BL} = \frac{rm - ra}{q - e}.$$

The rest of Barrow's derivation is a straightforward algebraic computation augmented by Rules 1–3 until a homogeneous linear equation in a and e is established. The paragraph has more than its fair share of typographical errors, which do not enhance its readability.

The final step of replacing e by t and a by m would seem to be the most mysterious part of the procedure, suggesting to the modern reader the taking of the limit as $e \to t$ and $a \to m$. This, of course, is not the case, as t and m are fixed finite values while e and a are infinitesimal variables. Barrow is, as indicated earlier,[124] using the supposed similarity of the triangles NRM and TPM and the consequent equation of proportionality, $e/a = t/m$.

---

[124]Cf. his comment following Rule 3 on page 91, above.

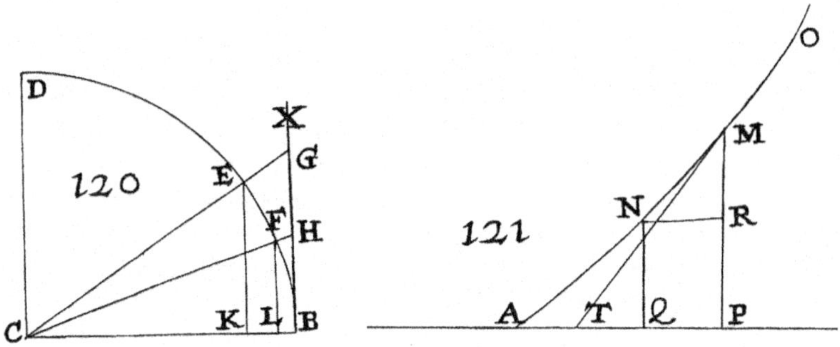

Fig. 2.36 Barrow's figures for tangent

The other examples of the algebraic determination of tangents given in the *Lectiones geometricæ* are of somewhat more interesting curves. The first, which we have discussed here, is at first sight a strange little curve.

2.3.7 Exercise For the sake of definiteness choose r to be 2 in Barrow's example, so that

$$f(q,m) = q^4 + q^2m^2 - 4m^2.$$

i. Graph $f(q,m) = 0$ on the qm-plane.
ii. Define $g(x,y) = f(2-y,x)$ and graph $g(x,y) = 0$ in the xy-plane.
iii. Writing $q = r\cos\theta$, $m = r\sin\theta$, express $f(q,m) = 0$ in polar coordinates and graph the resulting equation.

The result of the Exercise shows us that Barrow's Example I is a rotated and translated version of the polar curve $r = c\tan\theta$ for some constant c. His fifth example replaces r and θ by y and x: in other words, he shows the derivative to $y = c\tan x$ to be $y' = c\sec^2 x$.[125] In doing so, he introduced two figures — 120 and 121 reproduced in Fig. 2.36. The thing to notice is that his Fig. 121 is essentially the same as his Fig. 115 (in Fig. 2.34), albeit more carefully drawn: That N is on the curve and not on the tangent line is clearly evident in the new image; this was not at all evident in Fig. 115.[126]

In the *Lectiones geometricæ*, the illustrations are not presented in the text where they are referred to, but are collected in special folded sheets. Figures 115, 116, 120, and 121 are all on the same sheet in close proximity, with Fig. 115 situated directly above 120, which is directly to the left of 121. This effectively renders Fig. 115 redundant, a less well-drawn copy of 121.

The distinction between the point N on the curve and the point, say I, of intersection of the tangent line MT and the line NR, which was clear enough on Fermat's cluttered

[125]The other curves for which he finds tangents are two versions of the *folium of Descartes*, with equations $x^3 + y^3 = c$ and $x^3 + y^3 = cxy$, and the quadratrix.
[126]Cf. footnote 120.

diagram[127] is missing in Barrow's 115 and 116, but visible in Fig. 121. Like Fermat's reference to "ad-equate" in place of "equate", he was consciously assuming the difference between two quantities,

$$\frac{e}{a} = \frac{\text{NR}}{\text{MR}} \text{ and } \frac{\text{IR}}{\text{MR}} = \frac{\text{TP}}{\text{MP}} = \frac{t}{m},$$

to be negligible, i.e., infinitesimal. This makes NI/MR infinitesimal and NI thus to be an infinitesimal of higher order than NR and MR. This puts Rule 3, the identification of e/m with t/m, on par with Rule 1, the removal of higher order infinitesimals. The language of higher order infinitesimals would be developed by Leibniz. The importance of the higher infinitesimality of IN would emerge in the work of Lagrange.

A few parting words about Barrow might be in order. I begin with a quotation from Margaret Baron:

> In conclusion it is perhaps worth saying once again that Barrow's *Geometrical Lectures* should be viewed, not as an isolated study, but as the culmination of all the seventeenth-century geometrical investigations leading to the calculus. In this context the work represents the most detailed and systematic treatment of these properties of curves such as tangents, arcs, areas and so on, which, in the hands of Newton and Leibniz, led so rapidly to the invention of the calculus. By the use of modern notation, it is, of course, possible to transform the geometrical results arrived at by Barrow into standard differentials and definite integrals and Child has drawn up a formidable array of results which he obtained by so doing. Moreover, Barrow was able to integrate the concepts of time and motion with those of space in the manner suggested by Torricelli, Galileo and Roberval, and thus to move nearer to Newton's fluxions.[128]

Child's translation of Barrow's geometrical lectures was not undertaken so much to translate the work from Latin into English as to translate the geometric theorems presented therein into analytic terms and thereby prove that Barrow had pretty much invented the Calculus before Newton and Leibniz and to make the case that their works were highly dependent on Barrow's. Like any conclusion in the history of mathematics, this has been disputed by other historians.

Baron continues

> Mathematical invention is a process of continuous change and development rather than something which takes place at a given point in time, but if it be considered necessary to draw a line between those mathematicians of the seventeenth century who "had the calculus" and those who "had not" the line would inevitably exclude Barrow on the grounds that he exhibited no calcular rules and used no specialised notation or symbolism. The claim made by Child that Barrow privately made use of notation, rules and symbols and that he turned these over to Newton whilst preferring to publish his own work in purely geometrical language, cannot be considered seriously. Barrow was a skilful geometer, not only in the purely formal

[127] Figure 2.32. Fermat's and Barrow's labelling differ:

Barrow	M	N	R	P	T	–
Fermat	*R*	*N*	*E*	*D*	*B*	*I*

[128] Baron, *op. cit.*, p. 251.

> sense, but also in his intuitive appreciation, through the concepts of time and motion, of the properties of curves, tangents and areas. His approach to the study of curves was made possible by the acceptance, in his thinking processes, of Cavalierian indivisibles, and there is no evidence that he evolved, or indeed felt any need for, any kind of analytical procedure.[129]

The infinitely small had been around for some time in the form of *indivisibles*, the exact nature of which varied from person to person. Bonaventura Cavalieri (1598–1647), mentioned by Baron, was particularly adept at using them, but his key contribution based on indivisibles was in finding areas and, as much fun as it is, it is not at first sight really relevant here and, indeed, Cavalieri's infinitesimals are not particularly relevant to the present book. One aspect of his work, however, will be considered in the next chapter, in Sect. 3.2.5. For now, I must content myself with suggesting the reader look up some account of Cavalieri's method.[130]

2.3.3 Transition to Newton and Leibniz

Barrow, as Baron says, did not invent the Calculus. But perhaps one should draw two lines, on either side of Barrow separating the predecessors to Barrow — Fermat, Hudde, de Sluse, Torricelli, Roberval, Cavalieri, etc. — from Barrow and Barrow from his successors — Newton and Leibniz. Child did do the transformations referred to by Baron and found one can read into Barrow most of the rules which, analytically expressed, would be used by Newton and Leibniz in constructing the differentiation algorithm which turned the analytic *art* into *a calculus*, one so powerful it became *the Calculus*. Moreover, Barrow derived geometrically a version of the Fundamental Theorem of the Calculus by which the area and tangent problems are inverse to one another and, in the hands of Newton and Leibniz, the integral calculus became at least semi-algorithmic as well. Barrow, however, was a geometrician and, as evidenced by his remark on having had to be persuaded to include some examples of the analytic determination of tangents, was not interested in the analytic development of his results. This was where Newton and Leibniz came in. Historians debate on how much they owe to Barrow. Child, in his translation of Barrow's geometrical lectures and a translation of Leibniz's early mathematical manuscripts[131] attempted to prove that they both owed almost all to Barrow's work, but most historians consider this view extreme.

Grabiner explains the difference nicely:

---

[129] *Ibid.*, pp. 251–252.

[130] Excerpts from Cavalieri's work can be found in: David Eugene Smith (ed.), *A Source Book in Mathematics*, 1929 (reprinted: Dover Publications, Inc., New York, 1959, pp. 605–609); Struik, *op. cit.*, pp. 209–219; and Stedall (*op. cit.*, pp. 62–65. Accounts can also be found in Edwards (*op. cit.*, pp. 104–109) and Baron (*op. cit.*, pp. 122–135). By far the most complete discussion in English however is Kirsti Andersen, "Cavalieri's Method of Indivisibles", *Archive for History of Exact Sciences* 31(1985), pp. 291–367.

[131] J.M. Child (ed. and trans.), *The Early Mathematical Manuscripts of Leibniz*, The Open Court Publishing Company, Chicago and London, 1920.

> In the latter third of the seventeenth century, Newton and Leibniz, each independently, invented the calculus. By "inventing the calculus" I mean that they did three things. First, they took the wealth of methods that already existed for finding tangents, extrema, and areas, and they subsumed all these methods under the heading of two general concepts, the concepts which we now call **derivative** and **integral**. Second, Newton and Leibniz each worked out a notation which made it easy, almost automatic, to use these general concepts...Third, Newton and Leibniz each gave an argument to prove what we now call the Fundamental Theorem of the Calculus: the derivative and the integral are mutually inverse. Newton called our "derivative" a *fluxion* — a rate of flux or change; Leibniz saw the derivative as a ratio of infinitesimal differences and called it the *differential quotient*. But whatever terms were used, the concept of derivative was now embedded in a general subject — the calculus — and its relationship to the other basic concept, which Leibniz called the integral, was now understood.[132]

Another thing Newton and Leibniz did, not nearly as successfully, was to provide justifications for the ad-equations. Barrow's analytic determination of the tangent made a double advance on Fermat by explicitly introducing infinitesimals into the discussion, something we now recognise as the equivalent of taking limits, and by drawing attention to the *characteristic triangle* MTP or the "triangle" MNR from which we get the slope of the tangent. But he left unexplained why MNR could be taken as similar to MTP.

Once one has the characteristic triangle in mind I suppose it is inevitable to view the tangent line through a point P on a curve C as the limiting position of secant lines passing through P and a second point N on the curve as N nears P. Or, one might view the tangent as the secant line for N infinitesimally close to P. That the emphasis would fall on the slope of the line, i.e., on $\tan(\angle MTP)$ as opposed to its reciprocal $\cot(\angle MTP)$ as calculated by Barrow is perhaps a little less inevitable and may be attributable to the desire to represent the tangent line in functional form, $y = mx + b$, where m is the slope. Or, it may be due to the notion of rates of change and the habit of choosing $x(t) = t$ wherever possible. For whatever reason, before Newton and Leibniz any property of the tangent line was used to find it, and after Newton and Leibniz everyone calculated the slope of the tangent line via the difference quotient

$$\frac{\Delta y}{\Delta x} = \frac{f(x + \Delta x) - f(x)}{\Delta x},$$

for very small Δx. Exactly how Δx was equivalent to 0 was variously explained and not satisfactorily so until the 19th century.

2.3.4 *Newton*

Newton was the first to "invent" the Calculus. The story is well-known. He was studying in Cambridge when the plague arrived in England in 1665 and he retreated to his family farm in Woolsthorpe where he worked out the Calculus and Physics in

132Grabiner, "Changing concept...", *op. cit.*, p. 199.

one *annus mirabilis*. He then kept as quiet as possible about it for as long as possible. He did share a few results with colleagues, but it was decades before he published the Calculus in definitive Newtonian form. By then, of course, Leibniz had rediscovered everything and published it all.

A proper discussion of Newton can only be given by a dedicated Newton scholar. His publications do not reflect the time or even order of discovery. It is well-known, for example, that he used the Calculus to arrive at the results of his *Principia*, but there is no trace of this in the book, in which these results are established by classical geometrical methods. With respect to tangents, Baron informs us that

> In the first stages of Newton's investigations into the properties of curved lines he relied mainly on Descartes' tangent method. He took over Descartes' characteristic symbolism...as well as the method of equal roots. This he incorporated with *Hudde's Rule*.[133]

Once he began to develop the Calculus, however, he switched over to infinitesimals, and later to some vague notion of limit:

> It is well known that at first the method employed by Newton involved fixed infinitesimals. But in the Introduction to the *Quadratura curvarum*, published in 1704, Newton aimed to develop his theory without the use of infinitely small quantities. Both in his *Principia* and *Quadratura curvarum*, he used "prime and ultimate ratios," which involve the concept of limits, though in a form differing from that adopted by mathematicians later. These prime and ultimate ratios do not contemplate primarily one constant which one variable approaches. The prime and ultimate ratios are ratios of two quantities just springing into being or else vanishing. Only secondarily does Newton, in applying his theory to finding the fluxion[134] of x^n, for example, consider in the right member of his equations what we would call the limit of a ratio. Newton was really considering the ratio of two quantities, each of which was approaching the limit zero, rather than the limit of one quantity that was the ratio of two quantities.[135]

The works cited are the *Tractatus de quadratura curvarum* [*Treatise on the quadrature of curves*] (1704) and the *Philosophiæ naturalis principia mathematica* [*Mathematical principles of natural philosophy*] (1687), both usually referred to by the abbreviated titles given in the quotation. The *Quadratura curvarum* is Newton's most famous work on the Calculus and the *Principia*, of course, his most famous and important work.

Newton's first work on the Calculus to be disseminated was *De analysis per æquationes numeri terminorum infinitas* of 1669, not published until 1711. This dealt mainly with infinite series and term-by-term differentiation and integration and is of no importance for our present purposes. This was followed in 1671 by *De methodus fluxionum et serierum infinitorum*, first published in English translation by John Colson as *The Method of Fluxions and Infinite Series with its Application to the Geometry of CURVE-LINES* in 1736. The *Principia*, which could have been the

---

[133] Baron, *op. cit.*, p. 257.

[134] The fluxion is essentially the derivative of one of the variables with respect to time — more anon.

[135] Florian Cajori, "Newton's fluxions", in: David Eugene Smith (ed.), *Sir Isaac Newton, 1727–1927; A Bicentenary Evaluation of His Work*, The Williams & Wilkins Company, Baltimore, 1928, p. 193.

first publication of the Calculus, was published in 1687, but Newton replaced the Calculus, which he had used to obtain his results, by classical geometrical reasoning. Thus, the *Quadratura curvarum*, written a few years later and tacked on to Newton's *Opticks* became in 1704 his first published account of his treatment of the Calculus.

English translations of these works are all available online or in any university library, and substantial excerpts are to be found in most source books. I quote from an early translation of the *De methodus fluxionum*:

> Now those Quantities which I consider as gradually and indefinitely increasing, I shall hereafter call *Fluents*, or *Flowing Quantities*, and shall represent them by the final Letters of the Alphabet v, x, y, and z; that I may distinguish them from other Quantities, which in Equations are to be consider'd as known and determinate, and which therefore are represented by the initial Letters a, b, c &cc. And the Velocities by which every Fluent is increased by its generating Motion, (which I may call *Fluxions*, or simply Velocities or Celebrities,) I shall represent by the same Letters pointed thus $\dot{v}$, $\dot{x}$, $\dot{y}$, and $\dot{z}$. That is, for the Celerity of the Quantity v I shall put $\dot{v}$, and so for the Celebrities of the other Quantities x, y, and z, I shall put $\dot{x}$, $\dot{y}$, and $\dot{z}$ respectively.[136]

We can think of the fluents v, x, y, and z as functions $v(t)$, $x(t)$, $y(t)$ and $z(t)$ of time t, and their fluxions as their instantaneous rates of change: $\dot{v} = dv/dt$, $\dot{x} = dx/dt$, $\dot{y} = dy/dt$, and $\dot{z} = dz/dt$. Having introduced such, Newton next considers several problems, offering examples of each type with their solutions, and then follows up with a demonstration of the solution:

PROB. I.

The Relation of the Flowing Quantities to one another being given, to determine the Relation of their Fluxions.

SOLUTION.

1. Dispose the Equation, by which the given Relation is express'd, according to the Dimensions of some one of its flowing Quantities, suppose x, and multiply its Terms by any Arithmetical Progression, and then by $\frac{\dot{x}}{x}$. And perform this Operation separately for every one of the flowing Quantities. Then make the Sum of the Products equal to nothing, and you will have the Equation required.
2. EXAMPLE I. If the Relation of the flowing Quantities x and y be $x^3 - ax^2 + axy - y^3 = 0$; first dispose the Terms according to x, and then according to y, and multiply them in the following manner.

Mult	x^3	$-ax^2$	$+axy$	$-y^3$	$-y^3$	$+axy$	$-ax^2$ $+x^3$
by	$\frac{3\dot{x}}{x}$.	$\frac{2\dot{x}}{x}$.	$\frac{\dot{x}}{x}$.	0	$\frac{3\dot{y}}{y}$.	$\frac{\dot{y}}{y}$.	0
makes	$3\dot{x}x^2 - 2ax\dot{x} + a\dot{x}y$			$*$	$-3\dot{y}y^2 + a\dot{y}x$		$*$

The Sum of the Products is $3\dot{x}x^2 - 2ax\dot{x} + a\dot{x}y - 3\dot{y}y^2 + a\dot{y}x = 0$, which Equation gives the Relation between the Fluxions $\dot{x}$ and $\dot{y}$. For if you take x at pleasure, the Equation $x^3 - ax^2 + axy - y^3 = 0$ will give y. Which being determined, it will be $\dot{x} : \dot{y} :: 3y^2 - ax : 3x^2 - 2ax + ay$.[137]

---

[136] Isaac Newton (John Colson ed. and trans.), *The Method of Fluxions and Infinite Series with its Application to the Geometry of CURVE-LINES*, 1736, p. 20.

[137] *Ibid.*, p. 21.

Overall one should recognise the method from Calculus as differentiating $f(x(t), y(t)) = 0$ to obtain $df/dt(x(t), y(t)) = 0$ for $f(x, y) = x^3 - ax^2 + axy - y^3$ and then determining dx/dy. The reference to an arbitrary arithmetical progression indicates that he is relying on the methods of Hudde and de Sluse.

After presenting a few additional examples, he comes to the demonstration. Here he resorts to infinitesimals by introducing the "indefinitely small Quantity" o, which we may take to be the differential dt and considers the "Moments" $\dot{v}o$, etc., which we may consider to be the differentials $dv = v'(t)dt$, etc.

DEMONSTRATION *of the Solution.*

13. The Moments of flowing Quantities, (that is, their indefinitely small Parts, by the accession of which, in indefinitely small portions of Time, they are continually increased,) are as the Velocities of their Flowing or Increasing.
14. Wherefore if the Moment of any one, as x, be represented by the Product of its Celerity $\dot{x}$ into an indefinitely small Quantity o (that is, by $\dot{x}o$,) the Moments of the others v, y, z, will be represented by $\dot{v}o, \dot{y}o, \dot{z}o$; because $\dot{v}o, \dot{x}o, \dot{y}o$, and $\dot{z}o$ are to each other as $\dot{v}, \dot{x}, \dot{y}$, and $\dot{z}$.
15. Now since the Moments, as $\dot{x}o$ and $\dot{y}o$ are the indefinitely little accessions of the flowing Quantities x and y, by which those Quantities are increased through the several indefinitely little intervals of Time; it follows, that those Quantities x and y, after any indefinitely small interval of Time, become $x + \dot{x}o$ and $y + \dot{y}o$. And therefore the Equation, which at all times indifferently expresses the Relation of the flowing Quantities, will as well express the Relation between $x + \dot{x}o$ and $y + \dot{y}o$, as between x and y: So that $x + \dot{x}o$ and $y + \dot{y}o$ may be substituted in the same Equation for those Quantities, instead of x and y.
16. Therefore let any Equation $x^3 - ax^2 + axy - y^3 = 0$ be given, and substitute $x + \dot{x}o$ for x, and $y + \dot{y}o$ for y, and there will arise

$$\left.\begin{array}{llll} x^3 & + 3\dot{x}ox^2 & + 3\dot{x}^2oox & + \dot{x}^3o^3 \\ - ax^2 & - 2a\dot{x}ox & - a\dot{x}^2oo & \\ + axy & + a\dot{x}oy & + a\dot{y}ox & + a\dot{x}\dot{y}oo \\ - y^3 & - 3\dot{y}oy^2 & - 3\dot{y}^2ooy & - \dot{y}^3o^3 \end{array}\right\} = 0.$$

17. Now by Supposition $x^3 - ax^2 + axy - y^3 = 0$, which therefore being expunged, and the remaining Terms being divided by o, there will remain $3\dot{x}x^2 + 3\dot{x}^2ox + \dot{x}^3oo - 2a\dot{x}x - a\dot{x}^2o + a\dot{x}y + a\dot{y}x + a\dot{x}\dot{y}o - 3\dot{y}y^2 - 3\dot{y}^2oy - \dot{y}^3oo = 0$. But whereas o is supposed to be infinitely little, that it may represent the Moments of Quantities; the Terms that are multiply'd by it will be nothing in respect of the rest. Therefore I reject them, and there remains $3\dot{x}x^2 - 2a\dot{x}x + a\dot{x}y + a\dot{y}x - 3\dot{y}y^2 = 0$, as above in Examp. I.
18. Here we may observe, that the Terms that are not multiply'd by o will always vanish, as also those Terms that are multiply'd by o of more than one Dimension. And that the rest of the Terms being divided by o, will always acquire the form that they ought to have by the foregoing Rule: Which was the thing to be proved.[138]

The method here is virtually the same as Fermat's calculation of

$$\frac{f(A+E) - f(A)}{E}$$

[138] *Ibid.*, pp. 24–25.

and the subsequent elimination of all terms containing E, or Barrow's similar procedure. The big difference is the explicit reference to the infinitesimal nature of o as justification.

Newton gives several examples of the use of this technique before going on to discuss related problems, the third of which is

PROB. III.

To determine the Maxima *and* Minima *of Quantities.*

1. When a Quantity is the greatest or the least that it can be, at that moment it neither flows backwards or forwards. For if it flows forwards, or increases, that proves it was less, and will presently be greater than it is. And the contrary if it flows backwards, or decreases. Wherefore find its Fluxion, by Prob. I. and suppose it to be nothing.
2. EXAMP. I. If in the Equation $x^3 - ax^2 + axy - y^2 = 0$ the greatest Value of x be required; find the Relation of the Fluxions of x and y, and you will have $3\dot{x}x^2 - 2a\dot{x}x + a\dot{x}y - 3\dot{y}y^2 + a\dot{y}x = 0$. Then making $\dot{x} = 0$, there will remain $-3\dot{y}y^2 + a\dot{y}x = 0$ or $3y^2 = ax$. By the help of this you may exterminate either x or y out of the primary Equation, and by the resulting Equation you may determine the other, and then both of them by $-3y^2 + ax = 0$.[139]

His next problem is the construction of tangent lines.

PROB. IV.

To draw Tangents to Curves.

First Manner.

1. Tangents may be variously drawn, according to the various Relations of Curves to right Lines. And first let BD be a right Line, or Ordinate, in a given Angle to another right Line AB, as a Base or Absciss, and terminated at the Curve ED. Let this Ordinate move[140] through an indefinitely small Space to the place *bd*, so that it may be increased by the Moment *cd*, while AB is increased by the Moment B*b*, to which D*c* is equal and parallel. [See Fig. 2.37.] Let D*d* be produced till it meets with AB in T, and this Line will touch the Curve in D or *d*; and the Triangles *dc*D, DBT will be similar.[141] So that it is TB : BD :: D*c* (or B*b*) : *cd*.
2. Since therefore the Relation of BD to AB is exhibited by the Equation, by which the nature of the Curve is determined; seek for the Relation of Fluxions, by Prob. I. Then take TB to BD in the Ratio of the Fluxion of AB to the Fluxion of BD, and TD will touch the Curve in the point D.
3. EX. I. Calling AB = x, and BD = y, let their Relation be $x^3 - ax^2 + axy - y^3 = 0$. And the Relation of the Fluxions will be $3\dot{x}x^2 - 2a\dot{x}x + a\dot{x}y - 3\dot{y}y^2 + a\dot{y}x = 0$.

---

[139] *Ibid.*, p. 44.

[140] For Newton the curve is traced out as the point of intersection of two non-rotating lines moving along a pair of axes. The axes need not meet at right angles and the lines, called the *abscissa* and *ordinate*, need not be vertical and horizontal, but must remain parallel to the axes. The abscissa is parallel to the y-axis and its coordinate is x, while the ordinate is parallel to the x-axis with coordinate y.

[141] Newton is, of course, being only approximate here. If cd is the moment, d lies on the curve infinitesimally close to the tangent line, but not *on* the tangent line. The line through D and d will cross the curve not touch it.

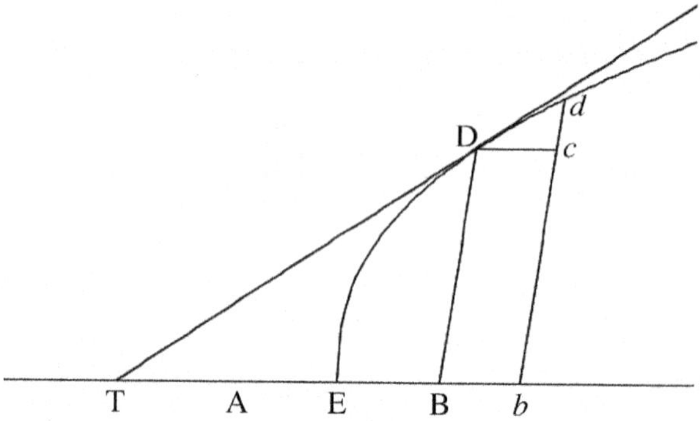

Fig. 2.37 Newton's tangent construction

So that $\dot{y} : \dot{x} :: 3xx - 2ax + ay : 3y^2 - ax ::$ BD (y) : BT. Therefore BT = $\frac{3y^3 - axy}{3x^2 - 2ax + ay}$. Therefore the point D being given, and thence DB and AB, or y and x, the length BT will be given, by which the Tangent TD is determined.[142]

I find it amusing to contemplate the full title of the English edition of the *De methodus fluxionum* and the implied promise of the last line. I present it below without the numerous variations in font size, letting the frequent changes in style testify to its garishness:

THE
METHOD of FLUXIONS
AND
INFINITE SERIES;
WITH ITS
Application to the Geometry of CURVE- LINES.

By the INVENTOR
Sir ISAAC NEWTON, *K^t.*
Late President of the Royal Society.

Translated from the AUTHOR'*s* LATIN ORIGINAL
not yet made publick.

To which is subjoin'd,
A PERPETUAL COMMENT upon the whole Work,
Consisting of
ANNOTATIONS, ILLUSTRATIONS, and SUPPLEMENTS,
In order to make this Treatise
A compleat Institution for the use of LEARNERS.

The "learners" in question are almost certainly not undergraduates. University instruction in mathematics of the day was simply not up to this level. Moreover, by all accounts, Newton was not a successful lecturer, having few students if any, often returning to his office early or occasionally speaking to an empty hall when students failed to show up. And one can see from the above passages, despite the care to logic,

---

[142]Newton, *op. cit.*, p. 46.

why students would have a hard time understanding him. Today we would begin with the simplest case of a curve given by the graph of a function $y = f(x)$, and form the difference quotient

$$\frac{f(x+o) - f(x)}{o}. \tag{2.33}$$

Newton started off with a curve $f(x, y) = 0$ and calculated the mysterious

$$\frac{f(x + \dot{x}o, y + \dot{y}o) - f(x, y)}{o},$$

which we would represent as

$$\frac{f\big(x(t+o), y(t+o)\big) - f\big(x(t), y(t)\big)}{o},$$

i.e., as

$$\frac{g(t+o) - g(t)}{o},$$

where $g(t) = f\big(x(t), y(t)\big)$. Unfortunately, he did not yet have functional notation and launched right into multivariable calculus.

As for tangents, starting with $y = f(x)$, after simplifying (2.33), expunging o, whether by ignoring infinitesimal differences *à la* Newton or taking the limit as we do today, we would get the slope $f'(x)$ of the tangent and be led in the two variable case to finding $dy/dx = (dy/dt)/(dx/dt)$. Newton, not having the single variable case causing one to standardise on the slope, was slightly inconsistent. Thus, for his Example I, in Problem I he calculated the slope's reciprocal $\dot{x}/\dot{y} = dx/dy$ and in paragraphs 1 and 2 of Problem IV he describes the process in terms of finding this reciprocal, and then, in considering this Example, immediately finds the slope $\dot{y}/\dot{x}$ itself. I want to add the simultaneous placement of d on the curve and the tangent line as another source of possible confusion, but his "learners" would be astronomers, physicists, mathematicians, and other learned scholars already familiar with ad-equality or the dismissal of infinitesimals, but not Newton's systematic method for solving problems involving curves and certainly not the Leibnizian d-notation, and this simultaneous occupation of two places by d would not have been confusing to them.

As already mentioned, the *De methodus fluxionum* was privately disseminated and a number of mathematicians learned the fluxional calculus before the book's publication in 1736. The first public announcement by Newton of the method was in 1687 in the first edition of the *Principia* and he still used infinitesimals. Within a few years he rejected them in favour of a not very clearly presented notion of limit in the *Quadratura curvarum*, composed in the years 1691–1692 and published in 1704 — his first publication on the Calculus. It was also the last to be written and the most widely read of his accounts, thus perhaps the definitive version of his theory. The second English translation of 1745 bears the subtitle "The Treatises themselves, translated into English, with a large Commentary; in which the Demonstrations are supplied where wanting, the Doctrine illustrated, and the whole accommodated to the Capacities of Beginners, for whom it is chiefly designed". Reading Newton's opening words confirms that it is the Commentary that is chiefly designed for beginners:

INTRODUCTION to the Quadrature of Curves.

1. I Consider[143] mathematical Quantities in this Place not as consisting of very small Parts; but as describ'd by a continued Motion. Lines are describ'd, and thereby generated not by the Apposition of Parts, but by the continued Motion of Points[144]; Superficies's[145] by the Motion of Lines; Solids by the Motion of Superficies's; Angles by the Rotation of the Sides; Portions of Time by a continual Flux: and so in other Quantities. These Geneses really take Place in the Nature of Things, and are daily seen in the Motion of Bodies. And after this Manner the Ancients, by drawing moveable right Lines along immoveable right Lines, taught the Genesis of Rectangles.
2. Therefore considering that Quantities, which increase in equal Times, and by increasing are generated, become greater or less according to the greater or less Velocity with which they increase and are generated; I sought a Method of determining Quantities from the Velocities of the Motions or Increments, with which they are generated; and calling these Velocities of the Motions or Increments *Fluxions*, and the generated Quantities *Fluents*, I fell by degrees upon the Method of Fluxions, which I have made use of here in the Quadrature of Curves, in the Years 1665 and 1666.
3. Fluxions are very nearly as the Augments of the Fluents generated in equal but very small Particles of Time, and, to speak accurately, they are the *first Ratio* of the nascent Augments; but they may be expounded by any Lines which are proportional to them.[146]
4. Thus if the Area's ABC, ABDG [See Fig. 2.38.[147]] be described by the Ordinates BC, BD moving along the Base AB with an uniform Motion, the Fluxions of these Area's shall be to one another as the describing Ordinates BC and BD, and may be expounded by these Ordinates, because that these Ordinates are as the nascent Augments of the Area's.[148]
5. Let the Ordinate BC advance from its Place into any new Place *bc*. Complete the Parallelogram BCE*b*, and draw the right Line VTH touching the curve in C, and meeting

---

[143]The paragraph opens with a normal sized "1", followed by a large drop cap "I", and "consider" subsequently capitalised. I decided this required too much effort to duplicate completely.

[144]Newton is here taking a very Aristotelian view of the line.

[145]I.e., surfaces.

[146]"First ratio" and "nascent" are not exactly defined here. Their meaning will emerge when examples are discussed. In a couple of pages the "first ratio" will be called the "prime ratio". My interpretation of this paragraph is that, for any fluents v, w, $\dot{v}/\dot{w} \approx \Delta v/\Delta w$, that is, $\dot{v}/\dot{w} \approx (\dot{v}o)/(\dot{w}o)$.

[147]This image is taken from: Florian Cajori, *A History of the Conceptions of Limits and Fluxions in Great Britain from Newton to Woodhouse*, The Open Court Publishing Company, Chicago and London, 1919, p. 42, and is a clean reproduction of Newton's original.

[148]In light of paragraph 3, he seems to be suggesting something like

$$\frac{d\int_a^x f(t)dt}{dx} \Big/ \frac{d\int_a^x g(t)dt}{dx} = \frac{f(t)}{g(t)},$$

where the curve AC*c* is given by $y = f(x)$ and GD*a* by $y = g(x)$. Paragraphs 3 and 4 are unhelpful in the extreme. He thus seems to be asserting the Fundamental Theorem of the Calculus via some clumsy reference to proportion.

Fig. 2.38 Newton's prime and ultimate ratios

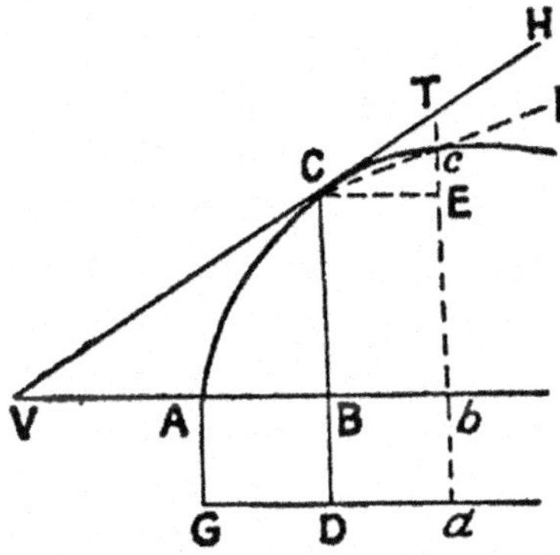

the two lines *bc* and BA produc'd in T and V: and B*b*, E*c*, and C*c* will be the Augments now generated of the Absciss AB, the Ordinate BC and the Curve Line AC*c*[149]; and the Sides of the Triangle CET are in the *first Ratio* of these Augments considered as nascent, therefore the Fluxions of AB, BC and AC are as the Sides CE, ET and CT of that Triangle CET, and may be expounded by these same Sides, or, which is the same thing, by the Sides of the Triangle VBC, which is similar to the Triangle CET.

6. It comes to the same Purpose to take the Fluxions in the *ultimate Ratio* of the evanescent Parts. Draw the right Line C*c*, and produce it to K. Let the Ordinate *bc* return into its former place BC, and when the Points C and *c* coalesce, the right Line CK will coincide with the Tangent CH, and the evanescent Triangle CE*c* in its ultimate Form will become similar to the Triangle CET, and its evanescent Sides CE, E*c* and C*c* will be *ultimately* among themselves as the sides CE, ET and CT of the other Triangle CET, are, and therefore the Fluxions of the Lines AB, BC and AC are in this same Ratio. If the Points C and *c* are distant from one another by any small Distance, the right Line CK will likewise be distant from the Tangent CH by a small Distance. That the right Line CK may coincide with the Tangent CH, and the ultimate Ratios of the Lines CE, E*c* and C*c* may be found, the Points C and *c* ought to coalesce and exactly coincide. The very smallest Errors in mathematical Matters are not to be neglected.[150]

The last sentence is occasionally interpreted as the complete and final rejection of infinitesimals on Newton's part. Newton is here fumbling for a notion of limit: What are the ratios in the triangle CE*c* when C and *c* finally coalesce? He expressed himself more clearly elsewhere. In the second edition of the *Principia* of 1713 we read:

> Quantities, as also ratios of quantities, which constantly tend toward equality in any finite time, and before the end of that time approach each other more nearly than [with] any given difference whatever, become ultimately equal...[151]
>
> The objection is that there is no ratio of evanescent quantities, which obviously, before they have vanished, is not ultimate; when they have vanished, there is none. But also by the same like argument it may be contended that there is no ultimate velocity of a body arriving

---

[149] Thus, these will be our dx, dy, and ds.

[150] John Stewart, *Sir Isaac Newton's Two Treatises of the Quadrature of Curves, and Analysis by Equations of an infinite Number of Terms, explained*, London, 1745, pp. 1–2. The second treatise referred to in the title is *De analysi*.

[151] This is remarkably close to our modern definition of "limit".

at a certain position; for before the body attains the position, this is not ultimate; when it has attained [it], there is none. And the answer is easy: By ultimate velocity I understand that with which the body is moved, neither before it arrives at the ultimate position and the motion ceases, nor thereafter, but just when it arrives; that is, that very velocity with which the body arrives at the ultimate position and with which the motion ceases. And similarly for the motion of evanescent quantities is to be understood the ratio of the quantities, not before they vanish, nor thereafter, but [that] with which they vanish. And likewise the first nascent ratio is the ratio with which they begin. And the prime and ultimate amount is to be [that] with which they begin and cease (if you will, to increase and diminish). There exists a limit which the velocity may attain at the end of the motion, but [which it may] not pass. This is the ultimate velocity. And the ratio of the limit of all quantities and proportions, beginning and ceasing, is equal...

The ultimate ratios in which quantities vanish, are not really the ratios of ultimate quantities, but the limits toward which the ratios of quantities, decreasing without limit, always approach; and to which they can come nearer than any given difference, but which they can never pass nor attain before the quantities are diminished indefinitely.[152]

Newton uses the word "limit" here in the sense of the word "bound". The curves familiar in those days were well-behaved with little oscillation. As one approached a point on a curve from the left or from the right, once one got close enough, the curve was monotone. Thus, as x approached a from the left, say, the values $f(x)$ either approached their limit from above or from below, but never passed it. Today, as with Darboux's function of Fig. 2.26, we drop the clause disallowing the quantities from attaining or surpassing their limits.

Newton clearly has a conception of limit, but has been having difficulty expressing it. He hasn't isolated its defining characteristic.

Following the introductory remarks in the *Quadratura curvarum*, Newton gives a few examples to illustrate his concepts. The one most anthologised is his differentiation of powers of x:

11. *Let the Quantity x flow uniformly,*[153] *and let it be proposed to find the Fluxion of* x^n. In the same Time that the Quantity x, by flowing, becomes $x + \text{o}$, the Quantity x^n will become $\overline{x+\text{o}}|^n$, that is, by the Method of infinite Series's,[154] $x^n + n\text{o}x^{n-1} + \frac{n^2-n}{2}\text{oo}x^{n-2} + \&c$. And the Augments o and $n\text{o}x^{n-1} + \frac{n^2-n}{2}\text{oo}x^{n-2} + \&c$. are to one another as 1 and $nx^{n-1} + \frac{n^2-n}{2}\text{o}x^{n-2} + \&c$.
    Now let these Augments vanish, and their ultimate Ratio will be 1 to nx^{n-1}.
12. By like ways of reasoning, the Fluxions of Lines, whether right or curve in all Cases, as likewise the Fluxions of Superficies's, Angles and other Quantities, may be collected by the Method of *prime* and *ultimate* Ratios. Now to institute an Analysis after this manner in finite Quantities and investigate the *prime* or *ultimate* Ratios of these finite Quantities when in their nascent or evanescent State, is consonant to the Geometry of the Ancients: and I was willing to show that, in the Method of Fluxions, there is no necessity of introducing Figures infinitely small into Geometry. Yet the Analysis

---

[152] Smith, *Source Book...*, *op. cit.*, pp. 617–618. Cf. also Struik, *op. cit.*, pp. 299–300.

[153] That is, let x be a constant multiple of time, so that $y = x^n$, being a function of time, is in fact a function of x.

[154] Newton had extended the Binomial Theorem to the case of arbitrary rational exponents. For n not a positive integer, however, the expansion is an infinite series. Thus, Newton is here differentiating x^n for arbitrary rational n. For more information, I refer the reader to Smoryński, *Treatise*.

> may be performed in any kind of Figures, whether finite or infinitely small, which are imagin'd similar to the evanescent Figures; as likewise in these Figures, which, by the Method of Indivisibles, use to be reckoned as infinitely small, provided you proceed with due Caution.[155]

I imagine the modern mathematician reacting to paragraph 11 in a manner not unlike the way one typically reacts to fingernails on a chalkboard. However lacking in rigour, Newton has nevertheless essentially defined the derivative of x^n as

$$\lim_{o\to 0} \frac{(x+o)^n - x^n}{o} = nx^{n-1}.$$

When I say this, I do so as a mathematician in identifying things that, however different, are abstractly the same. Newton did not have our conceptual framework. He did not have the concept of function. Thus, 11 does not read: Let $f(x) = x^n$; then $f'(x) = nx^{n-1}$. x^n was not a function of x, but another flowing quantity y varying with time and the two quantities were related by an equation $y = x^n$. They had fluxions $\dot{x}$ and $\dot{y}$, which we think of as dx/dt and dy/dt, as we regard x and y as functions of t. Time is somewhere behind the scenes in Newton, but not as an explicit variable t. The closest he comes to this is when he assumes x to "flow uniformly" — in essence making x stand in for t.

2.3.5 Leibniz

Leibniz independently discovered the Calculus in the 1670s, making his first discoveries around 1672 and publishing his first paper on the subject a little over a decade later in 1684. Where today mathematical papers have abstracts following the titles, in those days the fashion was to incorporate the abstract into the title: "Nova methodus pro maximis et minimis, itemque tangentibus, quæ nec fractas nec irrationales quantitates moratur, et singulare pro illis calculi genus"[156] ["A new method for maxima and minima as well as tangents, which is impeded neither by fractional nor irrational quantities, and a remarkable type of calculus for this"]. Leibniz, however, never published a systematic account of the whole, spreading his work out over numerous short papers and correspondence with others.

While Leibniz's early publications are important for the history of mathematics, his earlier unpublished manuscripts may offer more insight. From a manuscript of 1677 we get the following announcement, which may also serve as a partial review of the history of tangent finding methods:

---

[155] Stewart, *op. cit.*, p. 4.

[156] *Acta Eruditorum* 3 (1684), pp. 467–473. A German translation of this and several further papers of Leibniz was published by Gerhardt Kowalewski as number 162 in Ostwald's series of scientific classics in 1908. A partial English translation appears in Smith's source book, *op. cit.*, a full translation in Struik, *op. cit.*, and a nearly full translation in Stedall, *op. cit.*.

Fermat was the first to find a method which could be made general for finding the straight lines that touch analytical curves. Descartes accomplished it in another way, but the calculation that he prescribes is a little prolix. Hudde has found a remarkable abridgment by multiplying the terms of the progression by those of the arithmetical progression. He has only published it for equations in one unknown; although he has obtained it for those in two unknowns. Then the thanks of the public are due to Sluse; and after that, several have thought that this method was completely worked out.[157] But all these methods that have been published suppose that the equation *has been reduced* and cleared of fractions and irrationals; I mean of those in which the variables occur. I however have found means of obviating these useless reductions, which make the calculation increase to a terrible degree, and oblige us to rise to very high dimensions, in which case we have to look for a corresponding depression with much trouble; instead of all this, everything is accomplished at the first attack.[158]

This method has more advantage over all the others that have been published, than that of Sluse has over the rest, because it is one thing to give a simple abridgment of the calculation, and quite another thing to get rid of reductions and depressions. With respect to the publication of it, on account of the great extension of the matter which Descartes himself has stated to be the most useful part of Geometry, and of which he has expressed the hope that there is more to follow — in order to explain myself shortly and clearly, I must introduce some *fresh characters*, and give to them a *new Algorithm*, that is to say, altogether special rules, for their addition, subtraction, multiplication, division, powers, roots, and also for equations.[159]

The "fresh characters" are the differentials:

Explanation of the characters.

Suppose there are several curves, as CD, FE, HJ, connected with one and the same axis AB by ordinates drawn through one and the same point B, to wit, BC, BF, BH. The tangents CT, FL, HM to these curves cut the axis in the points T, L, M [See Fig. 2.39.]; the point A in the axis is fixed, and the point B changes with the ordinates. Let AB = x, BC = y, BF = w, BH = v; also let the ratio of TB to BC be called that of dx to dy, and the ratio of LB to BF that of dx to dw, and the ratio of MB to BH that of dx to dv. Then if, for example, y is equal to vw, we should say dvw instead of dy, and so on for all other cases. Let a be a constant straight line; then if y is equal to a, that is if CD is a straight line parallel to AB, dy or da will be equal to 0, or equal to zero. If the magnitude dx/dw comes out negative, then FL, instead of being drawn toward A, above B, will be drawn in the contrary direction, below B.[160]

Leibniz follows this with a list of the computation rules for differentials: for a constant,

$$\text{if } y = v \pm w \pm a \text{ then } dy = dv \pm dw$$
$$\text{if } y = avw \text{ then } dy = avdw + awdv,$$

[157]Newton had not yet published on the subject and consequently is not mentioned. It is not clear why Barrow is not mentioned. Child is of the opinion that Leibniz was hiding his dependence on Barrow, but it could also be the similarity of Barrow's analytical method to Fermat's.

[158]"Reductions" obviously refers to simplifying the equations for which tangents are sought. The process, especially in clearing surds, results in polynomials of higher degrees and the introduction of extra possibilities for the tangent. Presumably "depressions" refers at least in part to the elimination of false solutions. The point of this passage is that the method he is introducing is more direct and eliminates this excess work.

[159]Child, *Early Mathematical Manuscripts…*, *op. cit.*, pp. 131–132.

[160]*Ibid.*, pp. 132–133.

Fig. 2.39 Leibniz's diagram

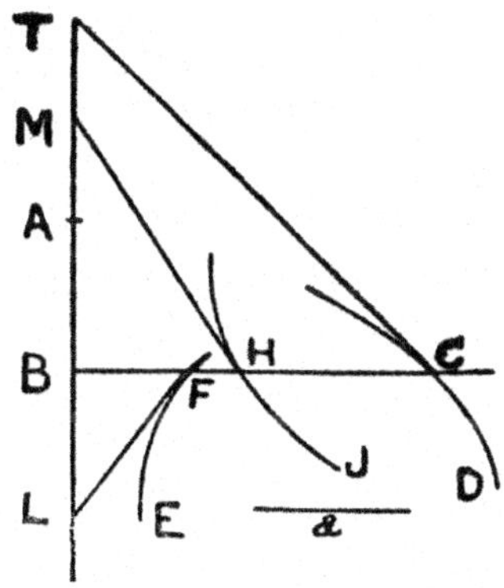

etc. The proofs for *infinitely small* quantities dx, dy are given in a revision of the manuscript written sometime in the period 1677–1680.

Edwards makes the interesting point that for Newton the fluxion $\dot{x}$, or derivative, was the same sort of quantity as x, while for Leibniz the fundamental notion was the differential dx (= Newton's $\dot{x}o$). Where we see a derivative, Leibniz saw "'merely' a geometrically significant quotient".[161] For Leibniz, once dx was given, dy was defined by setting dx/dy equal to the reciprocal of the slope of the tangent line.[162]

With Newton and Leibniz we finally have a calculus making routine the determination of the tangent for any analytic curve. As Leibniz emphasised, Hudde and de Sluse could do so for curves $f(x, y) = 0$ for polynomials $f(x, y)$. When f involved rational functions or roots, one had to do some preliminary work eliminating the fractions or roots under their method. This was no longer necessary with the Newton-Leibniz algorithm. Moreover, as new functions like $\sin x$ or $\ln x$ were differentiated, they could be combined with the old functions and the new combinations differentiated by the new method. A definite advance had been made with respect to *finding* tangents. But what about *defining* tangents? The passages quoted contain no definition.

Each author hinted at the modern definition. I have already quoted one translation of Newton above on page 107:

> If the points C and c are distant from one another by any small Distance, the right Line CK will likewise be distant from the Tangent CH by a small Distance. That the right Line CK may coincide with the Tangent CH…the Points C and c ought to coalesce and exactly coincide.

The meaning of this passage is perhaps more easily seen in Smith's translation:

---

[161] Edwards, *op. cit.*, p. 266. Edwards offers, incidentally, an excellent discussion of Leibniz's early papers as well as the published ones.

[162] Think of AB and BC as the x-and y-axes, respectively, in Fig. 2.39.

> If the points are distinct from each other by an interval, however small, the secant will be distant from the tangent by a small interval. That it may coincide with the tangent and the last ratio be found, the two points must unite and coincide altogether.[163]

A comparison with the Latin original reveals that Smith's is a loose translation, Newton himself referring to the specific points and lines of Fig. 2.38 as in Struik's translation. Nonetheless, he makes perfectly clear the fact that Newton is stating that the tangent is the limiting position of the secant lines as b moves closer and closer to B.

Leibniz unabashedly embraces the infinitely small and instead of referring to motion or limits appeals to infinitesimals. In his first published paper on the Calculus we read

> We have only to keep in mind that to find a *tangent* means to draw a line that connects two points of the curve at an infinitely small distance, or the continued side of a polygon with an infinite number of angles, which for us takes the place of the *curve*.[164]

A formal definition of the tangent line ought to be possible from either of these remarks. What is lacking is in Newton's case a clear and precise definition of limit and in Leibniz's case a solid justification of the use of infinitesimals. Leibniz attempted such a justification after receiving some harsh criticism, but his attempt was ultimately unsuccessful. Both approaches received stinging criticism in 1734 by George Berkeley, Bishop of Cloyne, (1685–1753) who, irked by mathematicians criticising theologians for their weak reasoning exposed the equal or even greater weakness of the reasoning resorted to by mathematicians in explaining the new calculus. Berkeley's tract, *The Analyst; or, a Discourse Addressed to an Infidel Mathematician*, is a carefully reasoned critique of the flaws in the arguments put forward by the mathematicians of the day.

2.3.6 Bumps in the Road

A discussion of Berkeley is not immediately relevant to our discussion of the Mean Value Theorem, but it is interesting, it sheds some light on the shortcomings of the work cited so far, and, of all the criticisms levelled against the Calculus, Berkeley's was the most influential. Hence I have decided to enter into a digression on Berkeley. The reader who is pressed for time may prefer to skip this and jump ahead to the next subsection on page 123, below.

Berkeley was certainly not the first to criticise analytical practice. "The philosopher Thomas Hobbes raised the first doubts, to be echoed by many others later, on the use of infinitely small or indivisible quantities".[165] In response to some criticisms by Bernard Nieuwentijt (1654–1718), Leibniz began

---

[163] Smith, *Source Book*, *op. cit.*, p. 617.

[164] Struik, *op. cit.*, p. 276.

[165] Stedall, *op. cit.*, p. 66. Stedall includes (p. 69) an excerpt from Hobbes, *Six lessons to the Professors of Mathematics*, 1656, p. 46, criticising in the plainest language the use of infinitesimals

When my infinitesimal calculus, which includes the calculus of differences and sums, had appeared and spread, certain over-precise veterans began to make trouble; just as once long ago the Sceptics opposed the Dogmatics, as is seen from the work of Empicurus [*sic*: Epicurus is meant.] against the mathematicians (i. e., the dogmatics), and such as Francisco Sanchez, the author of the book *Quod nihil scitur*, brought against Clavius; and his opponents to Cavalieri, and Thomas Hobbes to all geometers, and just lately such objections as are made against the quadrature of the parabola by Archimedes by that renowned man, Dethlevus Cluver. When then our method of infinitesimals, which had become known by the name of the calculus of differences, began to be spread abroad by several examples of its use, both of my own and also of the famous brothers Bernoulli, and more especially by the elegant writings of that illustrious Frenchman, the Marquis d'Hospital, just lately a certain erudite mathematician, writing under an assumed name in the scientific *Journal de Trevoux*, appeared to find fault with this method. But to mention one of them by name, even before this there arose against me in Holland Bernard Nieuwentijt, one indeed really well equipped both in learning and ability, but one who wished rather to become known by revising our methods to some extent than by advancing them.[166]

Though milder in tone than the comments by Hobbes and dripping with less sarcasm than Berkeley's tract, and as entertaining as it is, today such writing would be deemed inappropriate, particularly the *ad hominem* remark of the last cited sentence. But even the opening claim to martyrdom would raise the issue of paranoia, and an author would be ill-advised to submit such remarks for publication.

Berkeley too does not restrain himself from making personal attacks. Still, Berkeley does offer some cogent criticism, and overall Robert Woodhouse's assessment is probably fair:

The name of Berkeley has occurred more than once in the preceding pages: and I cannot quit this part of my subject without commending the analyst and the subsequent pieces, as forming the most satisfactory controversial discussion of pure science, that ever yet appeared: into what perfection of perspicuity and of logical precision, the doctrine of fluxions may be advanced, is no subject of consideration: But, view the doctrine as Berkeley found it, and its defects in metaphysics and logic are clearly made out.

If, for the purpose of habituating the mind to just reasoning, (and mental discipline is all the good the generality of students derive from the mathematics)[167] I were to recommend a book, it should be the *Analyst*. Even those who still regard the doctrine of fluxions as clearly and firmly established by their immortal inventor, may read it, not unprofitably, since, if it does not prove the cure of prejudice, it will be at least the punishment.[168]

Berkeley begins with a brief description of the method of fluxions as it is described in the *Quadratura curvarum*:

III. The Method of Fluxions is the general Key, by help whereof the modern Mathematicians unlock the secrets of Geometry, and consequently of Nature. And as it is that which hath enabled them so remarkably to outgo the Ancients in discovering Theorems and

(Footnote 165 continued)
by John Wallis (1616–1703) whose *Arithmetica infinitorum* of 1655 provided the spark that ignited Newton.

[166]Child, *Early Mathematical Manuscripts...*, *op. cit.*, pp. 145–146.

[167]And not even this when courses are watered down and only drill is offered.

[168]Robert Woodhouse, *The Principles of Analytical Calculation*, University of Cambridge Press, Cambridge, 1803, pp. xvii–xviii.

solving Problems, the exercise and application thereof is become the main, if not sole, employment of all those who in this Age pass for profound Geometers. But whether this Method be clear or obscure, consistent or repugnant, demonstrative or precarious, as I shall inquire with the utmost impartiality, so I submit my inquiry to your own Judgment, and that of every candid Reader. Lines are supposed to be generated by the motion of Points, Plains by the motion of Lines, and Solids by the motion of Plains. And whereas Quantities generated in equal times are greater or lesser, according to the greater or lesser Velocity, wherewith they increase and are generated, a Method hath been found to determine Quantities from the Velocities of their generating Motions. And such Velocities are called Fluxions: and the Quantities generated are called flowing Quantities. These Fluxions are said to be nearly as the Increments of the flowing Quantities, generated in the least equal Particles of time; and to be accurately in the first Proportion of the nascent, or in the last of the evanescent, Increments. Sometimes, instead of Velocities, the momentaneous Increments or Decrements of undetermined flowing Quantities are considered, under the Appellation of Moments.[169]

Berkeley launches his criticism on general epistemological principles:

IV. By Moments we are not to understand finite Particles. These are said not to be Moments, but Quantities generated from Moments, which last are only the nascent Principles of finite Quantities. It is said, that the minutest Errors are not to be neglected in Mathematics: that the Fluxions are Celebrities, not proportional to the finite Increments though ever so small; but only to the Moments or nascent Increments, whereof the Proportion alone, and not the Magnitude is considered. And of the aforesaid Fluxions there be other Fluxions, which Fluxions of Fluxions are called second Fluxions. And the Fluxions of these second Fluxions are called third Fluxions: and so on, fourth, fifth, sixth, *&c. ad infinitum*. Now as our Sense is strained and puzzled with the perception of Objects extremely minute, even so the Imagination, which Faculty derives from Sense, is very much strained and puzzled to frame clear Ideas of the least Particles of time, or the least Increments generated therein: and much more so to comprehend the Moments, or those Increments of the flowing Quantities in *statu nascenti*, in their very first origin or beginning to exist, before they become finite Particles. And it seems still more difficult, to conceive the abstracted Velocities of such nascent imperfect Entities. But the Velocities of the Velocities, the second, third, fourth and fifth Velocities, *&c*. exceed, if I mistake not, all Humane Understanding. The further the Mind analyseth and pursueth these fugitive Ideas, the more it is lost and bewildered; the Objects, at first fleeting and minute, soon vanishing out of sight. Certainly in any Sense a second or third Fluxion seems an obscure Mystery. The incipient Celerity of an incipient Celerity, the nascent Augment of a nascent Augment, *i. e.* of a thing which hath no Magnitude: Take it in which light you please, the clear Conception of it will, if I mistake not, be found impossible, whether it be so or no I appeal to the trial of every thinking Reader. And if a second Fluxion be inconceivable, what are we to think of third, fourth, fifth Fluxions, and so onward without end?[170]

Paragraph IV is a philosophical assault on the method of fluxions and instances his opposition to abstract entities and his belief that Geometry should deal only with the immediately perceivable:

...he rejects infinitesimals on the grounds that they are simply incomprehensible:

[169]George Berkeley, *The Analyst; or, a Discourse Addressed to an Infidel Mathematician.*, London, 1734, pp. 6–7.
[170]*Ibid.*, pp. 8–9.

> Axiom. No reasoning about things whereof we have no idea. Therefore no reasoning about Infinitesimals.
>
> Nor can it be objected that we reason about Numbers wch are only words & not ideas, for these Infinitesimals are words of no use if not supposed to stand for Ideas.
>
> Much less infinitesimals of infinitesimals &c.

> Berkeley argues here that we can frame no idea of infinitesimals and, thus, that there is no legitimate purpose served by introducing signs such as dx or $o\dot{x}$ into mathematical discourse. His criticisms clearly depend upon the "axiom" that no word is to be used without an idea. When Berkeley later repudiates this axiom it might be thought that he is no longer entitled to this kind of critique of infinitesimals. I think, however, that this conclusion is unduly hasty.[171, 172]

Infinitesimals were already being hotly debated before Berkeley wrote *The Analyst*. Newton had used them before replacing them by his references to nascent and evanescent augments and prime and ultimate ratios, finding, he believed, greater rigour in the replacement's implied limit concept. Others accepted them wholeheartedly, but disagreed on their exact nature. Nieuwentijt and Leibniz embraced them, but while Nieuwentijt was willing to accept infinitesimal objects infinitely small in comparison with finite numbers, he was unwilling to accept things infinitely small compared to them. Leibniz, on the other hand, had no qualms about positing a whole hierarchy of infinitesimals: There were first order infinitesimals, which were infinitely small with respect to finite numbers; then second order infinitesimals infinitely small with respect to first order infinitesimals; and so on. When Comparing finite quantities, one could ignore infinitesimal differences; when comparing first order infinitesimal quantities, one could ignore second and higher order infinitesimal differences; and so on. An exact account of infinitesimals was lacking and their use was informal, intuitive, and most decidedly non-rigorous.

Berkeley now carried his criticism of fluxions and their iterations over to infinitesimals and higher infinitesimals in paragraphs V and VI. His argument is again not that a justification for their use is lacking, but that a justification must be lacking because they make no sense:

> Now to conceive a Quantity infinitely small, that is, infinitely less than any sensible or imaginable Quantity, or than any the least finite Magnitude, is, I confess, above my Capacity. But to conceive a Part of such infinitely small Quantity, that shall be still infinitely less than it, and consequently though multiply'd infinitely shall never equal the minutest finite Quantity, is, I suspect, an infinite Difficulty to any Man whatsoever; and will be allowed such by those who candidly say what they think; provided they really think and reflect, and do not take things upon trust.[173]

---

171 Douglas M. Jesseph, *Berkeley's Philosophy of Mathematics*, University of Chicago Press, Chicago, 1993, pp. 158–159. I confess to having given this book only a quick and superficial reading, but it strikes me as offering an excellent in-depth discussion of Berkeley's criticism of the Calculus. Another source worthy of mention is Cajori, *A History of the Conceptions of Limits...*, *op. cit.*.

172 However "unduly hasty", the rejection of Berkeley's critique follows from Berkeley's own Lemma cited in paragraph XII of *The Analyst* — cf. p. 117, below.

173 Berkeley, *op. cit.*, p. 10.

Berkeley was no enemy of mathematics, nor even of the method of fluxions itself. His goal was not to criticise analysis *per se*, but to deflate the mathematicians who criticised theological argumentation by showing that the analysts were guilty of the same crimes against reason they accused the theologians of. The hint of this in the last clause of the above citation is replaced by more direct statements of this thesis in paragraphs VII and VIII, for example:

> VII. All these Points, I say, are supposed and believed by certain rigorous Exactors of Evidence in Religion, Men who pretend to believe no further than they can see. That Men, who have been conversant only about clear Points, should with difficulty admit obscure ones might not seem altogether unaccountable. But he who can digest a second or third Fluxion, a second or third Difference, need not, methinks, be squeamish about any Point in Divinity... But with what appearance of Reason shall any Man presume to say, that Mysteries may not be Objects of Faith, at the same time that he himself admits such obscure Mysteries to be the Object of Science?[174]

Berkeley is on firmer ground in paragraph IX when he attacks Newton's derivation of the product formula for differentiation, stated in terms of moments, in the *Principia*:

> The main Point in the method of Fluxions is to obtain the Fluxion or Momentum of the Rectangle or Product of two indeterminate Quantities. Inasmuch as from thence are derived Rules for obtaining the Fluxions of all other Products and Powers; be the Coefficients or the Indexes what they will, integers or fractions, rational or surd. Now this fundamental Point one would think should be very clearly made out, considering how much is built upon it, and that its Influence extends throughout the whole Analysis. But let the Reader judge. This is given for Demonstration. Suppose the Product or Rectangle AB increased by continual Motion: and that the momentaneous Increments of the Sides A and B are a and b. When the Sides A and B were deficient, or lesser by one half of their Moments, the Rectangle was $\overline{A - \frac{1}{2}a} \times \overline{B - \frac{1}{2}b}$, i.e. $AB - \frac{1}{2}aB - \frac{1}{2}bA + \frac{1}{4}ab$. And as soon as the Sides A and B are increased by the other two halves of their Moments, the Rectangle becomes $\overline{A + \frac{1}{2}a} \times \overline{B + \frac{1}{2}b}$ or $AB + \frac{1}{2}aB + \frac{1}{2}bA + \frac{1}{4}ab$. From the latter Rectangle subduct the former, and the remaining difference will be $aB + bA$. Therefore the Increment of the Rectangle generated by the intire [*sic*] Increments a and b is $aB + bA$. *Q.E.D.* But it is plain that the direct and true Method to obtain the Moment or Increment of the Rectangle AB, is to take the Sides as increased by their whole Increments, and so multiply them together, $A + a$ by $B + b$, the product whereof $AB + aB + bA + ab$ is the augmented Rectangle; whence if we subduct AB, the Remainder $aB + bA + ab$ will be the true Increment of the Rectangle, exceeding that which was obtained by the former illegitimate and indirect Method by the Quantity ab. And this holds universally be the Quantities a and b what they will, big or little, Finite or Infinitesimal, Increments, Moments or Velocities. Nor will it avail to say that ab is a Quantity exceeding small: Since we are told that *in rebis mathematics errores quàm minimi non sunt contemnendi.*[175, 176]

Newton's argument is a bit of sleight of hand, but not the good kind where he dazzles us with a clever trick we would never have thought of. No, he tries to pull the

[174] *Ibid.*, p. 12.

[175] This is the Latin original of Newton's remark cited above that "The very smallest Errors in mathematical Matters are not to be neglected".

[176] Berkeley, *op. cit.*, pp. 14–16.

wool over our eyes by calculating a different expression. Berkeley was not alone in criticising Newton on this; no less a mathematician than William Rowan Hamilton wrote later, in 1862, to no less an admirer of Newton than Augustus de Morgan:

> His mode of getting rid of *ab* appeared to me long ago (I must confess it) to involve so much of *artifice*, as to deserve to be called *sophistical*; although I should not like to say so publicly. He subtracts, you know $\left(A - \frac{1}{2}a\right)\left(B - \frac{1}{2}b\right)$ from $\left(A + \frac{1}{2}a\right)\left(B + \frac{1}{2}b\right)$; whereby, of course, *ab* disappears in the result. But by *what right*, or *what reason* other than to give an unreal air of *simplicity* to the calculation, does he *prepare* the *products* thus?[177]

Newton's trick survives today in the form of the following exercise.

2.3.8 Exercise Let f be a function defined in some interval containing the number a.

i. Show: If $f'(a)$ exists, then

$$\lim_{h\to 0} \frac{f(a+h) - f(a-h)}{2h} = f'(a).$$

ii. Show by example that

$$\lim_{h\to 0} \frac{f(a+h) - f(a-h)}{2h}$$

can exist even when $f'(a)$ does not.

So Newton is, in effect, calculating a generalised derivative which may exist when the derivative doesn't, but will always give the correct value when it does exist. He may not have provided any justification for his procedure, but some justification does exist.

Berkeley's next mathematical assault is a more direct frontal attack. He has just said that Newton's derivation of the product formula succeeds by ignoring an error and thus is in violation of one of Newton's own principles. He now claims that Newton's argument is self-contradictory and conclusions based on contradictions carry no weight.

> XII. From the foregoing Principle so demonstrated, the general Rule for finding the Fluxion of any Power of a flowing Quantity is derived. But, as there seems to have been some inward Scruple or Consciousness of defect in the foregoing Demonstration, and as this finding the Fluxion of a given Power is a Point of primary Importance, it hath therefore been judged proper to demonstrate the same in a different manner independent of the foregoing Demonstration. But whether this other Method be more legitimate and conclusive than the former, I proceed now to examine; and in order thereto shall premise the following Lemma. "If with a View to demonstrate any Proposition, a certain Point is supposed, by virtue of which certain other points are attained; and such supposed Point be it self afterwards destroyed or rejected by contrary Supposition; in that case, all the other Points, attained thereby and consequent thereupon, must also be destroyed and rejected, so as from thence forward to be no more supposed or applied in the Demonstration." This is so plain as to need no Proof.

---

[177] Smith, *Source Book*, *op. cit.*, p. 631.

XIII. Now the other Method of obtaining a Rule to find the Fluxion of any Power is as follows. Let the Quantity x flow uniformly, and be it proposed to find the Fluxion of x^n. In the same time that x by flowing becomes $x + o$, the Power x^n becomes $\overline{x+o}|^n$, i. e. by the Method of infinite Series $x^n + nox^{n-1} + \frac{nn-n}{2}oox^{n-2} + \&c.$ and the Increments o and $nox^{n-1} + \frac{nn-n}{2}oox^{n-2} + \&c.$ are one to another as 1 to $nx^{n-1} + \frac{nn-n}{2}ox^{n-2} + \&c.$ Let now the Increments vanish, and their last Proportion will be 1 to nx^{n-1}. But it should seem that this reasoning is not fair or conclusive. For when it is said, let the Increment vanish, *i. e.* let the Increments be nothing, or let there be no Increments, the former Supposition that the Increments were something, or that there were Increments, is destroyed, and yet a Consequence of that Supposition, *i. e.* an Expression got by virtue thereof, is retained. Which, by the foregoing Lemma, is a false way of reasoning. Certainly when we suppose the Increments to vanish, we must suppose their Proportions, their Expressions, and everything else derived from the Supposition of their Existence to vanish with them.[178]

Berkeley was right about the product formula yielding the formula for differentiating powers x^n — for positive integers n. For other rational exponents a bit more is required. Newton's new derivation is not an "inward Scruple or Consciousness of defect" about the earlier proof, but an attempt to unify the treatment of the differentiation of x^n for all rational n. This new derivation is far from rigorous and the necessary rigour was a long time coming.[179]

Berkeley's objection to the derivation is, however, completely off the mark. Newton in no way violates Berkeley's Lemma. He is not first taking o not to be 0 so he can divide by o and obtain a new equation, and then changing his mind and saying o is 0 in the resulting equation. Newton does indeed assume o is not 0 in calculating

$$\frac{(x+o)^n - x^n}{o} = nx^{n-1} + \frac{n(n-1)}{2}ox^{n-2} + \ldots,$$

but he now wants to claim that, the values on the left always equalling those on the right, the two expressions will share the same limit as $o \to 0$ and that the limit on the right can be calculated by setting o equal to 0 in *that* expression. He wants to say this, but the limit concept has not yet crystallised sufficiently for him to say this clearly.

Newton, Leibniz, and Berkeley agreed on the value of the Calculus and on the truth of its results. All three also believed that any result obtained by the method of fluxions or the use of infinitesimals could be rigorously verified by the old Greek methods. What they disagreed on was the justification of the new procedures. Newton was evidently not one to worry himself much on the matter. When, for example, he extended the Binomial Theorem from positive integral exponents to arbitrary rational exponents by guessing the form for an expansion of $\sqrt{1-x^2}$, he did not prove the result, but checked it by formally multiplying the resulting series by itself to get

---

[178] *Ibid*, pp. 19–21.

[179] I give a fairly complete account of the history of the Binomial Theorem in: Smoryński, *Treatise*, *op. cit.* It might be added that the two proofs are from two different works of Newton's and thus their simultaneous existence, even should the results have been on equal footing, would not necessarily have been proof of anything more than variety.

$1 - x^2$. He also formally applied the familiar old algorithm for finding square roots and obtained the same infinite series. As to the question of fluxions and its use of infinitesimals, he merely said that infinitesimals weren't needed, that they could be replaced by sufficiently small finite quantities, and mumbled something about "limits". He made no attempt at a rigorous justification.

Berkeley tried to explain the success by a theory of compensating errors. I confess to find this too absurd to have read this part of his essay, though it may well be the case that he is able to treat one or two special cases successfully.

Leibniz is the one who thought deeply about the matter. In one of his manuscripts only published posthumously, he wrote

> It has been proposed to me several times to confirm the essentials of our calculus by demonstrations, and here I have indicated below its fundamental principles, with the intent that any one who has the leisure may complete the work. Yet I have not seen up to the present anyone who would do it. For what the learned Hermann has begun in his writings, published in my defence against Nieuwentiit [*sic*], is not yet complete.
>
> For I have, beside the mathematical infinitesimal calculus, a method also for use in Physics, of which an example was given in the *Nouvelles de la République des Lettres*; and both of these I include under the Law of Continuity; and adhering to this, I have shown that the rules of the renowned philosophers Descartes and Malebranche were sufficient in themselves to attack all problems on Motion.
>
> I take for granted the following postulate:
>
> *In any supposed transition, ending in any terminus, it is permissible to institute a general reasoning, in which the final terminus may also be included.*
>
> For example, if A and B are any two quantities, of which the former is the greater and the latter is the less, and while B remains the same, it is supposed that A is continually diminished, until A becomes equal to B; then it will be permissible to include under a general reasoning the prior cases in which A was greater than B, and also the ultimate case in which the difference vanishes and A is equal to B. Similarly, if two bodies are in motion at the same time, and it is assumed that while the motion of B remains the same, the velocity of A is continually diminished until it vanishes altogether, or the speed of A becomes zero; it will be permissible to include this case with the case of the motion of B under one general reasoning. We do the same thing in geometry, when two straight lines are taken, produced in any manner, one VA being given in position or remaining in the same site, the other BP passing through a given point P, and varying in position while the point P remains fixed; at first indeed converging toward the line VA and meeting it in the point C; then, as the angle of inclination VCA [*sic*, this should read PCA] is continually diminished, meeting VA in some more remote point (C), until at length from BP, through the position (B)P, it comes to βP, in which the straight line no longer converges toward VA, but is parallel to it, and C is an impossible or imaginary point (Fig. 2.40). With this supposition it is permissible to include under some one general reasoning not only all the intermediate cases such as (B)P but also the ultimate case βP.
>
> Hence also it comes to pass that we include as one case ellipses and the parabola, just as if A is considered to be one focus of an ellipse (of which V is the given vertex), and this focus remains fixed, while the other focus is variable as we pass from ellipse to ellipse, until at length (in the case when the line BP, by its intersection with the line VA, gives the variable focus) the focus C becomes evanescent[180] or impossible, in which case the ellipse passes into a parabola. Hence it is permissible with our postulate that a parabola should

[180]Child adds a footnote here explaining that "evanescent" should be read "vanishing into the far distance".

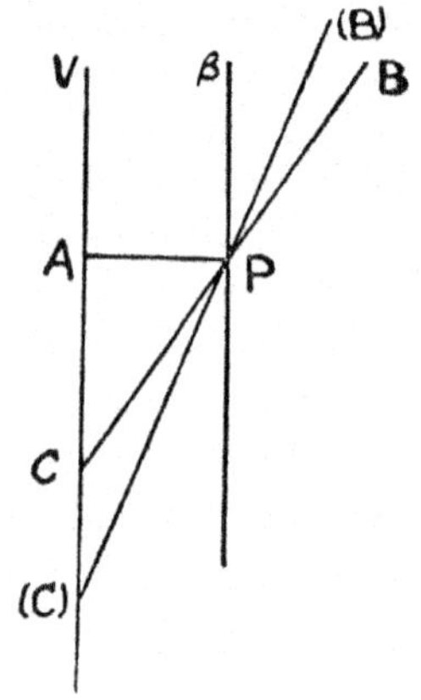

Fig. 2.40 Leibniz's diagram

> be considered with ellipses under a common reasoning. Just as it is common practice to make use of this method in geometrical constructions, when they include under one general construction many different cases, noting that in a certain case the converging straight line passes into a parallel straight line, the angle between it and another straight line vanishing.[181]

The Law of Continuity is not very clearly expressed. The simplest interpretation would be that, if a property held for all values of o as $o \to 0$ then it must also hold at 0. Obviously, this does not hold for all properties — for example for $P(o) : o \neq 0$. Presumably it holds in some sense for some crucial properties, but its value as a postulate rests on a delineation of the properties to which the Law applies. Without such the Law of Continuity can be no more than a heuristic useful in finding solutions, but not in establishing them.

In Leibniz's early publications he seemed to believe in the existence of infinitesimals, and a number of his followers accepted them. In a letter to François Pinsson published in 1701, he surprised many of his followers:

> For instead of the infinite or infinitely small, one takes quantities as large, or as small, as necessary in order that the error be smaller than the given error, so that one differs from Archimedes's style only in the expressions, which are more direct in our method and more conform to the art of invention.[182]

Another quote:

> It will be sufficient if, when we speak of infinitely great (or more strictly unlimited), or of infinitely small quantities (i. e., the very least of those within our knowledge), it is understood that we mean quantities that are indefinitely great or indefinitely small, i. e., as great as you please, or as small as you please, so that the error that any one may assign may be less than a certain assigned quantity. Also, since in general it will appear that, when any small error is assigned, it can be shown that it should be less, it follows that the error is absolutely nothing; an almost exactly similar kind of argument is used in different places by Euclid,

[181] Child, *Early Mathematical Manuscripts...*, *op. cit.*, pp. 146–148.

[182] Quoted in translation in H.J.M. Bos, *Differentials, Higher Order Differentials and the Derivative in the Leibnizian Calculus*, dissertation, Rijksuniversiteit te Utrecht, 1973, p. 73.

> Theodosius and others; and this seemed to them to be a wonderful thing, although it could not be denied that it was perfectly true that, from the very thing that was assumed as an error, it could be inferred that the error was non-existent. Thus, by infinitely great and infinitely small, we understand something indefinitely great, or something indefinitely small, so that each conducts itself as a sort of class, and not merely as the last thing of a class. If any one wishes to understand these as the ultimate things, or as truly infinite, it can be done, and that too without falling back upon a controversy about the reality of extensions, or of infinite continuums in general, or of the infinitely small, ay, even though he think that such things are utterly impossible; it will be sufficient simply to make use of them as a tool that has advantages for the purpose of the calculation, just as the algebraists retain imaginary roots with great profit. For they contain a handy means of reckoning, as can manifestly be verified in every case in a rigorous manner by the method already stated.[183]

Phrases like Newton's "come nearer than any given difference" (p. 106, above) or Leibniz's "one takes quantities as large, or as small, as necessary in order that the error be smaller than the given error" (in the quotation from the letter to Pinsson), especially the latter, are awfully close to the modern ϵ-δ definition of limit and one wonders how it took so long for that definition to crystallise. For it seems only first to have been used and almost expressed by Bolzano in 1816, then by Cauchy in 1821, finally to be expressed clearly and unambiguously in modern form sometime in the mid-19th century by Weierstrass and his students (like Heine, cited in Sect. 1). Did Berkeley's critique accelerate or decelerate the process?

The immediate effect of Berkeley's attack on infidel mathematicians and their limits and infinitesimals seems to have been beneficial. Two years after the appearance of *The Analyst* no fewer than four textbooks on the method of fluxions were published in England, more than all the previously existing expositions put together. Also, Colin Maclaurin's (1698–1746) later text of 1742 on fluxions was a response to Berkeley. And there were other published responses to Berkeley. On the continent the debate over infinitesimals raged on. While analysis continued to grow and develop, its foundations remained shaky.

Our modern foundation for the Calculus rests on two pillars — the familiar ϵ-δ definitions of limit (pointwise, uniform, etc.) and a characterisation of the real number field (completeness, archimedean order). The latter only began to emerge in the mid-1810s in the work of Bolzano and reached final form in 1872 with an explosion of papers offering different methods of reaching a common solution. The former, however, was seemingly within reach of everyone — or, at least, anyone who chanced upon the key phrases of Newton and Leibniz, who was not attracted by the lure of infinitesimals, and who knew to ignore everything else.

Perhaps, without a clear concept of the completeness of the real numbers, the problem was just too intractable. In any event, the serious attempts to found the Calculus rigorously in the 18th century went in a different direction. I would even venture to say they went astray in that the approaches were doomed to failure.

The two commonly cited attempts to found the Calculus without recourse to limits were the *residual analysis* of John Landen (1719–1790) and the formal power series of Joseph Louis Lagrange (1736–1813). Landen wrote two books on the subject, a

---

[183] Child, *Early Mathematical Manuscripts...*, *op. cit.*, p. 150.

short volume of 44 pages, *A Discourse Concerning the Residual Analysis* (1758), and a longer exposition, *The Residual Analysis* (1764). Lagrange also wrote two books on his approach, *Théorie des fonctions analytiques, contenant les principes du calcul différentiel dégagés de toute considération d'infiniment petits ou d'évanouissans, de limites ou de fluxions, et réduits a l'analyse algébrique des quantités finies* [*Theory of Analytic Functions*...] (1797), usually referred to as *Théorie des Fonctions*, and *Leçons sur le calcul des fonctions* [*Lessons on the Calculus of Functions*] (1806).

Landen's starting point was Newton's Binomial Theorem,

$$(1+x)^q = 1 + qx + \frac{q(q-1)}{2}x^2 + \frac{q(q-1)q-2)}{6}x^2 + \ldots,$$

for rational q. The right-hand-side generally requires $|x| < 1$ to guarantee convergence, a fact widely recognised but not emphasised. His approach was to consider and simplify an expression,

$$\frac{f(x) - f(y)}{x - y},$$

obtaining an equation that generally held for $x \neq y$ and claiming it must therefore hold for $x = y$ as well. Using a simple algebraic identity

$$\frac{u^{m/n} - v^{m/n}}{u - v} = u^{m/n-1} \frac{1 + \frac{v}{u} + \left(\frac{v}{u}\right)^2 + \ldots + \left(\frac{v}{u}\right)^{m-1}}{1 + \left(\frac{v}{u}\right)^{m/n} + \left(\frac{v}{u}\right)^{2m/n} + \ldots + \left(\frac{v}{u}\right)^{(n-1)m/n}} \tag{2.34}$$

and an assumed expansion

$$(1+x)^{m/n} = 1 + ax + bx^2 + cx^3 + \ldots,$$

he derived

$$\frac{m}{n}(1+x)^{m/n-1} = a + 2bx + 3cx^2 + \ldots$$

by algebraic manipulation. Multiplication of both sides by $1 + x$ allowed him to determine $a, b, c, \ldots$ in succession. Following this he proceeded to apply (2.34) to a variety of tangent and max/min problems. I forego discussion of these matters, referring the interested reader to the literature.[184]

Lagrange's approach was to assume every function f could be expanded into a power series:

$$f(x) = a_0 + a_1(x-a) + a_2(x-a)^2 + \ldots$$

[184]Landen's *Discourse* is available online and a print edition by Gale, 2010, of Book I of *The Residual Analysis* exists. Additionally, excerpts from *Discourse* are reproduced in Struik, *op. cit.*, pp. 386–388 and Stedall, *op. cit.*, pp. 398–401. I also refer to Smoryński, *Treatise*, *op. cit.*, pp. 148–151, for a detailed account of Landen's "proof" of the Binomial Theorem.

He defined the derivative by $f'(a) = a_1$, wrote f' as a power series

$$f'(x) = b_0 + b_1(x - a) + b_2(x - a)^2 + \ldots,$$

noted that $f''(a) = b_1$ and that f'' could be expanded into a power series itself, etc., and went on to calculate

$$f'(a) = a_1, \quad f''(a) = 2a_2, \quad f'''(a) = 6a_3, \ \ldots$$

Lagrange's approach was popular for a while,[185] but its limitations were startlingly revealed by Cauchy in his *Résumé des leçons* in 1823 where he produced an analytically expressible function that equalled its power series expansion at exactly one point and deduced from it that two distinct analytic functions could have the same power series expansions (though they could not, of course, both equal these expansions). Again I refer the interested reader to the literature.[186]

Landen's work is of fleeting importance, of interest today only as an example of the nonlinear development of mathematics: Mathematics is not an unbroken progressive development; it occasionally goes down blind alleys, and the residual analysis was one of them. Lagrange, though he based his approach on the false assumption that every function could be expanded into a power series and used this "fact" in proving theorems, produced results of lasting importance. These results required new proofs, but he made the discoveries and provided starting points for some of these new proofs. One of his results was the Mean Value Theorem, which contribution will be discussed in the next chapter. For now, we skip ahead to Lagrange's successors and the definition of the derivative.

2.3.7 *The Derivative Defined*

The modern definition of the derivative is given in terms of limits, which are themselves defined in terms of approximations. The first serious analyses of approxima-

---

185 His term "derivative" and notation f' for the derivative are still used today.

186 Both books by Lagrange are available online. English translations of excerpts from *Théorie des fonctions* can be found in Struik, *op. cit.*, pp. 389–391, and Stedall, *op. cit.*, pp. 404–406. Other discussions of Lagrange's approach can be found in Edwards, *op. cit.*, pp. 296–299, and Smoryński, *Formalism*, *op. cit.*, pp. 127–135. This last reference, incidentally, includes in Exercise 6.6 of Chapter II, pp. 184–185, an outline of Cauchy's result mentioned above.

tions were performed independently by Lagrange[187] and d'Alembert.[188] And three decades later, in his two books on founding the Calculus on power series, Lagrange presented such a treatment for these series.

Unlike Lagrange, d'Alembert was a firm believer in founding the Calculus on the notion of limit. In volume IX (1765) of Diderot's *Encyclopédie*, d'Alembert writes

> LIMIT (*Mathematics*). One says that a magnitude is the *limit* of another magnitude, when the second may approach the first more closely than by a given quantity, as small as one wishes, moreover without the magnitude which approaches being allowed ever to surpass the magnitude that it approaches; so that the difference between such a quantity and its *limit* is absolutely unassignable...
>
> The theory of *limits* is the foundation of the true justification of the differential calculus. *See* DIFFERENTIAL, FLUXION, EXHAUSTION, INFINITE. Strictly speaking, the *limit* never coincides, or never becomes equal to the quantity of which it is the *limit*, but the latter approaches it ever more closely, and may differ from it as little as one wishes.[189]

This is no improvement on Newton's proclamation about limits of half a century earlier.[190]

Oddly enough, despite his distaste for limits, Lagrange came closer to our modern definition of limit than d'Alembert even though both men all but proved certain limits to exist using their numerical analyses of convergence. Lagrange, in fact, came close to the definition of continuity:

> And, Lagrange said, "The course of the curve will necessarily be *continuous* from this point; thus it will, little by little, approach the axis before cutting it, and approach it, consequently, within a quantity less than any given quantity." This characterization of continuity appears geometric. But Lagrange rendered it algebraic: "So we can always find an abscissa h corresponding to an ordinate less than any given quantity; and then all smaller values of h correspond also to ordinates less than the given quantity." This is a far cry from "insensible degrees" or "infinitely small changes." But it is not far from this characterization of continuity at $h = 0$ to the Bolzano-Cauchy definitions of continuity in general. Even though Lagrange himself did not take his characterization to be the defining property of continuous function, he had for the first time stated, in terms of inequalities, what Cauchy and Bolzano later recognized as such.[191]

Taking their cue from Lagrange, Bolzano and Cauchy defined continuity, as we saw in the preceding section, and gave rigorous ϵ-δ proofs of limit theorems, albeit not always using this notation.

It is high time we defined the notions of limit and derivative.

---

[187] J.L. Lagrange, "Sur la résolution des équations numériques, et additions au mémoire sur la résolution des équations numériques", *Mémoires de l' Academie...Berlin* 23 (1767), pp. 311–352 and 24 (1768), pp. 111 –180; reprinted in volume 2 of *Oeuvres de Lagrange*, Gauthier-Villars, Paris, 1867–1882.

[188] "Réflexions sur les suites et sur les racines imaginaires", in: J. d'Alembert, *Opuscules mathématiques*, vol. 5, Briasson, Paris, 1768, pp. 171–215. An annotated English translation of the relevant portions can be found in Smoryński, *Treatise*, *op. cit.*, pp. 182–188.

[189] English translation from: Stedall, *op. cit.*, pp. 297–298. Stedall includes also excerpts on limits from Wallis, Newton, Maclaurin, and Cauchy.

[190] Cf. p. 95, above.

[191] Judith V. Grabiner, *Origins*, *op. cit.* p. 95.

2.3.9 Definition Let f be a function defined everywhere in an interval I with the possible exception of a point a, with $a \in I$. A number L is the *limit of f as x approaches a*, written

$$\lim_{x \to a} f(x) = L,$$

if, for any $\epsilon > 0$ there is a $\delta > 0$ such that for all $x \in I$,

$$0 < |x - a| < \delta \Rightarrow |f(x) - L| < \epsilon.$$

The main difference between this and the definition of the continuity of f at a is the clause $0 < |x - a|$, i.e., $x \neq a$, in the premise of the final implication. For, it is not assumed that $f(a)$ is defined, and, in any event, any possible value of f *at* a is irrelevant in determining how f behaves as x *approaches* a.

The quintessential functional limit is the derivative:

2.3.10 Definitions Let I be an interval, $a \in I$, and $f : I \to \mathbb{R}$. f is *differentiable at a* if

$$\lim_{h \to 0} \frac{f(a+h) - f(a)}{h} = \lim_{x \to a} \frac{f(x) - f(a)}{x - a}$$

exists. This limit, when it exists, is called the *derivative of f at a* and is denoted $f'(a)$. The function f' mapping a to $f'(a)$ is called the *derivative of f*.

The first to take the hint and give rigorous ϵ-δ proofs in print, as far as I know, was Bernard Bolzano in his *Der binomische Lehrsatz, und als Folgerung aus ihm der polynomische, und die Reihen, die zur Berechnung der Logarithmen und Exponentialgrößen dienen, genauer als bisher erwiesen*[192] [*The Binomial Theorem, and as a Consequence from it the Polynomial Theorem and the Series which serve for the Calculation of Logarithmic and Exponential Quantities, proved more strictly than before*]. He did not formulate it as explicitly as we do today, and it looks a lot like he is dealing with infinitesimals. However, if one looks carefully at the details of his treatment, one sees he is using standard ϵ-δ arguments.

Bolzano is quite adamant about his avoidance of infinitesimals:

> I have, not only in this work, already generally made the rule, by which instead of the so-called *infinitely small quantities* I have used with equal success the concept of *quantities which can be made smaller than any given value* or, (to avoid monotony, I sometimes call, albeit less accurately) the quantities *which can be made as small as one wishes*. Hopefully, one will not misjudge the difference between quantities of this sort and those which one otherwise imagines under the scope of the infinitely small. The demand, to imagine a quantity (I mean a variable one) which can always be made smaller than one has already taken it to be, and generally smaller than any given value, really contains nothing that can be offensive to anyone. Must one not see, rather, that frequently there are such quantities in space as well as in time? Against this the thought of a quantity which can not only be *assumed* smaller, but rather already *is* smaller than any mere given one but also [than] any *supposed*, i.e.,

[192] Prague, 1816; English translation in: Russ, *op. cit.*.

thinkable, quantity; should this not be contradictory? So reads the customary definition of the infinitely small.[193]

Convention. To indicate a quantity which can become smaller than any given value, we choose the symbols ω, Ω or something similar.[194]

The "something similar" is generally an ω or Ω with a superscript — a numeral written directly above the omega. The two variants have distinct uses: ω is a variable quantity we can make as small as we wish at will and Ω is a variable quantity that can be made as small as desired as a consequence of our choices for the sizes of various ω's. Having stated these conventions, he proves for these variable quantities versions of closure properties of infinitesimals used in proving various limit theorems. For example, the theorem to the effect that the limit of a sum is the sum of the limits, which traditionally followed from the fact that the sum of infinitesimals is itself infinitesimal, is rendered:[195]

Lemma. If each of the quantities $\omega, \overset{(1)}{\omega}, \overset{(2)}{\omega}, \ldots, \overset{(m)}{\omega}$ can become as small as one wishes while the (finite) number of them does not change, then their algebraic *sum* or *difference* is also a quantity which can become as small as one wishes, i.e.

$$\omega \pm \overset{(1)}{\omega} \pm \overset{(2)}{\omega} \pm \ldots \pm \overset{(m)}{\omega} = \Omega.$$

He proves this by showing the sum Ω can be made less than D (think: ϵ) in absolute value by choosing each ω less than $D/(m+1)$ (think: δ) in absolute value.

With respect to differentiation, in Sect. 23 he determines the derivative of $f(x) = x^n$ for arbitrary real a and $x > 0$:[196]

Lemma. The quantity

$$\frac{(x+\omega)^n - x^n}{\omega}$$

can be brought as close to the value nx^{n-1} as one wishes, if ω is taken small enough; n and x may denote whatever desired, if only x is not = 0; i.e.,

$$\frac{(x+\omega)^n - x^n}{\omega} = nx^{n-1} + \Omega.$$

His proof is not perfect: His treatment of the case for irrational n is garbled, but the argument is overall correct. He shows that the difference,

$$\frac{(x+\omega)^n - x^n}{\omega} - nx^{n-1},$$

can be made smaller than any given D ($= \epsilon$) by showing how small ω needs to be taken (i.e., he finds δ).

---

[193] *Ibid.*, p. v; Russ, p. 158.

[194] *Ibid.*, p. 15; Russ, p. 173.

[195] *Ibid.*, p. 15; Russ, p. 173.

[196] *Ibid.*, p. 20; Russ, p. 176.

The ω, Ω notation is a convenient shorthand for our use of δ, ϵ, respectively. However, when things get complicated it is not sufficient as dependencies can get confused. Bolzano's next big result, that, if $f_0, f_1, \ldots$ converges to f, then $f_0', f_1', \ldots$ converges to f', is simply false.

Bolzano does not offer a definition of the derivative nor even introduce the term in his first paper of 1816. In his later unpublished "Functionenlehre", he would do so, even defining one-sided derivatives. By then Cauchy had published his two famous textbooks on Analysis, treating differentiation in the *Résumé des leçons* in 1823.

As with continuity, Cauchy did not define differentiability at a point, but on an interval. And, as with continuity, he built some uniformity into the definition. Today we define differentiability on an interval as follows.

2.3.11 Definition Let I be an interval and $f : I \to \mathbb{R}$. f is *differentiable on* I if, for every $x \in I$, f is differentiable at x.

Cauchy's introduction of the derivative in the *Résumé des leçons* gives a relatively loose definition:

THIRD LESSON

Derivatives of Functions of a single Variable

WHEN the function $y = f(x)$ remains continuous between two limits of the variable x, and when one assigns to this variable a value between the limits at hand and confers an infinitely small increment to this variable, an infinitely small increment of the function itself is produced. Consequently, if one puts $\Delta x = i$, the two terms of the *ratio of differences*

$$\frac{\Delta y}{\Delta x} = \frac{f(x+i) - f(x)}{i}$$

will be infinitely small quantities. But, whereas when these two terms approach indefinitely and simultaneously the limit zero, the ratio itself will be able to converge towards another limit, either positive, or negative. This limit, when it exists, has a determinate value, for each particular value of x; but this varies with x. Thus, for example, if one takes $f(x) = x^m$, m denoting a whole number, the ratio of the infinitely small differences becomes

$$\frac{(x+i)^m - x^m}{i} = mx^{m-1} + \frac{m(m-1)}{1.2}x^{m-2}i + \ldots + i^{m-1}$$

and it has for a limit the quantity mx^{m-1}, that is to say, a new function of the variable x. It will be the same in general; however, the form of the new function which serves up the limit of the ratio $\frac{f(x+i)-f(x)}{i}$ depends on the form of the proposed function $y = f(x)$. To indicate this dependence, we give the new function the name of *derived function*, and we denote it, with the aid of an accent, by the notation

$$y' \text{ or } f'(x).$$

Several pages later, when it comes time to prove the theorem on which his proof of the Mean Value Theorem depends, he introduces ϵ and δ:[197]

[197] Cauchy, *Résumé*, *op. cit.*, p. 9. After making many of my translations from the *Résumé* for this book, I learned of a complete translation of the work by Dennis M. Cates. There are two versions of

> Denote by δ, ε, two very small numbers, the first being chosen of such kind which, for the numerical value[198] of i less than δ, and for any value whatsoever of x between the limits x_0, X, the ratio
> $$\frac{f(x+i)-f(x)}{i}$$
> always lies above $f'(x)-\varepsilon$, and below $f'(x)+\varepsilon$.[199]

The thing to note is that, given ϵ, the same δ is claimed to work for all x in the given interval:

2.3.12 Definition Let I be an interval and $f : I \to \mathbb{R}$. f is *uniformly differentiable on I* with *derivative* f' if for all $\epsilon > 0$ there is a $\delta > 0$ such that for all $x \in I$ and all h,

$$0 < |h| < \delta \ \&\ x+h \in I \Rightarrow \left|\frac{f(x+h)-f(x)}{h} - f'(x)\right| < \epsilon.$$

A differentiable function is readily seen to be continuous, but it does not imply the same for the derivative (Cf. Remark 2.2.6 or Exercise 2.3.20.). Cauchy's uniform differentiability, however, does.

2.3.13 Lemma *Let f be uniformly differentiable on I. Then: f' is uniformly continuous on I.*

Proof Let $\epsilon > 0$ be given, choose $\delta > 0$ so that for all $y \in I$,

$$0 < |h| < \delta \ \&\ y+h \in I \Rightarrow \left|\frac{f(y+h)-f(y)}{h} - f'(y)\right| < \frac{\epsilon}{2},$$

and note

$$\begin{aligned}|f'(x+h)-f'(x)| &= \left|f'(x+h) - \frac{f(x)-f(x+h)}{-h} - \frac{f(x+h)-f(x)}{-h} - f'(x)\right| \\ &\le \left|f'(x+h) - \frac{f(x)-f(x+h)}{-h}\right| + \left|\frac{f(x+h)-f(x)}{h} - f'(x)\right| \\ &< \frac{\epsilon}{2} + \frac{\epsilon}{2}.\end{aligned}$$

□

(Footnote 197 continued)
this translation, an expensive annotated edition, *A Guide to Cauchy's Calculus; A Translation and Analysis of Calcul Infinitesimal*, Fairview Academic Press, Walnut Creek (California), 2011, and a more affordable student edition, *Cauchy's Calcul Infinitesimal; A Complete English Translation*, same publisher, 2012. In comparing my translations with his, I find the differences minor and have kept my own. Nonetheless, I shall give page references to the less expensive copy which is now in my possession. The reference in the present case is to p. 7.

198 I.e., the absolute value.

199 Cauchy, *Résumé*, *op. cit.*, p. 27; Cates, *op. cit.*, p. 23.

2.3.14 Definition Let I be an interval and $f : I \to \mathbb{R}$. f is *continuously differentiable* on I if f is differentiable on I and f' is continuous on I.

An immediate corollary to Lemma 2.3.13 is that a uniformly differentiable function is continuously differentiable. Thus

uniform differentiability $\Rightarrow$ continuous differentiability $\Rightarrow$ differentiability.

The converse implications fail in general.

2.3.15 Exercise i. Show that Darboux's function,

$$f(x) = \begin{cases} x^2 \sin \frac{1}{x}, & x \neq 0 \\ 0, & x = 0, \end{cases}$$

is everywhere differentiable, but f' is not continuous at $x = 0$.
ii. Show that $f(x) = 1/x$ is continuously differentiable on $(0, 1)$, but f' is not uniformly continuous there, whence f is not uniformly differentiable on this interval. [In part i you may assume the usual rules for differentiation already to have been established.]

With a little more theory, it can be shown that the uniform continuity of f' on a closed, bounded interval $[a, b]$ entails the uniform differentiability of f on $[a, b]$.[200]

With three distinct candidates for a notion of differentiability to choose from — differentiability on an interval, continuous differentiability on an interval, and uniform differentiability on an interval — the question arises: Which notion is fundamental and which are variants — stronger forms or weaker generalisations of the concept? Obviously, the nomenclature gives away the conventional answer: Differentiability is the fundamental concept, while continuous and uniform differentiability are modifications. The reason for this is theoretical. In practice, thanks to the Mean Value Theorem, most results of interest depend only on differentiability, although this was not always the case.

In 1816 Bolzano's interest in differentiation was primarily in differentiating a few specific functions. His paper on the Binomial Theorem does not even refer to the derivative by name, much less establish any of its properties. In contrast, Cauchy set out immediately to provide rigorous proofs for those properties of the derivative Lagrange had discovered, and to derive some of his own. This does not include Lemma 2.3.13, which he was apparently unaware of. His statement of the Mean Value Theorem, for example, explicitly assumes the continuity of f' as an added assumption to the uniform differentiability of f. Indeed, his proof of the theorem from which he derived the Mean Value Theorem itself relies explicitly on the continuity of f' — as we will see in the next chapter. Many of the early proofs of theorems about differentiable functions were valid only for uniformly differentiable functions and, were it not for the validity of the Mean Value Theorem for the broader class of

[200] Cf. pp. 301–304, below, for details.

differentiable functions, we might well regard uniform differentiability today as the fundamental concept, terming it "differentiability" and calling differentiability itself by some derivative name like "weak differentiability" or "generalised differentiability".

Some of the more interesting of these theorems about differentiable functions will be discussed in the next two sections. For now we consider only a few simpler properties. In a course on the Calculus, after defining differentiation, one generally differentiates a few simple functions, derives the rules for differentiating functions built up from simpler ones by the arithmetic operations and composition, and then applies differentiation to find tangent lines and maxima and minima.

The student who has had an honest Calculus course in which he or she was required to perform some simple ϵ-δ proofs should have no difficulty proving some of the basic computation rules:

$$\begin{aligned}
&f'(x) = 0 \text{ for any constant function } f\\
&f'(x) = 1 \text{ for } f(x) = x\\
&(f \pm g)'(x) = f'(x) \pm g'(x)\\
&(cf)'(x) = cf'(x) \text{ for any constant } c\\
&(f \cdot g)'(x) = f'(x)g(x) + f(x)g'(x)\\
&(1/f)'(x) = \frac{-f'(x)}{f(x)^2}, \text{ whenever } f(x) \neq 0\\
&(f/g)'(x) = \frac{f'(x)g(x) - f(x)g'(x)}{g(x)^2}, \text{ whenever } g(x) \neq 0.
\end{aligned}$$

More difficult are the rules for differentiating the trigonometric, exponential, and logarithmic functions, the demonstrations of which few students can master in a first Calculus course. Likewise, the derivation of the Chain Rule,

$$(f \circ g)'(x) = f'(g(x)) \cdot g'(x),$$

and the existence assertion of the Inverse Function Theorem,

$$(f^{-1})'(y) = \frac{1}{f'(x)} = \frac{1}{f'(f^{-1}(y))}, \text{ when } y = f(x) \text{ is one-to-one and } f'(x) \neq 0,$$

may well require greater experience in formulating proofs. The reader who has just completed a Calculus course may find this a perfect occasion to review these proofs, carrying out the proofs for the easy algebraic rules cited above and looking up the demonstrations for the transcendental functions, the Chain Rule, and the Inverse Function Theorem in his or her Calculus text.[201] For my own part, I have resisted the temptation to repeat the proofs here as the details do not bear directly on our main topic, the Mean Value Theorem, which, believe it or not, we are rapidly approaching.

---

[201] Particularly nice proofs of the Chain Rule and the Inverse Function Theorem can be found in Jan Mikusiński and Piotr Mikusiński, *An Introduction to Analysis: From Number to Integral*, John Wiley & Sons, New York, 1993, pp. 123–124 and 132–133, respectively.

The value of having an official definition of the derivative is not so much in being able to prove that this or that function is differentiable with such and such a derivative. For a century and a half, from Newton to Cauchy, mathematicians could find derivatives without a formal definition or rigorously proven theorems about derivatives. Without such a definition, however, one could not specify exactly the conditions under which theorems held. In the words of Niels Henrik Abel (1802–1829), theorems could "admit exceptions", and it is precise definitions and strict rigour in proofs that explain and, with luck, delineate these exceptions.

Historically, the exceptions were rare. Just as one had had a paucity of curves for a couple of millennia, the new analysis had a restricted collection of functions. Initially they were algebraic or trigonometric and were quite smooth, continuous and differentiable at all but a few isolated points. As the concept of function crystallised and the stock of functions grew, it was realised that this was false. The first example was given in the 1830s by Bolzano in his "Functionenlehre",[202] wherein he constructed a continuous function which failed to have a derivative anywhere, though he only proved it to fail to have a derivative on a dense set of points. Bolzano's work, however, was unfinished and his result was not published until the 20th century. In the 1850s Bernhard Riemann (1826–1866) produced examples of functions which failed to have derivatives on dense sets of points, and around 1860 Charles Cellerier (1818–1889) produced a *(continuous) nowhere differentiable function*, as continuous functions possessing no derivatives at any point are now called. His example,

$$C(x) = \sum_{n=0}^{\infty} \frac{1}{a^n} \sin(a^n x),$$

where $a > 1000$ is an even integer, was not published until 1890. In the meantime, in 1872, Weierstrass startled the mathematical world when he announced in a lecture that

$$W(x) = \sum_{n=0}^{\infty} b^n \cos(a^n x\pi)$$

is nowhere differentiable when a is an odd positive integer, $0 < b < 1$, and $ab > 1 + \frac{3}{2}\pi$.[203] His proof was not published until 1895, by which time his followers had published a number of such examples. And by then Peano had published (1890) his continuous space-filling curve $\gamma_P(t) = \langle x_P(t), y_P(t)\rangle$ and announced that neither of its component functions x_P nor y_P was differentiable anywhere in their domain.[204]

2.3.16 Exercise Prove Peano's assertion.

Obviously, although one can construct these functions without a definition of differentiation, one would be hard put to prove their nowhere differentiability without

---

[202] Russ, *op. cit.*, pp. 487–489, 507–508.

[203] In one course, at least, given in 1874, he said that the conditions could be relaxed to a being an integer > 1 and $ab > 1$, but that the proof was more difficult under these more general conditions.

[204] Cf. Sect. 2.2.3, above.

such a definition. And one similarly needs such a definition to prove the various properties of derivatives as those used in the familiar procedures for finding tangents and solving maxima/minima problems. Our problem is, of course, to give a formal definition of the tangent to a curve in order to give scope and meaning to the Mean Value Theorem, and to justify the methods of finding maxima/minima and thereby lend validity to the proof of the Mean Value Theorem outlined in the Preface.

Let us begin with the merely technical problem of maxima and minima.

The crucial result is quite simple.

2.3.17 Lemma *Let $f : [a, b] \to \mathbb{R}$ assume a maximum or a minimum at $c \in (a, b)$ and assume $f'(c)$ exists. Then: $f'(c) = 0$.*

That some form of the Lemma holds is intuitively obvious. We have already quoted Newton (p. 102, above) on this. In Europe, the principle seems first to have been enunciated by Johannes Kepler (1571–1630). The principle, in fact, is said to be found in a work of Bhāskara II (1114–1185) dating from 1150.[205] However, prior to Cauchy there was no attempt to determine the conditions on f under which the conclusion held. For Cauchy, of course, f had to be uniformly differentiable in an interval and his result was not as general as that above.

The proof of Lemma 2.3.17 reduces to another pair of Lemmas.

2.3.18 Lemma *Let $f : I \to \mathbb{R}$ be differentiable at $c \in I$ and suppose $f'(c) > 0$. There is a $\delta > 0$ such that, for all $x \in I$,*

$$c - \delta < x < c \Rightarrow f(x) < f(c) \text{ and } c < x < c + \delta \Rightarrow f(c) < f(x).$$

Proof Let $\epsilon = f'(c)/2$ and choose $\delta > 0$ so that, for $x \in I$,

$$0 < |x - c| < \delta \Rightarrow \left| \frac{f(x) - f(c)}{x - c} - f'(c) \right| < \epsilon.$$

For $c < x < c + \delta$, one has

$$-\frac{f'(c)}{2} < \frac{f(x) - f(c)}{x - c} - f'(c) < \frac{f'(c)}{2},$$

whence

$$0 < \frac{f'(c)}{2} < \frac{f(x) - f(c)}{x - c}.$$

[205] Bibhutibhushan Datta and Avadesh Narayan Singh (Kripa Shankar Shukla, reviser), "Use of Calculus in Hindu mathematics", *Indian Journal of History of Science* 19, No. 2 (1984), pp. 95–109; here: p. 98. Mediæval Hindu mathematicians, particularly in the Kerala region, were several centuries ahead of the Europeans in many areas, including the beginnings of the infinitesimal calculus. In the last few decades some primary sources have been published in English translation, but not enough yet for one to develop an accurate picture of their state of knowledge. The internet is rife with references to the Hindu origins of the above Lemma and the Mean Value Theorem, but they tend to offer no details. We discuss the matter in greater detail in Sect. 2.3 of Chap. 3.

$x - c$ being positive, multiplication by it will preserve the inequality:

$$0 < f(x) - f(c), \text{ i.e. } f(c) < f(x).$$

For $c - \delta < x < c$, we again have

$$0 < \frac{f(x) - f(c)}{x - c},$$

but $x - c$ is negative, whence multiplication by it reverses the inequality:

$$0 > f(x) - f(c),$$

i.e. $f(c) > f(x)$. □

2.3.19 Lemma *Let $f : I \to \mathbb{R}$ be differentiable at $c \in I$ and suppose $f'(c) < 0$. There is a $\delta > 0$ such that, for all $x \in I$,*

$$c - \delta < x < c \Rightarrow f(c) < f(x) \text{ and } c < x < c + \delta \Rightarrow f(x) < f(c).$$

I leave the proof as an exercise to the reader.

Proof of Lemma 2.3.17. Let $c \in (a, b)$ be where f assumes a maximum in $[a, b]$. If $f'(c) > 0$, choose $\delta > 0$ according to Lemma 2.3.18. Because c is an interior point there is some $x \in I$ satisfying $c < x < c + \delta$. But for such x, $f(x) > f(c)$, contradicting the maximality of f at c. Similarly Lemma 2.3.19 tells us that $f'(c)$ cannot be less than 0. As $f'(c)$ exists by assumption, we must have $f'(c) = 0$.

The proof for f assuming the minimum value is similar. □

Note that Lemma 2.3.18, for example, says that if $f'(c) > 0$, then for x, y sufficiently close to c, one has

$$x < c < y \Rightarrow f(x) < f(c) < f(y).$$

One sometimes expresses this in words as *f is increasing at c*, which is not the same as saying that *f is increasing in a neighbourhood of c*,

$$x < y \Rightarrow f(x) < f(y)$$

for x, y sufficiently close to c, as the result of the following exercise shows.

2.3.20 Exercise Define

$$f(x) = \begin{cases} x + x^2 \sin\left(\dfrac{1}{x^2}\right), & x \neq 0 \\ 0, & x = 0. \end{cases}$$

i. Show directly using the definition of the derivative that $f'(0) = 1 > 0$.

ii. Show that $f'(1/\sqrt{2n\pi}) = 1 - 2\sqrt{2n\pi} < 0$ for positive integers n.
iii. Use Lemma 2.3.19 to conclude that, for any $\delta > 0$, there are $x, y \in (0, \delta)$ such that $x < y$ and $f(x) > f(y)$.

Lemma 2.3.17 is the key lemma needed to complete the proof of the Mean Value Theorem. However, we have raised the issue of the relation between the derivative's sign and whether or not the function is increasing in an interval and we might as well answer it here. One would expect, from Lemma 2.3.18 that, if the derivative is always positive in an interval, the function is increasing at every point and thus must be increasing in a more global sense. This is true, but the proof is subtle.

2.3.21 Corollary (Strictly Increasing Function Theorem) *Let $f : I \to \mathbb{R}$ be differentiable on the interval in question and suppose $f'(x) > 0$ for all $x \in I$. Then f is strictly increasing on I: for all $x, y \in I, x < y \Rightarrow f(x) < f(y)$.*

Proof We use the same sort of continuous induction used in the proof of the Intermediate and Extreme Value Theorems and the Uniform Continuity Theorem.

Let $x \in I$ be any element of I other than the right endpoint if I has one, and define

$$X = \{z \in I \,|\, x < z \,\&\, \forall y \in I(x < y < z \Rightarrow f(x) < f(y))\}.$$

By Lemma 2.3.18, X is nonempty: for some $\delta > 0$, $x + \delta \in X$.

If X is unbounded, then for all $y \in I$ there is some $z \in X$ such that $y < z$. If $x < y$, then $x < y < z$ and it follows that $f(x) < f(y)$.

If X is bounded, it has a least upper bound z_0. Let $x < y < z_0$ and choose $z \in X$ such that $y < z$. Then

$$x < y < z \in X \Rightarrow f(x) < f(y).$$

Either $z_0 \in I$ or $z_0 \notin I$.

If $z_0 \notin I$, then $z_0 > y$ for all $y \in I$ and we have shown that, for all $y \in I$, if $x < y$ then $f(x) < f(y)$.

Thus, assume $z_0 \in I$. Then $z_0 \in X$.

We first apply Lemma 2.3.18 to conclude $f(x) < f(z_0)$: Choose $\delta > 0$ according to the Lemma so that $x < z_0 - \delta$ and for all $y \in I$

$$z_0 - \delta < y < z_0 \Rightarrow f(y) < f(z_0). \tag{2.35}$$

But z_0 is the least upper bound of X, whence there is some $z \in X$ satisfying $y < z < z_0$. This means, for $x < z_0 - \delta < y < z < z_0$, we have $f(x) < f(y)$. Combined with (2.35) this yields $f(x) < f(z_0)$.

If z_0 is the right endpoint of I, we have shown, for all $y \in I$, that $x < y \Rightarrow f(x) < f(y)$.

If z_0 is an interior point of I, we apply the second part of Lemma 2.3.18 to obtain a contradiction. Choose δ so small that $(z_0, z_0 + \delta) \subseteq I$ and for all y,

$$z_0 < y < z_0 + \delta \Rightarrow f(z_0) < f(y)$$
$$\Rightarrow f(x) < f(y), \text{ by the above.}$$

If we choose δ small enough so that $z_0 + \delta$ is also an interior point, we see that $z_0 + \delta \in X$, contrary to the assumption that z_0 is an upper bound on X. □

By Corollary 2.3.21, we know that f is strictly increasing on (a, b) if $f'(x) > 0$ for all $x \in (a, b)$ and, similarly, f is strictly increasing on $[a, b]$ if $f'(x) > 0$ on $[a, b]$. In the latter case, the conclusion still holds for f continuous on $[a, b]$ if we weaken the differentiability requirement to assuming $f'(x)$ exists and is positive for all $x \in (a, b)$. For, if $f(a) > f(x)$ for some $a < x \in (a, b)$, one can choose $\delta > 0$ so small that, for $y \in (a, a + \delta)$, $|f(x) - f(a)| < \frac{1}{2}(f(a) - f(x))$. But then $x < y$ and $f(y) > f(x)$. Likewise, $f(x) < f(b)$ for $x < b$.

2.3.22 Remark The proof given of Corollary 2.3.21 is surprisingly complicated and it seems to have taken some time for mathematicians to realise that it was not an immediate consequence of Lemma 2.3.18. That something has to be added may be seen by considering the simple example,

$$f(x) = \begin{cases} x, & 0 < x < 1 \\ x - 1, & 1 < x < 2. \end{cases}$$

Here $f'(x) = 1 > 0$ everywhere in the domain of f, yet $f(1/2) > f(5/4)$. The problem here, of course, is that the domain of f is not one interval, but two disjoint intervals. That said, one might acknowledge that a proof is necessary but still question the need for anything as inelegant as the proof given here, with its cases and subcases. I have chosen the current proof using the Least Upper Bound Principle as a sort of induction principle because it fits in with earlier proofs along these lines and is thus a natural choice. There are slicker proofs. In the next chapter we will encounter Weierstrass's more direct proof, which proof relies on the Extreme Value Theorem, which we proved by appeal to the Least Upper Bound Principle. Weierstrass's proof evolved into a very simple proof by appeal to the Mean Value Theorem and will be given a few pages from now (p. 139, below). An alternative reduction not as dependent on the Least Upper Bound Principle is to assume f is uniformly differentiable. We will encounter this more than once in the next chapter when we discuss the history of the Mean Value Theorem.

2.3.23 Remark With Exercise 2.3.20 we showed that it does not follow from the assumption $f'(c) > 0$ that f is increasing in any neighbourhood around c. If, however, we assume f' to be continuous in some interval around c, then Lemma 2.2.17 (p. 62, above) tells us that $f'(x) > 0$ everywhere in some interval $(c - \delta, c + \delta)$, whence Corollary 2.3.21 tells us f is strictly increasing there.

The negative derivative has its analogous result:

2.3.24 Corollary (Strictly Decreasing Function Theorem) *Let $f : I \to \mathbb{R}$ be differentiable on the interval in question and suppose $f'(x) < 0$ for all $x \in I$. Then f is strictly decreasing on I: for all $x, y \in I$, $x < y \Rightarrow f(y) < f(x)$.*

This can be given a proof analogous to to that of Corollary 2.3.21 or it can be reduced to the application of that Corollary to $g(x) = -f(x)$. Again, if $I = [a, b]$ is closed and f is continuous on I, $f'(x)$ need only be assumed to exist and be negative for $x \in (a, b)$.

2.3.25 Exercise Paired with the Strictly Increasing Function Theorem is the *(Weakly) Increasing Function Theorem*: Let $f : I \to \mathbb{R}$ be differentiable in the interval in question and suppose $f'(x) \geq 0$ for all $x \in I$. Then f is increasing on I: for all $x, y \in I, x < y \Rightarrow f(x) \leq f(y)$.

i. Prove this.
ii. State and prove an analogous *(Weakly) Decreasing Function Theorem.*
iii. Prove the *Constant Function Theorem*: Let $f : I \to \mathbb{R}$ be differentiable in the interval in question and suppose $f'(x) = 0$ for all $x \in I$. Then f is constant on I.

[Darboux's function shows that the analogue to Lemma 2.3.18 for $f'(x) \geq 0$ fails. Thus, instead of modifying the proof of the Strictly Increasing Function Theorem, reduce the Increasing Function Theorem to it by considering the functions $f_n(x) = f(x) + x/n$. Use a similar function for ii. The same can be done for iii, or one can reduce it directly to i and ii.]

Getting back on track, I note that Lemma 2.3.17 is the obvious lemma to use to find the point on a smooth curve C of maximum distance from a given line in the proof of the Mean Value Theorem as outlined in the Preface. To apply it we need a formula for the distance from a point to a line. Such was given first by Ludwig Otto Hesse (1811–1874):

2.3.26 Lemma *Let the line L have the equation $Ax + By + C = 0$ and let $\langle\alpha, \beta\rangle$ be any point in the plane. The distance from $\langle\alpha, \beta\rangle$ to L is given by*

$$d_L(\alpha, \beta) = \frac{|A\alpha + B\beta + C|}{\sqrt{A^2 + B^2}}. \tag{2.36}$$

Proof There are three cases to consider.

Case 1. $B = 0$. Then $A \neq 0$ as otherwise the equation is $C = 0$ which either defines the plane or the empty set, in either case not a line. The line in question is the vertical one $Ax + C = 0$, i.e., $x = -C/A$. The distance from $\langle\alpha, \beta\rangle$ to L is measured horizontally:

$$d_L(\alpha, \beta) = \left| x - \frac{-C}{A} \right| = \frac{|Ax + C|}{|A|} = \frac{|Ax + By + C|}{\sqrt{A^2 + B^2}},$$

since $B = 0$.

Case 2. $A = 0$. Similar.

Case 3. $A, B \neq 0$. Let L denote the line given by the equation $Ax + By + C = 0$ and let P denote the point $\langle\alpha, \beta\rangle$. A simple translation of axes will not change L and P, just their representations. Write

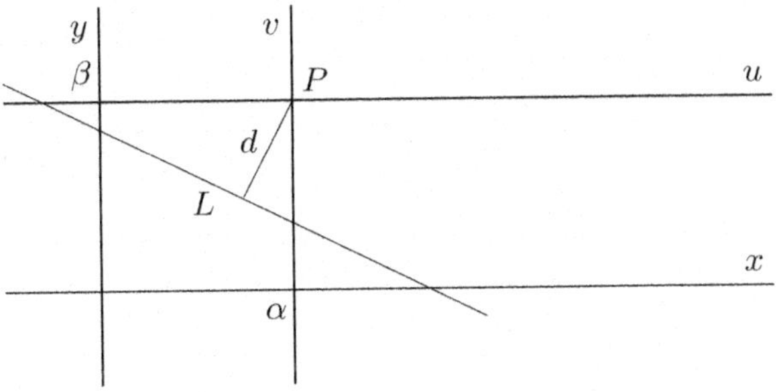

Fig. 2.41 Distance from P to L

$$x = u + \alpha,\ y = v + \beta \text{ i.e., } u = x - \alpha,\ v = y - \beta.$$

P acquires the uv-coordinates $\langle 0, 0\rangle$ and the equation of L is transformed into

$$L:\ A(u + \alpha) + B(v + \beta) + C = 0,$$

i.e.,

$$Au + Bv + (A\alpha + B\beta + C) = 0.$$

Write D for $A\alpha + B\beta + C$.

In terms of u, v, the distance from P to L is the distance from the origin to L, which is the distance d from P to the intersection of L and the line perpendicular to L connecting L to the origin. (See Fig. 2.41.)

The equation of this perpendicular is thus $Bu - Av = 0$. The intersection point is the simultaneous solution to the pair of equations

$$\begin{aligned} Bu - Av &= 0 \\ Au + Bv &= -D, \end{aligned}$$

and simple algebra tells us this is

$$u_0 = \frac{-AD}{A^2 + B^2},\quad v_0 = \frac{-BD}{A^2 + B^2}.$$

The distance from $\langle u_0, v_0\rangle$ to the uv-origin P is the square root of

$$u_0^2 + v_0^2 = \frac{A^2 D^2}{(A^2 + B^2)^2} + \frac{B^2 D^2}{(A^2 + B^2)^2} = \frac{(A^2 + B^2) D^2}{(A^2 + B^2)^2} = \frac{D^2}{A^2 + B^2}.$$

Taking the square root,

$$d_L(\alpha, \beta) = d = \frac{|D|}{\sqrt{A^2 + B^2}} = \frac{|A\alpha + B\beta + C|}{\sqrt{A^2 + B^2}}.$$

□

Recall the definition

$$d_\gamma(t) = d_L(\gamma(t)) = d_L(x(t), y(t))$$

for a given line L with equation $Ax + By + C = 0$ and a parametrisation $\gamma : I \to \mathbb{R} \times \mathbb{R}$ of a curve, $\gamma(t) = \langle x(t), y(t) \rangle$. If each of $x(t)$ and $y(t)$ is differentiable at some point t, then d_γ^2 is also differentiable at t:

$$f(t) = d_\gamma^2(t) = \frac{\big(Ax(t) + By(t) + C\big)^2}{A^2 + B^2},$$

whence

$$f'(t) = \frac{2\big(Ax(t) + By(t) + C\big)}{A^2 + B^2}\big(Ax'(t) + By'(t)\big).$$

2.3.27 Corollary *Let $\gamma : [a, b] \to \mathbb{R} \times \mathbb{R}$ be a continuous parametrisation of a curve that is not a straight line, with $\gamma(a) \neq \gamma(b)$, and suppose the component functions $x(t)$ and $y(t)$ are differentiable on (a, b). Let L be the line*

$$\big(y(b) - y(a)\big)\big(x - x(a)\big) = \big(x(b) - x(a)\big)\big(y - y(a)\big) \tag{2.37}$$

connecting the points $\gamma(a)$ and $\gamma(b)$. There is a $c \in (a, b)$ such that

$$\big(y(b) - y(a)\big)x'(c) = \big(x(b) - x(a)\big)y'(c). \tag{2.38}$$

If we consider only the case where $x(b) \neq x(a)$, we can rewrite (2.37) in the more familiar form,

$$\frac{y - y(a)}{x - x(a)} = \frac{y(b) - y(a)}{x(b) - x(a)},$$

and note that (2.38) can be rewritten

$$\frac{y'(c)}{x'(c)} = \frac{y(b) - y(a)}{x(b) - x(a)}.$$

Thus, since

$$\frac{dy}{dx}(t) = \frac{y'(t)}{x'(t)},$$

this tells us that the slope of the tangent line to the curve passing through $\gamma(c)$ equals the slope of the secant line connecting $\gamma(a)$ and $\gamma(b)$. Or, rather, it will tell us that as soon as we have formally defined "smooth" and "tangent", an easy enough but slightly subtle matter.

Proof of Corollary 2.3.27. By Lemma 2.2.31 there is a point $c \in (a, b)$ at which $d_\gamma(c)$ is maximum. For such c,

$$f'(c) = \frac{2\big(Ax(c) + By(c) + C\big)}{A^2 + B^2}\big(Ax'(c) + By'(c)\big) = 0. \tag{2.39}$$

Because the curve is not a straight line, $d_\gamma(c) > 0$ and the fraction in (2.39) is not 0. Thus we have

$$Ax'(c) + By'(c) = 0, \text{ i.e., } Ax'(c) = -By'(c).$$

But if we expand (2.37) we find that

$$A = y(b) - y(a), \quad B = -\big(x(b) - x(a)\big).$$

Thus

$$\big(y(b) - y(a)\big)x'(c) = \big(x(b) - x(a)\big)y'(c).$$

□

Ignoring the geometric interpretation, Corollary 2.3.27 is already quite strong, encompassing the classroom versions of the Mean Value Theorem and an often poorly motivated generalisation called the Cauchy Mean Value Theorem:

2.3.28 Corollary (Classroom Mean Value Theorem)[206] Let $f : [a, b] \to \mathbb{R}$ be continuous on $[a, b]$ and differentiable on (a, b). There is a $c \in (a, b)$ such that

$$f'(c) = \frac{f(b) - f(a)}{b - a},$$

or, if one prefers,

$$f(b) = f(a) + f'(c)(b - a).$$

Proof Apply Corollary 2.3.27 to $\gamma(t) = \langle t, f(t)\rangle$ □

2.3.29 Corollary (Cauchy Mean Value Theorem) *Let $f, g : [a, b] \to \mathbb{R}$ be continuous on $[a, b]$ and differentiable on (a, b). Assume $g(a) \neq g(b)$ and $g'(x) \neq 0$ for all $x \in (a, b)$. There is some $c \in (a, b)$ such that*

$$\frac{f'(c)}{g'(c)} = \frac{f(b) - f(a)}{g(b) - g(a)}.$$

Proof Apply Corollary 2.3.27 to $\gamma(t) = \langle f(t), g(t)\rangle$. □

2.3.30 Remark The assumption $g(a) \neq g(b)$ is redundant. It follows by Corollary 2.3.28 from the assumption that $g'(x) \neq 0$ for any $x \in (a, b)$. The latter assumption

---

[206]The use of the word "Classroom" here is a local one. The reader will not find it elsewhere in the literature and I introduce it merely to distinguish the theorem as stated from the myriad of forms of the Mean Value Theorem as the one familiar from the first year Calculus course. When the distinction is unimportant, I drop the adjective.

is not redundant, as evidenced by the functions $f(x) = x^2$, $g(x) = x^3$ on the interval $[-1, 1]$. for which the only value of c satisfying (2.38),

$$\big(f(1) - f(-1)\big)g'(c) = \big(g(1) - g(-1)\big)f'(c),$$

i.e.,

$$(1-1)3c^2 = (1-(-1))2c,$$

i.e., $0 = 4c$, is $c = 0$. But $g'(0) = 0$, whence the division yielding

$$\frac{f'(c)}{g'(c)} = \frac{f(b) - f(a)}{g(b) - g(a)}$$

cannot be performed. (Exercise: Examine the same pair of functions on $[-1, 2]$.)

Corollary 2.3.28 permits simpler proofs of Corollaries 2.3.21 and 2.3.24.

Simple proof of Corollary 2.3.21. Assume $f'(x) > 0$ everywhere in the interior of I and let $x, y \in I$ with $x < y$. By Corollary 2.3.28 there is some c with $x < c < y$ such that

$$f'(c) = \frac{f(y) - f(x)}{y - x}.$$

Thus $f(y) - f(x) = f'(c)(y - x) > 0$ since $y > x$ and $f'(c) > 0$. Thus $f(y) > f(x)$. □

The corresponding proof of Corollary 2.3.24 is similar, as are proofs of the weaker variants given in Exercise 2.3.25. Of these I only single out part iii for demonstration here:

2.3.31 Corollary (Constant Function Theorem) *Let $f : I \to \mathbb{R}$ be differentiable on the interval in question and suppose $f'(x) = 0$ for all $x \in I$. Then: f is constant on I.*

Proof Fix $a \in I$ and let $x \neq a$ be any other element of I. By Corollary 2.3.28 there is an element c between a and x such that

$$f(x) - f(a) = f'(c)(x - a) = 0 \cdot (x - a) = 0,$$

whence $f(x) = f(a)$. □

The classroom version of the Mean Value Theorem is the ultimate interest in the present work and we will begin its study in earnest in the next chapter. Up till now, however, we have been concerned with the general geometric form as introduced in the Preface, and we still have a few loose ends to clear up. These concern the meaning to be assigned to the words "smooth" and "tangent".

In the traditional Calculus course, one mainly treats curves of the form $y = f(x)$, i.e., the graphs of functions. Here the definition of the tangent to the curve at a point is fairly simple:

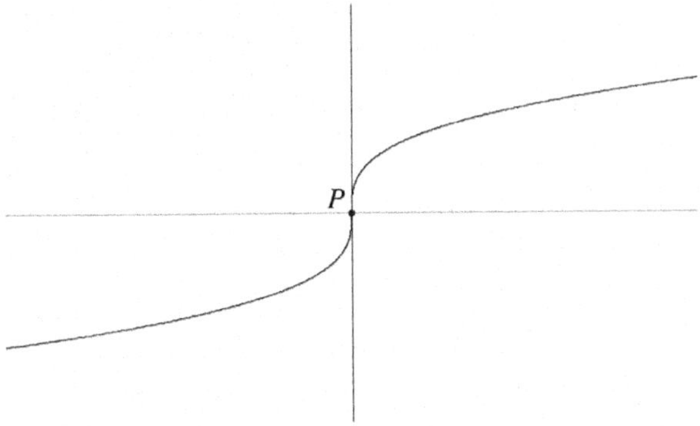

Fig. 2.42 An inverse cubic

2.3.32 Definition Let the function $y = f(x)$ be defined on the interval $[a, b]$ and let $c \in [a, b]$ and assume f is differentiable at c. The *tangent to the curve* $y = f(x)$ *at* $\langle c, f(c)\rangle$ (or, simply, *at* c) is the line with equation

$$y = f(c) + f'(c)(x - c). \tag{2.40}$$

This definition is not entirely satisfactory. Consider the inverse cubic function of Fig. 2.42. According to Definition 2.3.32 it will have no tangent at P because, at $c = 0$, $f'(c)$ is infinite. Yet as a curve, it is not much different from the graph of the cubic (see Fig. 2.30 on page 80, above), which has tangents everywhere. The present curve is the reflexion of a cubic across the line $y = x$ and the reflexion of the tangents to the cubic ought to be considered as tangents of the cube root. To do so, we need to allow infinite slopes, i.e., vertical lines, as tangents.

In doing this, some care is necessary. The inverse cubic is not much of a problem. For $y = x^{1/3}$, we have

$$f'(0) = \lim_{h\to 0} \frac{h^{1/3} - 0^{1/3}}{h} = \lim_{h\to 0} \frac{1}{h^{2/3}} = +\infty.$$

For a curve like the cycloid, which might also be considered to have a vertical tangent, we have, for example, distinct left- and right-sided limits at the cusps:

$$\lim_{h\to 0+} \frac{f(h) - f(0)}{h} = \lim_{t\to 0+} \frac{(1 - \cos t) - 0}{t - \sin t} = +\infty$$
$$\lim_{h\to 0-} \frac{f(h) - f(0)}{h} = \lim_{t\to 0-} \frac{(1 - \cos t) - 0}{t - \sin t} = -\infty.$$

Here, by $\lim_{h\to 0+}$ we mean to imply that the variable h is restricted to lying to the right of 0 in the definition:

$$\lim_{h\to 0+} g(h) = L$$

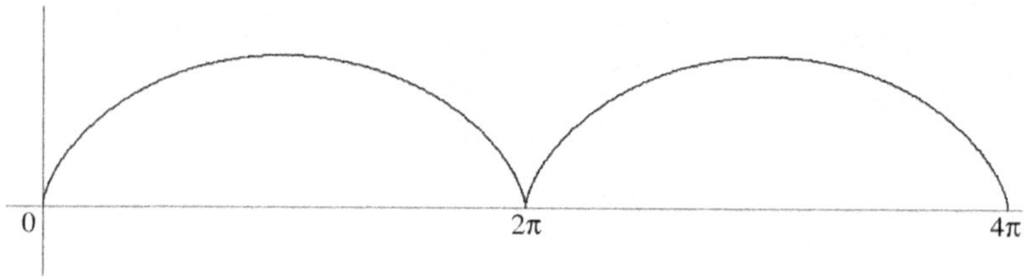

Fig. 2.43 Only one non-tangent

means that for any $\epsilon > 0$ there is some $\delta > 0$ such that

$$0 < h < \delta \Rightarrow \left|g(h) - L\right| < \epsilon.$$

Similarly, $\lim_{h \to 0-}$ requires the conclusion of the implication to hold only for $-\delta < h < 0$.

For the inverse cubic, we can say that the derivative exists and is $+\infty$, while for the cycloid we would have to say the derivative does not exist because a single two-sided limit does not exist.

2.3.33 Remark Actually, the situation is even more subtly complicated than this. Definition 2.3.32 allowed one to consider the derivative at the endpoint of an interval, which derivative, of course, depends on the limit of the difference quotient at the endpoint. This will necessarily be a one-sided limit, as our definition of limit required $f(x+h)$ to approach the candidate for the limit only for $x+h$ in the interval under consideration. Thus, if we define the curve C by,

$$\begin{array}{l} x(t) = t - \sin t \\ y(t) = 1 - \cos t \end{array}, \; t \in [0, 4\pi],$$

as in Fig. 2.43, we have $f'(0) = +\infty$, $f'(4\pi) = -\infty$, but $f'(2\pi)$ is undefined. This truncated cycloid does not have cusps at 0 and 4π, whence we can take the vertical lines there to be tangents (using the equations $x = 0$ and $x = 4\pi$ in place of (2.40)). We could extend the cycloid beyond the endpoints, thus restoring the cusps and losing the tangents, but we could also extend the curve differently, for example by adding a copy of C reflected across the x-axis, yielding "true" tangents at 0 and 4π, as in Fig. 2.44. Whether or not there is a tangent at 2π in this new figure will depend on how we define "tangent" for parametrically defined curves and how we parametrise the curve.

So, how do we define the notion of tangent for a parametrically definable curve? The classroom definition for the graph of a function $y = f(x)$ is unambiguous. We may express f in numerous ways, but when it comes to numerical values they all agree. For example, for $f(x) = \sin^2 x$, it makes no difference if we represent f by the expression $\sin^2 x$ or $1 - \cos^2 x$: Not only are $y = \sin^2 x$ and $y = 1 - \cos^2 x$ the same curve, but each value of x in the domain of f determines the same point

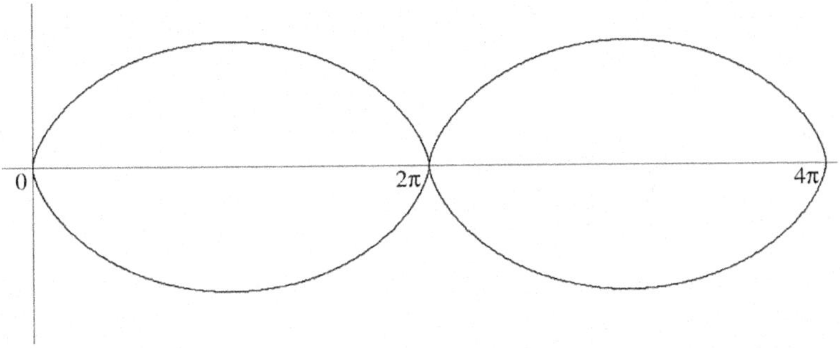

Fig. 2.44 Is there a tangent at 2π?

$\langle x, y\rangle$ on the graph. General curves have many different parametrisations that behave differently; the graph of a function is, essentially, a single canonical parametrisation of a curve:

$$\begin{matrix} x(t) = t \\ y(t) = f(t) \end{matrix}, t \in domain(f).$$

2.3.34 Exercise We can give two essentially different parametrisations of the curve of Fig. 2.44 as follows. First, let

$$f(t) = t - \sin t, \quad g(t) = 1 - \cos t$$

be the functions used in the usual parametrisation of the cycloid, and define γ_1, γ_2 as follows.

$$\gamma_1 : \quad x_1(t) = \begin{cases} f(t), & 0 \le t \le 4\pi \\ 4\pi - f(t - 4\pi), & 4\pi < t \le 8\pi, \end{cases}$$

$$y_1(t) = \begin{cases} g(t), & 0 \le t \le 4\pi \\ -g(t), & 4\pi < t \le 8\pi. \end{cases}$$

$$\gamma_2 : \quad x_2(t) = \begin{cases} f(t), & 0 \le t \le 4\pi \\ 4\pi - f(t - 4\pi), & 4\pi < t \le 8\pi, \end{cases}$$

$$y_2(t) = \begin{cases} g(t), & 0 \le t \le 2\pi \text{ or } 4\pi < t \le 6\pi \\ -g(t), & 2\pi < t \le 4\pi \text{ or } 6\pi < t \le 8\pi. \end{cases}$$

Graph these on your graphing calculator, watching the process slowly unfold. Viewing the function as defining the trajectory of a moving particle, which would you consider as representing a smooth motion? Which has two "bounces", i.e., cusps? What about the curve they trace out?

The point here is that functions are simpler to deal with than curves. We can easily unambiguously define differentiability for a function:

2.3.35 Definition A function $\gamma : I \to \mathbb{R} \times \mathbb{R}$ given by $\gamma(t) = \langle x(t), y(t)\rangle$ is *differentiable at a point* $c \in I$ if each of $x(t)$ and $y(t)$ is differentiable at c. We write

$$\gamma'(c) = \langle x'(c), y'(c)\rangle.$$

2.3.36 Exercise Which of the functions γ_1, γ_2 of Exercise 2.3.34 is differentiable at $t = 2\pi$? Find $\gamma'(2\pi)$ and $\gamma'(6\pi)$ for this function.

Defining the smoothness of a function is a matter of some delicacy, for there is not a unique notion of smoothness used in Analysis, nor two notions, nor three, ..., but an infinite number of levels of smoothness depending on how many times the function is differentiable and whether or not the last derivative is continuous. For our purposes we don't require much smoothness at all, but, as we want to conclude the existence of tangent lines from smoothness, we add an extra condition:

2.3.37 Definition A function $\gamma : I \to \mathbb{R} \times \mathbb{R}$ is *smooth* if for all $a, b \in I$ with $a < b$

i. γ is continuous on $[a, b]$,
ii. γ is differentiable on (a, b), and
iii. $\gamma'(c) \neq \langle 0, 0\rangle$ for any $c \in (a, b)$.

With respect to Exercise 2.3.34, I note that the function γ_2 is smooth, but γ_1 is not.

2.3.38 Definition A curve C is *smooth* if there is an interval I and a smooth function $\gamma : I \to \mathbb{R} \times \mathbb{R}$ such that $C = \gamma(I) = \{\gamma(t) \mid t \in I\}$.

According to this definition, the curve of Fig. 2.44 is smooth because of the smooth parametrisation γ_2 of Exercise 2.3.34. And the cube root, $y = x^{1/3}$ is just as smooth as the cube $y = x^3$, as one can see by comparing their respective parametrisations,

$$\gamma_1(t) = \langle t^3, t\rangle \quad \gamma_2(t) = \langle t, t^3\rangle,$$

with derivatives

$$\gamma_1'(t) = \langle 3t^2, 1\rangle \quad \gamma_2'(t) = \langle 1, 3t^2\rangle.$$

Conditions (i) and (ii) of the definition of smoothness are natural enough. Condition (iii) is explained by the following definition of a tangent line:

2.3.39 Definition Let $\gamma : [a, b] \to \mathbb{R} \times \mathbb{R}$ be differentiable at $c \in (a, b)$ with $\gamma'(c) \neq \langle 0, 0\rangle$. The *line tangent to* γ *at* c (or: *at* $\gamma(c)$) is the line L given, according to case, by the equation,

$$\begin{aligned}
&x = x(c), \text{ if } x'(c) = 0, \text{ or}\\
&y = y(c), \text{ if } y'(c) = 0, \text{ or}\\
&y = y(c) + \frac{y'(c)}{x'(c)}(x - x(c)), \text{ if } x'(c) \neq 0 \text{ and } y'(c) \neq 0.
\end{aligned}$$

Were it not for the common preference for writing the equation of a line in slope-intercept form, we could consolidate these cases in the single line,

$$\big(y - y(c)\big)x'(c) = y'(c)\big(x - x(c)\big),$$

or

$$y'(c)x - x'(c)y + y(c)x'(c) - x(c)y'(c) = 0.$$

And we see immediately why $\gamma'(c) = \langle 0, 0\rangle$ is unwelcome: The resulting equation is $0x + 0y + 0 = 0$, which defines the plane and not a line. (Or, using the equations of the Definition, one has two lines $x = x(c)$ and $y = y(c)$.)

A $\langle 0, 0\rangle$ derivative at some point on a differentiable curve may accompany a geometric tangent or it may not.

2.3.40 Exercise Define γ_1, γ_2 on $[-1, 1]$ by

$$\gamma_1(t) = \langle t^6, t^3\rangle$$

$$\gamma_2(t) = \begin{cases} \gamma_1(t), & t \le 0 \\ -\gamma_1(t), & 0 < t. \end{cases}$$

i. Show that $\gamma_1(t)$ is continuously differentiable on $[-1, 1]$ and parametrises the parabola $x = y^2$ over the y-interval $[-1, 1]$ and the curve has a tangent at $t = 0$ although $\gamma_1' = \langle 0, 0\rangle$.
ii. Show that $\gamma_2(t)$ is also continuously differentiable on $[-1, 1]$ with $\gamma_2'(0) = \langle 0, 0\rangle$, but γ_2 has a cusp at $t = 0$.

Referring once again to the functions of Exercise 2.3.34, we see that the vertical line $x = 2\pi$ is tangent to γ_2 at $t = 2\pi$ and $t = 6\pi$, while γ_1 has no tangent at these points. As for their common curve $C = \gamma_1(\,[0, 8\pi]\,) = \gamma_2(\,[0, 8\pi]\,)$, we would say that a line is tangent to C at a point P if it is tangent there with respect to some smooth parametrisation:

2.3.41 Definition A line L is *tangent to a curve* C at a point $P \in C$ if there is some smooth parametrisation $\gamma : I \to \mathbb{R} \times \mathbb{R}$ of C and some point $c \in I$ such that $\gamma(c) = P$ and L is tangent to γ at c.

With all of this we can now restate Corollary 2.3.27 in the form of the Mean Value Theorem as presented to the man-in-the-street back in the Preface.

2.3.42 Theorem (Mean Value Theorem; Geometric Form) *Let C be a smooth curve with distinct points P, Q on the curve. There is a point R on the curve at which the tangent line is parallel to the segment PQ.*

You will notice that the statement of the Theorem omits the mention that R lies "between" P and Q. A curve, as we have defined it, is just a set of points. It has no orientation of its own, hence no notion of betweenness. The orientation is determined

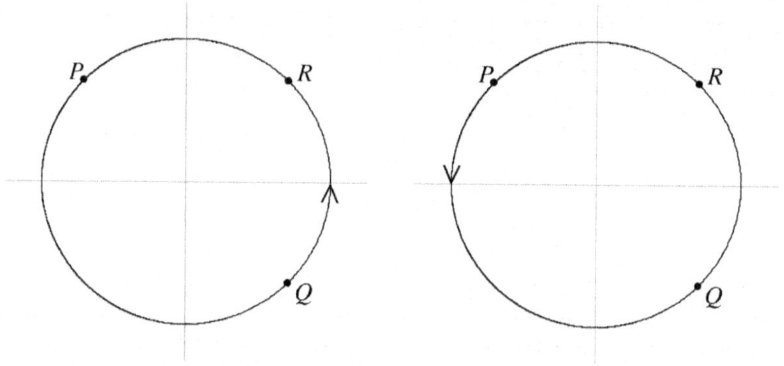

Fig. 2.45 Is R between P and Q?

not by the curve, but by the particular parametrisation chosen to demonstrate the smoothness of the curve. For example, we can parametrise the circle by

$$\gamma_1(t) = \langle \cos t, \sin t \rangle, \quad t \in [0, 2\pi]$$

or, say, by

$$\gamma_2(t) = \langle \cos t, \sin t \rangle, \quad t \in [\pi, 3\pi].$$

Consider the points

$$P = \langle -\sqrt{2}/2, \sqrt{2}/2 \rangle, \quad Q = \langle \sqrt{2}/2, -\sqrt{2}/2 \rangle, \quad R = \langle \sqrt{2}/2, \sqrt{2}/2 \rangle,$$

as in Fig. 2.45. In the first circle, $R = \gamma_1(\pi/4)$ does not lie between $P = \gamma_1(3\pi/4)$ and $Q = \gamma_1(7\pi/4)$ because, as one traces the curve going from $t = 0$ to $t = 2\pi$, R appears before both P and Q. In the second circle $R = \gamma_2(9\pi/4)$ occurs after $Q = \gamma_2(7\pi/4)$, but before $P = \gamma_2(11\pi/4)$, as the curve is traced, whence R lies between P and Q.

The matter appears to be one of unnecessary subtlety and is probably best ignored, i.e., resolved by introducing the parametrisation explicitly into the statement of the Theorem.

2.3.43 Theorem (Mean Value Theorem; Algebraic-Geometric Form) *Let $\gamma : I \to \mathbb{R} \times \mathbb{R}$ parametrise a smooth curve C and let $a, b \in I$ with $a < b$ and $\gamma(a) \neq \gamma(b)$. There is some $c \in (a, b)$ such that the line tangent to γ at c is parallel to the secant connecting $\gamma(a)$ and $\gamma(b)$.*

Basically, the only difference between this statement and that of Corollary 2.3.27 is that the Theorem does not exclude the trivial case in which that portion of the curve between $\gamma(a)$ and $\gamma(b)$ coincides with the secant line.

I leave the details of the proof to the reader.

We are almost finished with this chapter. What we haven't done to (at least: my) complete satisfaction is to justify the choices of definitions of a smooth curve and a tangent to the curve. This would be done by showing that the definitions agree with the intuitively defined tangents in all classical cases and do not generalise too much, defining tangents where they shouldn't be or accepting curves that are too un-curvelike. Now, this is a matter of some depth and subtlety and, even though we have been routinely straying somewhat from our ostensible purpose of discussing the Classroom Mean Value Theorem, I should at some point get back on track. Thus, I shall, for the most part, leave the reader to convince him- or herself that these definitions are, if not written in stone, at least reasonable. I do feel compelled, however, to address the issue of space-filling curves.

2.3.44 Exercise Let $\gamma(t) = \langle \cos t^4, \sin t^2 \rangle$ on $[.0001, 45]$. Drag out your TI-83 or TI-84, put it into parametric graphing mode, enter

```
X1T= cos(T^4)
Y1T= sin(T^2),
```

and set the window to

```
Tmin = .0001
Tmax = 45
Tstep = .05
Xmin =⁻2.35
Xmax = 2.35
Ymin =⁻1.55
Ymax = 1.55,
```

and graph the function. What do you see? Is this a smooth function?

It should not be revealing too much to say that what one sees on one's calculator tells one more about the resolution of the calculator's display than about the nature of the graph of γ. Graphing the function on the computer at higher resolution reveals a lot of white space. If one extends γ to larger and larger intervals, the space available for the image of γ fills in more and more, and the resolution of the graph may again be overtaken by the function. Nonetheless, there are points that γ will miss. For example, for no value of $t \in (-\infty, \infty)$ does $\gamma(t) = \langle 1, 1 \rangle$. To see this, assume by way of contradiction that

$$\gamma(t) = \langle \cos t^4, \sin t^2 \rangle = \langle 1, 1 \rangle.$$

Now,

$$\cos t^4 = 1 \Rightarrow \text{for some natural number } m,\ t^4 = 2m\pi$$
$$\sin t^2 = 1 \Rightarrow \text{for some natural number } n,\ t^2 = \frac{4n+1}{2}\pi.$$

Thus,

$$2m\pi = \left(\frac{4n+1}{2}\pi\right)^2,$$

i.e.,

$$8m\pi = (4n+1)^2\pi^2,$$

and, since $\pi \neq 0$,

$$\pi = \frac{8m}{(4n+1)^2},$$

contrary to the irrationality of π.

2.3.45 Exercise For γ as in Exercise 2.3.44,

i. show that the points $\langle 0, 0\rangle$, $\langle 1, -1\rangle$, $\langle -1, 1\rangle$, and $\langle -1, -1\rangle$ do not lie on the curve $\gamma([.0001, 45])$; and
ii. show that $\gamma'(t) \neq \langle 0, 0\rangle$ for $t \in [.0001, 45]$. Conclude that γ is smooth.

One can do much better. In general, a differentiable curve misses "most" points.

As the reader may remember, at the end of his paper Peano cited some conditions which, when assumed in addition to continuity, prevented a curve $\gamma : [0, 1] \to [0, 1] \times [0, 1]$ from being a space-filling curve. One was that the curve be the graph of a function $y = f(x)$ and the other was that the curve be of *bounded variation*. In both cases, the reason the curve could not fill the entire unit square was that it could be fit inside a set of arbitrarily small area. I have not seen Jordan's proof that a curve of bounded variation cannot be a space-filling curve, but can report that Robert Burckel and Caspar Goffman have published a fairly simple combinatorial proof of this result.[207] I will not prove the result in this generality, but will present instead a simpler proof that no continuously differentiable curve is a space-filling curve.

2.3.46 Theorem *Let $\gamma : [0, 1] \to [0, 1] \times [0, 1]$ be continuously differentiable on $[0, 1]$ and let $\epsilon > 0$. There is a set $X \subseteq [0, 1] \times [0, 1]$ of area $< \epsilon$ such that $\gamma([0, 1]) \subseteq X$. In other words, the curve $C = \gamma([0, 1])$ has zero area and thus cannot equal the entire square, which has area 1.*

Proof Let $\epsilon > 0$. Writing $\gamma(t) = \langle x(t), y(t)\rangle$, we are assuming $x'(t)$ and $y'(t)$ continuous, hence bounded on $[0, 1]$. Let $B > 0$ be a common bound on $|x'(t)|$, $|y'(t)|$ for $t \in [0, 1]$. For $s, t \in [0, 1]$, the Mean Value Theorem (Corollary 2.3.28) yields

$$x(s) - x(t) = x'(t_0)(s - t), \quad y(s) - y(t) = y'(t_1)(s - t),$$

for some $t_0, t_1 \in (0, 1)$, whence

$$\big|x(s) - x(t)\big|, \big|y(s) - y(t)\big| \leq B \cdot |s - t| < \frac{B}{n}, \text{ for } |s - t| < \frac{1}{n+1}, \tag{2.41}$$

[207] R.B. Burckel and C. Goffman, "Rectifiable curves are of zero content", *Mathematics Magazine* 44 (1971), pp. 179–180.

where $n > 1$ will be chosen shortly.

Observe

$$\gamma([0,1]) = \bigcup_{k=0}^{n-1} \gamma\left(\left[\frac{k}{n}, \frac{k+1}{n}\right]\right)$$
$$\subseteq \bigcup_{k=0}^{n-1} x\left(\left[\frac{k}{n}, \frac{k+1}{n}\right]\right) \times y\left(\left[\frac{k}{n}, \frac{k+1}{n}\right]\right),$$

and each rectangle

$$x\left(\left[\frac{k}{n}, \frac{k+1}{n}\right]\right) \times y\left(\left[\frac{k}{n}, \frac{k+1}{n}\right]\right)$$

has area less than

$$\frac{B}{n} \cdot \frac{B}{n} = \frac{B^2}{n^2} = \frac{B^2}{n} \cdot \frac{1}{n}.$$

Now choose $n > B^2/\epsilon$, so that $\epsilon > B^2/n$. Then $\gamma([0,1])$ is contained in a set of area less than

$$\sum_{k=0}^{n-1} \frac{B^2}{n} \cdot \frac{1}{n} = n \cdot \frac{B^2}{n} \cdot \frac{1}{n} = \frac{B^2}{n} < \epsilon.$$

□

Our definition of smoothness did not require continuous differentiability, and not all smooth curves are continuously differentiable, whence the above proof does not apply to them generally. It can be shown that a differentiable curve cannot be a space-filling curve, but I don't know any proof of comparable simplicity in the general case.[208] Thus I shall simply allow Theorem 2.3.46 to stand as an indication that some sort of differentiability is the appropriate condition to add to continuity to formally capture the intuitive notion of a smooth curve. The emergence of sharp formal concepts from vague intuitive ones is an interesting study, and we have been following it throughout this chapter, but it is time now to change gears. The concepts have been formalised and we wish to consider the Mean Value Theorem itself.

---

[208] Burckel and Goffman prove Theorem 2.3.46 for rectifiable curves. A curve is rectifiable just in case it is of bounded variation. Using (2.41) one easily shows continuously differentiable curves to be rectifiable, whence Theorem 2.3.46 is a special case of Jordan's result. Not every differentiable function, however, is rectifiable. Gelbaum and Olmsted, *op. cit.*, pp. 140–141, cite $x^2 \sin(1/x^2)$ as an example.

Chapter 3
The Mean Value Theorem

3.1 The Mean Value Theorem and Related Results

Today the name "Mean Value Theorem" is attached to a specific, precisely stated result, namely the Classroom Mean Value Theorem (Corollary 2.3.28 of the preceding chapter). Over history, however, there have been several variants of this result, some equally precise, that we would recognise as instances of the Mean Value Theorem. And even today there are variants with the words "Mean Value Theorem" in their names: the Cauchy Mean Value Theorem, the Extended Mean Value Theorem, and the Mean Value Theorem for Integrals, to name a few. The main purpose of the present book is to discuss the Classroom Mean Value Theorem as it is commonly, but no longer universally, presented in the introductory course in the Calculus. And the topic of the present chapter is the history of the development of this Theorem. Before beginning this history, however, it will be convenient to compile a list of those results that will pop up in one form or another in this history. This list includes those results on which proofs and attempted proofs of the Theorem have depended, variants of the Theorem, and certain applications thereof. This is the rôle of the present section.

3.1.1 Variants of the Mean Value Theorem

The name "Mean Value Theorem" will refer throughout the rest of this book to the Classroom Mean Value Theorem. For, the geometric significance of the Theorem, however intuitive and motivational it may be, pales in comparison to the algebraic/analytic applicability of the Theorem in the non-parametric directly functional case: $y = f(x)$.

The simplest variant of the Mean Value Theorem is a special case, the incorrectly eponymous Rolle's Theorem:

C. Smoryński, *MVT: A Most Valuable Theorem*,
DOI 10.1007/978-3-319-52956-1_3

3.1.1 Theorem (Rolle's Theorem) *Let* $f : [a, b] \to \mathbb{R}$ *be continuous on* $[a, b]$, *differentiable on* (a, b), *and suppose* $f(a) = f(b)$. *Then, for some* $c \in (a, b)$, $f'(c) = 0$.

Proof. If f is constant on $[a, b]$, then $f'(c) = 0$ for all $c \in (a, b)$, so assume f is not constant: for some $d \in (a, b)$, $f(d) \neq f(a)$. By the Extreme Value Theorem (Theorem 2.2.12 in Chap. 2), f assumes a maximum and a minimum value on $[a, b]$. At least one of these is not $f(a) = f(b)$ and thus occurs at some $c \in (a, b)$. But by Lemma 2.3.17 in Chap. 2, $f'(c) = 0$. □

The Mean Value Theorem reduces algebraically to Rolle's Theorem.

3.1.2 Theorem (Mean Value Theorem) *Let* $f : [a, b] \to \mathbb{R}$ *be continuous on* $[a, b]$ *and differentiable on* (a, b). *There is a real number* $c \in (a, b)$ *such that*

$$f'(c) = \frac{f(b) - f(a)}{b - a}.$$

Proof. Consider the auxiliary function,

$$\phi(x) = f(x) - f(a) - \frac{f(b) - f(a)}{b - a}(x - a).$$

ϕ is also continuous on $[a, b]$ and differentiable on (a, b). Moreover,

$$\phi(a) = f(a) - f(a) - \frac{f(b) - f(a)}{b - a}(a - a) = 0 - \frac{f(b) - f(a)}{b - a}0 = 0$$
$$\phi(b) = f(b) - f(a) - \frac{f(b) - f(a)}{b - a}(b - a) = f(b) - f(a) - \big(f(b) - f(a)\big) = 0.$$

Rolle's Theorem applies: for some $c \in (a, b)$, $\phi'(c) = 0$. But

$$0 = \phi'(c) = f'(c) - 0 - \frac{f(b) - f(a)}{b - a}(1 - 0) = f'(c) - \frac{f(b) - f(a)}{b - a},$$

whence

$$f'(c) = \frac{f(b) - f(a)}{b - a}.$$ □

The textbook may or may not offer some motivation behind this choice of ϕ. The occasional failure to provide motivation for the choice of ϕ has been noted, e.g., in several papers of a volume, *Selected Papers on Calculus*,[1] an anthology drawn from *The American Mathematical Monthly* and *Mathematics Magazine* aimed at teachers and students coming to grips with the Calculus. About ϕ we read

[1]Tom M. Apostol, Hubert E. Chrestenson, C. Stanley Ogilvy, Donald E. Richmond, and N. James Schoonmaker (eds.), *Selected Papers on Calculus*, The Mathematical Association of America, 1969.

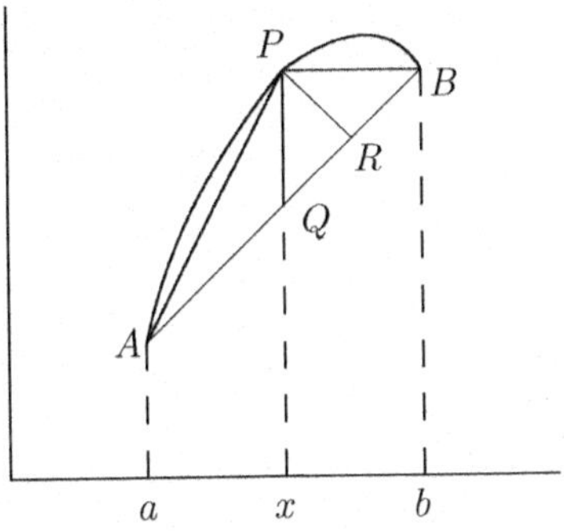

Fig. 3.1 Where ϕ comes from

> This is a formidable expression whose origin is puzzling until it is pointed out as the difference PQ of the ordinate of a point on the graph of $f(x)$ and the ordinate of the secant line for the same x.[2] [See Fig. 3.1, which is borrowed from Yates's paper, below.]

> The usual proofs of the first and extended mean value theorems involve the process of applying Rolle's theorem to functions happily designed to yield the desired conclusions. Frequently, no mention is made of how these functions are discovered.[3]

> In all of the textbooks on elementary and advanced calculus with which the author is acquainted, the various mean-value theorems and Taylor series with a remainder are arrived at by setting up a function judiciously and then applying Rolle's Theorem. In many cases the student justifiably may get the feeling that this "suitable" function is pulled out of the proverbial hat.[4]

Neither the validity of the proof nor the simplicity thereof is being called into question here. The objection is pædagogical, the desire that the choice of ϕ appear natural and not be presented as a *deus ex machina*. This point might more properly be discussed in the next chapter on Calculus reform, but we have already alluded to it in the preceding chapter and this might be a better place to discuss it. The function ϕ is not just something that is found by trial and error or by a flash of insight; it arises quite naturally in a number or ways, any one of which could profitably be presented in class.

The simplest explanation of the choice of ϕ is probably that cited by Yates, above. The equation of the line connecting the points $\langle a, f(a)\rangle$ and $\langle b, f(b)\rangle$ (points A and B in Fig. 3.1) is

$$y = f(a) + \frac{f(b) - f(a)}{b - a}(x - a)$$

and ϕ is obtained simply by subtracting this function from $f(x)$. Because the two curves meet in $\langle a, f(a)\rangle$ and $\langle b, f(b)\rangle$ one will have $\phi(a) = \phi(b) = 0$. One has a

[2] R.C. Yates, "The law of the mean", in: Apostol *et al.*, *op. cit.*, p. 195; reprinted from *The American Mathematical Monthly* 66 (1959), pp. 579–580.

[3] Louis C. Barrett and Richard A. Jacobson, "Extended laws of the mean", in: Apostol *et al.*, *op. cit.*, p. 198; reprinted from *The American Mathematical Monthly* 67 (1960), pp. 1005–1007.

[4] Murray R. Spiegel, "Mean value theorems and Taylor series", in: Apostol *et al.*, *op. cit.*, p. 204; reprinted from *Mathematics Magazine* 29 (1956), pp. 263–266.

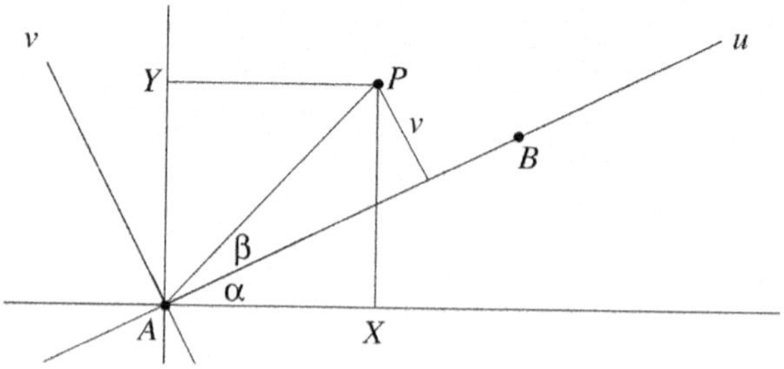

Fig. 3.2 Explaining ϕ through rotation

function to which one can apply Rolle's Theorem, and if one does so, the Mean Value Theorem falls out.

This may still have the appearance of an accident, but a more natural reason for considering ϕ can be given. I refer the reader to Fig. 3.1. If one were to rotate the figure to make AB parallel to the x-axis, Rolle's Theorem would apply,[5] yielding a point P with a horizontal tangent. Rotating back into the original orientation the tangent at P remains parallel to AB. Now, one doesn't have to actually carry out the rotation. If P' is any other point on the curve on the same side of AB as P, one can drop the vertical from P' to a point Q' on AB and a perpendicular to a point R' on AB. The angles $\angle PQR$ and $\angle P'Q'R'$ will be equal, as will be the right angles $\angle PRQ$ and $P'R'Q'$, whence the triangles PQR and $P'Q'R'$ are similar and

$$\frac{PQ}{P'Q'} = \frac{PR}{P'R'}.$$

Hence PR is maximised when PQ is maximised and the function ϕ suggests itself.

Nevertheless, one can in fact carry out the rotation without too much computational difficulty. It is convenient first to translate the axes to place $\langle a, f(a)\rangle$ at the origin:

$$X = x - a, \quad Y = y - f(a).$$

Then AB is a line through the origin of slope

$$\tan\alpha = \frac{f(b) - f(a)}{b - a}.$$

Letting AB be the u-axis of a uv-pair of axes with origin at A, we calculate the v-coordinate of the point P on the curve with xy-coordinates $\langle x, f(x)\rangle$ by referring to Fig. 3.2. We have

$$\frac{v}{\sqrt{X^2+Y^2}} = \sin\beta = \sin(\theta - \alpha),$$

[5] Well, almost: The rotated curve may no longer be the graph of a function. However, one can still follow the argument as a heuristic to obtain ϕ and then apply the proof of Theorem 3.1.2 as above.

where θ is the angle between AP and the X-axis

$$= \sin\theta\cos\alpha - \cos\theta\sin\alpha$$
$$= \frac{Y}{\sqrt{X^2+Y^2}}\cdot\frac{b-a}{K} - \frac{X}{\sqrt{X^2+Y^2}}\cdot\frac{f(b)-f(a)}{K},$$

where $K = \sqrt{\big(f(b)-f(a)\big)^2 + (b-a)^2}$. Multiplying by $K\sqrt{X^2+Y^2}$, we have

$$\begin{aligned} Kv &= Y\cdot(b-a) - X\cdot\big(f(b)-f(a)\big)\\ &= \big(f(x)-f(a)\big)(b-a) - (x-a)\big(f(b)-f(a)\big). \end{aligned}$$

It doesn't require any great inspiration to see that dividing by $b-a$ will yield ϕ. And, v being thus a constant multiple of ϕ, its absolute value will be maximised when that of ϕ is maximised.

There are other ways of obtaining ϕ. One approach is to note that for points P on the curve, the area of the triangle APB is proportional to the height PR above the base AB. Thus, the height PR is maximised when the area is maximised. From Analytic Geometry one might remember the formula for the area as $\frac{1}{2}\big|D(x)\big|$ for the determinant

$$D(x) = \begin{vmatrix} x & f(x) & 1\\ a & f(a) & 1\\ b & f(b) & 1\end{vmatrix},$$

where $P = \langle x, f(x)\rangle$. Now $D(a) = D(b) = 0$, whence Rolle's Theorem applies and $D'(c) = 0$ for some $c \in (a,b)$. To calculate $D'(c)$, first expand $D(x)$:

$$\begin{aligned} D(x) &= x\begin{vmatrix} f(a) & 1\\ f(b) & 1\end{vmatrix} - f(x)\begin{vmatrix} a & 1\\ b & 1\end{vmatrix} + \begin{vmatrix} a & f(a)\\ b & f(b)\end{vmatrix}\\ &= \big(f(a)-f(b)\big)x - f(x)(a-b) + \big(af(b) - bf(a)\big). \end{aligned}$$

Then differentiate

$$D'(x) = f(a) - f(b) - f'(x)(a-b).$$

Setting $D'(c) = 0$ now yields the Mean Value Theorem.

Of the papers quoted above, Yates presents the proof as just described, while Barrett and Jacobson use the parallelogram of which APB is half and note that a variant yields the Cauchy Mean Value Theorem as well. The method of proof is due to Peano (1884), who will be discussed in Sect. 3.9, below.

I am, as already confessed, not one to remember formulæ and this dragging in of what many students will consider a forgotten obscurity may be less pædagogically advisable than introducing ϕ with some minimal explanation. It is true that such determinants will prove necessary later in the course when functions of several variables are discussed, but this occurs sufficiently later that introducing the determinant here will provide no reinforcement value.

Finally, Spiegel, in the third paper cited above, notes that one can solve for ϕ: One wants a function ϕ satisfying $\phi(a) = \phi(b)$ to which one can apply Rolle's Theorem and yet is simply related to f. Such a function ϕ can be found by assuming it to be of the form

$$\phi(x) = f(x) - (A + Bx), \tag{3.1}$$

and, setting $\phi(a) = \phi(b)$,

$$\begin{aligned} \phi(a) &= f(a) - A - Ba \\ \phi(b) &= f(b) - A - Bb, \end{aligned} \tag{3.2}$$

whence

$$0 = f(a) - f(b) - Ba + Bb.$$

We immediately conclude

$$B = \frac{f(b) - f(a)}{b - a}.$$

With the proper choice of A, we can make $\phi(a) = 0$: Setting $\phi(a) = 0$ in (3.2) yields

$$A = f(a) - \frac{f(b) - f(a)}{b - a} a.$$

Thus (3.1) yields

$$\begin{aligned} \phi(x) &= f(x) - \left(f(a) - \frac{f(b) - f(a)}{b - a} a + \frac{f(b) - f(a)}{b - a} x \right) \\ &= f(x) - f(a) - \frac{f(b) - f(a)}{b - a} (x - a). \end{aligned}$$

Returning from the question of proof to that of theorem, we find two variants of the Mean Value Theorem often presented shortly after the Theorem itself. The first of these is the Cauchy Mean Value Theorem, also known as the *Extended Mean Value Theorem*.

3.1.3 Theorem (Cauchy Mean Value Theorem) *Let $f, g : [a, b] \to \mathbb{R}$ be continuous on $[a, b]$ and differentiable on (a, b). There is some $c \in (a, b)$ such that*

$$g'(c)\big(f(b) - f(a)\big) = f'(c)\big(g(b) - g(a)\big). \tag{3.3}$$

If $g'(x)$ is never 0 on (a, b), then, in fact,

$$\frac{f'(c)}{g'(c)} = \frac{f(b) - f(a)}{g(b) - g(a)}. \tag{3.4}$$

As in proving the Mean Value Theorem, one usually proves this by introducing an auxiliary function,

$$\phi(x) = \big(f(x) - f(a)\big)\big(g(b) - g(a)\big) - \big(g(x) - g(a)\big)\big(f(b) - f(a)\big),$$

and notes that $\phi(b) = \phi(a) = 0$, whence Rolle's Theorem yields some $c \in (a, b)$ such that

$$0 = \phi'(c) = f'(c)\big(g(b) - g(a)\big) - g'(c)\big(f(b) - f(a)\big),$$

thus yielding (3.3). And, as in our earlier proof of this theorem (Lemma 2.3.29 of page 139, above), if $g'(x)$ is never 0 on (a, b), then $g(b) \neq g(a)$ and we can perform the divisions to conclude (3.4).

Again, the choice of ϕ is a little mysterious at first, but if one assumes $g'(x)$ never to equal 0, one can divide by $g(b) - g(a)$ and use

$$\phi_1(x) = f(x) - f(a) - \frac{f(b) - f(a)}{g(b) - g(a)}\big(g(x) - g(a)\big),$$

which is more directly analogous to the earlier auxiliary function ϕ used in the proof of the Mean Value Theorem: the linear terms $b - a$ and $x - a$ have simply been replaced by $g(b) - g(a)$ and $g(x) - g(a)$, respectively.

One can also use a determinant to produce the needed auxiliary function:

3.1.4 Exercise Define

$$D(x) = \begin{vmatrix} g(x) & f(x) & 1 \\ g(a) & f(a) & 1 \\ g(b) & f(b) & 1 \end{vmatrix},$$

and use D to give a proof of the Cauchy Mean Value Theorem.

An alternative proof of the Cauchy Mean Value Theorem reduces it to the Mean Value Theorem by appeal to the Inverse Function Theorem. When $g'(x)$ is assumed never equal to 0, then, by the Mean Value Theorem, g is one-to-one. Being continuous it has, by the Extreme Value Theorem, minimum and maximum values m and M, respectively. $g : [a, b] \to [m, M]$ has an inverse $h : [m, M] \to [a, b]$ and we can apply the Mean Value Theorem to $F(x) = f(h(x))$ to derive (3.4). Now, g is either strictly increasing or strictly decreasing. This is obvious if g' is continuous. In the more general case we need the following Lemma of Gaston Darboux from his paper "Mémoire sur les fonctions discontinues"[6] cited earlier.

3.1.5 Lemma (Intermediate Value Theorem for Derivatives) *Let $g : [\alpha, \beta] \to \mathbb{R}$ be continuous and differentiable on $[\alpha, \beta] \subseteq (a, b)$ and suppose $g'(\alpha) < d < g'(\beta)$. Then, for some $c \in (\alpha, \beta)$, $g'(c) = d$.*

---

[6] Darboux, *op. cit.*, pp. 109–110.

Proof. Define $h(x) = g(x) - dx$. h is also continuous and differentiable on $[\alpha, \beta]$. By continuity it assumes a minimum at some $c \in [\alpha, \beta]$.

Now $h'(\alpha) = g'(\alpha) - d < 0$, whence Lemma 2.3.19 in Chap. 2 tells us $h(\alpha) > h(x)$ for all x in some interval $(\alpha, \alpha + \delta)$. Thus $h(\alpha)$ is not the minimum value of h.

Likewise, $h'(\beta) = g'(\beta) - d > 0$ and Lemma 2.3.18 in Chap. 2 implies the minimum does not occur at β.

Hence the minimum occurs at $c \in (\alpha, \beta)$ and Lemma 2.3.17 in Chap. 2 applies: $h'(c) = 0$. But $h'(c) = g'(c) - d = 0$ implies $g'(c) = d$. □

It follows, since $g'(x)$ is never 0 in (a, b) that one cannot have $g'(\alpha) < 0 < g'(\beta)$ or $g'(\beta) < 0 < g'(\alpha)$ (applying the Lemma to $-g$) for $\alpha < \beta$ in (a, b). Thus either $g'(x)$ is always positive and g is strictly increasing (by Corollary 2.3.21 in Chap. 2), or g' is always negative and g is strictly decreasing (by Corollary 2.3.24 in Chap. 2).

Assume for the sake of definiteness that g is strictly increasing and observe

$$\begin{aligned}\frac{f(b) - f(a)}{g(b) - g(a)} &= \frac{f(h(M)) - f(h(m))}{M - m}\\ &= \frac{F(M) - F(m)}{M - m}\\ &= F'(m_0), \quad \text{for some } m_0 \in (m, M)\\ &= f'(h(m_0)) \cdot h'(m_0), \quad \text{by the Chain Rule}\\ &= f'(c) \cdot \frac{1}{g'(c)},\end{aligned}$$

for $c = h(m_0)$ by the Inverse Function Theorem. The proof for g strictly decreasing merely switches m and M at their various occurrences.

Incidentally, the Cauchy Mean Value Theorem is aptly named: Cauchy himself proved it, though not in the same way and under the stronger assumption that f, g were uniformly differentiable.

Our next variant of the Mean Value Theorem is usually termed *Taylor's Theorem with the Lagrange Form for the Remainder* and is sometimes even called *Taylor's Theorem.*[7] Brook Taylor (1685–1731), of course, proved nothing at all like this, although the Theorem does have some bearing on the series named after him. The result was actually proven by Lagrange, under very stringent requirements and by a different method of proof.

---

[7]G.H. Hardy (1877–1947) (*A Course of Pure Mathematics*, 10th edition, Cambridge University Press, Cambridge, 1952, p. 256) gives "Taylor's Theorem" and the *General Mean Value Theorem* as names for this variant. George B. Thomas, Jr. (1914–2006) (*Calculus*, 2nd edition, Addison-Wesley Publishing Company, Inc., Reading (Mass.), 1961, p. 149) calls it the *Extended Mean Value Theorem*. This last name is now more commonly applied to the Cauchy Mean Value Theorem, which I have also seen referred to as the *Generalised Mean Value Theorem*. To avoid confusion, it is best to use a more descriptive name for each theorem. The eponymous "Cauchy Mean Value Theorem" is not very descriptive, but it is unambiguous and the attribution is correct. One could try to be more descriptive with names like "Two-Function Mean Value Theorem" or "Parametric Form of the Mean Value Theorem", but I doubt such clumsy names would catch on. And, as for Theorem 3.1.6, I think "Higher Order Mean Value Theorem" most suitable.

3.1.6 Theorem (Higher Order Mean Value Theorem) *Let $n \geq 1$ be a positive integer and let $f : [a, b] \to \mathbb{R}$ be n times continuously differentiable on $[a, b]$ and $n + 1$ times differentiable on (a, b). There is some $c \in (a, b)$ such that*

$$f(b) = f(a) + \frac{f'(a)}{1!}(b-a) + \frac{f''(a)}{2!}(b-a)^2 + \\ \ldots + \frac{f^{(n)}(a)}{n!}(b-a)^n + \frac{f^{(n+1)}(c)}{(n+1)!}(b-a)^{n+1}.$$

If one defines "0 times continuous differentiability" to mean continuity, the Mean Value Theorem becomes a special case of the Theorem after it has been modified to allow $n = 0$.

One proves this today in analogy to the proofs of the Mean and Cauchy Mean Value Theorems by choosing the right auxiliary function:

$$\phi(x) = \left(f(b) - \sum_{k=0}^{n} \frac{f^k(x)}{k!}(b-x)^k\right) - \left(\frac{b-x}{b-a}\right)^n \left(f(b) - \sum_{k=0}^{n} \frac{f^k(a)}{k!}(b-a)^k\right).$$

3.1.7 Exercise Prove Theorem 3.1.6 by showing $\phi(a) = \phi(b) = 0$ and then applying Rolle's Theorem.

Most textbooks using this choice of ϕ as the auxiliary function to be used in proving Theorem 3.1.6 present it with no explanation. It is sufficiently similar to the auxiliary functions used in proving the Mean Value Theorem and the Cauchy Mean Value Theorem, and the proof itself is short enough, that no explanation of where it came from may seem to be necessary. But it differs enough from what one might expect, for example by including factors $b - x$ instead of $x - a$, that one would like the choice motivated or explained. At the moment all I can say is that ϕ yields the result in a single step, and the other proofs I am aware of require iterated applications of Rolle's Theorem[8] or of the Cauchy Mean Value Theorem.[9]

With the Higher Order Mean Value Theorem, we have completed our discussion of direct variants of the Mean Value Theorem. There are, of course, other direct variants in higher dimensions, but these lie beyond the scope of this book.

3.1.2 The Mean Value Theorem and Integration

There are two results about integration that bear directly on the Mean Value Theorem. These are the *Fundamental Theorem of the Calculus* and the *Mean Value Theorem for Integrals*.

---

[8]For example that in Spiegel, *op. cit.*, pp. 205–206.

[9]Cf. Sect. 3.5 on Cauchy's contributions, below.

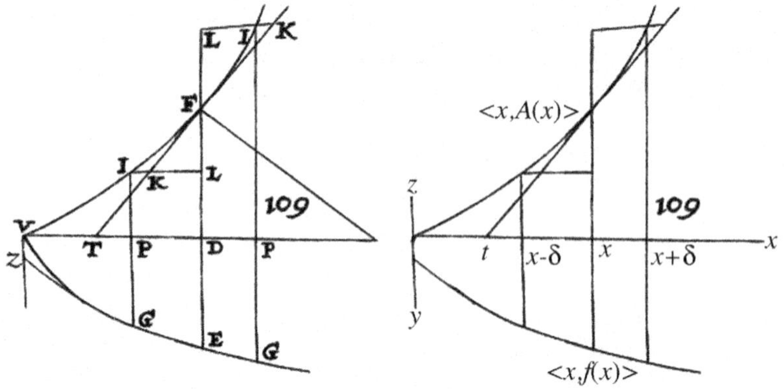

Fig. 3.3 Barrow's diagram

The Fundamental Theorem of the Calculus is, like the Mean Value Theorem, not a theorem so much as a cluster of theorems. It has a geometric *ur*form preceding the invention of the Calculus, and a heuristic analytic form used from the time of Newton and Leibniz on down to Cauchy, who gave the first formal definition of the integral and proved the result rigorously. Today the preferred formulation is as a pair of theorems.

Dirk Struik summarises the emergence of the Fundamental Theorem of the Calculus as follows:

> The next step [after some small success at finding inverse-tangents] was the recognition that finding quadratures and solving inverse-tangent problems were identical propositions — in other words, the discovery that the integral calculus is the inverse of the differential calculus. Torricelli came to this understanding in his case of generalized parabolas and hyperbolas, satisfying the equation $x\,dy = ky\,dx$... James Gregory (1638 – 1675), the great Scottish mathematician who died so young, seems to have been the first to see the proposition in its generality, though still in a geometric manner... We then find the fundamental theorem in the *Lectiones geometricae* (London, 1670) by Isaac Barrow ... The method is thoroughly geometrical, and this makes it not easy to recognize the importance of Barrow's results.[10]

Barrow's version of the Fundamental Theorem is given in paragraph XI of Lecture X and the geometric style of proof is rather opaque to one approaching the subject from an analytic viewpoint. His proof is probably more intelligible if we explain it analytically. To this end, I reproduce his Fig. 109 (Fig. 3.3) in both its original and an analytically relabelled form.

The graph depicts two curves, ZGEG and VIFI. Thinking of the horizontal line passing through V as the x-axis, with origin at, say, V, the first curve represents the graph of a strictly increasing function $y = f(x)$, the y-axis being the half-line extending VZ downward. Simultaneously one imagines the second curve graphed in the xz-plane superimposed on the first graph, where the positive z-axis is the upper half of the prolongation of ZV from the origin V. The second curve represents the

---

[10]Struik, *op. cit.*, p. 253.

area $z = A(x) = \int_0^x f(x)\,dx$. Thus, if we think of D as at distance x from the origin, the point E is $\langle x, f(x)\rangle$ and F is $\langle x, A(x)\rangle$.

The point T on the x-axis with x-coordinate t is chosen so that

$$x - t = \frac{A(x)}{f(x)} \tag{3.5}$$

and the line FT is drawn. Barrow shows this line to be the tangent to the area curve at F by showing every point on the line other than F to lie strictly to the right of the curve VIFI, i.e., $z = A(x)$. To this end, choose a point P on the x-axis to the left or right of D at $x \pm \delta$. For the sake of definiteness, we consider $x + \delta$. Let

$$\mathrm{I} = \langle x+\delta, A(x+\delta)\rangle, \quad \mathrm{L} = \langle x, A(x+\delta)\rangle, \quad \mathrm{G} = \langle x+\delta, f(x+\delta)\rangle$$

and K the point on the line FT of height $A(x+\delta)$ above the x-axis.

The triangles KLF and TDF are similar, whence

$$\frac{\mathrm{KL}}{\mathrm{TD}} = \frac{\mathrm{LF}}{\mathrm{DF}}.$$

Thus

$$\begin{aligned}\mathrm{KL} = \mathrm{TD}\cdot\frac{\mathrm{LF}}{\mathrm{DF}} &= (x-t)\frac{A(x+\delta)-A(x)}{A(x)}\\ &= \frac{A(x)}{f(x)}\cdot\frac{A(x+\delta)-A(x)}{A(x)} = \frac{A(x+\delta)-A(x)}{f(x)}.\end{aligned}$$

Now

$$A(x+\delta) - A(x) = \text{area of DPEG} > \mathrm{DP}\cdot\mathrm{DE}$$

since $f(x)$ is increasing. Thus

$$\mathrm{KL} = \frac{A(x+\delta)-A(x)}{f(x)} > \frac{\mathrm{DP}\cdot\mathrm{DE}}{f(x)} = \frac{\mathrm{DP}\cdot f(x)}{f(x)} = \mathrm{DP}.$$

But DP = LI, so K lies strictly to the right of I on the curve $z = A(x)$.

The argument for the point P at $x - \delta$ is similar and we see that the line TKFK touches the curve at F, i.e., TKFK is the tangent to the curve in the familiar geometric sense.

For good measure, Barrow also includes Fig. 110 (our Fig. 3.4) which, if flipped horizontally, is the corresponding figure for the case of a strictly decreasing function f.

This is not quite the familiar result learned in the Calculus course. For one thing it does not say explicitly that $dA(x)/dx = f(x)$, but determines the tangent line by finding the geometrically more natural subtangent TD rather than the slope, a common practice in those days — as we saw in the last chapter. I haven't looked

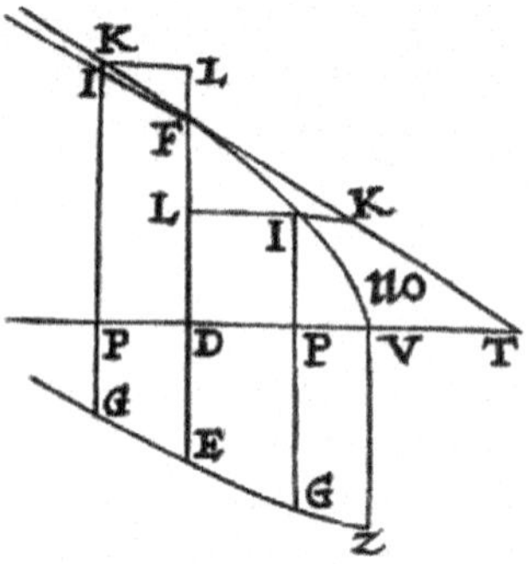

Fig. 3.4 Barrow's diagram for a decreasing function

into this, but I would hazard a guess that emphasis on slopes rather than subtangents emerged from comparisons of rates of change — Newton's $\dot{y}/\dot{x}$ or Leibniz's dy/dx. Geometrically, there is no reason to emphasise the ratio $\dot{y}/\dot{x}$ or Leibniz's dy/dx over $\dot{x}/\dot{y}$ or Leibniz's dx/dy until the functional dependence of y on x came to the fore. In any event, from the defining equation for t (namely, (3.5)), one readily calculates

$$f(x) = \frac{A(x)}{x-t} = \text{slope of TKFK}.$$

And under Newton and Leibniz, this slope would become the derivative.

Also, in the modern formulation, f is not required to be monotone. This would require one to repeat the proof at a maximum or minimum: On one side of D, K would be on the right of the tangent, and on the other K would be to the left — for small enough δ (so that $x \pm \delta$ remained in an interval in which f was monotone). This argument, of course, fails for Darboux's function of Fig. 2.2.27 in Chap. 2.

Newton and Leibniz did not add to the theory behind the Fundamental Theorem of the Calculus, but they transformed it from a geometrical theorem into an analytical tool. For nearly a century and a half, from Leibniz's first publications until Cauchy's rigorisation of Analysis, mathematicians accepted that area was given by integration and defined the integral by antidifferentiation. It was in Cauchy's *Résumé des leçons* that the integral of a continuous function as the limit of a sum was first precisely defined and the Fundamental Theorem of the Calculus was first rigorously proven.

Cauchy's definition of the integral differs from the modern one in two respects. First, and most importantly, he defined the integral directly only for functions uniformly continuous[11] on closed, bounded intervals. Bernhard Riemann (1826–1866) later analysed the situation and characterised the more general class of integrable functions for which the necessary limit existed.

The limit in question is the familiar limit as $\max\{x_{i+1} - x_i \mid i = 0, 1, \ldots, n-1\}$ goes to 0 of sums

[11] Recall that Cauchy's definition of continuity agreed with our modern definition of uniform continuity; he had no concept of ordinary continuuity.

$$\sum_{i=0}^{n-1} f(x_i^*)(x_{i+1} - x_i), \tag{3.6}$$

where $a = x_0 < x_1 < \ldots < x_{n-1} < x_n = b$. Cauchy, in defining $\int_a^b f(x)dx$ took x_i^* to be x_i while Riemann allowed x_i^* to be any element of $[x_i, x_{i+1}]$. For continuous functions, Cauchy's proof that the limit of the sums existed contained within it a proof that the more general *Riemann sums* (3.6) where x_i^* is not restricted to being an endpoint also converged to the same limit. Thus, my above pronouncement that the important difference between Cauchy's and Riemann's definitions of the integral is Cauchy's restriction to the (uniformly) continuous case.

As Cauchy's proof of the existence of the integral of any continuous[12] function on a closed, bounded interval has no bearing on the Mean Value Theorem, I postpone its discussion here and refer the interested reader to pages 281–284 or to the literature.[13]

Cauchy devotes several lessons of the *Résumé des leçons* to the integral. Lesson 21 defines the definite integral, proves its existence and cites

$$\int_{x_0}^{X} a\,dx = a(X - x_0)$$

as an example of the definite integral. Lesson 22 calculates a few definite integrals and states some immediate properties of the definite integral, the most immediately relevant being his formula (19):

$$(19) \qquad \int_{x_0}^{X} f(x)dx = (X - x_0)f[x_0 + \theta(X - x_0)], \text{ where } 0 \leq \theta \leq 1.$$

This is known as the Mean Value Theorem for Integrals.

3.1.8 Theorem (Mean Value Theorem for Integrals) *Let $f : [a, b] \to \mathbb{R}$ be continuous. There is a $c \in (a, b)$ such that*

$$\int_a^b f(x)dx = f(c)(b - a).$$

Proof. Because f is continuous it attains, by the Extreme Value Theorem, minimum and maximum values m and M, respectively, at some points $\alpha, \beta \in [a, b]$. Now any Riemann sum satisfies

[12]Recall the result first published by Heine in 1872: on a closed, bounded interval $[a, b]$ continuity implies uniform continuity.

[13]Cauchy's proof is quite interesting and intelligible. It rests on Note II of Cauchy, *Cours*, *op. cit.*, and is given in full in Lesson 21 of Cauchy, *Résumé*, *op. cit.* English translations can be found in Bradley and Sandifer, *op. cit.*, pp. 291–307 for the Note and Stedall, *op. cit.*, pp. 440–444 for the integral itself. Cates, *op. cit.*, includes two of the notes from the *Cours* in the *Résumé*, the relevant portion of Note II being given on pp. 188–189.

$$\sum_{i=0}^{n-1} m(x_{i+1} - x_i) \leq \sum_{i=0}^{n-1} f(x_i^*)(x_{i+1} - x_i) \leq \sum_{i=0}^{n-1} M(x_{i+1} - x_i),$$

i.e.,

$$m(b-a) = m\sum_{i=0}^{n-1}(x_{i+1} - x_i) \leq \sum_{i=0}^{n-1} f(x_i^*)(x_{i+1} - x_i)$$
$$\leq M\sum_{i=0}^{n-1}(x_{i+1} - x_i) = M(b-a). \tag{3.7}$$

Thus, taking the limit,

$$m \leq \frac{1}{b-a}\int_a^b f(x)dx \leq M.$$

If $m = M$, then $f(c) = m = M$ for all $c \in [a, b]$ and one can choose c to be any element of (a, b). If $m < M$, then by the Intermediate Value Theorem there is some c strictly between α and β such that

$$f(c) = \frac{1}{b-a}\int_a^b f(x)dx. \qquad \square$$

3.1.9 Example To see that the continuity of f is required, consider the simple example,

$$f(x) = \begin{cases} x, & 0 \leq x \leq 1 \\ 1 + x, & 1 < x \leq 2. \end{cases}$$

The function is integrable in the modern sense and one has

$$\begin{aligned}\int_0^2 f(x)dx &= \int_0^1 x\,dx + \int_1^2 (1+x)dx \\ &= \frac{x^2}{2}\bigg|_0^1 + \left(x + \frac{x^2}{2}\right)\bigg|_1^2 \\ &= \frac{1}{2} - 0 + \left(2 + \frac{4}{2} - 1 - \frac{1}{2}\right) = 3.\end{aligned}$$

And the average value of f is thus

$$\frac{\int_0^2 f(x)dx}{2-0} = \frac{3}{2},$$

but the range of f is

$$f\big([0, 1]\big) \cup f\big((1, 2]\big) = [0, 1] \cup (2, 3]$$

and $3/2$ is not in the range of f, whence the Mean Value Theorem for Integrals fails for f.

Today we would use the Fundamental Theorem to derive the Mean Value Theorem for Integrals. Cauchy does the opposite in Lesson 26 of the *Résumé des leçons*.[14]

3.1.10 Theorem (Fundamental Theorem of the Calculus) *Let* $f : [a, b] \to \mathbb{R}$ *be continuous and define*

$$F(x) = \int_a^x f(x)dx.$$

F is differentiable for all $x \in [a, b]$ *and* $F'(x) = f(x)$.

Proof. Observe, for $x, x + h \in [a, b]$,

$$\begin{aligned} F(x+h) - F(x) &= \int_a^{x+h} f(t)dt - \int_a^x f(t)dt = \int_x^{x+h} f(t)dt \\ &= (x + h - x)f(c) = h \cdot f(c), \end{aligned}$$

for some c between x and $x + h$. Thus

$$\frac{F(x+h) - F(x)}{h} - f(x) = f(c) - f(x).$$

By the continuity of f, $f(c) - f(x)$ can be made as small as we please by choosing h sufficiently small, i.e., given $\epsilon > 0$ there is $\delta > 0$ such that for $h < \delta$,

$$\left| \frac{F(x+h) - F(x)}{h} - f(x) \right| = |f(c) - f(x)| < \epsilon. \qquad \square$$

As I said, today we can reverse the dependence and derive the Mean Value Theorem for Integrals from the Fundamental Theorem: For $F(x) = \int_a^x f(t)dt$, we have

$$\begin{aligned} \int_a^b f(x)dx &= F(b) = F(b) - 0 = F(b) - F(a) \\ &= F'(c)(b-a), \text{ applying the Mean Value Theorem to } F \\ &= f(c)(b-a). \end{aligned}$$

[14] An English translation of this part of Lesson 26 is given in Stedall, *op. cit.*, p. 446 and Cates, *op. cit.*, p. 100.

Some authors refer to the Fundamental Theorem of the Calculus as the *First* Fundamental Theorem of the Calculus and pair it with the corollary:

3.1.11 Theorem (Second Fundamental Theorem of the Calculus) *Let $f : [a, b] \to \mathbb{R}$ be continuous and let F be any antiderivative of f, i.e., $F'(x) = f(x)$ for all $x \in [a, b]$. Then*

$$\int_a^b f(x)dx = F(b) - F(a).$$

Proof.[15] Let $G(x) = \int_a^x f(t)dt$. By the Fundamental Theorem, G is an antiderivative of f, and, as we just saw,

$$\int_a^b f(x)dx = G(b) - G(a).$$

If F is any antiderivative of f on $[a, b]$, then

$$(F - G)'(x) = F'(x) - G'(x) = f(x) - f(x) = 0.$$

But, using the Mean Value Theorem, we have shown in Corollary 2.3.31 in Chap. 2, above, that a function with derivative equal to 0 throughout an interval must be constant. Thus, for some C, $F(x) - G(x) = C$, i.e., $F(x) = G(x) + C$ and

$$F(b) - F(a) = \big(G(b) + C\big) - \big(G(a) + C\big) = G(b) - G(a) = \int_a^b f(x)dx. \quad \square$$

Alternate Proof. Partition the interval $a = x_0 < x_1 < \ldots < x_{n-1} < x_n = b$ and observe

$$\begin{aligned} F(b) - F(a) &= F(x_n) - F(x_{n-1}) + F(x_{n-1}) - F(x_{n-2}) + \ldots + F(x_1) - F(x_0) \\ &= \sum_{i=0}^{n-1} \big(F(x_{i+1}) - F(x_i)\big) \\ &= \sum_{i=0}^{n-1} (x_{i+1} - x_i)F'(x_i^*), \quad x_i < x_i^* < x_{i+1}, \end{aligned}$$

by the Mean Value Theorem,

$$= \sum_{i=0}^{n-1} f(x_i^*)(x_{i+1} - x_i).$$

---

[15] Cauchy gives this proof in Lesson 26 following his proof of the First Theorem. Oddly enough Stedall does not translate this part of the lesson. Thus I refer the interested reader to the original French of Cauchy, *Résumé*, *op. cit.*, pp. 102–104, or to Cates, *op. cit.*, pp. 102–103.

Thus $F(b) - F(a)$ is a Riemann sum for the given partition. But when we let $\max\{x_{i+1} - x_i \mid i = 0, 1, \ldots n-1\}$ go to 0, the Riemann sums tend to the integral. Thus

$$F(b) - F(a) = \int_a^b f(x)dx. \qquad \square$$

Both proofs yield the Second Theorem as a corollary to the Mean Value Theorem. Conversely, the Mean Value Theorem for continuously differentiable functions can be derived from the Second Fundamental Theorem and the Mean Value Theorem for Integrals: Suppose f' is continuous. Then

$$\begin{aligned} f(b) - f(a) &= \int_a^b f'(x)dx, \text{ by Theorem 3.1.11} \\ &= f'(c)(b-a), \end{aligned}$$

for some $c \in (a, b)$ by the Mean Value Theorem for Integrals, the proof of which depended on the Intermediate Value Theorem and not the Mean Value Theorem.

There is one final result of Cauchy's concerning integration that should be mentioned here, in part because its proof is not too dissimilar to the alternate proof of the Second Fundamental Theorem of the Calculus and in part because it depends on a Mean Value Theorem — the Mean Value Theorem for *Areas*. This is his proof in Lesson 23, two lessons before that in which he proves the Fundamental Theorems, that the area under a continuous curve is given by the definite integral.

> Imagine now that, the limit X is superior to x_0, the function $f(x)$ is positive from $x = x_0$ to $x = X$, x, y designate rectangular coordinates, and A the surface[16] included on the one hand between the x-axis and the curve $y = f(x)$, on the other between the ordinates $f(x_0), f(X)$.[17] This surface, which has for base the length $X - x_0$ reckoned over the x-axis, will be a mean[18] between the areas of the two rectangles constructed over the base $X - x_0$ with the respective heights equal to the least and the greatest ordinates above the various points of this base. It will thus be equivalent to a rectangle constructed over a mean ordinate represented by an expression of the form $f[x_0 + \theta(X - x_0)]$; so that we have
>
> (8) $\qquad A = (X - x_0)f[x_0 + \theta(X - x_0)],$
>
> θ denoting a number inferior[19] to unity. If we divide the base $X - x_0$ into very small elements $x_1 - x_0\, x_2 - x_1\, \ldots\, X - x_{n-1}$, the surface A finds itself divided into corresponding elements of which its values will be given by equations similar to that of formula (8). We thus have again
>
> (9) $A = (x_1 - x_0)f[x_0 + \theta_0(x_1 - x_0)] + (x_2 - x_1)f[x_1 + \theta_1(x_2 - x_1)] + \ldots$
>
> $\qquad \ldots\ldots\ldots\ldots\ldots + (X - x_{n-1})f[x_{n-1} + \theta_{n-1}(X - x_{n-1})],$

---

[16] I.e., two dimensional region.

[17] I.e., between the vertical lines $x = x_0$ and $x = X$. The f's are clearly typos.

[18] A *mean* of a set S of numbers is, according to Cauchy (cf. Note II of Bradley and Sandifer, *op. cit.*, pp. 291 – 307, and Cates, *op. cit.*, p. 188 – 189), any number lying between $\min(S)$ and $\max(S)$.

[19] Cauchy was a little sloppy distinguishing between $\leq$ and $<$.

$\theta_0, \theta_1 \ldots \theta_{n-1}$ denoting numbers inferior to unity. If then in this last equation we make the numerical values[20] of the elements of $X - x_0$[21] decrease indefinitely, we obtain, in passing to the limit,

$$(10) \qquad A = \int_{x_0}^{X} f(x)dx.^{22}$$

The argument does not prove conclusively that the area trapped between the limits a and b is the definite integral $\int_a^b f(x)dx$ because there is no accompanying proof that the figure has any area at all. And there is no such proof because he hasn't defined the geometric notion of area. However, he has shown the next best thing: If we can define area in such a manner that the area of a subset X of a given set Y does not exceed that of Y (provided both areas exist), the area of a union $X \cup Y$ of two sets intersecting in a set of area 0 is the sum of the areas of the sets X, Y, and the areas of rectangles are computed in the usual manner, then the only possible value for the area under a continuous curve is given by the definite integral.

3.1.3 Applications of the Mean Value Theorem

In Sect. 3.5, below, we will discuss Cauchy's proofs of the variants of the Mean Value Theorem given in the *Résumé des leçons*. It was in this book that the Mean Value Theorem was made a cornerstone of the Calculus, as Cauchy applied it repeatedly in providing rigorous proofs of many of the basic theorems of the Calculus. We have just seen its use in proving the Fundamental Theorem of the Calculus, and in Chap. 2 we used it to obtain quick proofs of some seemingly simple results, mainly Corollaries 2.3.21, 2.3.24, and 2.3.31 in Chap. 2, which, simple as they are, deserve, along with the results of Exercise 2.3.25 in Chap. 2, to be restated here with some fanfare:

3.1.12 Theorem (Strictly Increasing Function Theorem) *Let $f : [a, b] \to \mathbb{R}$ be continuous, differentiable on (a, b), and suppose $f'(x) > 0$ for all $x \in (a, b)$. Then f is strictly increasing on $[a, b]$,*

$$x < y \Rightarrow f(x) < f(y).$$

3.1.13 Theorem (Increasing Function Theorem) *Let $f : [a, b] \to \mathbb{R}$ be continuous, differentiable on (a, b), and suppose $f'(x) \geq 0$ for all $x \in (a, b)$. Then f is increasing on $[a, b]$: for all $x, y \in [a, b]$,*

$$x < y \Rightarrow f(x) \leq f(y).$$

---

[20] I.e., absolute values.

[21] I.e., all the differences $x_{i+1} - x_i$.

[22] Cauchy, *Résumé*, *op. cit.*, p. 91; Cates, *op. cit.*, p. 91.

3.1.14 Theorem (Strictly Decreasing Function Theorem) *Let $f : [a, b] \to \mathbb{R}$ be continuous, differentiable on (a, b), and suppose $f'(x) < 0$ for all $x \in (a, b)$. Then f is strictly decreasing on $[a, b]$*

$$x < y \Rightarrow f(x) > f(y).$$

3.1.15 Theorem (Decreasing Function Theorem) *Let $f : [a, b] \to \mathbb{R}$ be continuous, differentiable on (a, b), and suppose $f'(x) \leq 0$ for all $x \in (a, b)$. Then f is decreasing on $[a, b]$: for all $x, y \in [a, b]$,*

$$x < y \Rightarrow f(x) \geq f(y).$$

3.1.16 Theorem (Constant Function Theorem) *Let $f : [a, b] \to \mathbb{R}$ be continuous, differentiable on (a, b), and suppose $f'(x) = 0$ for all $x \in (a, b)$. Then f is constant on $[a, b]$: for all $x, y \in [a, b]$, $f(x) = f(y)$.*

The direct proofs of these results given on pages 131–136 were a bit complicated and indirect. The proofs via appeal to the Mean Value Theorem are more straightforward and unified: one takes $x < y$ and finds $c \in (x, y)$ such that $f(x) - f(y) = (x - y)f'(c)$. Then

$$f(x) \begin{Bmatrix} > \\ \geq \\ = \\ \leq \\ < \end{Bmatrix} f(y) \text{ iff } f'(c) \begin{Bmatrix} > \\ \geq \\ = \\ \leq \\ < \end{Bmatrix} 0.$$

Theorem 3.1.13 has been put forward as a replacement for the Mean Value Theorem in the first year Calculus course.[23] The first four of these theorems are easily shown equivalent. The equivalences 3.1.12⇔3.1.14 and 3.1.13⇔3.1.15 follow by reducing each theorem applied to a function f to the equivalent applied to $g(x) = -f(x)$. The reader has demonstrated the implication 3.1.12⇒3.1.13 in Exercise 2.3.25 back in Chap. 2, and the converse implication is simple enough: Let $f'(x) > 0$ on an interval I and let $a, b \in I$ with $a < b$.

$$\forall x\, f'(x) > 0 \Rightarrow \forall x\, f'(x) \geq 0,$$

whence for any c Theorem 3.1.13 yields

$$a < c < b \Rightarrow f(a) \leq f(c) \leq f(b).$$

[23]Cf., e.g., Thomas W. Tucker, "Rethinking rigor in Calculus: the role of the Mean Value Theorem", *The American Mathematical Monthly* 104 (1997), pp. 231–240. We discuss this paper in Chap. 4, below.

But, if $f(a) = f(b)$, it follows that f is constant on $[a, b]$, whence $f'(c) = 0$ for all $c \in (a, b)$, contrary to the assumption that $f'(x) > 0$ for all $x \in I$. Thus

$$a < b \Rightarrow f(a) < f(b).$$

Moreover, Theorem 3.1.16 follows readily from any of the first four theorems. For, their all being equivalent, Theorems 3.1.13 and 3.1.15 will both hold. If $f'(x) = 0$ for all $x \in I$, then $f'(x) \geq 0$ and

$$x < y \Rightarrow f(x) \leq f(y),$$

but we also have $f'(x) \leq 0$ for all $x \in I$, whence

$$x < y \Rightarrow f(y) \leq f(x).$$

Thus

$$x < y \Rightarrow f(x) \leq f(y) \& f(y) \leq f(x) \Rightarrow f(x) = f(y).$$

Simple as it is, Theorem 3.1.16 is supremely important. It embodies the use of the Mean Value Theorem in proving the Second Fundamental Theorem of the Calculus, by which any two antiderivatives to a function differ by a constant. And this is what allows us to evaluate definite integrals by searching for antiderivatives. I.e., the algorithmic nature of integration, insofar as it is algorithmic, derives from this Theorem — and thus from the Increasing Function Theorem.

Related to the various Increasing, Decreasing, and Constant Function Theorems are two more, initially less impressive results that should be mentioned here. The first is the irrelevantly named *Racetrack Principle*.

3.1.17 Theorem (Racetrack Principle) *Let f, g be continuous on $[a, b]$ and differentiable on (a, b). Suppose $f'(x) \leq g'(x)$ for all $x \in (a, b)$. Then, for all $x \in [a, b]$, $f(x) - f(a) \leq g(x) - g(a)$.*

One motivates this by saying that, if two cars are racing around a track and the first car is always moving at least as fast as the second, then, at the end of a specified amount of time, the first car will have travelled at least as far as the second. The Racetrack Principle follows easily from the Increasing Function Theorem: for $h(x) = g(x) - f(x)$, one has $h'(x) = g'(x) - f'(x) \geq 0$ for all $x \in (a, b)$, whence h is weakly increasing:

$$\begin{aligned} x \in [a, b] &\Rightarrow a \leq x \Rightarrow h(a) \leq h(x) \\ &\Rightarrow g(a) - f(a) \leq g(x) - f(x) \\ &\Rightarrow f(x) - f(a) \leq g(x) - g(a). \end{aligned}$$

3.1.18 Exercise The Racetrack Principle is stated for weak inequality. Formulate and prove a corresponding principle using the strict inequality.

The second result has been called the *Mean Value Inequality* and the *Law of Bounded Change*.

3.1.19 Theorem (Mean Value Inequality) *Let f be continuous on $[a, b]$ and differentiable on (a, b). Suppose $m \leq f'(x) \leq M$ for all $x \in (a, b)$. Then, for all $x \in [a, b]$,*

$$m(x-a) \leq f(x) - f(a) \leq M(x-a).$$

Proof. Let $h(x) = Mx - f(x)$. Then $h'(x) = M - f'(x) \geq 0$, whence h is increasing:

$$\begin{aligned} x \in [a, b] &\Rightarrow a \leq x \Rightarrow h(a) \leq h(x) \\ &\Rightarrow Ma - f(a) \leq Mx - f(x) \\ &\Rightarrow f(x) - f(a) \leq Mx - Ma = M(x-a). \end{aligned}$$

Similarly, one shows $m(x-a) \leq f(x) - f(a)$. □

The Increasing Function Theorem, Racetrack Principle, and Mean Value Inequality have each been proposed as replacements for the Mean Value Theorem in the first year Calculus course on the grounds that they yield most of the applications of the Mean Value Theorem as easily as does the Mean Value Theorem and they are more intuitive and thus more readily absorbed by students of the course. This is really a point for discussion in the next chapter and I only mention it now to explain why such apparent trivialities are emphasised here.

3.1.20 Digressive Rant (on Labelling) *I do not care for the name "Racetrack Principle", nor, indeed, for the habit of naming every result in sight, and, insofar as the issue of pædagogy has been raised, I think this the perfect opportunity to speak out. First, we name theorems to single them out as milestones, telling students that such results are important. A name says "Remember this". When I was first starting out, I recall reading articles bemoaning the fact that students were getting by on rote memorisation, and now we have reversed positions and are actively encouraging students to memorise more and more.*

Be that as it may, perhaps more deplorable is the lack of any rational naming policy in mathematics. The result is a multiplicity of names for some results and objects, duplicate names for disparate objects and results, inaccurate eponymous names, and misleading or irrelevant names. Just about every eponymous name is inaccurate: Pythagoras was certainly not the first to discover the Pythagorean Theorem, Rolle proved nothing like Rolle's Theorem, Pell had little to do with the Pell equation, Riemann sums can already be found in Cauchy, Cauchy convergence had been used by Euler, da Cunha, and Bolzano before Cauchy introduced it etc. etc. Some mathematical theorems are correctly named after their discoverers, and some are named as tributes, but often these names are based on nothing more than errors of attribution.

Unless, however, a name is so firmly entrenched in the literature that changing it is likely to do the student a disservice (e.g., calling the Pythagorean Theorem by a name the student will never see again), expositors ought to think carefully on their nomenclature. Already widely used names are preferable to novelties, unless the novel name is so perfectly succinct and descriptive as to be recognisably superior to anything else in use. And a mathematically or historically descriptive name is better than a fleetingly fashionable metaphor. The "Mean Value Theorem for Integrals" is ideal: The value $f(c)$ *in the equation*

$$f(c) = \frac{1}{b-a}\int_a^b f(x)dx$$

is indeed an average, or mean, of the function on the given interval. The "Mean Value Theorem for Derivatives", being, as we saw, intimately connected with the Mean Value Theorem for Integrals, is a good name. Its abbreviated form, the "Mean Value Theorem", is less descriptive but betrays its history and the leading role it has played among various mean value theorems. And it is more descriptive than the old-fashioned "Law of the Mean", which might remind one more of the Central Limit Theorem in Probability Theory than of the Mean Value Theorem.

Because of the histories of the concepts, I have no objection to the names "Monte Carlo Methods" or the "Monty Hall Problem" in Probability Theory, but the "Race-track Principle" has no justification whatsoever. It is pandering to the audience, an insulting declaration that the student is incapable of appreciating the result on its own grounds and announcing that the expositor is condescending to speak down to the level of his students or readers. The "Mean Value Inequality" is, to anyone familiar with any of the Mean Value Theorems, an appropriate designation and even mildly suggestive. Likewise the "Law of Bounded Change" refers, if not very precisely, to what the theorem is about.

3.1.21 Exercise Grab a Calculus textbook and go through it, deciding which theorems are well-named and which are stupidly named. Are there any unnamed theorems that could be given useful names? Do this with older and newer textbooks. Has the situation always been the same or has there been a change, and, if the latter is the case, what is the trend?

Now that I've got that off my chest, it is time to get back on track — to return to our discussion of applications of the Mean Value Theorem. In addition to furthering theory by providing rigorous proofs of results such as those we have been discussing, the Mean Value Theorem is also useful in applications. The two types of applications stressed in the Calculus course are in estimating error and establishing L'Hôpital's Rule for the evaluation of certain limits.

Error estimates can be straightforward applications of the Mean Value Inequality, or they can be the more subtle ϵ-δ calculations the students should have struggled with, with only partial success, earlier in the course. The point now is not, of course, to give ϵ-δ proofs of continuity — to apply the Mean Value Theorem in the first place one must assume much more than mere continuity. Rather, it is to show how close the independent variable must be to some specified value (within δ) in order to limit the error (less than ϵ) in the resulting value of the dependent variable. In the spirit of the "Racetrack Principle", we might call it the "Wobbly Joystick Problem": Determine the allowable amount of play (up to δ) a control device may have and still perform reliably (to within ϵ).

Insofar as this is a book about the Mean Value Theorem, some examples of the Wobbly Joystick Problem ought to be discussed. However, insofar as this is a book about the theory of the Calculus and not about the Calculus per se, such examples would not move our overall discussion forward unless for the sake of the next chapter I wanted to demonstrate as well how the Increasing Function Theorem et alia also solved the problem. But I think that, even there the most die-hard fan of the Mean Value Theorem will be willing to concede the applicability of the stated alternative principles. Hence we shall skip this problem and jump ahead to L'Hôpital's Rule.[24]

L'Hôpital's Rule was discovered by Johann Bernoulli, who along with his brother Jakob read Leibniz's first publications on the Calculus and contributed greatly to the early development of the field. Guillaume François Antoine Marquis de l'Hôpital (1661–1704), whose name is often translated into English as l'Hospital, was a French aristocrat and mathematician who wrote the first printed textbook on the Calculus, *Analyse des infiniment petits pour l'intelligence des lignes courbes* (1696) [*Analysis of the Infinitely Small for the Understanding of Curves*]. L'Hôpital and Bernoulli had an arrangement whereby the former paid the latter and acquired the right to include the latter's results in his textbook. Although he acknowledged the authorship of many of the results in general terms, he did not specify individually which results were his own and which were Bernoulli's. Thus, one of Bernoulli's contributions has come down to us as *L'Hôpital's Rule*.

By way of introduction to the Rule, I quote (in translation from a later edition of l'Hôpital's book[25]) l'Hôpital's statement of the problem, his solution, and his examples of its application.

SECTION IX.

Solution to some problems which depend on the preceding methods.

PROPOSITION I.

PROBLEM.

163. LET *a curved line* AMD (*Fig.* 130. *Pl.* 7 [See Fig. 3.5.]) (AP = x, PM = y, AB = a) *be such that the value of the applicand*[26] y *is expressed by a fraction, of which the numerator & denominator each becomes zero when* x = a, *that is, when the point* P *falls upon the given point* B. *We want the value of the applicand* BD *at that time.*

Suppose given two curved lines ANB, COB, which have the line AB as a common axis, & which are such that the applicand PN expresses the numerator, & the applicand PO the

[24] A good recent account of L'Hôpital's Rule, which I only learned about after completing this book, is: Horst Struve and Ingo Witzke, "Die Regel von l'Hôpital", *Elemente der Mathematik* 69 (2014), pp. 118–129.

[25] Marquis de l'Hôpital, *Analyse de infiniment petits*, Paris, 1768.

[26] The French original is *appliquée*. Struik, *op. cit.*, translates this more mathematically as "ordinate".

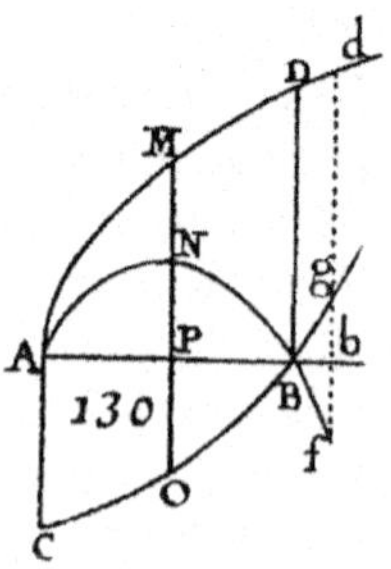

Fig. 3.5 L'Hôpital's diagram

denominator of the general fraction which always matches PM: so that PM $= \frac{\text{AB} \times \text{PN}}{\text{PO}}$. It is clear that the two curves will intersect at the point B; by the supposition that PN & PO each becomes zero, when the point P falls on B. That being granted, if one imagines an applicand *bd* infinitely close to BD, & which intersects the curved lines ANB, COB at points f, g; one has $bd = \frac{\text{AB} \times bf}{bg}$, which (*Art.* 2.) does not differ from BD. There is thus no question other than that of finding the ratio of bg to bf. Now it is obvious that the segment AP becoming AB, the applicands PN, PO become null, & that AP becoming Ab, they become bf, bg. Whence it follows that the applicands bf, bg themselves are the differentials[27] of the applicands on B & b with respect to the curves ANB, COB; & consequently if one takes the differential of the numerator, & divides by the differential of the denominator, after making x = a = Ab or AB, one will have the sought after value of the applicand *bd* or BD. Which was to be found.

EXAMPLE I.

164. LET $y = \frac{\sqrt{2a^3x - x^4} - a\sqrt[3]{aax}}{a - \sqrt[4]{ax^3}}$. It is clear that when $x = a$, the numerator & the denominator of the fraction each becomes equal to zero. Therefore one will take the differential $\frac{a^3dx - 2x^3dx}{\sqrt{2a^3x - x^4}} - \frac{aadx}{3\sqrt[3]{axx}}$ of the numerator, & one then will divide by the differential $-\frac{3adx}{4\sqrt[4]{a^3x}}$ of the denominator, that is, after setting $x = a$, one will divide $-\frac{4}{3}adx$ by $-\frac{3}{4}dx$; which gives $\frac{16}{9}a$ for the sought after value of BD.

EXAMPLE II.

165. LET $y = \frac{aa - ax}{a - \sqrt{ax}}$. One gets $y = 2a$, when $x = a$.

We can resolve this example without the need of the differential calculus, in this way.

[27]The French is *différence*, which translates more literally to difference. "Differential" conveys the infinitesimality of the difference more clearly.

> Having removed the incommensurables,[28] one obtains $aaxx + 2aaxy - axyy - 2a^3x + a^4 + aayy - 2a^3y = 0$, which being divided by $x - a$, will reduce to $aax - a^3 + 2aay - ayy = 0$; & substituting a for x, it comes as before to $y = 2a$.[29]

Some clarification may be in order. L'Hôpital is still in that transitional period between geometry and analysis. His statement of the problem and determination of the solution are semi-geometric, while the examples he gives are presented algebraically/analytically. His diagram, like Barrow's Fig. 109 (our Fig. 3.3), does not follow our modern conventions. The curves ANB, COB, and AMD represent the graphs of the functions $f(x)$, $g(x)$, and $f(x)/g(x)$, respectively. Taking AB for the x-axis, today we would read $f(x)$ as positive between A and B, $g(x)$ as negative there, and wonder why AMD is positive throughout when the ratio is clearly negative. His convention, like Barrow's, was different and we can again imagine the line AC extended in both directions as representing the positive y-axis above AB and the positive z-axis below AB with the graphs of $y = f(x)$ and $z = g(x)$ thus drawn in the same picture. To this he adds the graph of $y = f(x)/g(x)$.

The points P, B, b[30] denote points with x-coordinates $x, a, a+h$, respectively, where h is an infinitesimal. The y-coordinates of N, B, f are the values $f(x), f(a)$, $f(a+h)$, respectively; the z-coordinates of O, B, g are $g(x)$, $g(a)$, $g(a+h)$, respectively; and the y-coordinates of M, D, d are $f(x)/g(x), f(a)/g(a), f(a+h)/g(a+h)$, respectively. L'Hôpital's argument is essentially that, by the infinitesimality of h, BD and bd can be ad-equated:

$$\begin{aligned}
\mathrm{BD} \sim \mathrm{bd} &= \frac{f(a+h)}{g(a+h)} = \frac{f(a+h)-f(a)}{g(a+h)-g(a)}, \quad \text{since } f(a) = g(a) = 0 \\
&= \frac{f(a+h)-f(a)}{h} \cdot \frac{h}{g(a+h)-g(a)} \\
&= \frac{f(a+h)-f(a)}{h} \Big/ \frac{g(a+h)-g(a)}{h} \\
&\sim \frac{f'(a)}{g'(a)},
\end{aligned}$$

because h is taken to be infinitesimal.

---

28 I.e., having solved for $\sqrt{ax}$ and squared the result.

29 L'Hôpital, *op. cit.*, pp. 206–208. The typography of the original is inconsistent and I have tried to preserve some of the flavour of the standards, or lack thereof, in the original. The use of italics is inconsistent, the letters b, d, f, g being italicised in the text but not in the diagram and the equation $x = a$ occurring in both forms in the text. The treatment of the initial words in articles 163, 164, and 165 varies. In 163, the initial letter is a large drop cap, only the size reproduced here, the second letter is an ordinary capital, and the rest of the word (*Soit* in the French) is given in small caps. In 164, the initial letter is equally large, but rests on the base line, and the rest of the word is in small caps. And in 165 the initial letter is still larger than normal, but not as large as in the previous two articles, it is not vertically centred on the line, but is not a drop cap. Again the rest of the word is given in small caps.

30 Should I follow the lead of his diagram and write b, f, g, d for the points or follow the text and write b, f, g, d? I have decided to use plain text as I use f, g for the functions.

In the modern Calculus course, L'Hôpital's Rule is proven as an application of the Cauchy Mean Value Theorem. One would expect to find this proof in the *Résumé des leçons* and, indeed, the proof is there — but not where one might expect. Cauchy's initial discussion in Lesson 6 precedes his presentation of the Mean Value Theorem. The formulation is purely analytic, but it is still basically the same as l'Hôpital's:

> 4.th Problem. *We want the true value of a fraction of which the two terms are functions of the variable x, in the case where we assign to this variable a particular value, for which the fraction is presented by the indeterminate form* $\frac{0}{0}$.
>
> *Solution.* Let $s = \dfrac{z}{y}$ be the proposed fraction, y and z designating functions of the variable x, and suppose that the particular value $x = x_0$ reduces this fraction to the form $\frac{0}{0}$, that is to say, that it makes y and z vanish. If we represent by Δx, Δy, Δz the infinitely small increments and simultaneously by x, y, z the three variables, we have, for any value of x whatever,
>
> $$s = \frac{z}{y} = \lim \frac{z + \Delta z}{y + \Delta y},$$
>
> and, for the particular value $x = x_0$,
>
> $$s = \lim \frac{\Delta z}{\Delta y} = \frac{dz}{dy} = \frac{z'}{y'}. \tag{3}$$
>
> Thus, the value sought of the fraction s or $\dfrac{z}{y}$ coincides generally with the ratio $\dfrac{dz}{dy}$ or $\dfrac{z'}{y'}$.
>
> *Examples.* We have, for $x = 0$, $\dfrac{\sin x}{x} = \dfrac{\cos x}{1} = 1$, $\dfrac{l(1+x)}{x} = \dfrac{1}{1+x} = 1$[31]; for $x = 1$, $\dfrac{l(x)}{x-1} = \dfrac{1}{x} = 1$, $\dfrac{x-1}{x^n-1} = \dfrac{1}{nx^{n-1}} = \dfrac{1}{n}$; &c....[32]

Today, we would state and prove the result analytically, bringing out more explicitly the assumptions made about f, g and replacing the appeal to infinitesimals by reference to limits.

3.1.22 Theorem (L'Hôpital's Rule; Urform) *Let f, g be differentiable at a, with $f(a) = g(a) = 0$. Suppose $g'(a) \neq 0$. Then*

$$\lim_{x \to a} \frac{f(x)}{g(x)} = \frac{f'(a)}{g'(a)}. \tag{3.8}$$

Proof. Because $g'(a) \neq 0$, it follows that $g(a+h) \neq 0$ for h sufficiently small. Observe,

$$\begin{aligned}\lim_{x \to a} \frac{f(x)}{g(x)} &= \lim_{h \to 0} \frac{f(a+h)}{g(a+h)} = \lim_{h \to 0} \frac{f(a+h) - f(a)}{g(a+h) - g(a)}, \quad \text{since } f(a) = g(a) = 0 \\ &= \lim_{h \to 0} \left(\frac{f(a+h) - f(a)}{h} \cdot \frac{h}{g(a+h) - g(a)} \right)\end{aligned}$$

---

[31] Cauchy uses $l(x)$ for $\ln x$.

[32] Cauchy, *op.cit.*, p. 24.; Cates, *op. cit.*, pp. 20–21.

$$= \lim_{h\to 0} \frac{f(a+h)-f(a)}{h} \Big/ \lim_{n\to 0} \frac{g(a+h)-g(a)}{h}$$
$$= \frac{f'(a)}{g'(a)}. \qquad \square$$

Bernoulli and l'Hôpital would not have specified the differentiability of f, g at a as all functions were believed differentiable at all but some isolated points. Cauchy may have believed the same, but his definition of differentiability gives uniform differentiability in a neighbourhood. Thus, he would have cited the more stringent condition that f, g were uniformly differentiable in some interval containing a.

The Urform of L'Hôpital's Rule is, as the name implies, not the result one learns and applies in the Calculus.

3.1.23 Example Consider

$$\lim_{x\to 0} \frac{\sin x - x}{x^2}.$$

Here, $f(x) = \sin x - x$, $g(x) = x^2$, $a = 0$ and, indeed, $f(a) = g(a) = 0$. We want to say

$$\lim_{x\to 0} \frac{\sin x - x}{x^2} = \frac{f'(0)}{g'(0)}.$$

But $g'(x) = 2x$ and thus $g'(0) = 0$ and Theorem 3.1.22 does not apply. However, $f'(x) = \cos x - 1$ and $f'(0) = 0$ as well, and we can look at

$$\lim_{x\to 0} \frac{f'(x)}{g'(x)} = \frac{f''(0)}{g''(0)} = \frac{-\sin 0}{2} = 0.$$

Can we conclude

$$\lim_{x\to 0} \frac{f(x)}{g(x)} = \lim_{x\to 0} \frac{f'(x)}{g'(x)} = 0?$$

3.1.24 Exercise Analyse the problem for $f(x) = \sin x - x$, $g(x) = x\cos x - x$.

Cauchy handles ratios like those of the Example and the Exercise in an appendix to the *Résumé des leçons*. First, he derives, for functions f, F continuously differentiable on $[x_0, X]$ the formula

$$(1) \qquad \frac{f(X)-f(x_0)}{F(X)-F(x_0)} = \frac{f'[x_0+\theta(X-x_0)]}{F'[x_0+\theta(X-x_0)]},$$

where θ is some number "inferior to unity" and $F(x)$ is strictly[33] increasing or strictly decreasing on (x_0, X). Following this he generalises the Urform of L'Hôpital's Rule:

[33]Cauchy says "always". His statement of the conditions are slightly inadequate: In his later book, *Leçons sur le calcul differentiel*, de Bure, Paris, 1829, p. 33, he adds the condition that $F'(x)$ not change sign in the interval, whence $F'(x)$ is never 0 on $[x_0, X]$.

If one makes $X = x_0 + h$, Eq. (1) will become

$$(2) \qquad \frac{f(x_0+h)-f(x_0)}{F(x_0+h)-F(x_0)} = \frac{f'[x_0+\theta h]}{F'[x_0+\theta h]}.$$

This last, which includes, as a particular case, Eq. (6)[34] of the 7.th Lesson, is susceptible of several important applications; we will demonstrate this through a few remarks.

Imagine first that the functions $f(x)$ and $F(x)$ both vanish for $x = x_0$ and set, for brevity, $\theta h = h_1$. In this case, we obtain from formula (2)

$$(3) \qquad \frac{f(x_0+h)}{F(x_0+h)} = \frac{f'(x_0+h_1)}{F'(x_0+h_1)},$$

h_1 being a quantity of the same sign as h, but of a lesser numerical value.[35] If the functions

$$f(x), f'(x), f''(x), \ldots f^{(n-1)}(x),$$
$$F(x), F'(x), F''(x), \ldots F^{(n-1)}(x),$$

all vanish for $x = x_0$, and remain continuous, as well as[36] $f^{(n)}$ and $F^{(n)}$, between the limits $x = x_0$ and $x = x_0 + h$; then, on supposing each of the functions

$$F(x), F'(x), F''(x), \ldots F^{(n-1)}(x)$$

is always increasing or always decreasing[37] from the first limit to the second,[38] and designating by $h_1, h_2, \ldots h_n$ the quantities of the same sign, but of which the numerical values become progressively smaller, one would obtain, with Eq. (3), a series of similar equations the gathering of which comprise the formula

$$(4) \qquad \frac{f(x_0+h)}{F(x_0+h)} = \frac{f'(x_0+h_1)}{F'(x_0+h_1)} = \frac{f''(x_0+h_2)}{F''(x_0+h_2)} = \ldots = \frac{f^{(n)}(x_0+h_n)}{F^{(n)}(x_0+h_n)}.$$

If, in the formula (4), we content ourselves to equate the first fraction to the last, the equation at which one arrives can be written as follows

$$(5) \qquad \frac{f(x_0+h)}{F(x_0+h)} = \frac{f^{(n)}(x_0+\theta h)}{F^{(n)}(x_0+\theta h)},$$

θ being a number inferior to unity. Finally, if in Eq. (5) we substitute for the finite quantity h an infinitely small quantity, denoted by i, we have

$$(6) \qquad \frac{f(x_0+i)}{F(x_0+i)} = \frac{f^{(n)}(x_0+\theta i)}{F^{(n)}(x_0+\theta i)}.$$ [39]

Cauchy's Eq. (6) is equivalent to

$$\lim_{h\to 0} \frac{f(x_0+h)}{F(x_0+h)} = \lim_{h\to 0} \frac{f^{(n)}(x_0+h)}{F^{(n)}(x_0+h)}.$$

---

[34] I.e., the Mean Value Theorem.

[35] I.e., absolute value.

[36] I assume "as well as" refers only to remaining continuous and not to vanishing at x_0.

[37] See footnote 33, above.

[38] I.e., from x_0 to $x_0 + h$.

[39] Cauchy, *op. cit.*, pp. 162–164; Cates, *op. cit.*, pp. 171–172.

For those unbelievers in infinitesimals, if such creatures still exist, instead of substituting i for h in Cauchy's formula (5), simply take the limits in (5) as $h \to 0$.

What Cauchy has proven in this excerpt can be summed up, using the notation dominant in the present book, as follows:

3.1.25 Theorem (L'Hôpital's Rule; Cauchy Form) *Let f, g be functions defined on an interval $[a, b]$ and suppose $f, f', \dots, f^{(n-1)}, g, g', \dots, g^{(n-1)}$ are continuously differentiable on $[a, b]$, and $f^{(n)}$, $g^{(n)}$ are continuous there. Assume*

$$f(a) = f'(a) = \dots = f^{(n-1)}(a) = g(a) = g'(a) = \dots = g^{(n-1)}(a) = 0.$$

Assume further that each of $g^{(k)}(x)$ for $k = 1, 2, \dots, n-1$ is never 0 on (a, b) and that

$$\lim_{x\to a} \frac{f^{(n)}(a)}{g^{(n)}(a)}$$

exists. Then

$$\lim_{x\to a} \frac{f(x)}{g(x)} = \lim_{x\to a} \frac{f^{(n)}(x)}{g^{(n)}(x)}. \tag{3.9}$$

If $g^{(n)}(a) \neq 0$, then by the continuity of $f^{(n)}$ and $g^{(n)}$ on $[a, b]$, one can replace the right-hand side of (3.9) by the limit and write

$$\lim_{x\to a} \frac{f(x)}{g(x)} = \frac{f^{(n)}(a)}{g^{(n)}(a)}.$$

The Cauchy Form is still not the modern form of L'Hôpital's Rule, but it is powerful enough to handle the ratios of Example 3.1.23 and Exercise 3.1.24, as the reader can readily verify.

Cauchy's Form is slightly less general than the modern form of the Rule because his proofs of the Mean Value Theorem and the Cauchy Mean Value Theorem assume the derivatives to be continuous. A modern version of Theorem 3.1.25 would not require $f^{(n)}$ and $g^{(n)}$ to be continuous on $[a, b]$, and would only require these two functions to exist on (a, b). But one might not bother stating such a form in a modern textbook because one would state and prove a version of the Rule capable of iteration, thus allowing the calculation of limits like those of the aforementioned Example and Exercise by repeatedly applying the Rule instead of the Cauchy Form of L'Hôpital's Rule.

3.1.26 Theorem (L'Hôpital's Rule; Modern Form) *Let f, g be continuous on $[a, b]$ and differentiable on (a, b). Assume $f(a) = g(a) = 0$ and $g'(x) \neq 0$ for any $x \in (a, b)$. Then*

$$\lim_{x\to a} \frac{f(x)}{g(x)} = \lim_{x\to a} \frac{f'(x)}{g'(x)},$$

provided the right-hand limit exists.

Proof. I give a rigorous ϵ-δ proof. Let

$$L = \lim_{x \to a} \frac{f'(x)}{g'(x)}$$

and let $\epsilon > 0$. Choose $\delta > 0$ so that for all $x \in [a, b]$,

$$0 < |x - a| < \delta \Rightarrow \left| \frac{f'(x)}{g'(x)} - L \right| < \epsilon.$$

Choose x so that $0 < |x - a| < \delta$ and observe

$$\frac{f(x)}{g(x)} = \frac{f(x) - f(a)}{g(x) - g(a)} = \frac{f'(c)}{g'(c)}$$

for some $c \in (a, x)$ by the Cauchy Mean Value Theorem. But

$$\begin{aligned} \left| \frac{f(x)}{g(x)} - L \right| &= \left| \frac{f(x)}{g(x)} - \frac{f'(c)}{g'(c)} + \frac{f'(c)}{g'(c)} - L \right| \\ &\leq \left| \frac{f(x)}{g(x)} - \frac{f'(c)}{g'(c)} \right| + \left| \frac{f'(c)}{g'(c)} - L \right| \\ &< 0 + \epsilon = \epsilon, \end{aligned}$$

since $0 < |c - a| < |x - a| < \delta$. □

For convenience I have stated only the simple case in which the right-sided limit is desired (i.e., the case where the limit is taken at the left endpoint of an interval). I leave it to the reader to provide the proper formulations of the result for left-sided and two-sided limits.

3.1.27 Exercise Repeat Exercise 3.1.24 using Theorem 3.1.26.

3.1.28 Exercise Verify the x-coordinate of the point E of Figs. 2.1 and 2.2 on page 10 to be $2/\pi$.

3.2 Precursors to the Mean Value Theorem

3.2.1 Generalities

Historians and mathematicians hold different world views. Whereas mathematicians look for patterns and identify disparate objects that behave alike in certain contexts, ignoring as irrelevant those differences that only manifest themselves elsewhere, historians practise caution and hold as incomparable those things occurring in different times and places, taking what similarities there are as "accidental" or "superficial".

They warn against *Whig history* and its practice of *precursorism*, or, as it is more negatively pronounced, *precursoritis*.

Whig history is the tendency to view the past as an unbroken path to the present. Precursorism is the search for early signs of present-day knowledge in past work. Taken to the extreme, it is both bad and misleading. But extreme rejection can be bad and misleading as well.

Probably the most egregious error of precursorism concerns the *Ishango bone*, a 20000 year old bone with what appears to be tally marks scratched on it. The marks are grouped into three rows on two sides of the bone. In the first row the marks appear in groups of 9, 19, 21, and 11; in the second row the groups contain 19, 17, 13, and 11 scratches; and on the third row on the opposite side of the bone occur groups of 7, 5, 5, 10, 8, 4, 6, and 3 marks, respectively. The existence of such groupings suggests that the bone is more than a mere tally stick and the question of how much more has been raised and answers proposed. The description, which I have seen in print, of the bone as the "Ishango abacus" is truly absurd — at least until someone explains how to perform simple sums with it.[40]

One of the most notorious disputes between historians and mathematicians concerns the use of the term "geometric algebra" coined by Hieronymus Georg Zeuthen (1830–1920), a Danish mathematician, initially to describe the mathematics of Book II of Euclid's *Elements*.[41] In 1975, Sabetai Unguru, working in the Department of the History of Science at the University of Oklahoma made the unfortunate mistake of attacking the nomenclature by declaring the subject not to be algebra. He was soundly beaten down by three prominent mathematicians, each of whom had made important contributions to the history of mathematics as well as to mathematics itself: Bartel Leendert van der Waerden (algebra), Hans Freudenthal (topology), and André Weil (number theory). The language of the mathematicians was a bit strong; simply explaining what mathematicians mean by "algebra" and that geometric algebra fell under this description should have sufficed.

In discussing precursors to the Mean Value Theorem, I would hope to steer a course midway between the errors of over-interpretation and of failure to recognise the obvious. I can pinpoint results that we now recognise as implicitly embodying elements of the Mean Value Theorem and to make a case for their precursorness which the reader will have to evaluate for him- or herself.

In enumerating examples of precursors to the Mean Value Theorem I have been more rather than less inclusive. My criteria are that a result ought to be recognisably an instance of some generality of one of the versions of the Mean Value Theorem or it

---

[40] I forget where I came across the reference to the "Ishango abacus", but a Google search yielded interpretations of it as a calendar and even as proof of knowledge of prime numbers. A brief but thoughtful (and less fanciful) account of the Ishango bone and other archæological mathematical artifacts can be found in: George Gheverghese Joseph, *The Crest of the Peacock; Non-European Roots of Mathematics*, Princeton University Press, Princeton, 2011, pp. 30*ff*.

[41] A brief, informative description of Zeuthen's geometric algebra by his compatriot Jesper Lützen can be found in the biographical entry (pp. 574–578) in: Joseph W. Dauben and Christoph J. Scriba (eds.), *Writing the History of Mathematics: Its Historical Development*, Birkhäuser Verlag, Basel, 2002.

ought to exhibit the main principle on which the Mean Value Theorem depends. There is room for disagreement here. The more analytically oriented reader, for example, may feel the Geometric Form of the Mean Value Theorem insufficiently analytic in itself and may wish to disqualify the results of Apollonius, Archimedes, Valerio, and Cavalieri for that reason. The results of the Merton scholars and Rolle may be deemed too trivial and incapable of leading on their own to greater generality. For the most part I shall pass over such matters in silence, the exception being the work of Bhāskara II, which is currently widely touted without being given any account and must therefore be examined with some care.

3.2.2 *Conic Sections in Classical Greece*

The simplest instance of the geometric Mean Value Theorem occurs when the curve is a circle. If A and B are two points on the circle, the perpendicular bisector of the chord AB will intersect the circle at two points C and C' on opposite sides of AB. The perpendiculars to this bisector CC' at the points C and C', respectively, are tangent to the circle and parallel to AB. This follows easily from two propositions of Euclid's *Elements*:

> III.3. *If in a circle a straight line through the centre bisect a straight line not through the centre, it also cuts it at right angles; and if it cut it at right angles, it also bisects it.*
>
> III.16. *The straight line drawn at right angles to the diameter of a circle from its extremity will fall outside the circle, and into the space between the straight line and the circumference another straight line cannot be interposed...*

If AB is a diameter, III.16 immediately implies the perpendiculars to CC' at C and C' are tangents to the circle and, their being perpendicular to CC', they are thus parallel to AB.

If AB is not a diameter, one first applies III.3 to conclude that the diameter passing through the midpoint of AB is the perpendicular bisector CC' of AB, whence III.16 again applies.

This might seem too narrow and special to be considered a precursor to the geometric Mean Value Theorem, but it does generalise. According to Thomas Heath, Archimedes informs us that Aristæus and Euclid proved the result for parabolas. Archimedes tells us this in his "Quadrature of the parabola", one of two works in which Archimedes presented not only the proofs of his results, but also the methods he used to find them.

The "Quadrature of the parabola" has as its main goal the determination of the area of parabolic segments. It is prefaced with a letter to his colleague Dositheus outlining the contents of the paper and announcing that "Prefixed are, also, the elementary propositions in conics which are of service in the proof".[42] The first of these reads

[42] Heath, *Works of Archimedes*, *op. cit.*, p. 234.

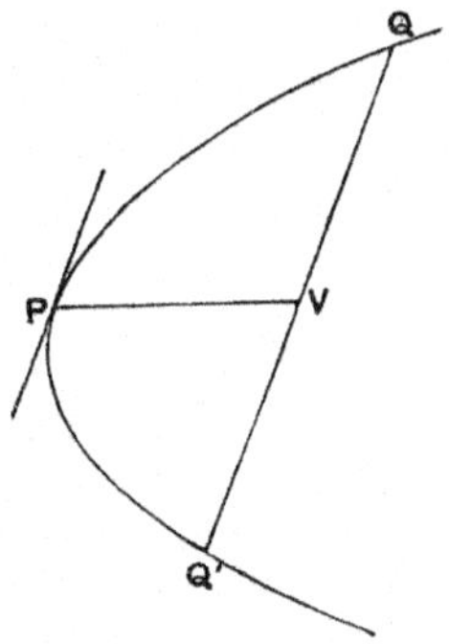

Fig. 3.6 Archimedes's diagram

Proposition 1.

If from a point on a parabola a straight line be drawn which is either itself the axis or parallel to the axis, as PV, [See Fig. 3.6.*] and if QQ' be a chord parallel to the tangent to the parabola at P and meeting PV in V,*

$$QV = VQ'.$$

Conversely, if $QV = VQ'$, the chord QQ' will be parallel to the tangent at P.[43]

Archimedes offers no proof of this, noting only that it and the next two propositions "are proved in the elements of conics", which Heath informs us means that the result was proven by Aristæus and Euclid in their now lost treatises on conics.[44]

We do have a proof, not only of this proposition, but also of its generalisation to conic sections in the *Conics* of Apollonius of Perga. He begins, in modern fashion, by generalising the concept of diameter in Definition 4:

> 4. Of any curved line which is in one plane I call that straight line the diameter which, drawn from the curved line, bisects all straight lines drawn to this curved line parallel to some straight line; and I call the end of that straight line (the diameter) situated on the curved line the vertex of the curved line, and I say that each of these parallels is drawn ordinatewise to the diameter.[45]

One diameter is special:

> 7. And I call that straight line the axis of a curved line... which being a diameter of the curved line... cuts the parallel straight lines at right angles.[46]

3.2.1 Remark That the "diameters" of families of parallel chords are straight lines is a property of conics not shared by general curves. For curves for which the chords exist, we can define the diameter corresponding to a family of parallel chords to be

[43] *Ibid.*

[44] *Ibid.*, p. 235.

[45] Apollonius, *Conics*, in: Mortimer J. Adler (ed.), *Great Books of the Western World*, vol. 11, Encyclopædia Britannica, Chicago, 1952, p. 604.

[46] *Ibid.* The missing words are for the case of an hyperbola, the diameters of which are defined separately in Definition 5.

the locus of midpoints of the chords. The diameter, so defined, need not be a straight line. For example, consider the graph of the cubic function $y = (x + 2)x(x - 1)$ on $[-1, \sqrt{2}]$ and the horizontal chords determined by the lines $y = 0, y = 1, y = 2$. These intercept the curve at the points:

$y = 0$: $x = 0, 1$
$y = 1$: $x = -.44504186791263, 1.2469796037175$
$y = 2$: $x = -1, \sqrt{2}$.

The midpoints of these chords are

$$\langle .5, 0\rangle, \ \langle .400968867902, 1\rangle, \ \langle -.479289321881, 2\rangle,$$

respectively, which are clearly not collinear.

For conic sections, however, the diameters are straight lines and Apollonius proves the following in Book I of the *Conics*

Proposition 32

If a straight line is drawn through the vertex of a section of a cone, parallel to an ordinate, then it touches the section, and another straight line will not fall into the space between the conic section and this straight line.[47]

Apollonius proves the result first for parabolas and then gives a combined proof for ellipses and hyperbolas. The proof uses an algebraic characterisation of the parabola, so I might as well cheat and give a modern proof. In fact, I will prove the relevant portion of Proposition 1 of Archimedes. To this end, let a parabola be defined by a given focus F and directrix L. The axis of the parabola, i.e., the perpendicular line drawn from F to L will serve as the x-axis. The vertex of the parabola, i.e., the intersection of the axis and the parabola, will be the origin O. If we choose OF as a unit, the directrix becomes the line $x = -1$, the focus the point $\langle 1, 0\rangle$, and any point $\langle x, y\rangle$ on the parabola will satisfy

$$x + 1 = \sqrt{(x - 1)^2 + y^2},$$

i.e.,

$$(x + 1)^2 = (x - 1)^2 + y^2. \tag{3.10}$$

(See Fig. 3.7.)

Simplifying (3.10) successively yields

$$\begin{aligned} x^2 + 2x+1 &= x^2 - 2x + 1 + y^2 \\ 4x &= y^2. \end{aligned} \tag{3.11}$$

[47] *Ibid.*, p. 638.

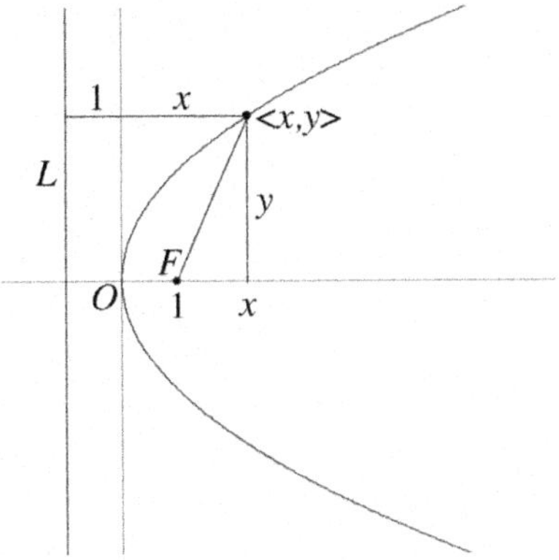

Fig. 3.7 Parabolic equation

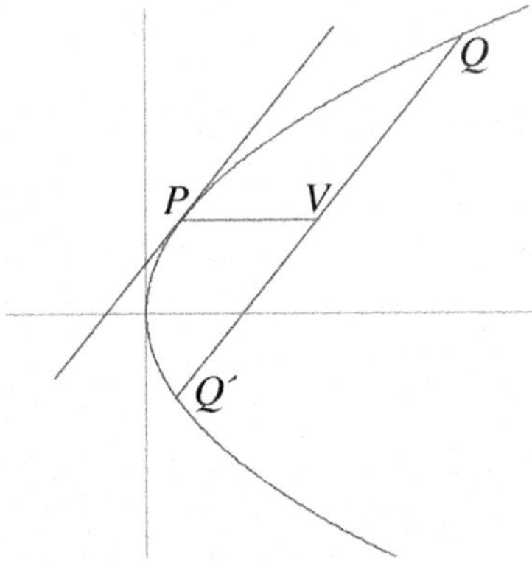

Fig. 3.8 Modified Archimedean diagram

Now, if $y = mx + b$ is any non-vertical chord, we can solve for the endpoints Q, Q' of the chord as in Fig. 3.8, by plugging $mx + b$ in for y in (3.11):

$$4x = (mx + b)^2,$$

i.e.,

$$m^2x^2 + (2mb - 4)x + b^2 = 0,$$

and the quadratic formula yields

$$\begin{aligned} x &= \frac{4 - 2mb \pm \sqrt{(2mb - 4)^2 - 4m^2b^2}}{2m^2} \\ &= \frac{2 - mb}{m^2} \pm \frac{\sqrt{(2mb - 4)^2 - 4m^2b^2}}{2m^2}. \end{aligned}$$

The x-coordinate of the midpoint V of QQ' is thus

$$x_0 = \frac{2 - mb}{m^2}.$$

And the y-coordinate is

$$y_0 = mx_0 + b = m\frac{2 - mb}{m^2} + b = \frac{2}{m} - b + b = \frac{2}{m}.$$

The equation of the parallel PV to the x-axis is thus $y = 2/m$.

The slope of the tangent to the parabola $4x = y^2$ at P is easily calculated by differentiation:

$$4 = 2yy',$$

i.e.,

$$\begin{aligned} y' &= \frac{2}{y} \\ &= \frac{2}{2/m} \quad \text{at } P = \langle x_0, 2/m \rangle \\ &= m. \end{aligned}$$

Thus, the tangent to the parabola at P has the same slope as the original chord and is thus parallel to the chord.

The case of a vertical line being trivial and no chord having slope $m = 0$, we have finished the proof.

I have proven the less general result of Archimedes rather than that of Apollonius for two reasons. One was that it saved me from having to prove the parabola to have linear diameters. The other is that Archimedes has another interesting observation to make.

> DEF. "In segments bounded by a straight line and any curve I call the straight line the **base**, and the **height** the greatest perpendicular drawn from the curve to the base of the segment, and the **vertex** the point from which the greatest perpendicular is drawn."

Proposition 18.

> *If Qq be the base of a segment of a parabola, and V the middle point of Qq, and if the diameter through V meet the curve in P, then P is the vertex of the segment.*[48]
>
> For Qq is parallel to the tangent at P [Prop. 1]. Therefore, of all the perpendiculars which can be drawn from points on the segment to the base Qq, that from P is the greatest. Hence, by the definition, P is the vertex of the segment.[49]

---

[48] Swapping q for Q' a copy of Fig. 3.6 is reproduced in the paper at this point.

[49] Heath, *op. cit.*, pp. 246–247.

3.2.2 Exercise This last part of the argument for Proposition 18, that no other point on the parabola can be farther from Qq than P requires a small argument to show that our definition of tangent in terms of slope entails the property taken by Archimedes and Apollonius to define tangency: No other line passing through P can fit between the tangent line and the parabola.
i. Prove that for any point P on the parabola $4x = y^2$ there is only one line $y = mx + b$ passing through P which has no other intersection with the parabola. [Hint. Solve $4x = (mx + b)^2$ and use the vanishing of the discriminant to express b, x, y successively in terms of m.]
ii. Assume the arc of QPQ' of the parabola (of Fig. 3.6) to be continuously parametrised by $\gamma(t)$ and that P' is a point on the tangent line more distant from QQ' than P. Use d_γ and the Intermediate Value Theorem to derive a contradiction.

Now Archimedes does not come out and say that the tangent to a curve at a vertex relative to a given base is parallel to the base, but, given his characterisation of the tangent, the realisation is implicit — as is the generality of the result. And it is this implicit realisation more than the explicit statement of the result in the parabolic case that earns his work its star status among the precursors to the modern Mean Value Theorem. However, it must be reported that he does not seem to have applied this principle to his spiral — the other curve for which he found the tangent at one point: no comparable result appears in his work *On Spirals*. It could be that the question never occurred to him, or that, lacking a construction determining where the parallel tangent touched the spiral, he felt the result to be of no interest or not properly established.

3.2.3 Example Consider the Archimedean spiral defined parametrically by

$$x = \theta \cos\theta, \quad y = \theta \sin\theta, \quad 0 \le \theta \le 2\pi.$$

Given two points $\langle \theta_i \cos\theta_i, \theta_i \sin\theta_i \rangle$ on the spiral for $i = 0, 1$, the slope of the chord connecting them is

$$\frac{\Delta y}{\Delta x} = \frac{\theta_1 \sin\theta_1 - \theta_0 \sin\theta_0}{\theta_1 \cos\theta_1 - \theta_0 \cos\theta_0},$$

while the slope of the tangent to the spiral at a point $\langle \theta \cos\theta, \theta \sin\theta \rangle$ is

$$\frac{dy}{dx} = \frac{\sin\theta + \theta\cos\theta}{\cos\theta - \theta\sin\theta}.$$

Algebraically, constructing the tangent line parallel to the chord means solving the equation

$$\frac{\sin\theta + \theta\cos\theta}{\cos\theta - \theta\sin\theta} = \frac{\theta_1 \sin\theta_1 - \theta_0 \sin\theta_0}{\theta_1 \cos\theta_1 - \theta_0 \cos\theta_0},$$

for θ in terms of θ_0, θ_1. Even for simple θ_0, θ_1, we might only be able to solve this approximately. For example, if $\theta_0 = \pi, \theta_1 = 3\pi/2$, the equation becomes

$$\frac{\sin\theta + \theta\cos\theta}{\cos\theta - \theta\sin\theta} = \frac{\frac{3\pi}{2}(-1) - \pi\cdot 0}{\frac{3\pi}{2}\cdot 0 - \pi(-1)} = -\frac{3}{2}.$$

The solution θ between π and $3\pi/2$ is

$$\theta = 3.975994055 \text{ radians} = 227.8076787^\circ,$$

which is not a familiar angle easily related to $\pi = 180^\circ$ or $3\pi/2 = 270^\circ$.

3.2.3 The Analytic Approach in Mediæval India

The Indians were several centuries ahead of the Europeans in some areas of mathematics, but they are nonetheless under-reported in histories of the field. The main reason for this is the difficulty of determining just what they knew. Even when claims were made on behalf of Indian mathematics by sympathetic Europeans, those of a more Eurocentric disposition have expressed doubts, sometimes very strongly, about the interpretation of the works cited or about the originality thereof, occasionally positing hidden European roots. The historical expositor who is not a specialist with a working knowledge of Sanskrit cannot confidently write on the subject. So he will of necessity not give Indian mathematics more than a brief, superficial coverage.

In Europe the Calculus arose in the 17th century largely because of the new symbolism, which made the discovery and description of new curves much easier. The mathematical landscape broadened considerably. The old geometric problems of determining areas and tangents were translated into algebra and their solution fueled the growth. In India, the Calculus started to emerge from trigonometry in connexion with problems from astronomy. A millennium passed between the earliest beginnings at the hands of Āryabhaṭa (*c.* 499 A.D.) and what P.P. Divakaran has justly called "the first textbook of calculus", namely the *Yuktibhāṣā* of Jyeṣṭhadeva (*c.* 1530). My impression is that mediæval Indian work on the Calculus was restricted to trigonometric functions, much as Torricelli's was limited to "generalized parabolas and hyperbolas, satisfying the equation $x\,dy = ky\,dx$".[50]

Because of the debate over Indian accomplishments in mathematics, two Indian scholars, Bibhutibhushan Datta (1888–1958) and Avadesh Narayan Singh (1901–1954), laid plans to publish three volumes of source material. Unfortunately, their third volume which was to include the Calculus never appeared. They did, however, write a short paper on the subject, later revised by Kripa Shankar Shukla (*1918) and published in 1984.[51]

---

[50] As quoted on page 160, above, from commentary of Dirk Struik.

[51] Datta and Singh, *op. cit.*

If there is little doubt today that the Indian tradition encompassed some aspects of the Calculus, this has been a hard-won realisation. Datta and Singh begin their paper citing an example:

A Controversy

Attention was first drawn to the occurrence of the differential formula

$$\delta(\sin\theta) = \cos\theta\,\delta\,\theta$$

in Bhāskara II's (1150) *Siddhānta Śiromaṇi* by Pandit Bapu Deva Sastri[52] in 1858. The Pandit published a summarised translation of the passages which involve the use of the above formula. His summary was defective in so far as it did not bring into prominence the idea of the infinitesimal increment which underlies Bhāskara's analysis. Without making clear to his readers, the full significance of Bhāskara's result, the Pandit made the mistake of asserting — what was plain to him — that Bhāskara was fully acquainted with the principles of the differential calculus.

The Pandit was adversely criticised by Spottiswoode [*sic*], who without consulting the original on which the Pandit based his conclusions, remarked (1) that Bapu Deva Sastri had overstated his case in saying that Bhāskarācārya[53] was fully acquainted with the principles of the differential calculus, (2) that there was no allusion to the most essential feature of the differential calculus, viz. the infinitesimal magnitudes of the intervals of time and space therein employed, and (3) that the approximative character of the result was not realized.[54]

So, what did Bāpūdeva and Spottiswoode say on the matter? Their papers are short and out of copyright, so we might as well quote them in full.

Bhāskarā's knowledge of the Differential Calculus.—By BAPU DEVA SHASTRI, *Professor of Mathematics and Astronomy in the Government Sanskrit College, Benares.*

To the Editor of the Asiatic Society's Journal.

SIR,— It appears to be generally believed that the principle of the Differential Calculus was unknown to the ancient Hindu mathematicians. Allow me to correct this impression by the following statement regarding what Bháskarácharya has written on the subject.

Bháskarácharya says that "the difference between the longitudes of a planet found at any time on a certain day and at the same time on the following day is called its rough motion during that interval of time; and that its *Tátkálika* motion is its exact motion."

The *Tátkálika* or instantaneous motion of a planet is the motion which it would have in a day, had its velocity at any given instant of time remained uniform. This is clear from the meaning of the term *Tátkálika* and it is plain enough to those who are acquainted with the principles of the Differential Calculus that this *Tátkálika* motion can be no other than the differential of the longitude of a planet. This *Tátkálika* motion is determined by Bháskarácharya in the following manner.

---

[52]"Pandit" and "Sastri" (or "Shastri") are titles singling Bapu Deva out as a man of learning. Thus one will often see his name in shorter form as Bapu Deva or even as Bapudeva, with or without various accents. Other variants I have come across are Bapu Deba and Bapu Deo. I have seen him indexed in books under "Deva" and "Sastri". Insofar as the most authoritative source I have gives "Bapudeva" as one word, I have indexed him under that name.

[53]Usually spelled Bhaskaracharya with or without the accents, "Bhāskarācārya" means "Bhāskara the Great Teacher" and is thus another name for Bhāskara II.

[54]Datta and Singh, *op. cit.*, p. 95.

"Suppose, x, x' = the mean longitudes of a planet on two successive days;

y, y' = the mean anomalies;

u, u' = the true longitudes and

a = the eccentricity or the sine of the greatest equation of the orbit.

Then, $x' - x$ = the mean motion of the planet, $y' - y$ = the motion of the mean anomaly and $u' - u$ = the true motion of the planet." Now according to Bháskarácárya, the equation of the orbit on the

first day $= \dfrac{a.\ \sin y}{\text{Rad}}$,[55] and

that on the next day $= \dfrac{a\ \sin y'}{\text{Rad}}$;

$$\therefore \quad u = x \pm \frac{a.\ \sin y}{\text{Rad}}, \ldots\ldots\ldots\ldots\ldots \quad (1)$$

$$\text{and} \quad u' = x' \pm \frac{a.\ \sin y'}{\text{Rad}};$$

$$\therefore \quad u' - u = x' - x \pm \frac{a(\sin y' - \sin y)}{\text{Rad}} \ldots\ldots \quad (2).$$

Now, in order to know the instantaneous value of $u' - u$, it is necessary first to know the instantaneous value of the *Bhogya-khaṇḍa* or the difference between two successive sines given in *Tables of sines*. Thus, suppose the sines of the arcs 0, A, $2A$, $3A$, &c. are given in the *Tables of sines*, then $\sin A - \sin 0$, $\sin 2A - \sin A$, $\sin 3A - \sin 2A$, &c. are the *Bhogya-khaṇḍas*.

"These are not equal to each other but gradually decrease, and consequently while the increase of the arc is uniform, the increment of the sine varies"—on account of the deflection of the arc. Hence the difference between any two successive sines is not the *Tátkálika Bhogya-khaṇḍa*; but if the arc instead of being deflected be increased in the direction of the tangent then the increase which would take place in the sine is the *Tátkálika Bhogya-khaṇḍa* i.e., the instantaneous motion of the sine.

Thus in the accompanying diagram [Fig. 3.9] suppose the arc $Df = A$, then, $\sin Af - \sin AD = fg - DE =$ fm, the *Bhogya-khaṇḍa* of the sine DE; but this is not the *Tátkálika Bhogya-khaṇḍa* of that sine. If the arc AD instead of being deflected towards f, be increased in the direction of the tangent so that $DF = Df = A$; then $FG - DE = Fn$, which would be the *Tátkálika Bhogya-khaṇḍa* of the sine DE i.e., the instantaneous motion of that sine."

Bháskarácárya has determined that "the *Tátkálika Bhogya-khaṇḍa* varies as the cosine of arc, i.e., when arc = 0, its cosine equals the radius, and A = the *Tátkálika Bhogya-khaṇḍa*. And, as the arc increases, the cosine and the *Bhogya-khaṇḍa* decrease. Hence, if y be any given arc, the *Tátkálika Bhogya-khaṇḍa* answering to it will be found by the following proportion.

As, R (or the cosine of an arc = 0.)

: The *Tátkálika Bhogya-khaṇḍa* (= A.)

:: Cosine y.

: *Tátkálika Bhogya-khaṇḍa* of $\sin y$.

[55] Rad = radius of the circle.

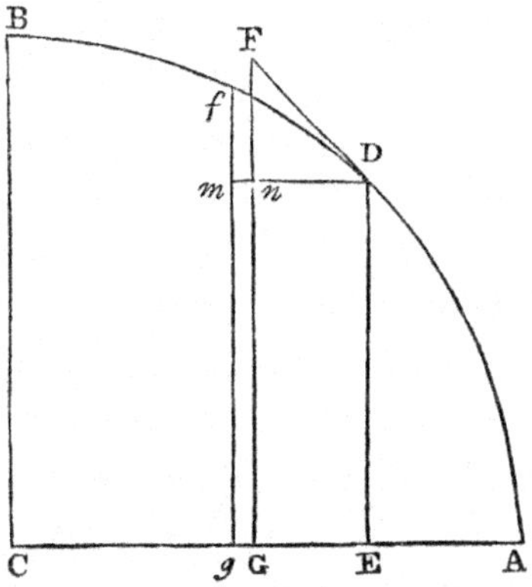

Fig. 3.9 Illustration from Bāpudēva

$$\therefore \textit{Tátkálika Bhogya-khaṇḍa} = \frac{\text{A.}\cos y.}{\text{R}}.\text{''}$$

The reason of the above proportion can be easily understood from the two similar triangles DCE and DFn in the above diagram.

"After having thus determined the *Tátkálika Bhogya-khaṇḍa*, the instantaneous value of $\sin y' - \sin y$ is found by the following proportion.

As $A : \frac{A.\ \cos y}{\text{R}} :: y' - y : \frac{\cos y\ \times\ (y' - y)}{\text{R}}$ (= the instantaneous value of $\sin y' - \sin y$.)

By substituting the instantaneous value of $\sin y' - \sin y$ in the Eq. (2), the instantaneous value of $u' - u$, the true motion of the planet will be found: that is,

$$u' - u = x' - x \pm \frac{a.\ \cos y}{\text{R}} \cdot \frac{y' - y}{\text{R}} \quad \ldots\ldots \quad (3)$$

This is the instantaneous motion of the planet."

This is the way in which Bháskarácharya determined the instantaneous motion of the sun and the moon.

Equation (3) is just the differential of Eq. (1). As,

$$d\,(u) = d\,(x \pm \frac{a.\ \sin y}{\text{R.}});$$

$$\text{or } du = d\,x \pm \frac{a}{\text{R}} \cdot \frac{\cos y}{\text{R}} \cdot dy;$$

which is similar to Eq. (3).

Now the term *Tátkálika* applied by Bháskarácharya to the velocity of a planet, and his method of determining it, correspond exactly to the differential of the longitude of a planet and the way for finding it. Hence it is plain that Bháskarácharya was fully acquainted with the principle of the Differential Calculus. The subject, however, was only incidentally and briefly treated of by him; and his followers, not comprehending it fully, have hitherto neglected it entirely.

I have the honor to be,
Your obedient servant,
BAPU DEVA SHASTRI,

4th May, 1858.[56]

Spottiswoode read Bāpūdeva's paper with interest and responded, his response introduced by the editor:

ART. VII. — *Note on the supposed Discovery of the Principle of the Differential Calculus by an Indian Astronomer. By* W. SPOTTISWOODE, Esq. *Communicated by the Director.*

In the number of the Journal of the Asiatic Society of Bengal last received, No. III., of 1858, is a short article by Bapu Deva Shastri, Professor of Mathematics and Astronomy at the Government College of Benares, in which he has undertaken to show, that Bháskarácháryа, an astronomer who flourished at Ujjain in the twelfth century, was fully acquainted with the principle of the Differential Calculus, one of the most important discoveries of the last century in Europe.

As this would have been a very remarkable circumstance in the history of astronomical science, it was obviously a matter of more than ordinary interest to have the accuracy of Professor Bapu Deva's statement carefully tested, and I therefore applied to our colleague, Mr. William Spottiswoode, who is well known as a mathematician,[57] for his opinion; the answer with which he has favoured me will, I doubt not, be thought by the Society worthy of being communicated to the public through our Journal, especially as, whilst it shows that Bapu Deva's statement is not correct to its whole extent, yet it does full justice to Bháskarácháryа's penetration and science, and acknowledges that his calculations bear a very remarkable analogy to the corresponding processes in modern mathematical astronomy.

12, James Street, Buckingham Gate,
London, May 5, 1859.

MY DEAR MR. WILSON,

[56]Bapu Deva Shastri, "Bháskará's knowledge of the Differential Calculus", *Journal of the Asiatic Society of Bengal* 27 (1858), pp. 213 – 216. Bāpūdeva was one of the first scholars to be familiar with both Indian and Western mathematics and astronomy, and, as a member of the faculty of the Government Sanskrit College in Benares from 1842, his competence in that language can also be assumed. A brief paragraph on him as an historian of Indian science can be found in: Radha Charan Gupta, "India", in: Joseph W. Dauben and Christoph J. Scriba (eds.), *Writing the History of Mathematics: Its Historical Development*, Birkhäuser, Basel, 2002; here: p. 312.

[57]Spottiswoode (1825–1883) was a solid choice to ask for an opinion on the matter. As a mathematician he was known for his work on determinants, including the first elementary treatise on the subject, *Elementary Theorems Relating to Determinants* (1851). In 1853 he was made a Fellow of the Royal Society, serving as its president from 1879 until his death. He was also president of the London Mathematical Society from 1870 to 1872 and a member of various other learned societies.

In his obituary in *The Times* we read "Mr. Spottiswoode was nearly as good a linguist as he was a mathematician, and so accomplished an Oriental scholar was he that he was urged to undertake an edition of a great work in Indian Astronomy, on which he contributed a paper to the Journal of the Asiatic Society". The work in question was the *Sūrya Siddhānta*, probably the most important early work of Indian Astronomy. Spottiswoode's report, giving a chapter by chapter account of the work, was published in 1863.

I have read Bapu Deva Shastri's letter on Bháskarácháry a's mode of determining the instantaneous motion of a planet, with great interest, and think that we are much indebted to him for calling our attention to so important an element in the old Indian methods of calculation. It still, however, seems to me, that he has overstated the case, in saying that "Bháskarácháry a was fully acquainted with the principle of the Differential Calculus." He has undoubtedly conceived the idea of comparing the successive positions of a planet in its path, and of regarding its motion as constant during the interval, and he may be said to have had some rudimentary notion of representing the arc of a curve by means of auxiliary straight lines. But on the other hand, in the method here given, he makes no allusion to one of the most essential features of the Differential Calculus, viz., the infinitesimal magnitude of the intervals of time and space therein employed. Nor indeed is anything specifically said about the fact that the method is an approximative one.

Nevertheless, with these reservations, it must be admitted, that the penetration shown by Bhaskara, in his analysis, is in the highest degree remarkable; that the formula which he establishes (Eq. 3, p. 216),[58] and his method of establishing it, bear more than a mere resemblance — they bear a strong analogy — to the corresponding process in modern mathematical astronomy; and that the majority of scientific persons will learn with surprise the existence of such a method in the writings of so distant a period and so remote a region.

With many thanks for communicating the paper to me,

I remain, very sincerely yours,

H.H. Wilson, Esq. W. Spottiswoode.

P.S. I may perhaps add, that if —

$$x, x' \text{ be the mean longitudes,}$$
$$y, y' \text{ be the mean anomalies,}$$
$$u, u' \text{ be the true anomalies}$$

of a planet on two successive days; and a the eccentricity, or sine of the greatest equation of the orbit; then $(u' - u)$, or the true motion of the planet,[59]

$$= x' - x\,a \pm (\sin . y' - \sin . y).$$

And Bhaskara's method consists in showing, that the "instantaneous" value of $\sin . y' - \sin . y$ (or the value which it would have if the velocity of the planet had remained uniform during the day) is $(y' - y) \cos . y$. His formula therefore becomes —

$$u' - u = x' - x \pm (y' - y)\, a \cos . y.$$

And the corresponding formula in modern analysis is —

$$\begin{aligned} du &= d(x \pm a \sin . y) \\ &= dx \pm a \cos . y\, dy. \end{aligned}$$

W.S.[60]

Is this "adverse" criticism, as Datta and Singh have labelled it in the citation given above on page 189? My copy of *The Concise Oxford Dictionary* offers the following definition of "adverse"

[58] I.e., formula (3) of page 191, above.

[59] The a in the formula should follow the symbol $\pm$.

[60] *Journal of the Royal Asiatic Society of Great Britain and Ireland*, 17 (1860), pp. 221–222.

ă'dverse *a.* Contrary, hostile (*to*); hurtful, injurious, (*to*); (arch.) placed opposite; hence ~- LY (-slĭ) *adv.* [ME, f. OF *advers* f. L *adversus* p.p. of AD(*vertere vers-* turn)]

I don't think anyone can call Spottiswoode's remarks hostile, hurtful, or injurious. Looking up "contrary" one finds a number of alternative definitions, mostly fairly strongly negative, only "opposite" being correctly applicable here, for Bāpūdeva believes he has proven Bhāskara II to have been fully aware of "the principle of the Differential Calculus" and Spottiswoode believes proof is lacking. It is odd that Datta and Singh should characterise Spottiswoode's remarks so strongly negatively, as they themselves acknowledge that Bāpūdeva has not proven his case.

Spottiswoode is, however, guilty of a lack of clarity of his own, when for example he says of Bhāskara II that "he may be said to have had some rudimentary notion", when he should have said that *on the basis of the information related by Bāpūdeva*, Bhāskara II "may [only] be said to have had some rudimentary notion". Datta and Singh were correct in criticising Spottiswoode for drawing such conclusions without first checking the *Siddhānta Śiromaṇi*.

To set the record straight, Datta and Singh repeat a bit of the proof, pausing to note that in 932 already Mañjula[61] had stated that

$$\sin w' - \sin w = (w' - w)\cos w \tag{3.12}$$

holds (approximately) when $w' - w$ is very small, but the earliest proof they had was given by Bhāskara II. They give his proof and a quote from the book:

> "The difference between the longitudes of a planet found at any time on a certain day and at the same time on the following day is called its (*sphuṭa*) *gati* (true rate of motion) for that interval of time."
>
> "This is indeed rough motion (*sthūlagati*). I now describe the fine (*sūkṣma*) instantaneous (*tāt-kālika*) motion. The *tāt-kālika-gati* (instantaneous motion) of a planet is the motion which it would have, had its velocity during any given interval of time remained uniform."
>
> During the course of the above statement, Bhāskara II observes that the *tāt-kālika-gati* is *sūkṣma* ("fine" as opposed to rough), and for that the interval must be taken to be very small, so that the motion would be very small. This small interval of time has been said to be equivalent to a *kṣaṇa* which according to the Hindus is an infinitesimal interval of time (immeasurably small).[62]

Even after including the derivation of (3.12), Datta and Singh acknowledge that this is not a convincing proof that Bhāskara II knew the basics of Differential Calculus. The proof offered places him somewhere between, say, Fermat and Leibniz. However, they mitigate their acknowledgement:

> If the above were the only result occurring in Bhāskara II's work, one would be justified in not accepting the conclusions of Pandit Bapu Deva Sastri. There is however other evidence in Bhāskara II's work to show that he did actually know the principles of the differential calculus. This evidence consists partly in the occurrence of the two most important results of the differential calculus:

[61] The name is usually rendered Muñjāla and Mañjula is either a typographical error or a variant.

[62] Datta and Singh, *op. cit.*, p. 97. But see footnote 97, below.

> (i) He has shown that when a variable attains the maximum value its differential vanishes.
> (ii) He shows that when a planet is either in apogee or in perigee the equation of the centre vanishes, hence he concludes that for some intermediate position the increment of the equation of centre (i.e., the differential) alos [*sic*] vanishes.
>
> The second of the above results is the celebrated Rolle's Theorem, the mean value theorem of the differential calculus.[63]

And these, of course, are why Bhāskara II and the *Siddhānta Śiromaṇi* are of interest here — for, Datta and Singh state that (*i*) and (*ii*) "occur in the *Golādhyāya, Spaṣṭādhikāra vāsanā* of the *Siddhānta Śiromaṇi*".[64] They also mention that this occurrence was first noted by Sudhakara Dvivedi,[65] who succeeded Bāpūdeva as Professor of Mathematics and Astronomy at Benares in 1889.

It is most desirable to include here passages from Bhāskara II on these points, particularly the passage demonstrating (*ii*). This turns out not to be an easy task, for reasons I shall get to later. First I wish to continue the story of the struggle for recognition of Indian achievements in the field of Differential Calculus, particularly as it relates to Bhāskara II.[66]

In 1931 in the Annual Report of the *Deutsche Mathematiker-Vereinigung* [German Mathematical Society] there appeared a short article by Prabodh Chandra Sengupta,[67] written to supplement an earlier article by Edmund Hoppe[68] on the history of the Calculus up to Newton and Leibniz. Sengupta wished to add the contributions of early Indian mathematicians to the story. The paper was accompanied by a nay-saying response:

> The question to what extent the concepts, which in further elaboration lead to the idea of differentiation, were already known by the Indians, particularly by Bhâskara, has been earlier discussed in the literature. The publisher of Bhâskara's Siddhânta-Śiromani 1866,[69] Bâpu Deva Śâstrin, already in 1858 went so far, on account of Bhâskara's computation of the so-called tâtkâlika-motion (instantaneous motion) of a planet, to ascribe to this a complete insight into the principle of the Differential Calculus. However W. Spottiswoode has already come out against this in a notice in the Journal of the Royal Asiatic Society, vol. 17 (1860), where he acknowledges, that Bhâskara indeed calculated for the difference of the sine of an angle under a change in angles and hereby replaced $\sin y_1 - \sin y$ by $(y_1 - y)\cos y$, that however there is no mention of passage to a limit, thus the principally important, methodological concept of the Differential Calculus is missing. Also one cannot

[63] *Ibid.*, pp. 98 – 99.

[64] Cf. page 201, below, for an explanation.

[65] *Ibid.*, p. 104.

[66] Those papers from Datta and Singh on down also discuss integration, but our central concern being the Mean Value Theorem I concentrate on differentiation.

[67] P.C. Sengupta, "History of the infinitesimal calculus in ancient and medieval India", *Jahresbericht der deutschen Mathematiker-Vereinigung* 40 (1931), pp. 223–227.

[68] E. Hoppe, "Zur Geschichte der Infinitesimalrechnung bis Leibniz und Newton", *Jahresbericht der deutschen Mathematiker-Vereinigung* 37 (1928), pp. 148–186.

[69] The date refers to a Sanskrit edition of the *Siddhānta Śiromaṇi* and the *Vāsanābhāṣya* published in 1866. The work was revised by Candradeva and Gaṇapatideva Śāstri and published in 1891 and 1929, respectively.

> at all be convinced by him, because he had in mind in his problem a time interval of a day's length.[70]

Whereas Spottiswoode's criticism had been spot-on — Bāpūdeva had neglected to back up his assertions — this new critique overlooked the fact that Sengupta had cited chapter and verse.

Sengupta published an expanded version of his paper in Calcutta the following year.[71] The new version begins with an explanation of its purpose:

> 1. Dr. Hoppe in his article in a recent issue of the "Jahresbericht der Deutschen Mathematiker-vereinigung" has traced the history of the Infinitesimal Calculus prior to the time of Leibnitz and Newton. Unfortunately he has put on record no account of the work in this line done by the ancient Indian astronomers and mathematicians. It is proposed to present some facts in regard to this subject and to claim for the Indian mathematicians their due place in the history of the origin and development of the Infinitesimal Calculus.
>
> ...
>
> In what follows we shall try to develop historically how the necessity spoken of above gave rise first to the idea of differentiation and the use of infinitesimal triangles, and led also to a process very akin to integration. Some of these results arrived at by Indian mathematicians have already been brought to light by European researchers. In Ball's History of Mathematics, the author says in reference to Bhāskara, "Amongst the trigonometrical formulæ there is one which is equivalent to the equation,
>
> $$d\,(\sin\theta) = \cos\theta\, d\theta."$$
>
> His authority on this point is Delambre, I. 456. Some more of the results treated later on were first brought to public notice by the late MM. Bāpudev Śāstrī, Spottiswoode, the then Astronomer-Royal,[72] called it an over statement of a case. Dr. B. N. Seal in his "Positive Sciences of the Ancient Hindus"[73] has tried to meet Spottiswoode's views but has not made

---

[70]C.H. Müller, "Bemerkung zu vorstehenden Aufsatz des Herrn P.C. Sengupta", *Jahresbericht der deutschen Mathematiker-Vereinigung* 40 (1931), p. 227.

[71]P.C. Sengupta, "Infinitesimal calculus in Indian mathematics — its origins and development", *Journal of the Department of Letters (Calcutta University)* 22 (1932), pp. 1–17. (The page numbering of the papers in the volume suggest they were published separately and then collected together: they all begin on page 1.).

[72]Spottiswoode was never Astronomer Royal.

[73]Brajendranath Seal, *The Positive Sciences of the Ancient Hindus*, Longmans, Green and Co., London, 1915, pp. 76–80, defends the claim that Bhāskara II was acquainted with "the principle of the Differential Calculus". He points out that the astronomical time units in the *Siddhānta Śiromaṇi* go down to one *truṭi*, which equals 1/33750 of a second. He says

> Bháskara, in computing the "instantaneous motion" of a planet, compares its *successive* positions, and regards its motion as constant during the interval (which of course cannot be greater than a Tru*t*i of time, though it may be indefinitely less). This Tátkálika motion is no other than the differential of the planet's longitude, and Bápudeva Śástrí claims that both the conception of the instantaneous motion and *the method of determining it* plainly show that Bháskara was acquainted with the principle of the Differential Calculus... And in the passage in which Bháskara describes the process, he distinguished between Sthúla-gati and Súkshma-gati (velocity roughly measured, and measured accurately, *i.e.* by reference to indefinitely small quantities; for Súkshma, as we have seen, has always a reference to the A*n*u, the indefinitely small); indeed, he expressly mentions that the Sthúla-gati takes only

the point sufficiently clear whether Indian astronomers had a clear notion of the principle and methods of the Differential Calculus.[74]

Unfortunately, Sengupta does not make a good case for Bhāskara II's grasp of the "principle of the Differential Calculus", and makes no mention of Rolle's Theorem. He refers to "'instantaneous' or …daily motion", "'instantaneous' daily motion", and even "instantaneous daily motion", supporting Spottiswoode and Hoppe in their contention that a limit is not being taken.

And, Sengupta's first attempt to display an instance of differentiation is indeed no more than an approximation based on tables. Before discussing this I should say that the Indian sine was not the modern sine, but the product of the sine and the radius of a given circle, thus $R\sin\theta$ as opposed to $\sin\theta$, and $R\cos\theta$ as opposed to $\cos\theta$. And tables of sines would be given for specific convenient radii. This had been the practice in the Greek tables of chords made by Hipparchus and Ptolemy, and it continued with respect to sines and cosines in India, the Arab world, and Europe well past Copernicus. In India, 3438 was a common choice[75]; Brahmagupta, whom Sengupta cites, uses 3270.

As for the tables themselves, the Indians knew the Addition and Subtraction Formulæ for sines and cosines, as well as the Half Angle Formulæ, and their early tables listed sines and cosines at intervals of $3\frac{3}{4}^\circ = 225'$.

With this in mind, we can quote Sengupta:

There are indeed many methods by which this relation, *i.e.*, the equation $d\,(\sin\theta) = \cos\theta\, d\theta$ may have been recognised. Here is the most probable way by which this was recognised.[76]

In the figure given below [Fig. 3.10] let XOY be a quadrant of a circle and XP any arc, $\angle\text{POX} = \theta$. Measure PP′ such that the arc = 225′. Now PP′ $= \dfrac{225' \times R}{3438'}$, where $R = \text{OX}$. Let the figure be completed as shown.

$$\text{Here PM} = R\sin\theta,\ \text{PN} = R\cos\theta,$$
$$\text{P}'\text{R} = \text{Tabular difference of sines at the arc XP.}$$

The Δs PP′R and PON are similar. OP : PN = P′P : P′R

$$\therefore \text{The Tabular difference of sines} = \text{P}'\text{R} = \frac{\text{P}'\text{P} \times \text{PN}}{\text{OP}} = \frac{R\cos\theta \times 225'}{3438}.$$

But according to Brahmagupta 225 : 3438 = 214 : 3270[77]

(Footnote 73 continued)
Sthúla-kála (finite time) into consideration, and that the determination of the Tátkáliki-gati (Súkshma-gati) must have reference to the moment, which is an indefinitely small quantity of time, being, of course, smaller than his unit, the Tru*t*i. [pp. 77 – 79.]

[74] Sengupta, "Infinitesimal calculus", *op. cit.*, pp. 1–2.

[75] Identifying 2π with $360^\circ = 360$ degrees $= 360$ degrees $\times$ 60 minutes/degree $= 21600$ minutes $= 21600'$, we have $21600/(2\pi) \approx 3437.746771 \approx 3438$.

[76] This is not exactly what one wants to read in a proof that the Indians had a knowledge of differentiation.

[77] I.e., sin 225′ is approximately 214/3270. The fractions agree to five decimal places.

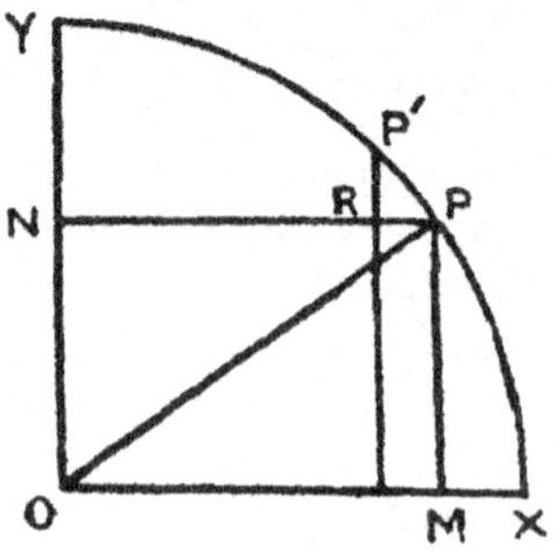

Fig. 3.10 Sengupta's diagram

∴ the tabular difference of sines

$$= \frac{R \cos\theta \times 225}{3438} \text{ or } \frac{R \cos\theta \times 214}{3270}.$$

Sengupta[78] has thus derived[79]

$$\frac{R \sin(\theta + 225')}{3438} - \frac{R \sin\theta}{3438} \approx \frac{R \cos\theta \times 225'}{3438},$$

i.e.,

$$\Delta \sin\theta \approx \cos\theta \cdot \Delta\theta. \tag{3.13}$$

Now the identification of the arc PP′ with the undrawn line segment PP′ is approximate for $\Delta\theta = 225'$. Was it obvious that a smaller choice of $\Delta\theta$ gave an even better approximation[80] and would in the limit yield

$$d \sin\theta = \cos\theta \, d\theta? \tag{3.14}$$

Following his presentation of the argument, Sengupta states, "We shall have ample proofs later on that the ancient Indian astronomers could really follow such a method". I'm not sure whether this comment refers only to the derivation of (3.13) or if it is intended to include (3.14) as well. In either case, he has thus far only presented a plausibility argument in favour of an early intuitive understanding of differentiation on the part of the Indians.

Sengupta cites some passages from the *Grahagaṇita*[81] in support of Bhāskara II's grasp of differentiation. The first, from VIII, 37, is a reference to the formula for instantaneous daily motion that one would naturally obtain by differentiation, and

---

[78]Sengupta, "Infinitesimal calculus", *op. cit.*, pp. 6–7. I have taken a few liberties with the notation, writing ∠ for the angle instead of <, and italicising those occurrences of "R" where it stands for a radius and not the point R of the diagram.

[79]The table would give 3438 sin α, whence one multiplies by $R/3438$ to get $R \sin \alpha$.

[80]But see pages 202–203, below.

[81]Cf. page 192, below, for an explanation.

the second is a slight modification of the proof in VIII, 39, of a formula going back to Brahmagupta (628 A.D.). It too uses (3.13) to replace the tabular difference by a derivative. Finally, he includes a long extract from the *Golādhyāya*, VIII, 69–74, again doing the calculation for $\Delta\theta = 225'$, and utilising (3.13).

In a very short passage (2 pages) in a more philosophical work, Susantha Goonatilake touches on Bhāskara II and the Differential Calculus, citing the now familiar[82]

$$\sin y' - \sin y = (y - y') \cos y.$$

He reminds us that instantaneous velocity is measured in trms of the *truti*, an "infinitesimal" unit of time, and continues:

> He also knew that the differential dy/dx vanishes when the variable is at a maximum. His approach had elements of the "mean value theorem," which is obtained from Rolle's Theorem (1691). Thus he stated in his *Graha ganita, Spasthadhikara*, "Where the planet's motion is an extremum, the motion is stationary; at the commencement and the end of retrograde motion, the apparent motion of the planet vanishes."[83]

"Elements of the mean value theorem" are not exactly the Mean Value Theorem itself. Still, if Bhāskara II recognised that $dy/dx = 0$ at a maximum, he certainly had the key element in the proof of the Mean Value Theorem in its modern analytic form. This would put him analytically on par with Archimedes's possession of the geometrical principle behind his Proposition 18. I use the conditional because it is not clear from the few scattered quotations we have cited exactly what Bhāskara II stated, or whether he was aware that some general principle was involved or he merely observed in specific cases something *we* recognise as instances of general principles. And, of course, is the Mean Value Theorem or even Rolle's Theorem to be found among these principles?

The reader may well be wondering at this point why I don't just quote Bhāskara II. To explain why, it may be a good idea to pause and explain a few things about Hindi mathematical literature. Early on the tradition was oral. Major works were composed in verse, which serves as an aid to memorisation. The same was, of course, true in Europe as well. Whereas in Europe the tradition of verse largely died out in scientific works, in India it lasted well past the rise of the written word. Primary scientific works were recited in verse and memorised by students. The advantage of verse as a memory aid is countered somewhat by the ofttimes cryptic nature of the verse. Thus the verses would be memorised by the students and then explained in commentary by their teachers. An oversimple example might be to consider "FOIL" as the verse to be memorised. The commentary would then be the explanation that "FOIL" stands for "first, outer, inner, last" and is a mnemonic for matching the terms when multiplying two binomials. The commentary could also include justification for the rule by appeal to the distributive law. Hindi verse often just cited the rules to use. Justification — proofs and explanation of principles — were reserved for commentaries. The

[82] There is a typographical error here: The prime on the right-hand-side of the equation is on the wrong occurrence of y; compare (3.12).

[83] Susantha Goonatilake, *Towards a Global Science: Mining Civilizational Knowledge*, Indiana University Press, Bloomington, 1998, p. 135.

Siddhānta Śiromaṇi was written in verse, and Bhāskara II accompanied it with his own *vāsanā* or commentary, the *Vāsanābhāṣya*.[84]

The *Siddhānta Śiromaṇi* is divided into four parts: (1) *Līlāvatī* on arithmetic and geometry, (2) *Bījagaṇita* on algebra, (3) *Golādhyāya* on trigonometry and spherical trigonometry, and (4) *Grahagaṇita* on planetary motion.[85] Translation of the work into English has been piecemeal: The first two parts were translated and published by Henry Thomas Colebrooke in 1817.[86] The third part was published in 1861 in a translation undertaken by Lancelot Wilkinson, a British official in India who had furthered Bāpūdeva's education and career, and revised by Bāpūdeva.[87] Finally, there are at least two English translations of the *Grahagaṇita* of the last and the present centuries. Colebrooke's translation, concerning as it does, only the most elementary portions of Bhāskara II's treatise, would not seem to have much to offer to our discussion were it not for the fact that there are many online references to a commentary on the *Līlāvatī* by a later scholar named Parameśvara and a statement by him therein of the Mean Value Theorem. I've not seen this commentary and can neither confirm nor deny these rumours.

Datta and Singh refer us to the *Golādhyāya, Spaṣṭādhikāra vāsanā*, which would be the *vāsanā* or commentary on the chapter *Spaṣṭādhikāra* of the *Golādhyāya*. None of the chapter titles given in the Wilkinson/Bāpūdeva edition is identified as *Spaṣṭādhikāra*, but the only chapter therein not given a Sanskrit name is the 5th chapter, covering much the same material as the 8th chapter of the *Grahagaṇita*, which is called *Spaṣṭādhikāra* in each of the two editions of that book at my disposal. Thus, Chap. V of the *Golādhyāya* and VIII of the *Grahagaṇita*, to which Sengupta refers are the places to look for any trace of the Mean Value Theorem in the *Siddhānta Śiromaṇi*.

Datta and Singh direct us to the *vāsanā* of Chap. V of the *Golādhyāya* for Dvivedi's points (*i*) and (*ii*). Now, the translation by Wilkinson and Bāpūdeva gives only the verses and not the commentary. Moreover, it was published decades ahead of Dvivedi's observation, whence one cannot expect Rolle's Theorem to be flagged for us by Wilkinson and Bāpūdeva in their annotations. Bearing in mind that "the principle of the Differential Calculus ... was only incidentally and briefly treated by him;

[84]For a fuller account of the tradition of Sanskrit verse I refer the reader to Kim Plofker, "Sanskrit mathematical verse", in: Eleanor Robson and Jacqueline Stedall (eds.), *The Oxford Handbook of the History of Mathematics*, Oxford University Press, Oxford, 2009.

[85]The *Līlāvatī* and *Bījagaṇita* are not universally considered part of the *Siddhānta Śiromaṇi*. Cf. David Pingree, "Bhāskara II", in: Charles Coulston Gillispie (ed.), *Dictionary of Scientific Biography*, vol. 2, Charles Scribner's Sons, New York, p. 115. Pingree places the *Grahagaṇita* (aka *Gaṇitadhyāya*) before the *Golādhyāya* and describes the latter as "to a large extent an expansion and explanation of the *Gaṇitadhyāya*" (pp. 117–118). In addition to these parts, Bhāskara II wrote a commentary, the *Vāsanābhāṣya* or *Mitākṣarā*, on the *Siddhānta Śiromaṇi*.

[86]Henry Thomas Colebrooke, *Algebra, with Arithmetic and Mensuration, from the Sanscrit of Brahmegupta and Bhascara*, John Murray, London, 1817.

[87]Lancelot Wilkinson and Pandit Bápú Deva S'ástrí, *Hindu Astronomy II. The Siddhánta* Sīromani, Calcutta, 1861. This is a companion volume to *Hindu Astronomy I. The Surya-Siddhánta*, translated by Bāpūdeva and published the previous year. The two books have been reprinted recently in single-volume editions, and can also be downloaded from the Internet in a single file.

and his followers, not comprehending it fully, have hitherto neglected it entirely", it is quite conceivable one could read this translation and not recognise Rolle's Theorem when (and if) it occurs. It is even conceivable that one afflicted with precursoritis can read it into the text where it does not in fact occur. And, of course, there is the possibility that one will not be able to settle the matter because the cited passage is not in the verses but only in the as yet untranslated *vāsanā*.

The two translations of the *Grahagaṇita* at my disposal are also not completely satisfactory. The first, by Arka Somayaji,[88] was published in 1980 and is available online. This translation covers the verses, but replaces Bhāskara II's *vāsanā* by his own modern commentary replete with diagrams and the use of the Differential Calculus. The commentary thus explains Bhāskara II's astronomy, but interprets the mathematics. Still, if we accept that Bhāskara II had something like the Differential Calculus at his command, such a modern commentary should make it easy to find those passages where he uses it. Specifically, if some instance of the Mean Value Theorem can be found in or read into the *Grahagaṇita*, Somayaji's translation and commentary ought to facilitate finding it. Unfortunately, what Somayaji supplies is partially taken away by bad printing. There are numerous blank spaces in the book, as if someone had gone over it with the opaque white fluid formerly applied by typists to cover their mistakes. There are empty spaces where, presumably, expressions typeset in Sanskrit were meant to be inserted, there is one label for a figure that isn't given, and, most seriously, there are gaps in mathematical formulæ, usually denominators going missing. Somayaji's translation promises to be difficult, occasionally impossible, to read.

The second translation is a religious work by Bimala Prasada Siddhānta Sarasvatī edited by Danivir Goswami.[89] This edition contains the Sanskrit text in poor photocopied reproduction, a very attractively typeset Bengali translation, and a decently typeset English translation from the Bengali by Pinaki Talukdar. There is one annotation that ends (on page 360) with "But we know" followed by a blank space covering the first quarter of the next page. In several places, passages are given only in Sanskrit and Bengali, unaccompanied by English renditions. And, following an annoying habit practised by many Indian translators of Indian mathematics, many of the technical terms are transliterated and not translated. For example:

> Now the *śīghraphala* of *gati* is stated. The *śīghraphala* of the planet should be subtracted from 90. We should taken [*sic*] the *jyā* for the remaining number. It should be multiplied by *śīghrakendra gati* and divided by *śīghra karna*. The resulting figure should be subtracted form [*sic*] the *śīghra ucchagati*. The resultant will be the *śīghra sphatagati* of the planet.[90]

There is a glossary of several hundred such untranslated terms, but they do not cover everything. That *jyā* is not listed is perhaps not a serious omission as most general histories of mathematics inform us that it is the sine. The phrase *śīghra*

[88] D. Arkasomayaji, *Siddhānta Śiromaṇi of Bhāskarācārya; English Exposition and Annotation in the light and language of modern Astronomy*, Rashtriya Sanskrit Sansthan, New Delhi, 1980.

[89] Bimala Prasada Siddhānta Sarasvatī (trans.) and Danavir Goswami (ed.), *The Sūrya-siddhānta and Siddhānta-śiromaṇi*, Gopsons Papers Ltd., Noida, 2007.

[90] *Ibid.*, p. 388.

sphatagati, however, requires a bit of detective work. It appears to be a typo for *śīghra sphuṭagati* and the glossary tells us that *śīghra* means "fast motion" and *sphuṭagati* (= *spaṣṭagati*) means "true motion of a planet".[91] As for *śīghra ucchagati*, the glossary provides no clue. In short, with respect to English, the book is only a partial translation of the *Grahagaṇita*.

I have not been able to identify any passage in either the translated verse or the translated *vāsanā* in the Sarasvatī/Goswami edition suggestive of the Mean Value Theorem. Nor have I found the phrase cited by Goonatilake. The matter requires a careful reading of the entire text by one who has mastered all the untranslated technical terms. This edition does, however, offer some evidence that Bhāskara II knew the difference between approximate and exact values. Verse 36 of Chap. VIII reads

> The difference in position of the corrected planetary position for one day is the instantaneous corrected motion of the planet.[92]

At first glance this reads as if it corroborates Spottiswoode's assertion that Bhāskara II, in taking differences for a period equalling one day and calling that "instantaneous", is not dealing with infinitesimals and limits at all, but only with approximations. In the *vāsanā*, however, we find explicit acknowledgement of the approximate nature of results:

> Calculations can be made based on this movement (taking it to be a uniform movement, which is not actually the case). This is the approximate movement calculated.[93]

We also find in the verses of Chap. IX a couple of references to the improvement of results by successive approximations. Verse 4 reads

> Assume the true planet to be the mean; compute the *mandaphala* and *śīghraphala* and apply them inversely, we have an approximation of the mean planets. Treating this as mean planets, again obtaining the *mandaphala* and *śīghraphala* and applying them inversely (i.e. add in negative and subtract in positive) and repeating the process until constant values are obtained, we have by this method of successive approximation the required mean planet.[94]

Again, in verse 19, we read

> Computation of rising times of the zodiacal signs would be approximate because we consider big arcs of the ecliptic. For much closer calculation (using the above method) one should use the rising times of *hora* (hours) or *drekkāna*.[95]

Sadly, Bhāskara II's commentary on these verses is given only in Sanskrit and Bengali. However, his comments are short and probably do not reveal more than what is obvious: He knew he was dealing with approximations and that approximations could be improved by dealing with finer divisions of time.

In Chap. I, verse 6 reads

---

[91] Datta and Singh, page 194, above, translate *(sphuṭa) gati* as "true rate of moton".

[92] Sarasvatī and Goswami, *op. cit.*, p. 378.

[93] *Ibid.*, pp. 382 – 383.

[94] *Ibid.*, p. 395.

[95] *Ibid.*, p. 410.

> The scholars name that book a *siddhānta* text which deals with various measures of time, ranging from *truṭi* up to the great flood at the end of a *kalpa*...[96]

Verses 16–18 and their accompanying *vāsana* offer details of the units of measurement:

> 100 *truṭi* = 1 *tatpara*
>
> 30 *tatparas* = 1 *nimeṣa* (time taken to blink an eyelid)
>
> 18 *nimeṣas* = 1 *kāṣṭhā*
>
> 30 *kāṣṭhas* = 1 *kalā*
>
> 30 *kalās* = 1 *ghaṭī*
>
> 2 *ghaṭīs* = 1 *muhūrta* = 1 *kṣaṇa*
>
> 30 *muhūrtas* = 1 *dina* (day).

If one performs the arithmetic, one finds there are 33750 *truṭis* to our temporal second. To the average person this is infinitesimal.[97]

In his popular account, *The Crest of the Peacock*, George Gheverghese Joseph sums up the claims on behalf of Bhāskara II. Following mention of the derivative of the sine, he announces

> It may seem far-fetched to claim, on this evidence alone, that Bhaskaracharya was one of the first mathematicians to conceive of the differential calculus, but there is further evidence to be found in his *Siddhanta Siromani*:
>
> 1. In computing the instantaneous motion of a planet, the time interval between successive positions of the planet was no greater than a *truti*, or 1/33 750 of a second, and his measure of velocity was expressed in this 'infinitesimal' unit of time.
> 2. Bhaskaracharya was aware that when a variable attains the maximum value, its differential vanishes.
> 3. He also showed that when a planet is either at its furthest from the Earth or at its closest, the equation of the centre vanishes. He therefore concluded that for some intermediate position the differential of the equation of the centre is equal to zero.
>
> In the third observation above there are traces of the "mean value theorem", which today is usually derived from Rolle's theorem (1691).[98]

---

[96] *Ibid.*, p. 189.

[97] Speaking of infinitesimals, on page 194, above, we quoted Datta and Singh on the *kṣaṇa* being an infinitesimal amount of time. At only 30 *kṣaṇas* per day, this makes a *kṣaṇa* equal to 48 min, which is hardly an infinitesimal length of time. A smaller *kṣaṇa* occurs in the mensuration of the ancient physicists: 1 *kṣaṇa* = 2/45 of a second, small but still not infinitesimal. The *truṭi*, though actually finite, would at first sight seem a better candidate for infinitesimality — as it was described by Goonatilake in a passage cited on page 198, above. However, words are often used in a number of ways: K. Ramasubramanian informs us (private correspondence), "The word most commonly employed — in the technical as well as classical Sanskrit literature — to refer to the 'instant of time' is *kṣaṇa* and NOT *truṭi*, which is just a very small unit of time compared to a day. For instance, if one were to say 'I am leaving this moment itself', the usage is: *ahaṃ asminneva kṣaṇe prasthāsyāmi*. No one would use the word *truṭi* as it DOES NOT mean an instant.".

[98] Joseph, *op. cit.*, p. 409.

The first of these assertions goes back to Seal's book of 1915[99] and bears on Bhāskara II's notion of velocity of a planet. Joseph's items 2 and 3 are just the claims (*i*) and (*ii*) of Datta and Singh.

The first point attempts to establish Bhāskara II's possession of the notion of the derivative through his consideration of motion during an infinitesimal interval of time. It would come close — only 1/33750 away from infinitesimality — if it were indeed true. I cannot verify the basic assertion. I have found the *truṭi* mentioned in the *Grahagaṇita*, but not in connexion with instantaneous motion, which, as I have cited, Bhāskara II defines to be the motion a planet undergoes between two successive days. This is exactly what Spottiswoode said. And some modern critics agree. I quote P.P. Divakaran:

> The credibility of the evidence depends on the interpretation of a particular piece of phraseology in their writings, evocatively rendered in [27][100] as "instantaneous velocity", a concept which has a long and contentious history in Europe going back at least to Aristotle. If this rendering is correct, these allusions will constitute an anticipation of the Newtonian view of dynamics as calculus, with position as a function of the primordial variable, time, and 'velocity' as the derivative. The texts however do not appear to support such a sharp reading. In all of the instances cited, the variations of planetary parameters being discussed are over an interval of time, generally one day. The phrase "instantaneous velocity" is used as the translation of *tātkālika gati*; but *tatkāla* has the literal (and, in the context, natural) meaning of "that (designated) *kāla*" and *kāla* itself is most commonly used for an interval of time.[101] Besides, the passages make it clear that the corrections discussed refer to changes over a day, differences rather than differentials.[102]

The linguistic argument has been disputed by K. Ramasubramanian:

> Here it must be pointed out that, *kāla* is a generic term which just means "Time" and is not specific to either intervals or instants as such — not to speak of the other connotations it has.[103]

This, however, does not blunt the force of the main point, that the time period in question is a day, not a *truṭi*, thus, as Divakaran says, a difference rather than a differential.

---

[99] Cf. footnote 73, above.

[100] The reference is to: K. Ramasubramanian and M.D Srinivas, "Development of Calculus in India", in: C.S. Seshadri (ed.), *Studies in the History of Indian Mathematics*, Hindustan Book Agency, New Delhi, 2010.

[101] Divakaran adds a footnote here: "The more precise word for 'instant' is *kṣaṇa* which in fact occurs in one passage but not as a qualifier of "velocity". For the record I add that *tātkālika gati* as an unbroken phrase does not actually occur in any of these passages." Ramasubramanian disputes this linguistically, stating that "whether the compound is 'broken' or 'unbroken' [it] gives the same meaning" in Sanskrit, and adds "I may also point out here, that the use of the word *tatkāla* to refer to 'that instant of time' is not something that is unique to Indian astronomy. It is quite common in other disciplines of study as well. For instance, in Indian Logic *Nyāya-śāstra*, people quite often employ the term *tatkāla* to refer to that instant of time". See also footnote 97 on page 203, above, regarding *kṣaṇa*.

[102] P.P. Divakaran, "Notes on *Yuktibhāsā*: the birth of calculus", to appear, p. 46.

[103] Private correspondence.

This is for planets. The moon's motion is larger to the naked eye and requires special consideration. Ramasubramanian and Srinivas cite verses 36–38 of the *Spaṣṭādhikāra* of the *Grahagaṇita*:

> The true daily motion of a planet is the difference between the true planets on successive days. And it is accurate (*sphuṭa*) over that period. The *koṭiphala* (Rcosine of anomaly) is multiplied by the rate of motion of the *manda*-anomaly (*mṛdu-kendra-bhukti*) and divided by the radius. The result added or subtracted from the mean rate of motion of the planet, depending on whether the anomaly is in *Karkyādi* or *Mṛgādi*, gives the true instantaneous rate of motion (*tātkālikī manda-sphuṭagati*) of the planet.
>
> In the case of the Moon, the ending moment of a *tithi*[104] which is about to end or the beginning time of a *tithi* which is about to begin, are to be computed with the instantaneous rate of motion at the given instant of time. The beginning moment of a *tithi* which is far away can be calculated with the earlier [daily] rate of motion. This is because Moon's rate of motion is large and varies from moment to moment.[105]

There is, of course, the presence of the phrase *tātkālikī manda-sphuṭagati* incorporating the oft-cited *tātkālikī gati*. The more pertinent point is the phrase "from moment to moment".[106] Ramasubramanian and Srinivasi explain

> Here, Bhāskara explains the distinction between the true daily rate of motion and the true instantaneous rate of motion. The former is the difference between the true longitudes on successive days and it is accurate as the rate of motion, on the average, for the entire period. The true instantaneous rate of motion is to be calculated from the Rcosine of the anomaly (*koṭiphala*) for each relevant moment.[107]

They proceed to quote Bhāskara's *vāsanā*:

> The true daily velocity is the difference in minutes etc., between the true planets of today and tomorrow, either at the time of sunrise or mid-day or sunset. If tomorrow's longitude is smaller than that of today, then we should understand the motion to be retrograde. It is said "over that period". This only means that, during that intervening period, the planet is to move with this rate [on the average]. This is only a rough or approximate rate of motion. Now we shall discuss the instantaneous rate of motion... In this way, the *manda*-corrected true instantaneous rate of motion (*tātkālikī manda-parisphuṭagati*) is calculated. In the case of Moon, this instantaneous rate of motion is especially useful... Because of its largeness, the rate of motion of Moon is not the same every instant. Hence, in the case of Moon, the special [instantaneous] rate of motion is stated.
>
> Then, the justification for the correction to the rate of motion (*gati-phala-vāsanā*)... The rate of motion of the anomaly is the difference in the anomalies of today and tomorrow. That should be multiplied by the [current] Rsine-difference used in the computation of Rsines and divided by 225. Now, the following rule of three to obtain the instantaneous Rsine-difference: If the first Rsine-difference 225 results when the Rcosine is equal to the radius, then how

---

[104]The authors explain that the "*Tithi* is the time taken by the Moon to lead the Sun exactly by 12° in longitude.

[105]Ramasubramanian and Srinivas, *op. cit.*, p. 21. The bracketed insertion was made by the authors.

[106]There is a lot of leeway in translating and the issue being the correctness of the use of words like "instantaneous" and "moment", it seems appropriate to note that the word "moment" here is the common choice of Ramasubramanian/Srinivas, Somayaji, and Sarasvati/Goswami in their independent translations of this passage.

[107]Ramasubramanian and Srinivas, *op. cit.*, p. 21.

> much is it for the given Rcosine. In this way, the Rcosine is to be multiplied by 225 and divided by the radius. The result is the instantaneous Rsine-difference and that should be multiplied by the rate of motion in the anomaly and divided by 225...[108]

Their comment on this passage,

> Thus, Bhāskara is here conceiving also of an instantaneous Rsin-difference, though his derivation of the instantaneous velocity is somewhat obscure,[109]

brings us to the next critical issue — 225′.

The Indian route to the Calculus was via tables of trigonometric functions. The sines and cosines of 90°, 60°, 45°, and 30° were known. The trigonometric Addition and Subtraction Formulæ were also known and sines and cosines of 75° and 15° could thus be added to the tables. Using the Half-Angle Formula one could then obtain the sines and cosines of $7\frac{1}{2}^{\circ}$ and finally of $3\frac{3}{4}^{\circ} = 225'$. The basic Indian trigonometric table evaluated sines and cosines between 0° and 90° in intervals of 225′. Not counting 0°, this gave tables of 24 sines or 24 cosines. Such tables appear in astronomical works from the *Sūrya Siddhānta* down to the *Siddhānta Śiromaṇi*. The calculations in the *Siddhānta Śiromaṇi* are based on this table. Divakaran criticises this. The classical method of approximating the circumference of a circle was to use the perimeters of inscribed or circumscribed polygons. If one starts with a hexagon having 6 sides, and successively doubles the number of sides four times, one will arrive at a 96-sided polygon. For some time Indian scholars assumed the regular 96-gon was the circle. He cites Kim Plofker's translation from a work of Bhāskara I (*fl. c.* 629):

> It is proper to say that a unit arc can be equal to its chord; even someone ignorant of treatises knows this; that a unit arc can be equal to its chord has been criticised by precisely this [master[110]]
>
> But we say: An arc equal to a chord exists. If an arc could not be equal to a chord then there would never be steadiness at all for an iron ball on level ground. Therefore, we infer that there is some spot by means of which that iron ball rests on level ground. *And that spot is the ninety-sixth part of the circumference.*[111]

Concerning Bhāskara II, Divakaran adds

> Five hundred years after Bhāskara I, we have another example of the staying power of the 96-fold division, this time in the work of his even more illustrious namesake, Bhāskara II.[112]

The quarter circle will consist of 96/4 = 24 arcs of length 225′. Bhāskara II finds the surface area of a hemisphere by adding the areas of 24 trapezia identified with flattened surface elements and then declaring the formula for the surface area, having approximately verified it numerically using the sine tables.

---

[108] *Ibid.*, pp. 22–23. The bracketed insertions were made by the Ramasubramanian and Srinivas.

[109] *Ibid.*, p. 23.

[110] The master referred to is Āryabhaṭa (*476) who made the first steps toward the Calculus *c.* 499. This bracketed insertion was made by Plofker.

[111] Divakaran, *op. cit.*, p. 38. In a footnote Divakaran notes that the emphasis is his and adds, "It is ironic that this appeal to physics, so very rare in Indian astronomy, should be such an absurdity".

[112] *Ibid.*, p. 39.

At one point in the explanation, Bhāskara does say that more chords will result in more annular regions but there is no evidence that he actually resorted to a finer division; that would have required a finer sine table and a more accurate π. And there is of course not the faintest suggestion of making an infinitely fine division. How then did he arrive at the neat formula to which his numerical answer was only an approximation? In looking for a plausible answer, it is perhaps useful to remember that all kinds of correct but illegitimate results for circles and spheres can be derived just from looking at polygons and polyhedra if only Bhāskara I's fallacy — that $\sin \pi/n = \pi/n$ for some n — held. The perimeter of a regular n-gon for example would be $2nR\sin(2\pi/2n) = 2nR\pi/n = 2\pi R$, independent of n, and the area of a regular $2n$-gon would be, similarly, πR^2 where R is the radius of the circumscribing circle. A circle would really be indistinguishable from a polygon.[113]

Ramasubramanian defends[114] Bhāskara II on this point:

> We wish to note here that in the *Jyotpatti* section[115] of SS [*Siddhānta Śiromaṇi*], Bhāskara II calculates the value of Sine 1 degree, accurately to five decimal places, which is clearly different from the arc of 1 degree. He also (like his predecessor Bhāskara I) uses the value of π to be approximately 3.1416, which would correspond to approximating the circle by a polygon of not less than 1024 sides![116]

There would, of course, be no point to calculating the sine of $1°$ if the circle were a regular 96-sided polygon. The obvious explanations of the use of $225'$ are not that Bhāskara II considers the arc indivisible, but that i. it is a traditional value and ii. one has to stop somewhere and an increment of $225'$ gave acceptable results. Overlapping or, perhaps, reinforcing these reasons was the fact that the *Siddhānta Śiromaṇi* was a textbook giving instructions for calculations and even offering some exercises.

Bhāskara II was aware of the difference between approximate and exact results and that successive approximations could be made. We have already seen some evidence of this. Another instance, given in verses 82–83 of Chap. X of the *Grahagaṇita*, is as follows:

> Assume the sine of the distance from the horizon in time (*unnatakaalajya*) be the required *hriti* (*ishta hriti*) in the first place. Multiply this by 12 times shadow, and divide the product by the square of the hypotenuse of the shadow. This result gives us the approximate value of the sine of declination ($\sin\delta$). Using the approx $\sin\delta$ determine the *dujya, kujya, charajya*...and *chara*. Use the *chara* to find out the corrected value of required *hriti* (*ishta hriti*). Multiply this by $\sin\delta$ determined previously and divide the product by the assumed required *hriti* (*ishta hriti*) to give us the nearer approximation of sine of declination ($\sin\delta$). Repeat this process till a stationary value has been reached. That would be the corrected value of sine of declination ($\sin\delta$).[117]

---

113 *Ibid.*, p. 40. I might perhaps add that even in modern Nonstandard Analysis, where one does more-or-less identify a circle with an n-gon for infinite n, one does not get an actual identity. Different infinite values of n can be invoked and the polygons will differ. The measurements also differ, but they do so infinitesimally and differ from a unique real number only infinitesimally.

114 Private correspondence.

115 In Wilkinson and Śástrí, *op. cit.*, this section is labelled "Appendix. On the construction of the canon of sines" and occupies pp. 263 – 265.

116 Private correspondence. The reference to "Sine" is to the Indian sine, which is the sine multiplied by the radius.

117 Sarasvatī and Goswami, *op. cit.*, p. 520. The ellipsis was in the original and does not indicate an omitted passage.

To understand completely what is being described requires a bit more translation and an explanation of the astronomical setting of the calculation. Fortunately for the writer, this is all beside the point, which is that Bhāskara II explicitly calls for the repetition of a procedure until a "stationary value" is reached. This is not, of course, a limit as such might never be reached, but, up to a given degree of accuracy, he expects such a value.

There is no doubt that the Calculus began to arise in India long before the European development got under way. What is in question is Bhāskara II's grasp of the subject and his supposed knowledge of "the principle of the Differential Calculus", whatever that might mean. The Calculus derives its name from, and Newton and Leibniz are credited with it for the invention of, the algorithms devised by them to calculate the slopes of tangents and the areas under curves. If we take these rules as the principles of the Calculus, in particular if we take the rules of differentiation to be the "principle of the Differential Calculus", then we must deny the Indians credit for inventing the Calculus. They did not have these rules. They, in fact, did not need them, as they never considered such large classes of functions as confronted the Europeans. However, as much fun as the methodical differentiation and integration can be, and as empowering as it is for the student of the Calculus, the fact that there are books dedicated to the history of the Calculus in Europe before Newton and Leibniz tells us that the specific algorithms of the Newton-Leibniz calculus are not the essence of the Calculus. Spottiswoode found this essence to lie in the notions of limit or infinitesimal; and, speaking more directly of the mechanism of their use, Divakaran[118] cites *local linearisation* — basically, the identification of infinitesimal arcs and line segments — to be essential. He finds the beginnings of this in the work of Āryabhaṭa, but says that the hint was not taken until the arrival of Mādhava of Saṇgamagrāma in the 14th century. Mādhava's works do not survive, but reports by his students and grandstudents do and within a few generations there appeared the *Yuktibhāṣā* of Jyeṣṭhadeva, a genuine textbook of the Calculus with aspects of the Differential and Integral Calculus worked out in detail not hinted at in the *Siddhānta Śiromaṇi*. Thus, for Spottiswoode and Divakaran, Bhāskara II did not have the Calculus at his disposal. But, as Spottiswoode says with respect to Bhāskara II's version of the differential of the sine, "the formula which he establishes … and his method of establishing it, bear more than a mere resemblance — they bear a strong analogy — to the corresponding process in modern mathematical astronomy". I must leave open the question of whether or not Bhāskara II was "fully acquainted with the principle of the Differential Calculus". Even if we do not accept this, we can look to see if his strong analogy extends to cover the vanishing of the differential at a local maximum and some form of the Mean Value Theorem itself. In other words, does the *Siddhānta Śiromaṇi* contain anything which *we* would recognise as necessarily instances of these results?

Joseph's points 2 and 3 are Dvivedi's observations (*i*) and (*ii*) as cited by Datta and Singh, who refer us to the *vāsanā* of Chap. V of the *Golādhyāya*. As I said earlier, Wilkinson translates only the verses and not the *vāsanā*. Nonetheless, we can hope

[118] Divakaran, *op. cit.*

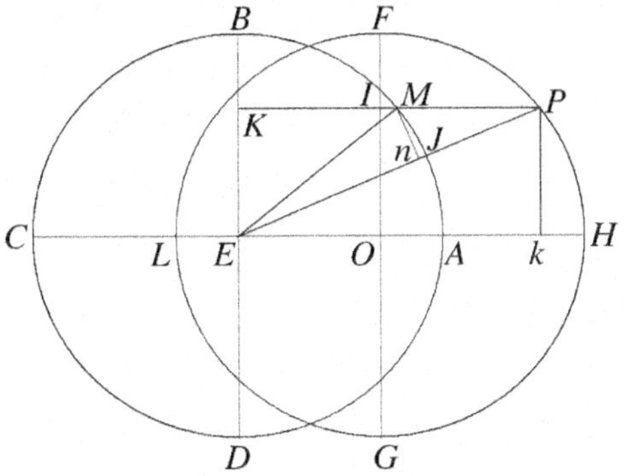

Fig. 3.11 The eccentric model

that some trace of it will appear in the verses about which the *vāsanā* is written. The chapter in question is quite short and in skimming over it, one immediately recognises something of what one is looking for in verse 39:

> 39. The mean motion of a planet is also its true motion when the planet reaches that point in the excentric cut by the transverse diameter which passes through the centre of the concentric: and it is where the planet is at that point that the amount of equation is at its maximum.[119]

To see to what extent this embodies (*i*) and (*ii*) or 2 and 3, we have to explain the astronomical terms — mean motion, true motion, excentric,[120] concentric, equation (of centre).

In ancient Indian astronomy planetary motion was described either by means of eccentric circles or epicycles. The eccentric theory is easier to explain, so I will base our discussion on it. The idea is very simple: The Earth is the centre of the universe and the moon, sun, stars and planets travel around the earth in circular orbits. As the motions of the planets exhibit some irregularity, some device must be employed to account for this. In the eccentric theory one moves the centre of a planetary orbit some distance from the Earth. This gives two circles — the *concentric circle* centred at the Earth, and the *eccentric circle* centred at some distance from the Earth.

In Fig. 3.11, the point E represents the Earth, O the centre of the planet's orbit, $ABCD$ the concentric circle representing the orbit had it indeed been centred at the Earth, and $HFLG$ the actual orbit. That is, $ABCD$ is the concentric circle and $HFLG$ the eccentric one. The points H and L are the *apses*, H being the higher apsis and L the lower apsis. The higher apsis is that point in the orbit at which the planet is at its farthest from the Earth and is also called the *apogee*, while the lower apsis is that point at which the planet is nearest the Earth and is called the *perigee*.

The point P on the diagram represents a planet revolving on its eccentric around the Earth. Its position is called the *true planet* and its motion the *true motion*. Its rotational speed is assumed constant. If we imagine another planet M beginning its revolution on the concentric circle at point A when P is at H, and assign to M the

[119] Wilkinson and S'ástrí, *op. cit.*, p. 149.

[120] The modern spelling is "eccentric" and will be used below.

same rotational speed as P, then M and P will always lie on a line parallel to the *line of apses CLEOAH*. The position of M is the *mean planet* and its motion the *mean motion*.

The point J on the diagram is where the planet P appears to be on the concentric when viewed from the Earth. The length of the arc MJ is called the *equation of centre*. If one drops the perpendicular from M to the point n on the line PE connecting the planet to the Earth, one arrives at a segment Mn, which is readily calculated by a little trigonometry:

$$\begin{aligned}\frac{Mn}{PM} &= \sin \angle MPn = \sin \angle KPE \\ &= \sin \angle kEP = \sin \angle HEP \\ &= \sin \theta,\end{aligned}$$

where θ is the longitude of P, taking the half-line EH as 0° longitude.[121] But $PM = HA = OE$ is a constant a, which we call the *eccentricity* of the eccentric circle. Thus

$$Mn = a \sin \theta.$$

At H and L, θ is 0° and 180°, respectively, whence $Mn = 0$, i.e., n lies on the concentric and coincides with M and J: The equation of centre is thus 0.

This is the first half of (*ii*) or 3, and is stated already in verse 21 of Chap. V:

> 21. The lower apsis is at a distance of six signs[122] from the higher apsis: when the planet is in either its higher or lower apsis, then its true place coincides with its mean place, because the line of the hypothenuse falls on the mean place of the planet in the concentric.[123]

Verse 39 is almost the conclusion of (*ii*) or 3, as we shall see shortly. The claim is that verse 39 *follows from* verse 21 and that this implication is Rolle's Theorem (Dvivedi) or, at least, has "traces of the 'mean value theorem'" (Joseph). But this premise and conclusion are separated by a number of verses which do not establish a connexion between the two.

Rolle's Theorem tells us that because the equation of centre vanishes at the apses H and L of Fig. 3.11 there will be some points on the arcs HFL and LGH where the derivative of the equation of centre vanishes. Verse 39 is slightly stronger than this conclusion: it says that, in fact, the maximum of the equation of centre occurs at these points — whence (*i*) and 2 tell us the derivative of the equation is 0. Moreover, 39 tells us explicitly where these maxima occur and is not a mere existence result. The implication, verse 21 ⇒ verse 39, is thus not Rolle's Theorem nor does it rely on Rolle's Theorem: Verse 39 is easily established directly if one applies a little Differential Calculus.

---

[121] Bhāskara II measures the longitude from a star in the constellation Aries, but any fixed direction will do.

[122] I.e., signs of the Zodiac. The point is that the apses are at opposite ends of a diagonal of the eccentric.

[123] Wilkinson and S'ástrí, *op. cit.*, p. 143.

The natural modern approach to establishing verse 39 is simply to solve

$$\frac{d\theta}{dt} = \frac{d\theta}{d\alpha} \cdot \frac{d\alpha}{dt} = 0,$$

where $\alpha = \angle MEA$, i.e., since $d\alpha/dt$ is a nonzero constant, to calculate $d\theta/d\alpha$ and determine where it is 0. In theory this is quite simple.

$$Ek = KP = KM + MP = R\cos\alpha + a,$$

where $R = EM$ is the radius of the concentric $ABCD$, and a is the eccentricity. And

$$Pk = R\sin\alpha.$$

Thus

$$\begin{aligned}\sin\theta = \frac{Pk}{EP} &= \frac{Pk}{\sqrt{Pk^2 + Ek^2}}\\ &= \frac{R\sin\alpha}{\sqrt{R^2\sin^2\alpha + (R\cos\alpha + a)^2}}\\ &= \frac{R\sin\alpha}{\sqrt{R^2\sin^2\alpha + R^2\cos^2\alpha + 2aR\cos\alpha + a^2}}\\ &= \frac{R\sin\alpha}{\sqrt{R^2 + 2aR\cos\alpha + a^2}}.\end{aligned}$$

It follows that

$$\theta = \sin^{-1}\left(\frac{R\sin\alpha}{\sqrt{R^2 + 2aR\cos\alpha + a^2}}\right).$$

3.2.4 Exercise Find when $d\theta/d\alpha = 0$.

I have given this as an exercise because it is too messy for my taste. In its place, I prefer a simpler, if less motivated approach.[124] Note that the equation of motion is

$$\text{arc } MJ = \text{arc } MA - \text{arc } JA = R\alpha - R\theta,$$

whence

$$\frac{d\,\text{arc } MJ}{dt} = R\frac{d\alpha}{dt} - R\frac{d\theta}{dt} = R\left(\frac{d\alpha}{dt} - \frac{d\theta}{dt}\right).$$

To determine when this derivative vanishes, consider the triangle EMP. The angle $\angle MEP$ equals $\alpha - \theta$ and angle $\angle MPE$ equals θ. Applying the Law of Sines to this triangle (the existence of which triangle requires $\alpha > 0$), we have

[124] In line with the discussion on pp. 152–156, above, I am duly ashamed of myself for applying such a trick.

$$\frac{\sin(\alpha-\theta)}{a}=\frac{\sin\theta}{R}.$$

Differentiation yields

$$\frac{\cos(\alpha-\theta)}{a}\cdot\frac{d(\alpha-\theta)}{dt}=\frac{\cos\theta}{R}\cdot\frac{d\theta}{dt}.$$

Now

$$\begin{aligned}\frac{d\,\text{arc}\,MJ}{dt}=0\Rightarrow\frac{d\alpha}{dt}&=\frac{d\theta}{dt}\\ \Rightarrow 0&=\frac{\cos\theta}{R}\cdot\frac{d\theta}{dt}\\ \Rightarrow \cos\theta&=0 \text{ or } \frac{d\theta}{dt}=0.\end{aligned}$$

Since $d\alpha/dt$ is a nonzero constant, the latter alternative cannot occur when the derivatives of the two angles are equal. It follows that $\cos\theta=0$, i.e., θ is 90° or 270°. And, conversely, at these angles the derivative of the equation of centre is 0.

Thus we have the conclusion of (*ii*) and 3: When θ is 90° or 270°, i.e., when P lies on the transverse axis of the concentric, the derivative of the equation of centre vanishes. As for Bhāskara II's verse 39, we know that the equation of centre vanishes at 0° and 180°, whence the equation must have maxima somewhere on the arcs HFL and LGH. From (*i*) and 2 we know that the derivative vanishes at such maxima. But the only points where the derivative vanishes are those just cited, whence the maxima occur there.

It would appear that verse 39 points implicitly to an instance of the principle that a derivative vanishes at a local maximum, i.e., some justification of point (*i*) is given. Perhaps the *vāsanā* is more explicit about this. Bāpūdeva is more explicit in his own footnoted comment on the verse:

> Now, as the difference between the true and mean motions is called the GATIPHALA, by cancelling therefore, p_2-p_1, p_3-p_2, &c. the parts of the true motions which are equal to the mean motion, the remaining parts e_2-e_1, e_3-e_2 &c. will evidently be the GATIPHALAS that is the differences between two successive amounts of equation are the GATIPHALAS. Thus, it is plain that the GATIPHALA entirely depends upon the amount of equation, but as the amount of equation increases, so the GATIPHALA is decreased and therefore when it is a maximum, the GATIPHALA will indifintely [*sic*] be decreased i. e. will be equal to nothing. Now as the amount of equation becomes a maximum in that place where the transverse diameter of the concentric circle cuts the excentric, ...the GATIPHALA, therefore becomes equal to nothing...[125]

The question is: how much of Bhāskara II's understanding is in this comment? Did Bāpūdeva paraphrase or replace the original justification?

Whether Bhāskara II knew the beginnings of the Differential Calculus or can only be credited with having had something strongly analogous I cannot decide from

[125] Bapudeva, *op. cit.*

my superficial inspection of the *Siddhānta Śiromaṇi*. For the sake of argument I am willing to concede to him knowledge of differentiation and even point (*i*). But I have been unable to find any evidence for the claim that Bhāskara II knew some instance of the Mean Value Theorem or even Rolle's Theorem. There is a claim, however, that Parameśvara states a version of the Mean Value Theorem in his commentary on the *Līlāvatī*.

There is a looseness in the use of the term "Mean Value Theorem" that renders such a claim totally uninformative. I've not seen Parameśvara's commentary on the *Līlāvatī* and cannot say what form his statement of the Mean Value Theorem took, nor for that matter can I yet verify that he made such a statement at all. If, indeed, he stated some version in response to his reading of Bhāskara II, then Bhāskara II genuinely belongs in the history of the Mean Value Theorem and not merely, as I fear may be the case, to the history of the historiography thereof.

As for Parameśvara, his rumoured involvement with the Mean Value Theorem extends beyond his commentary on the *Līlāvatī*. Typical of the online references is the following citation by J.J. O'Connor and E.F. Robertson on the MacTutor History of Mathematics Website:

> One of Paramesvara's most remarkable mathematical discoveries, no doubt influenced by Madhava, was a version of the mean value theorem. He states the theorem in his commentary *Lilavati Bhasya* on Bhaskara II's *Lilavati*. There are other examples of versions of the mean value theorem in Paramesvara's work which we now consider.
>
> The *Siddhantadipika* by Paramesvara is a commentary on the commentary of Govindasvami on Bhaskara I's *Mahabhaskaiya*. Paramesvara gives some of his eclipse observations in this work including one made at Navaksetra in 1422 and two made at Gokarna in 1425 and 1430. This work also contains a mean value type formula for inverse interpolation of the sine.

I have not seen the *Siddhānta Dipikā*, but there are accounts of this work by those who have seen it. Radha Charan Gupta (*1935) cites[126] the "mean-value-type formula"

$$R\sin(x+\theta) \approx R\sin x + (\theta/R).R\cos(x+\theta/2), \tag{3.15}$$

which is a very good approximation when θ is small enough. It misses being an instance of the Mean Value Theorem itself in not being an equality, but only an approximation. And, as Gupta points out, although there is an exact application of the Mean Value Theorem yielding (3.15),[127] Parameśvara's derivation of (3.15) is a more mundane application of a trigonometric identity and the familiar limit $\sin x/x \to 1$ for $x \to 0$. For,

---

[126]R.C. Gupta, "A mean-value-type formula for inverse interpolation of the sine", *The Mathematics Education* 10, no. 1 (1976), pp. 17–20; here: p. 17.

[127]One has

$$\sin(x+z) = \sin x + z\cos(x+cz),$$

where

$$c = \frac{1}{2} + \frac{z}{24}\cot x + \ldots$$

$$\begin{aligned}\sin(x+\theta)-\sin x &= \sin\left(x+\frac{\theta}{2}+\frac{\theta}{2}\right)-\sin\left(x+\frac{\theta}{2}-\frac{\theta}{2}\right)\\ &= \sin\left(x+\frac{\theta}{2}\right)\cos\frac{\theta}{2}+\sin\frac{\theta}{2}\cos\left(x+\frac{\theta}{2}\right)-\\ &\quad\left(\sin\left(\frac{x+\theta}{2}\right)\cos\frac{\theta}{2}-\sin\frac{\theta}{2}\cos\left(x+\frac{\theta}{2}\right)\right)\\ &= 2\sin\frac{\theta}{2}\cos\left(x+\frac{\theta}{2}\right).\end{aligned}$$

But for θ small, $\sin\theta/2 \approx \theta/2$, whence

$$\sin(x+\theta)-\sin x \approx 2\cdot\frac{\theta}{2}\cdot\cos\left(x+\frac{\theta}{2}\right)=\theta\cos\left(x+\frac{\theta}{2}\right). \tag{3.16}$$

Equation (3.15) follows after adjusting for the difference in radii.[128]

While the formula (3.15) is close to a special instance of the Mean Value Theorem, its approximative nature makes it more closely related to the Mean Value Inequality.[129] But like all the other results listed above and below as precursors, it is not the Mean Value Theorem itself, nor is it a special case thereof.

3.2.5 Exercise For z small, in addition to

$$f(x+z)\approx f(x)+zf'\left(x+\frac{z}{2}\right),$$

we have

$$f(x+z)\approx f(x)+zf'(x) \quad\text{and}\quad f(x+z)\approx f(x)+zf'(x+z).$$

Using a graphics calculator compare the graphs of

$$y=f(x+1),\ y=f(x)+f'\left(x+\frac{1}{2}\right),\ y=f(x)+f'(x),\ y=f(x)+f'(x+1)$$

for

i. $f(x)=\sin x$

ii. $f(x)=x^3$.

---

[128] In (3.15), x, θ measure arcs on a circle of radius R. The arcs on the corresponding circle of radius 1 are $x/R, \theta/R$, respectively. Thus the instance (3.15) follows from the instance of (3.16) for $x/R, \theta/R$.

[129] Albeit much sharper: Cf. the next two exercises.

3.2.6 Exercise Let $h > 0$, f twice continuously differentiable on $[x, x+h]$, and assume $f''(x) \neq 0$. Choose c, θ such that

$$f(x+h) = f(x) + hf'(x+ch), \quad 0 < c < 1$$
$$f(x+h) = f(x) + hf'(x) + \frac{h^2}{2}f''(\theta), \quad x < \theta < x+h.$$

Simple algebra yields

$$\frac{f'(x+ch) - f'(x)}{h} = \frac{1}{2}f''(\theta).$$

Using this show $\lim_{h\to 0} c = 1/2$.[130]

3.2.7 Exercise Let $f(x) = Ax^2 + Bx + C$ with $A \neq 0$. Show directly that, if

$$f(x+h) = f(x) + hf'(x+ch),$$

then $c = 1/2$. Interpret the results of Apollonius and Archimedes in the light of this information.

3.2.8 Exercise Let $f(x) = x^3$. Find c satisfying

$$f(0+h) = f(0) + hf'(0+ch).$$

We will return to the subject of these exercises in Sect. 3.12.2, below. Now, however, we move on to the next in our list of precursors.

[130] At the moment I don't know whom to credit this result to. The earliest relevant result I came across was a pair of exercises in H. Laurent, *Traité d'analyse*, Gauthier-Villars, Paris, 1885, p. 96. Exercise 17 on that page says that, if

$$f(x+h) = f(x) + hf'(x) + \ldots + \frac{h^n}{n!}f^{(n)}(x+ch),$$

then $c \approx 1/(n+1)$. He assumes c to be given as a power series in h and suggests calculating the terms. Exercise 18 covers the special case where $n = 1$, but does not specify how smooth f is assumed to be. The result is also given as an application of an infinite series representation of c in Exercise 55, page 124, in Joseph Edward, *Differential Calculus with Applications and Numerous Examples: An Elementary Treatise*, Macmillan and Co., London, 1886. The calculation of the first few terms of this series was then given as Exercise 56. In the third edition, *An Elementary Treatise on the Differential Calculus with Applications and Numerous Examples*, Macmillan and Co., London, 1896, this calculation was moved to the body of the text (p. 103). A devastating contemporary review of the second edition can be found in: Peter Duren, ed., *A Century of Mathematics in America, Part III*, American Mathematical Society, Providence, 1989, pp. 111–117.

3.2.4 The Mean Speed Theorem in Mediæval Europe

We are on firmer ground when it comes to the next precursor to the Mean Value Theorem. It has been thoroughly documented by Marshall Clagett, who has collected and translated the relevant papers and provided them with extended commentary in 256 pages of his source book, *The Science of Mechanics in the Middle Ages*.[131]

A short summary of Clagett's 256 pages is as follows: Aristotle and others considered *dynamics* — the relation between force and motion or cause and effect. In the 13th century Gerard of Brussels made an important contribution to the founding of *kinematics* — the quantitative study of motion — when he considered the motion of a moving line segment. In the following century the scholars of Merton College, Oxford, laid the foundations of the field of kinematics. The crowning achievement of their work was the *Mean Speed Theorem*, credited to William of Heytesbury for the earliest written statement of it in his *Regule solvendi sophismata* [*Rules for solving sophisms*] (*c.* 1335) and possibly the first proof in a subsequent work, *Probationes conclusionum* [*Proofs of conclusions*]. Other Merton scholars, Richard Swineshead (*fl. c.* 1340–1351) and John of Dumbleton (†*c.* 1349), gave additional proofs. Around 1350, the kinematics of the Merton school had reached Paris and Florence, where Nicole Oresme (1320–1382) and Giovanni di Casali, *aka* Johannes de Casali, (*c.* 1320 – *after* 1374), wrote tracts on the subject. They were followed by Blasius of Parma (*c.* 1345–1416) whose work on kinematics was posthumously published in 1482 and again in 1486 and 1505. The way was thus paved for Galileo Galilei (1564–1642) whose treatment in the *Dialogues Concerning Two New Sciences* has somewhat overshadowed the work of his predecessors.

In modern mathematical terms, the Mean Speed Theorem is an unimpressive special case of the Mean Value Theorem for Areas or Integrals, and Galileo's final result on free falling bodies a trivial instance of the Second Fundamental Theorem of the Calculus. In terms of our understanding of the physical world, however, it was notable progress. I quote Clagett:

> It is worth noting that, while the *mathematical* work of Gerard of Brussels proved to be the principal point of departure for the kinematic section of the earliest of the Merton treatises, namely, the *Proportions of Velocities in Movements* of Thomas Bradwardine, of perhaps equal importance and influence was a *philosophical* current. And indeed it was philosophers trained in mathematics who were responsible for the investigations into kinematics at Oxford and Paris. The philosophical problem which gave stimulus to kinematics was the problem of how qualities (or other forms) increase in intensity, e.g., how something becomes hotter, or whiter. In the technical vocabulary of the schoolmen, this was called the problem of the "intension and remission of forms," intension and remission simply meaning the increasing and decreasing of the intensity of qualities or other forms. The solution of this problem worked out by the philosopher Duns Scotus during the early years of the fourteenth century assumed a *quantitative* treatment of variations in the intensities of qualities suffered by bodies. It was accepted by the successors of Scotus, and it perhaps influenced the view held

[131] Marshall Clagett, ed., *The Science of Mechanics in the Middle Ages*, The University of Wisconsin Press, Madison, 1959. The three most important items were reprinted in Edward Grant, ed., *A Source Book in Medieval Science*, Harvard University Press, Cambridge (Mass.), 1974, items 41–43 (pp. 234–253) under the heading "Physics", subheading "Kinematics".

> by some of the men of Merton that the increase or decrease of qualitative intensity takes place by the addition or subtraction of degrees *(gradus)* of intensity. With this approach to qualitative changes accepted, the Merton schoolmen applied various numerical rules and methods to qualitative variations and then by analogy to kindred problems of motion in space *(motus localis)*. Needless to say, the discussion of variations of quality and velocity by the Oxford schoolmen (and their medieval successors) was almost entirely hypothetical and *not rooted in empirical investigations;* nor was it framed in such a way as to admit of such investigations. In fact, in medieval kinematics as well as dynamics all of the quantitative statements relative to pretended physical variables are in terms of general proportionality expressions; *and the proportionality constants, which can only be determined by experiment, are never found.*[132]

Their study acquired the appellation the *latitude of forms*, the word "latitude" indicating a quantitative change, or increment, and "form" some quality capable of numerical change. The latitude of forms was most famously applied to motion and questions of distance, speed, and acceleration. There is a definite non-numerical, qualitative character to their study of a quantitative subject: They discussed *uniform motion*, i.e., motion with constant velocity, and distinguished it from *difform motion* with its variable velocity. In difform motion they distinguished *uniformly difform motion* (constant acceleration) from *difformly difform motion* (variable acceleration). There was even *uniformly difformly difform motion* in which the acceleration changed uniformly.

The Merton Mean Speed Theorem asserts that the result of uniformly difform motion taken over a given finite interval of time can be achieved in the same period of time by a uniform motion at the mean velocity, the mean velocity being the instantaneous velocity achieved midway through the time period. The first, somewhat wordy, statement of the result was given by Heytesbury:

> In this connection, it should be noted that just as there is no degree of velocity by which, with continuously uniform motion, a greater distance is traversed in one part of the time than in another equal part of the time, so there is no latitude (i.e., increment, *latitudo*) of velocity between zero degree [of velocity] and some finite degree, through which a greater distance is traversed by uniformly accelerated motion in some given time, than would be traversed in an equal time by a uniformly decelerated motion of that latitude. For whether it commences from zero degree for from some [finite] degree, every latitude, as long as it is terminated at some finite degree, and as long as it is acquired or lost uniformly, will correspond to its mean degree [of velocity]. Thus the moving body acquiring or losing this latitude uniformly during some assigned period of time, will traverse a distance exactly equal to what it would traverse in an equal period of time if it were moved uniformly at its mean degree [of velocity].[133]

The passage reads as if some argument is being presented ("just as ...For ...Thus"), but it is all pure conclusion. First, as regards uniform motion, he remarks that if velocity is constant, one travels equal distances in equal times. This is an easy consequence of the formula,

$$Distance = Rate \times Time,$$

[132]Clagett, *op. cit.*, pp. 205–206.

[133]Clagett, *op. cit.*, p. 270. The bracketed insertions are Clagett's.

we learned in elementary school. The factor *Rate* is the same constant velocity during both intervals, and *Time* is the *duration* of time, which, of course, is the same for equal intervals.

The second point Heytesbury makes is that the distance travelled is the same for uniformly accelerated and decelerated motions provided the extremes of velocity are equal and the motions are of equal duration of time. Thus, for example, an automobile uniformly accelerating from 0 to 60 in 10 s will cover the same ground as one decelerating from 60 to 0 in the same period of time. Today we could imagine illustrating this by filming the car and running the film backwards and forwards in the projector at the same speed. The ground covered by the car in the given amount of time will literally be seen to be the same, although the direction of travel will be reversed. Heytesbury, however, does not have a motion picture camera at his command, and justifies his conclusion by appeal to the Mean Speed Theorem: In either case, the distance covered equals that given by uniform motion with velocity given by the means of the extremes; but the extremes are the same albeit given in reverse order (0 and 60 vs. 60 and 0 in our example).

The paragraph then finishes with an explicit statement of the Mean Speed Theorem. If a body with initial velocity v_0 is uniformly accelerated or decelerated over a period of time until a final velocity v_f is reached, the overall distance travelled will be the same as that of a body travelling for the same period of time at velocity v_m occurring at the middle of the given period of time. Moreover, he shows $v_m = (v_0 + v_f)/2$.

In the *Regule* Heytesbury does not prove the Mean Speed Theorem itself, but he does draw some conclusions and he gives an over-elaborate proof that $v_m = (v_0 + v_f)/2$ using the sum of an infinite geometric progression. A proof is given in the *Probationes* which is believed to have been written by him. The proof is mildly obscure, but the statement of the theorem is a model of clarity:

> EVERY increment of velocity (*latitudo motus*) uniformly acquired or lost will correspond to its mean degree [of velocity]. This means that a moving body uniformly acquiring or losing that increment will traverse in some given time a magnitude completely equal to that which it would traverse if it were moving continuously through the same time with the mean degree [of velocity].[134]

In applying the word "clarity" to this, I mean that the statement is clear modulo the definitions of the basic terms — "velocity", "velocity uniformly acquired" (i.e., uniform acceleration), and "mean degree". The terms are clear enough to us, and Heytesbury probably shared our *intuitive* understanding of the concepts, but his formal definitions differed from ours and it requires greater effort for the modern mind to follow his reasoning than to apply some Calculus and prove the result outright. Moreover, it is the result and not the argument for it that anticipates the Mean Value Theorem. Hence I shall cheat and present a modern proof, referring the curious reader to Clagett's book for various proofs by the Merton scholars.

To a modern mathematician, uniform acceleration means that velocity assumes the form $v(t) = \alpha + \beta t$ for some constants α and β, whence the total distance travelled

[134] *Ibid.*, p. 284.

in the interval from $t = a$ to $t = b$ is

$$\begin{aligned}\int_a^b (\alpha + \beta t)dt &= \alpha t + \beta \frac{t^2}{2}\bigg|_a^b \\ &= \alpha t\bigg|_a^b + \beta \frac{t^2}{2}\bigg|_a^b \\ &= \alpha(b-a) + \beta\left(\frac{b^2}{2} - \frac{a^2}{2}\right) \\ &= \alpha(b-a) + \beta\frac{b+a}{2}(b-a) \\ &= (b-a)\left(\alpha + \beta\frac{a+b}{2}\right), \end{aligned} \tag{3.17}$$

the last expression being the distance travelled in the interval from a to b at velocity

$$\alpha + \beta\frac{a+b}{2} = \frac{(\alpha + \beta a) + (\alpha + \beta b)}{2},$$

i.e., the mean of the initial $(\alpha + \beta a)$ and the final $(\alpha + \beta b)$ velocities, or the velocity at the median time $(a+b)/2$. Writing $c = (a+b)/2$, (3.17) assumes the form

$$\int_a^b f(t)\, dt = (b-a)f(c) \tag{3.18}$$

for linear f, i.e., a trivial special case of the Mean Value Theorem for Integrals.

To the extent that the modern proof relies on the Mean Value Theorem (in determining the form of the velocity function and in evaluating the definite integral), the result, a special case of the Mean Value Theorem for Integrals, is not truly impressive. One can avoid this dependence by integrating directly using approximating sums, or, given the special nature of the functions involved (constant and linear functions), one can simply appeal to Elementary Geometry to establish the result. Heytesbury and the other Merton scholars did neither, but argued numerically.

Clagett begins his final chapter on kinematics as follows:

> NOT long after the Merton kinematics had reached its maturity, its distinctive vocabulary and principal theorems began to spread throughout Europe. Their passage to Italy and France took place around A.D. 1350. In the course of this passage a significant event took place, the application of graphing or coördinate techniques (or more exactly two-dimensional figures) to the English concepts dealing with qualities and velocities. In the sense that this new system represented the functions implicit in the concepts of uniform velocity and acceleration, it resembled the analytic geometry of the seventeenth century; but it did not translate algebraic expressions as such into geometric curves, and vice versa. Hence we cannot yet call it analytic geometry.
>
> The basic idea of the system is simple. Geometric figures, particularly areas, can be used to represent the quantity of a quality. Extension of the quality in a subject is to be represented by a horizontal line, while the qualitative intensities at different points in the subject are to

> be represented by perpendiculars erected on the extension or subject line... In the case of motion, the line of extension represents time, and the line of intensity, velocity.[135]

This coordinate system was first used either in Italy by di Casali or in Paris by Oresme — the latter's work is not precisely dated. It was a step up from cartography and astronomy, in which places or stars were located by longitude and latitude, in that a functional relationship — the intensity of a quality at a given time — was graphed. As Clagett emphasises, however, without the expressive symbolism of algebra, it fell far short of Analytic Geometry, graphing only straight lines and not being capable of making finer divisions within the class of difformly difform motions.

Oresme's exposition is deemed the superior of the two and his name is usually discussed in this matter. His important work on the subject has a title something like *De configurationibus qualitatum*[136] [*On the configurations of qualities*].

Oresme's geometric representation proceeds as follows. One draws one horizontal line, basically our modern x-axis, to represent some interval of time. At each point in the interval one imagines a vertical line placed to represent the velocity of the given body at that moment of time, i.e., to represent the *instantaneous velocity* — whatever that may be. Today we define instantaneous velocity to be the limit of the average velocities over ever smaller intervals, with average velocity being defined by the distance-rate-time formula. It is not clear to me what Oresme and the Merton scholars meant by instantaneous velocity, or indeed if they agreed on what was meant. Reading scholastic mathematics tends to make me drowsy, so I cannot say whether or not they offered a definition or took the notion as primitive. I can say that they viewed time, velocity, and distance as continua, but that there was no universal agreement on the nature of the continuum, and without such I don't see how to treat instantaneous velocity other than as a primitive notion, something assumed to exist.

Uniform motion was readily defined. A motion is uniform just in case velocity is constant. Graphically, this means that the straight lines representing the individual velocities are all the same length, and the upper extremities fall into a horizontal straight line. The total distance travelled, by the distance-rate-time equation, thus equals the area of the rectangle whose base equals the duration of the motion and whose height is the common velocity.

Uniform acceleration was formally defined by Heytesbury as motion for which the differences between final and initial velocities over equal periods of time were equal:

$$v(t_1 + \Delta t) - v(t_1) = v(t_2 + \Delta t) - v(t_2)$$

for all t_1, t_2, and Δt. Heytesbury and others also used another property of uniform acceleration: The ratios of differences in velocities is proportional to the ratios of differences in time:

---

[135] *Ibid.*, p. 331.

[136] Clagett, *op. cit.*, pp. 339–340, gives a long footnote explaining the numerous variations of the title in manuscript copies found in various European libraries.

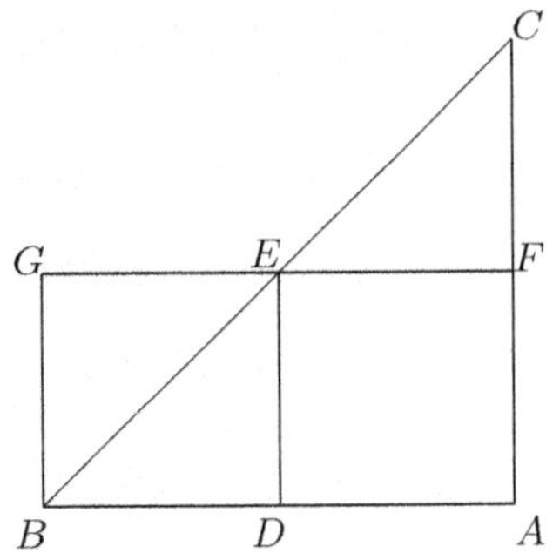

Fig. 3.12 Mean speed for a triangle

$$\frac{v(t_4) - v(t_3)}{v(t_2) - v(t_1)} = \frac{t_4 - t_3}{t_2 - t_1}, \text{ or } \frac{v(t_4) - v(t_3)}{t_4 - t_3} = \frac{v(t_2) - v(t_1)}{t_2 - t_1}.$$

This is established by a simple arithmetic argument when $(t_4 - t_3)/(t_2 - t_1)$ is rational, but requires limits or the Eudoxian theory of proportion in the general case. I am too ignorant of mediæval European mathematics to say whether or not they were aware of this.

Oresme seems to define uniform acceleration geometrically.[137] His definition refers more generally to "quality" and not "velocity", and is given in two stages. First, for a zero initial or terminal velocity:

> I.8. Concerning a right triangular quality. Every quality imaginable by a triangle having a right angle upon the base can be imagined by every triangle having a right angle upon the same base and by no other figure. For that such a quality is imaginable by such a triangle is obvious from the preceding chapter, because some quality can be proportional in intension to such a right triangle in altitude [i.e., some quality can be related to a triangle whose varying altitude represents the varying intension]. This quality is commonly called a "quality uniformly difform terminated at zero degree" (*non gradum*) [of intension].[138]

What this means in modern language is that the graph of uniformly accelerated motion starting or terminating at rest is a straight line with one end lying on the x-axis. The figure trapped between the line and the x-axis over the given interval is a right triangle. (See Fig. 3.12.) Once again, distance is taken to be (this time without justification) the area of the triangle.

In I.10 Oresme discusses the more general "quadrangular quality" — uniformly accelerated motion neither beginning nor ending at rest. In this case, the graph is a straight line and the figure trapped between the line and the x-axis (and I suppose I should mention the initial and final latitudes) is a quadrilateral. Total distance covered is again the area of the figure.

After a lot of discussion he reaches the Mean Speed Theorem.

[137] I am relying on Clagett's translation of excerpts and there may well be a more direct definition in terms of proportionality in the untranslated parts.

[138] *Ibid.*, p. 350. Bracketed insertions are Clagett's.

> III.7. On the measure of difform qualities and velocities. Every uniformly difform quality [in a subject] is just as great as would be a quality in the same or equal subject uniform at that degree [of intensity] of the middle point of the same subject; and I understand this [to be so] if the quality is linear…
>
> Let there be a quality imaginable by a triangle *ABC*, which is uniformly difform, and is terminated at zero degree in point *B* (see Fig. 6.5B[139]); and let *D* be the middle point of the subject line. The degree of this midpoint, or its intension, is imagined by the line *DE*. Hence the quality which is uniform at degree *DE* throughout the whole subject is imaginable by a quadrangle *AFGB*, as is clear from the tenth chapter of Part I. And it is evident by the twenty-sixth [proposition] of the first [book] of Euclid, that the two small triangles *EFC* and *EGB* are equal. Therefore the larger triangle *BAC*, which designates the quality uniformly difform, and the quadrangle *AFGD*, which would designate the quality uniform at the degree of the middle point, are equal. Hence the qualities imaginable by a triangle of this kind and a quadrangle are equal; and this was proposed.[140]

I will spare the reader Oresme's proof in the quadrangular case.

One might feel a bit reluctant to accept so special a case of the Mean Value Theorem for Areas as an instance of the Mean Value Theorem for Integrals, and thus as a precursor of the Mean Value Theorem itself. This reluctance, at least on my part, stems from the coincidental nature of the proof; no general principle is involved, just the observation that the median velocity happens to agree with the velocity at the median time in the case of uniform acceleration, something that is clearly not the case with more general motions. However, although it is such a special case, established serendipitously by a simple observation, the Merton Mean Speed Theorem, via the representation of every continuous function as the uniform limit of piecewise linear functions, readily yields the Mean Value Inequality for Integrals of continuous functions.[141] Thus, trivial as it is, the Mean Speed Theorem deserves serious consideration as a precursor to the Mean Value Theorem.

3.2.5 Valerio and Cavalieri

The 16th century brought great change to mathematics in Europe. It was from the seeds planted in this century that the Analytic Geometry and the Calculus of the 17th century sprouted. In the mid-16th century the solutions of the cubic and biquadratic equations propelled Europe into the forefront of algebraic research. By the end of the century much of the use of awkward (to the modern reader) linguistic expressions in algebra had been replaced by symbolism. And the rediscovery of Archimedes led to a renewed interest in the two chief technical problems that would later form the nucleus of the Calculus — finding tangents and finding areas. One of the masters of the Archimedean mathematics was Luca Valerio (1552–1618), called by Galileo

---

[139] I.e., Fig. 3.12.

[140] Clagett, *op. cit.*, pp. 358–359. "*AFGD*" should be "*AFGB*" in the sentence second to last.

[141] For details, cf. the very readable: Ádám Besenyei, "Lebesgue's road to antiderivatives", *Mathematics Magazine* 86 (2013), pp. 255–260.

"the Archimedes of our age". Valerio expanded Archimedes's results, streamlined the method of proof by exhaustion, and brought a greater universality to these results than did Archimedes. Unfortunately for his reputation, Valerio was the last of the Archimedean scholars. Mathematicians of the day were impatient with the Greek approach, which led to rigorous proofs of results once one knew what the results were, but hid the methods followed to obtain these results. Consequently, the emphasis changed to heuristics and the hunt for methods of discovery. One of the leaders of this new mathematics was Cavalieri, a generation younger than Valerio. Both men made what we now recognise as contributions to the Geometric Form of the Mean Value Theorem. A disciple of Galileo, Cavalieri inherited the latter's respect for Valerio and we can probably safely assume from the similarity of their approaches to this theorem that his result derives from Valerio's. However, Cavalieri's work is generally better known, so we shall consider him first.

In his biographical entry on Cavalieri in the *Dictionary of Scientific Biography*, historian Ettore Carruccio writes

> In proposition I of Book I of the *Geometria*, we find in geometric form the theorem of mean value, also known as the Cavalieri theorem. The theorem is presented as the solution of the following problem: Given a plane curve, provided with a tangent at every point and passing through two points A and B, to find a straight line parallel to AB and tangent to the curve at some point on the curve between A and B. Analytically we have: If the real function $f(x)$ of the real variable x is continuous in the interval (a, b) and at every point within this interval it is differentiable, at least one point ζ exists with $a < \zeta < b$, so that
>
> $$\frac{f(b) - f(a)}{b - a} = f'(\zeta).$$

Kirsti[142] Andersen, another authority on Cavalieri, makes a more modest claim on Cavalieri's behalf:

> I only consider it a precursor and not an early version of the mean value theorem, because it is based on geometrical ideas very different from the concepts underlying the mean value theorem.[143]

Andersen, however, provides a reference to Lucio Lombardo-Radice, another Cavalieri scholar, who evidently would agree with Carruccio.

Cavalieri was one of the important precursors to the European discovery of the Calculus; his work on integration is reported on in every major work on the history of the Calculus,[144] and excerpts from his *Geometria indivisibilibus continuorum nova quadam ratione promota* (1635; 2nd ed. 1653) can be found in English translation

---

[142] Ettore Carruccio, "Cavalieri, Bonaventura", in: Charles Coulston Gillispie (ed.), *Dictionary of Scientific Biography*, vol. 3, Charles Scribner's Sons, New York, p. 152. Note that Carruccio has not stated the analytic form correctly: f must be continuous on the closed interval $[a, b]$ and not merely on (a, b).

[143] Kirsti Andersen, "Cavalieri's method of indivisibles", *Archive for History of Exact Sciences* 31 (1985), pp. 291–367; here: p. 299.

[144] Cf. e.g., Edwards, *op. cit.*, pp. 104–109 and Baron, *op. cit.*, pp. 122–135.

in the various source books.[145] These references, however, tend to pass over his precursive contribution to mean value theory in silence. We are not, however, facing the same complete lack of information we were up against in the case of Bhāskara II. Carruccio quotes chapter and verse, which happen to be in Latin — but, Kirsti Andersen has written a long paper on Cavalieri's *Geometria* and her report, in English, touches on Cavalieri's version of the Mean Value Theorem in sufficient detail to tell us what the result was if not the justification he gave for asserting it.

To Cavalieri, his version of the Mean Value Theorem was not an end in itself, but an important lemma underlying his method of integration. His method is summed up in what is termed *Cavalieri's Principle*:

> The following extract, known as Cavalieri's theorem, is from the *Geometria Indivisibilibus*, Book VII, Theorem 1, Proposition 1.
>
> Any plane figures, constructed between the same parallels, in which [plane figures] any straight lines whatever having been drawn equidistant from the same parallels, the included portions of any straight line are equal, will also be equal to one another; and any solid figures, constructed between the same parallel planes, in which [solid figures] any planes whatever having been drawn equidistant from the same parallel planes, the plane figures of any plane so drawn included within these solids, are equal, the [solid figures] will be equal to one another.[146]

I have included Smith's introductory comment referring to the result as "Cavalieri's theorem" precisely because it is not the mean value theorem in any form, which Carruccio says is known as "the Cavalieri theorem". Mathematical nomenclature, in this most exact of sciences, may well be the least precise of all the exact sciences.

The Mean Value Theorem, in Cavalieri's form, comes in setting up the figure. In the two-dimensional case it tells us that any plane figure can be trapped between two parallel lines. Needless to say, his proof would not have been rigorous by today's standards, which require precise definitions and which rely on the rigorously proven Extreme Value Theorem. I doubt that Cavalieri offered anything like an acceptable definition of a "plane figure", but he did offer a definition of a tangent line:

> I say that a straight line touches a curve situated in the same plane as the line when it meets the curve either in a point or along a line and when the curve is either completely to the one side of the meeting line [in the case when the meeting is a point] or has no parts on the other side of it [in the case when the meeting is a line segment].[147]

This definition of tangent does not agree with our current definition. It does not allow tangents like those through the point P in Figs. 2.30 and 2.31 of page 80, but does allow tangents to coincide with the curve for a while as in Fig. 2.32 on page 80. It also allows infinitely many tangents at a corner or a cusp — any line passing through the point C and not the interior of the segment AB in Figs. 5 and 6 of page 3

[145] Smith, *Source Book*, *op. cit.*, pp. 605–609; Struik, *op. cit.*, pp. 209–219; and Stedall, *op. cit.*, pp. 62–65.

[146] Smith, *Source Book*, *op. cit.*, p. 605. The quotation, given in small print, is part of Smith's introductory comment, and the bracketed insertions are Smith's.

[147] Andersen, *op. cit.*,p. 297. Bracketed insertions are Andersen's.

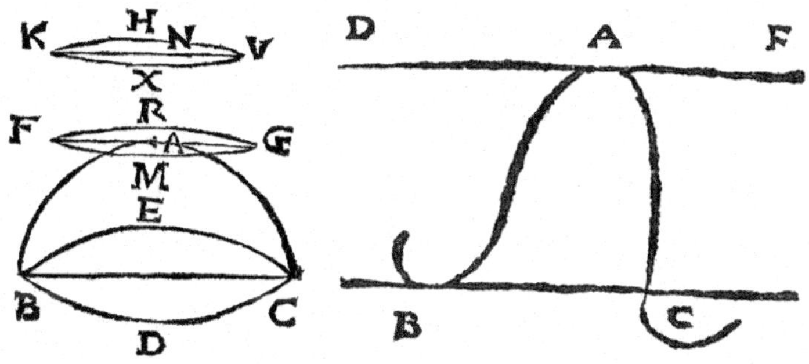

Fig. 3.13 Cavalieri's diagrams

is a tangent. The lines *AC* and *BC* are tangents in Fig. 1.5 in Chap. 1, but apparently not in Fig. 1.6 in Chap. 1. This latter failure is, however, probably a mere oversight.

The choice of definition is dictated by Cavalieri's Principle. He wanted to view any plane figure as being composed of parallel line segments trapped between two *opposite tangents* parallel to the segments. To this end he chose a line segment he called a *regula* to determine the direction of parallelism and then proved in Lemma 3 of Book VII his geometric version of the Mean Value Theorem:

> If a curved line is situated in one plane and if a straight line meets it in either two points, two line segments, or in a line segment and a point, then we can draw another straight line parallel to the previous line which touches the part of the curve situated between the two mentioned meetings.[148]

Proposition I of Book I also stated the result, but for solid figures bounded by parallel planes as well as plane figures bounded by parallel lines. Two illustrations from the 1653 edition of Cavalieri's book are pictured in Fig. 3.13. The diagram on the left in the figure is the illustration for the more general Proposition I and simultaneously depicts both cases. The disk *KHVX* is the planar regula for the solid figure *ABDCE* with tangent disc *RFMG* at point *A*, while the line *KV* is the regula for the plane curve *BAC* with tangent *FG* at point *A*. The diagram on the right is self-explanatory and is basically the modern illustration of Rolle's Theorem, albeit primitively drawn — something one might see on a blackboard, but not in a modern text.

The existence of the opposite tangents parallel to a given regula is now intuitively obvious. Given a regula and a plane figure, choose a point *P* in the interior of the figure and draw the line parallel to the regula passing through *P* as in Fig. 3.14.

It will enter and exit the figure at points *A*, *B*. Apply the Lemma to the two curves of the boundary on either side of *AB*. The parallels promised by the Lemma are Cavalieri's opposite tangents.

Obviously a rigorous proof of this would require precise definitions of "plane figure", "boundary", "curve", etc., along with the isolation of properties like continuity and smoothness, and theorems like the Extreme Value Theorem — all of which

---

[148] *Ibid.*, p. 298.

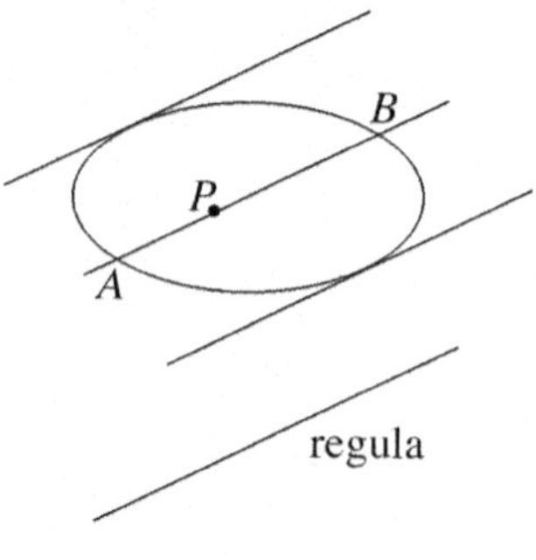

Fig. 3.14 Opposite parallels

came some centuries after Cavalieri. Nonetheless, his approach has great heuristic value and, although Cavalieri's Principle needed a bit more to determine areas of complex figures, it played a major rôle in the pre-Newtonian development of the Integral Calculus in Europe.

His version of the Mean Value Theorem — essentially the geometric form we introduced back in Chap. 1 and proved in Chap. 2 — was less influential. This is probably due in part to the fact that, other than to conclude the existence of opposite parallels, which was intuitively obvious in the cases considered, he did not demonstrate the usefulness of the result. Indeed, its use was to treat opposite parallels of plane and solid figures as two- and three- dimensional analogues to the endpoints of an interval. And, given his generosity of allowing corners and cusps to have tangents, his result might actually obscure what is useful about the Geometric Form of the Mean Value Theorem as proven in Chap. 2.

We began our discussion of Cavalieri with two views of his result and I think we are now in position to say that Carruccio's estimation of what Cavalieri accomplished in this matter is an overstatement. As we saw in Chap. 2, the Geometric Form of the Mean Value Theorem does indeed yield the analytic version we are calling the Classroom Mean Value Theorem. But this implication is not immediate; it requires knowledge that the slope of a tangent line is the derivative of the given function at a point on the curve. Cavalieri does not connect the two, as emphasised by Andersen when she says his result "is based on geometrical ideas very different from the concepts underlying the mean value theorem".[149] In the cases where the derivative does exist, however, Cavalieri's tangents coincide with the modern ones. So should we consider Cavalieri's result a variant of the Mean Value Theorem or, as Andersen feels, merely a precursor to the Mean Value Theorem?

More clear-cut as not being more than a precursor to the Mean Value Theorem is Valerio's result. In 1582 Valerio published a book[150] in which he proved in some generality that, given a bounded convex figure in the plane and a point exterior to the figure, a tangent line could be drawn from the point to the figure. The problem goes

---

[149]Cited on page 244, above.

[150]Lucæ Valerii, *Subtilium indagationum liber primus*, Zannettum, Rome, 1582. Cf. p. 5 for his result.

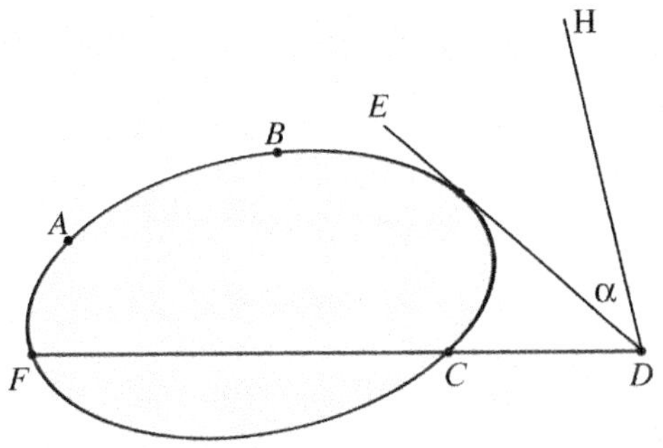

Fig. 3.15 Modified Valerian diagram

back to the Greeks who solved special cases of it by explicit constructions. When a construction was not known they did not assert its validity. Thus, for example, there is no mention by Archimedes of the problem as it relates to his spiral in *On spirals*. Valerio had no such qualms about constructions and proved the general result.

Valerio's proof in simple modern terms proceeds as follows. Let a bounded convex figure *ABC* be drawn and let *D* be a point exterior to the figure. Extend the line *DC*. Either it is the tangent sought or it cuts through the figure to a point *F* as in Fig. 3.15. By the boundedness and convexity of *ABC* and the exteriority of *D* there is a line *DH* passing through *D* which has no intersection with *ABC* no matter how far extended. Consider a line *DE* initially laid over *DH*, but allowed to rotate in the direction of *DC* with pivot at *D*. By the Least Upper Bound Principle, there is an angle α that is the least upper bound of all those angles for which *DE* does not intersect *ABC*. The claim is that at this angle the line *DE* is tangent to the curve *ABC*.

The argument at this point would correspond to the application of Lemma 2.2.17 in Chap. 2: If *DE* does not meet *ABC* then the distance of *DE* from *ABC* is positive and would remain so for some interval $(\alpha, \alpha + \epsilon)$, contrary to the choice of α. And if *DE* cuts the figure, there are points of the curve on either side of *DE*, hence one at least would lie on the line given by some angle $\beta < \alpha$, again contradicting the choice of α. Hence *DE* is tangent to *ABC* at some point of intersection.

I haven't read Valerio's full book, so I can't say if his definition of tangent was as broad as Cavalieri's, nor if he allowed linear segments of his curves or if he ruled them out. He did allow such to be the base of a figure trapped between a convex curve and one of its chords.

The proof as just sketched is not as rigorous as that of the Geometric Form of the Mean Value Theorem we gave in Chap. 2, but it is clear that a similar rigorisation is possible. It is also clear that Valerio was in no position to provide modern rigour. His actual proof — recall I said I was presenting his proof in modern terms — differed slightly. He began with an oversimplified[151] diagram (Fig. 3.16). The line *MN* represents the time it takes for *DE* to rotate from *DH* to *DC* and *L* represents the instant *DE* touches the curve at *B*, i.e., it corresponds to our angle α. *HK* is an arc

[151] Mathematically oversimplified: had I thought of the floral touch in the upper left before starting on this book, the graphics here-in would have been quite different.

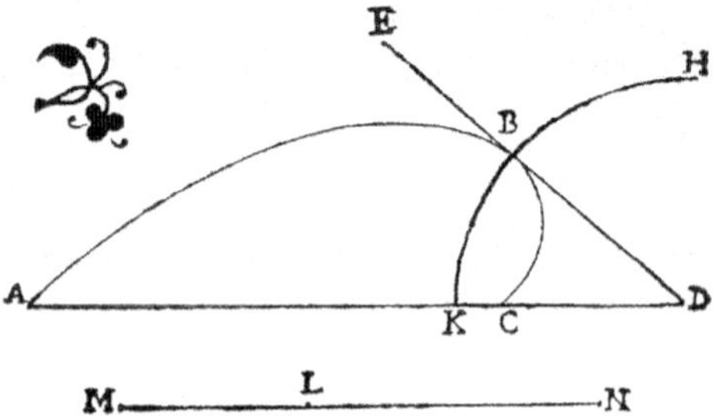

Fig. 3.16 Valerio's diagram

of a circle centred at D and is used to represent the rotation of DE around D. That HK intersects ABC at the point of tangency, that B is this point, and that A, C, D are collinear is a misleading pictorial simplification never made use of in the proof.

What this result has to do with the Geometric Form of the Mean Value Theorem is that Valerio adds a remark that the same statement holds if instead of rotating the line one movies it in a straight line towards the figure. He adds that if the figure is given by a convex curve with a straight line as base and the exterior line is parallel to the base, the tangent line opposite the base will be at maximum distance from the base. He calls this distance the *altitude* of the figure.

3.2.6 *Rolle*

Rolle's Theorem (Theorem 3.1.1), being a special case of the Mean Value Theorem as well as a lemma used in the modern proof of the Theorem, would naturally be a precursor to the Mean Value Theorem had it in fact preceded rather than followed the Mean Value Theorem chronologically. Rolle, however, did prove a precursor to Rolle's Theorem and he must thus be mentioned here.

Smith introduces Rolle's actual theorem as follows:

> Writers on the history of mathematics of the early part of the present [20th] century did not know where in the writings of Michel Rolle the theorem named after him could be found — the theorem according to which $f'(x) = 0$ has at least one real root lying between two consecutive real roots of $f(x) = 0$. One historian went so far as to express the opinion that the theorem is wrongly attributed to Rolle. Finally, in 1910, the theorem was found in a little-known book of Rolle, entitled, *Démonstraton d'une Methode pour résoudre les Egalitez de tous les degreez; suivie de deux autres Méthodes, dont la première donne les moyens de resoudre ces mêmes égalitez par la Geometrie, et la seconde, pour resoudre plusieurs questions de Diophante qui n'ont pas encore esté resoluës.*[152]

[152] Smith, *Source Book*, *op. cit.*, p. 253. The title translates roughly to *Demonstration of a method for solving equations of all degrees; followed by two other methods, of which the first gives the means of solving these same equations by geometry, and the second, for solving many Diophantine equations which have not previously been solved.*

If Smith's inclusion of a short list of European libraries in possession of copies of Rolle's book is any indication, the book was rare as well as little-known. Today, however, we have the relevant excerpts from the book in English translation in Smith's *Source Book*.[153]

There is also a more detailed historical account of Rolle's work given by June Barrow-Green.[154] She begins her account with a statement of the modern form of Rolle's Theorem and continues,

> It is clear from the language of functions and derivatives that the theorem is now presented as a theorem of calculus. Its importance lies in the fact that it is needed in the proof of the mean value theorem and for establishing the existence of Taylor series. When Michel Rolle (1652 – 1719) made the first statement of this theorem in 1690, however, Taylor series had not yet been discovered and calculus itself was in its infancy. Moreover, Rolle was deeply suspicious of its methods. His theorem first appeared not in the context of calculus at all but of equation solving.[155]

A problem much considered at the time was the location of the roots of polynomials. In 1690 Rolle published his most famous work, the *Traité d'algebre*, in which he explained his "method of cascades" for isolating the roots of polynomials. A *cascade* is essentially a generalisation of a formal version of the derivative, and Rolle must have known this as Leibniz had been publishing on the Calculus for half a decade by the time Rolle's book appeared. Rolle, however, did not hold the Calculus in high regard, in coming years criticising "the infinitesimal calculus for its lack of rigour and, as he believed, its propensity for error".[156] His definition of a cascade was purely algebraic, as was his proof of Rolle's Theorem for polynomials the following year when he published it in answer to the apparent criticism that his *Traité d'algebre* lacked a proof that his method works.

Rolle's method of finding the roots of a polynomial was a modified bisection method. The bisection method is brute simplicity. If $P(a)$ and $P(b)$ have opposite signs, one calculates $P\left(\frac{a+b}{2}\right)$. The result is either 0 or has a sign differing from one of $P(a), P(b)$. Ignoring the labour involved in calculating numerous values of $P(x)$, the hard part, at least conceptually, is determining a, b to begin with. One can get rough estimates of lower and upper bounds for any possible root without too much difficulty. Rolle used his cascades to get a finer division, separating the different roots of P. These separators were the roots of the cascade, i.e., the derivative, P'. Of

---

153 *Ibid.*, pp. 253–260.

154 June Barrow-Green, "From cascades to calculus: Rolle's theorem", in: Robson and Stedall, *op. cit.*, pp. 737–754.

155 *Ibid.*, pp. 737–738. The remarks about Taylor series are a little misleading. The general Taylor formula (1715) might not have been stated by 1690, but the Taylor series for the most important functions were known: the binomial series, trigonometric functions and their inverses, as well as e^x and $\ln x$.

156 *Ibid.*, p. 739.

course, this requires one to find the roots of P', for which one needed those of P''. Then the roots of P''', etc.

One example Rolle used, in modern notation, is

$$P(x) = x^4 - 24x^3 + 198x^2 - 648x + 473 = 0.$$

Successive differentiation (occasionally removing common factors from coefficients)[157] yields

$$\begin{aligned} P'(x) &= 4x^3 - 72x^2 + 396x - 648 = 0 \\ P''(x) &= 6x^2 - 72x + 198 = 0 \\ P'''(x) &= 4x - 24 = 0. \end{aligned}$$

Now $P'''(x) = 0$ has the trivial root $x = 6$. Through other means he knows all roots of P'' to lie between 0 and 13, thus the distinct real roots of P'', if they exist, are in the intervals $[0, 6]$ and $[6, 13]$. To the nearest integer he estimates these roots to be 4 and 7. All roots of P' lie between 0 and 163, yielding the intervals $[0, 4]$, $[4, 7]$, and $[7, 163]$ as the homes for the distinct roots of P'. To the nearest integers these are 3, 6, 9. This gives him the intervals $[0, 3]$, $[3, 6]$, $[6, 9]$ and $[9, 649]$ in which to search for the roots of P — which he finds to be 1, 6, 8 and 10, the first exact and the other three approximate.

3.2.9 Exercise Explore Rolle's example on a graphing calculator.

The crucial step in justifying the procedure is proving that distinct roots of $P(x)$ are indeed separated by roots of $P'(x)$. That is, the crucial step in justifying Rolle's method is proving Rolle's Theorem for polynomials.

Rolle's proof was not clearly presented, as emphasised by Barrow-Green:

> Today, interpreting Rolle's 'cascades' as 'derivatives' it is not difficult to understand why his method works. Rolle, however, neither used nor trusted calculus. Considering his method algebraically, it is not at all obvious what is happening. In the *Algebre* Rolle gave no clue as to any theoretical underpinning and introduced the idea of multiplying by an arithmetic progression without giving any reason for it. Nor did he prove that the roots of each cascade are limits for [i.e., separate the roots of] the previous equation. The latter in particular is not easy to see — it relies on some clever algebraic manipulation — and the fact that it was hidden from the reader in the *Algebre* is one of the reasons that Rolle realized the necessity of bringing out his *Demonstration*.[158]

---

[157] Why he didn't carry out the simplification process further and take

$$\begin{aligned} P'(x) &= x^3 - 18x^2 + 99x - 162 = 0 \\ P''(x) &= x^2 - 12x + 33 = 0 \\ P'''(x) &= x - 6 = 0 \end{aligned}$$

is unclear.

[158] Barrow-Green, *op. cit.*, p. 742.

Now the obvious proof of Rolle's Theorem, be it in the polynomial case or in the more general case of differentiable functions, is the modern one. And there is no reason not to assume Rolle was aware of this. The Extreme Value Theorem was still over a century away from being proven, but so was the Intermediate Value Theorem on which his procedure depended. Any number of results were not yet established but were accepted, e.g., that a polynomial of degree n had n roots, some possibly repeated and not all necessarily real. And the vanishing of the derivative at a local maximum was "in the air". Anyone familiar with Fermat's method of finding maxima and minima as well as the Differential Calculus would recognise this principle as underlying Fermat's method and, as we saw on page 102, above, Newton had stated this explicitly in 1671 in *De methodus fluxionum*. Moreover, Kepler had even earlier enunciated the principle. The problem was that Rolle distrusted the Calculus and had to use another method to establish his principle.

For polynomials there are other proofs. Setting aside Rolle's distrust of the Calculus for a moment, we can offer a different proof as follows. Let $a < b$ be successive roots of the polynomial P:

$$P(a) = P(b) = 0 \text{ and } \forall x(a < x < b \Rightarrow P(x) \neq 0).$$

Write

$$P(x) = (x - a)^k (x - b)^m Q(x),$$

where k, m are the multiplicities of a, b as roots of P. Note that for any $x \in [a, b]$, $Q(x) \neq 0$.

Now

$$P'(x) = k(x - a)^{k-1}(x - b)^m Q(x) + \\ m(x - a)^k (x - b)^{m-1} Q(x) + (x - a)^k (x - b)^m Q'(x),$$

whence

$$P'(x) = (x - a)^{k-1}(x - b)^{m-1}\big[k(x - b)Q(x) + m(x - a)Q(x) + (x - a)(x - b)Q'(x)\big].$$

Define

$$R(x) = k(x - b)Q(x) + m(x - a)Q(x) + (x - a)(x - b)Q'(x). \tag{3.19}$$

Now

$$R(a) = k(a - b)Q(a), \quad R(b) = m(b - a)Q(b)$$

have opposite signs because $Q(a)$ and $Q(b)$ must have the same sign, else, by the Intermediate Value Theorem, $Q(x)$ would have a zero in (a, b). Applying the Intermediate Value Theorem to R, there is some $c \in (a, b)$ such that $R(c) = 0$. But then

$$P'(c) = (c-a)^{k-1}(c-b)^{m-1}R(c) = 0.$$

The appeal to the Intermediate Value Theorem cannot be eliminated, but for polynomials the derivative can be defined formally without appeal to limits or infinitesimals, and the product formula can be established purely algebraically, thus allowing the above proof to be carried out in algebra augmented by the Intermediate Value Theorem — which Rolle did not dispute, having applied it freely.

Rolle's definition of a cascade is slightly more general than formal differentiation. One starts with a polynomial of degree n,

$$P(x) = a_0 + a_1x + \ldots + a_nx^n,$$

and an $(n+1)$-term arithmetic progression $a, a+c, a+2c, \ldots, a+nc$ and multiplies the coefficients of P by the terms of the progression:

$$P^*(x) = aa_0 + (a+c)a_1x + \ldots + (a+nc)a_nx^n. \qquad (3.20)$$

If $a = 0$, each term of the polynomial has a factor x and it can be divided out, yielding a new P^* of degree $n-1$:

$$\begin{aligned} P^*(x) &= ca_1 + 2ca_2x + \ldots + nca_nx^{n-1} \\ &= cP'(x). \end{aligned}$$

In applications Rolle chose $a = 0, c = 1$, so that his cascade P^* agreed with our derivative P'.

The algebraic rules for calculating derivatives are easily verified algebraically:

3.2.10 Lemma *Let P, Q be polynomials, $r \in \mathbb{R}$,*

i. $(rP)' = rP'$
ii. $(P+Q)' = P' + Q'$
iii. $(P \cdot Q)' = P' \cdot Q + P \cdot Q'$.

Proof. Write

$$P(x) = \sum_{k=0}^{n} a_kx^k, \quad Q(x) = \sum_{k=0}^{m} b_kx^k.$$

i. Observe

$$rP(x) = r\sum_{k=0}^{n} a_kx^k = \sum_{k=0}^{n} ra_kx^k,$$

whence

$$(rP)'(x) = \sum_{k=0}^{n} kra_kx^{k-1} = r\sum_{k=0}^{n} ka_kx^{k-1} = rP'(x).$$

ii. By assigning 0 coefficients to powers of x greater than the degree of P or Q, whichever is of lower degree, in the polynomial of lower degree we can assume the two polynomials to have the same number of terms and observe

$$\begin{aligned}\sum_{k=0}^{\max\{m,n\}} k(a_k+b_k)x^{k-1} &= \sum_{k=0}^{\max\{m,n\}} ka_kx^{k-1} + \sum_{k=0}^{\max\{m,n\}} kb_kx^{k-1}\\ &= \sum_{k=0}^{n} ka_kx^{k-1} + \sum_{k=0}^{m} kb_kx^{k-1},\end{aligned}$$

whence $(P+Q)'(x) = P'(x) + Q'(x)$.

iii. By induction on n. By i, the lemma holds for $n = 0$.

Suppose iii holds for P of degree n and consider the product $\left(P(x) + a_{n+1}x^{n+1}\right)Q(x)$.

$$\begin{aligned}\left(\left(P(x)+a_{n+1}x^{n+1}\right)Q(x)\right)' &= \left(P(x)Q(x)+a_{n+1}x^{n+1}Q(x)\right)'\\ &= \left(P(x)Q(x)\right)' + \left(a_{n+1}x^{n+1}Q(x)\right)', \text{ by ii}\\ &= P'(x)Q(x) + P(x)Q'(x) + \left(a_{n+1}x^{n+1}Q(x)\right)', \quad (3.21)\end{aligned}$$

by the induction hypothesis, but

$$\begin{aligned}\left(a_{n+1}x^{n+1}Q(x)\right)' &= a_{n+1}\left(\sum_{k=0}^{m} b_kx^{k+n+1}\right)', \text{ by i}\\ &= a_{n+1}\left(\sum_{k=0}^{m}(k+n+1)b_kx^{k+n}\right)\\ &= a_{n+1}\sum_{k=0}^{m} kb_kx^{k+n} + a_{n+1}\sum_{k=0}^{m}(n+1)b_kx^{k+n}\\ &= a_{n+1}x^{n+1}\sum_{k=0}^{m} kb_kx^{k-1} + a_{n+1}(n+1)x^n\sum_{k=0}^{m} b_kx^k\\ &= a_{n+1}x^{n+1}Q'(x) + (n+1)a_{n+1}x^nQ(x).\end{aligned}$$

Plugging this last into (3.21) yields

$$\begin{aligned}\left(\left(P(x)+a_{n+1}x^{n+1}\right)Q(x)\right)' &= P'Q + PQ' + a_{n+1}x^{n+1}Q' + (n+1)a_{n+1}x^nQ\\ &= \left(P' + (n+1)a_{n+1}x^n\right)Q + \left(P + a_{n+1}x^{n+1}\right)Q'\\ &= \left(P + a_{n+1}x^{n+1}\right)'Q + \left(P + a_{n+1}x^{n+1}\right)Q'.\end{aligned}$$

□

With this Lemma established, the proof of Rolle's Theorem for polynomials is thus rendered almost purely algebraic, the only non-algebraic component being the appeal to the intermediate value property for polynomials.

Rolle's proof does not quite go like this, but is perhaps best motivated by thinking about this proof. He began with the case in which there were no multiple roots and a, b were successive roots. He then wrote

$$P(x) = (x-a)(x-b)Q(x) = \big(x^2 - (a+b)x + ab\big)Q(x)$$

and considered

$$T(x) = \big(2x - (a+b)\big)Q(x),$$

which we would recognise as resulting from $R(x)$ of (3.19) by setting $k = m = 1$ and ignoring the final term. He then proved that $T(b)$ was "measured by" $(b-a)$, i.e., that $b-a$ was a factor of $T(b)$. He did so without explicitly noting that

$$T(b) = \big(2b - (a+b)\big)Q(b) = (b-a)Q(b).$$

Likewise $T(a) = (a-b)Q(a)$ is "measured by" $a-b$. In the case of n distinct real roots, he takes $Q(x)$ to be the product of monomials $x-r$ for roots r of P other than a, b. Having already shown, in a passage omitted from but referred to in Smith's *Source Book*, that $T(a)$ and $T(b)$ have opposite signs, he applies the Intermediate Value Theorem to conclude $T(c) = 0$ for some c between a and b. Exactly how he concludes $P'(c) = 0$ is unclear to me, but it is clear to us that if we restore the term $(x-a)(x-b)Q'(x)$ to $T(x)$ we get $R(x)$ and application of the Intermediate Value Theorem to R instead of to T yields $P'(c) = 0$ for some c between a and b — as we did earlier.

Without reference to the Calculus, which explains the steps of his proof, Rolle's demonstration comes across, in Barrow-Green's words, as "some clever algebraic manipulation", an unmotivated display of technical skill rather than a simple idea routinely worked out.

Moreover, Rolle muddies the waters by introducing general cascades and giving the argument in part for them and in part for the specific cascade we call the formal derivative. The final result is only established for derivatives, but he is aware of greater generality to its validity:

3.2.11 Exercise Let P, Q be polynomials and consider the cascades P^*, Q^*, etc., based on the arithmetic progression $a, a+c, a+2c, \ldots$, with $c \neq 0$

i. Show: $P^* = aP + cxP'$.
ii. Show: $(PQ)^* = P^*Q + PQ^* - aPQ$.
iii. Show: If $\alpha < \beta$ are consecutive *positive* roots of a polynomial R, then there is some γ between α, β such that $R^*(\gamma) = 0$. [Suggestion: Assume for simplicity that α, β are not multiple roots and $R = (x^2 - (\alpha+\beta)x + \alpha\beta)Q(x)$.]
iv. Show: The result of iii fails for $R(x) = x^2 - 2$, $a = 3$, $c = -1$.

Barrow-Green offers some summary comments:

> Clearly Rolle knew that his theorem was true for general arithmetic series... In this sense his original theorem was more general than its modern counterpart, because it was not restricted only to derivatives. At the same time it was also more restricted because it applied only to polynomial functions. Rolle gave no indication that he considered this theorem to have any more importance than his other results, despite the fact that it is clearly a cornerstone of his method of cascades.[159]

The extra generality is of dubious value and accentuates the extent to which Rolle's theorem is not Rolle's Theorem. His theorem is about the use of cascades to separate the roots of polynomials and not about the tangent to a curve being horizontal at some point between two such roots. The equivalence of his result with a special case of Rolle's Theorem is an accident, a consequence of the fact that the formal derivative happens to be an example of a cascade.

3.2.7 Taylor's Theorem

By "Taylor's Theorem" we do not mean here the result often referred to as Taylor's Theorem, namely Taylor's Theorem with the Lagrange Form for the Remainder (Theorem 3.1.6), nor a similar Taylor's Theorem with the Cauchy Form for the Remainder, neither of which predate the Mean Value Theorem and are thus not precursors thereto. The former, which we have termed the Higher Order Mean Value Theorem, was proven alongside the Mean Value Theorem by Lagrange, and the latter was later proven by Cauchy. Lagrange assumed Taylor's Theorem before proving his result, while Cauchy proved his as a lemma in proving Taylor's Theorem.

Properly speaking, Taylor's Theorem is the erroneous proposition that every function can be expanded into a power series around any fixed argument. The result simply does not hold in such generality. Fairly early on it was recognised that the expansion need not always converge and, in 1821, Cauchy produced an example for which the expansion did converge but not to the given function. The two theorems with the correctly eponymous remainders are tools for establishing the validity of Taylor's Theorem in certain cases.

Despite these caveats, Taylor's Theorem counts as a precursor to the Mean Value Theorem for a couple of reasons. First, it was a precursor to the correct Taylor's Theorems with the Lagrange and Cauchy Forms for the Remainder, the former of which directly generalises the Mean Value Theorem. Second, it was his work on Taylor series expansions that led Lagrange to the Mean Value Theorem.

The quintessential power series is the summation of the infinite geometric progression. If a is a fixed constant and r a fixed ratio with $|r| < 1$, we know

$$a + ar + ar^2 + \ldots = \frac{a}{1-r}. \tag{3.22}$$

---

[159] Barrow-Green, *op. cit.*, p. 744.

For $r = 1/4$ this is all but explicit in Archimedes's quadrature of the parabola.[160] And the general summation (3.22) was carried out with some rigour for $0 < r < 1$ by Oresme in the 14th century. However, this instance does not have the flavour of Taylor's Theorem which finds the series from the function. Oresme goes in the opposite direction, from the series to a closed form for the function of which it is an expansion.

The real breakthrough came with the Indians, particularly with Mādhava who was active in the latter half of the 14th and first quarter of the 15th centuries, one generation, say, after Oresme. Mādhava knew the infinite series expansions of the sine, cosine, and inverse tangent functions, series which would not be discovered in Europe until after the middle of the 17th century by Newton, James Gregory, Nicolaus Mercator (*c.* 1619–1687), and others.[161] The most important and influential of these was Newton, who gave power series expansions of numerous functions and applied such expansions, e.g., in solving differential equations.

At first the expansions of functions into power series was performed by a variety of *ad hoc* techniques. Newton, for example, obtained his binomial series,

$$(1+x)^\mu = \sum_{k=0}^{\infty} \binom{\mu}{k} x^k, \quad \binom{\mu}{k} = \frac{\mu(\mu-1)\cdots(\mu-k+1)}{k!}, \tag{3.23}$$

for rational values of μ, essentially by guessing the form of the coefficients. He first did this for $f(x) = (1-x^2)^{1/2}$ in connexion with the quadrature of the circle, and verified the result by multiplying the series by itself, obtaining $1 - x^2$. He also applied the familiar square root algorithm to $1 - x^2$ formally and derived the same series. Expansions of rational functions can be had by long division, thus, e.g.,

$$\frac{1}{1+x^2} = 1 - x^2 + x^4 - x^6 + \ldots$$

Integration yields

$$\tan^{-1} x = \int_0^x \frac{dx}{1+x^2} = x - \frac{x^3}{3} + \frac{x^5}{5} - \frac{x^7}{7} + \ldots$$

[160] Heath, *Works of Archimedes*, *op. cit.*, pp. 249–251.

[161] In line with traditional European historiography of Indian science, when Charles Matthew Whish (1794–1833), a civil servant of the East India Company brought the work of Mādhava to the attention of his fellow Europeans in the 1820s, doubts were raised: Indian knowledge of such series was recent and derivative, somehow having been learned from the Europeans. By the time Whish's paper was posthumously published in 1834 his belief in the originality of the Indian work had been eroded. Cf. U.K.V. Sarma, Vanishri Bhat, Venketeswara Pai, and K. Ramasubramanian, "The discovery of Mādhava series by Whish: an episode in historiography of science", to appear.

A similar trick yields

$$\ln(1+x) = \int_0^x \frac{dx}{1+x} = x - \frac{x^2}{2} + \frac{x^3}{3} - \frac{x^4}{4} + \ldots$$

The inverse sine was a bit trickier in that long division had to be replaced by application of the Binomial Theorem:

$$\begin{aligned}\sin^{-1} x &= \int_0^x \frac{dx}{\sqrt{1-x^2}} = \int_0^x (1-x^2)^{-1/2} dx \\ &= \int_0^x \sum_{k=0}^{\infty} \binom{-1/2}{k} (-x^2)^k dx \\ &\vdots \\ &= x - \frac{x^3}{6} + \frac{3x^5}{40} + \frac{5x^7}{112} + \ldots\end{aligned}$$

Eventually a number of mathematicians — Newton, Taylor, Johann Bernoulli, Leibniz, Abraham de Moivre (1667–1754), and probably Gregory[162] — discovered that, if f has an expansion,

$$f(x) = \sum_{k=0}^{\infty} a_k (x-a)^k, \tag{3.24}$$

around a point a, then the general coefficient of the series was given by

$$a_k = \frac{f^{(k)}(a)}{k!}, \tag{3.25}$$

the k-th derivative of f at a, divided by $k!$. Of these discoverers, Taylor is best known, and a series of the form (3.24) has come to be known as a *Taylor series*, or the *Taylor expansion of f around a*. And the assertion that functions have such expansions is called *Taylor's Theorem*.

I shall not present Taylor's proof here. It has been translated from the Latin of his *Methodus incrementorum* of 1715 and anthologised[163] and described elsewhere in the literature.[164] I shall only state that Taylor gave a direct derivation of the expansion of a function into a power series, establishing the formula (3.25) in the process. The proof was formally correct, but invalid in that Taylor paid no attention to the

[162] Cf., e.g., Giovanni Ferraro, *The Rise and Development of the Theory of Series Up to the Early 1820s*, Springer Science+Business Media, LLC, New York, 2008, p. 87. But see also Edwards, *op. cit.*, pp. 287–291.

[163] Struik, *op. cit.*, pp. 328–333; Stedall, *op. cit.*, pp. 201–206.

[164] Cf., e.g., Edwards, *op. cit.*, pp. 287–289; Smoryński, *Formalism*, *op. cit.*, pp. 125–127; Smoryński, *Treatise*, *op. cit.*, pp. 121–127.

problem of convergence, taking, as Felix Klein later wrote, "a transition to the limit of extraordinary audacity".[165]

A simpler determination of the coefficients of the Taylor expansion of a function around $a = 0$,

$$f(x) = \sum_{k=0}^{\infty} \frac{f^{(k)}(0)}{k!} x^k, \tag{3.26}$$

was given by Colin Maclaurin in his *A treatise on fluxions* of 1742. Unlike Taylor, who presumed to show the expansion existed, Maclaurin presupposed the existence of an expansion,

$$f(x) = a_0 + a_1 x + a_2 x^2 + \ldots$$

He noted that[166]

$$f(0) = a_0 + a_1 \cdot 0 + a_2 \cdot 0 + \ldots = a_0.$$

Differentiating term-by-term, he concluded

$$f'(x) = a_1 + 2a_2 x + 3a_3 x^2 + \ldots,$$

whence

$$f'(0) = a_1 + 2a_2 \cdot 0 + 3a_3 \cdot 0 + \ldots = a_1.$$

Similarly,

$$f''(x) = 2a_2 + 3 \cdot 2a_3 x + \ldots,$$

whence

$$f''(0) = 2a_2, \quad \text{i.e., } a_2 = \frac{f''(0)}{2}.$$

And so on.

In Maclaurin's honour, the series (3.26) is called a *Maclaurin series*, or the *Maclaurin expansion of f*.

Functions generated from polynomials, trigonometric functions, and the exponential and logarithmic functions via the algebraic operations and composition — in short, those functions given by elementary expressions of the Calculus — tend to be very well-behaved. Except for isolated points they have derivatives of all orders and one can assign to such functions f the Taylor expansions (3.24) with coefficients (3.25) or the Maclaurin expansion (3.26). This is unproblematic. The problem is convergence. It had been known since Oresme or earlier that the geometric series,

$$1 + x + x^2 + x^3 + \ldots,$$

---

165 Struik, *op. cit.*, p. 332.

166 The interested reader can look up Maclaurin's own words in the anthologies: Struik, *op. cit.*, pp. 338–341; Stedall, *op. cit.*, pp. 206–207.

fails to converge for $x \geq 1$. And it was no accident that Newton restricted x to having absolute value less than 1 for general values of μ in his binomial series (3.23). But it was assumed that the Taylor expansion of a function f around a point a did converge for x sufficiently close to a, and that it converged to $f(x)$. These two assumptions would be refuted by Cauchy in the early 1820s, but not until after Lagrange would attempt to found the Calculus upon them.

3.3 Lagrange and the Mean Value Theorem

With no knowledge of the relation between differentiability and the existence of tangents, the key analytic content of our modern Mean Value Theorem did not initially follow from the result of Valerio and Cavalieri and I have assigned to them and their result a mere precursor status. Lagrange transcends this, having stated and proved the analytic result under powerfully restrictive conditions — but conditions broad enough to cover the functions of his day. Where today we prove the Classroom Mean Value Theorem (Corollary 2.3.28 in Chap. 2 and Theorem 3.1.2) for functions continuous on the closed interval and differentiable on the open interval, or, in some classrooms, for continuously differentiable functions, Lagrange proved his result for *real analytic functions* — functions admitting expansion into Taylor series.

Like many of his contemporaries, Lagrange was concerned with the foundations of the Calculus. His *Thèorie des fonctions analytiques* of 1797 began with his critique of the existing attempts to lay a foundation for the Calculus:

> The first geometers who applied the Differential Calculus — Leibnitz, both Bernoullis, L'Hopital, etc. — grounded it on the consideration of infinitely small quantities of various orders, and on the assumption that one can consider and handle as equal [those] quantities which differ from each other only in infinitesimal amounts. They were satisfied through this Calculus to arrive quickly and surely at correct results, and on that account did not linger further thereby, to demonstrate the grounds for this. Euler, D'Alembert, and others who followed them tried to fill this gap in that they showed in particular applications that the differences which one took to be infinitely small must throughout be nothing other than nulls, and that their ratios (the single quantities, which actually arise in this Calculus) are no more than limits of the ratios of finite or unbounded differences.
>
> One must admit however that this concept, as correct as it may itself be, is clearly not sufficient to serve as the principle of a science, whose reliability is supposed to be grounded on evidence...[167]

With respect to Newton's kinematic approach, he points out that the Differential and Fluxional Calculi differ only in their metaphysics and cites as an advantage of

---

[167] Johann Philipp Grüson (trans.), *Lagrange's Theorie der analytischen Funktionen*, F.T. Lagarde, Berlin, 1798, pp. 3–4. As I know no French and some German, I have taken the liberty of translating from the German translation of *Théorie des fonctions analytiques*, which appeared a year after the original.

the latter the fact that "every man has or believes he has a concept of velocity".[168] But

> ...on the other hand one must confess that one has not even a proper clear concept of the velocity of a point in each instant, if this velocity is variable; and one can see from Maclaurin's scholarly work on fluxions, how difficult it is to prove the method of fluxions rigorously, and how many special tricks one must apply in order to show the various parts of this method and its proof.[169]

And Landen's residual calculus he finds to be awkward and unnatural. Finally, he announces his own approach:

> In one of the memoirs of the Berlin Academy from the year 1772 I maintained that the development of functions in a series contained the true principles of the Differential Calculus, and, to be sure, independently of the consideration of the infinitely small quantities or limits. I proved through this theory, the theorem of Taylor, which one can see as the main principle of this Calculus, and which one had previously known only through the help of this Calculus, or through the consideration of infinitely small differences.[170]

In the interim, he added, Arbogast had drawn the same conclusion and had presented a development of Analysis along the same lines. Arbogast, however, had not yet published his treatment and Lagrange "through certain special circumstances" needed the development of the principles of Analysis, so he returned to his earlier ideas.

Lagrange's approach depended on several assumptions we now reject, most important of which is that every function can be expanded into a power series,

$$f(x) = \sum_{k=0}^{\infty} a_k x^k \quad \text{or} \quad f(x) = \sum_{k=0}^{\infty} a_k (x-a)^k \quad \text{or} \quad f(x+i) = \sum_{k=0}^{\infty} a_k i^k.$$

While, as a matter of applied mathematics he considered the question of how many terms of the series were needed to calculate f to a desired degree of accuracy, he did not explain an equation like

$$f(x) = \sum_{k=0}^{\infty} a_k x^k$$

in terms of limits. He, in fact, did not explain what was meant; he merely assumed it was a sum and dealt with infinite sums accordingly. He assumed, without justification, that the usual rules of algebra could be applied and that his results for power series carried over automatically to real functions.

The problem with his approach is best explained by looking at the modern algebraic treatment of *formal power series*. A formal power series, which we might write as an infinite sum,

---

[168] *Ibid.*, p. 5.

[169] *Ibid.*, pp. 5–6.

[170] *Ibid.*, pp. 7–8.

$$\sum_{k=0}^{\infty} a_k x^k,$$

is not viewed algebraically as a function $f : \mathbb{R} \to \mathbb{R}$, but is formally defined as a sequence of coefficients,

$$f = \langle a_0, a_1, a_2, \ldots \rangle.$$

The algebraic operations on formal power series are formally defined. Addition is defined componentwise,

$$\langle a_0, a_1, a_2, \ldots \rangle + \langle b_0, b_1, b_2, \ldots \rangle = \langle a_0 + b_0, a_1 + b_1, a_2 + b_2, \ldots \rangle;$$

and multiplication is defined by *convolution*,

$$\langle a_0, a_1, a_2, \ldots \rangle * \langle b_0, b_1, b_2, \ldots \rangle = \langle c_0, c_1, c_2, \ldots \rangle,$$

where

$$c_k = \sum_{j=0}^{k} a_j b_{k-j},$$

so as to agree with the Cauchy product of the two series.

3.3.1 Exercise Thinking of $\langle a_0, a_1, a_2, \ldots \rangle$, $\langle b_0, b_1, b_2, \ldots \rangle$ as defining functions

$$f(x) = \sum_{k=0}^{\infty} a_k x^k, \quad g(x) = \sum_{k=0}^{\infty} b_k x^k,$$

determine the first few coefficients of

$$f(g(x)) = \sum_{k=0}^{\infty} c_k x^k.$$

Bearing in mind that power series can be differentiated termwise, the algebraists also give a formal definition of the derivative of a formal power series in the obvious way:

$$D(\langle a_0, a_1, a_2, \ldots \rangle) = \langle a_1, 2a_2, 3a_3, 4a_4, \ldots \rangle. \tag{3.27}$$

In the traditional post-Calculus course on Abstract Algebra one works through the proof that the formal power series form what is called a commutative ring with unit element, i.e., one proves the commutative, associative, and distributive laws, that $\langle 0, 0, 0, \ldots \rangle$ is an additive identity, $\langle -a_0, -a_1, -a_2, \ldots \rangle$ is the additive inverse of $\langle a_0, a_1, a_2, \ldots \rangle$, and that $\langle 1, 0, 0, 0, \ldots \rangle$ is the multiplicative identity. One might also determine the multiplicatively invertible elements (those for which $a_0 \neq 0$) and find their multiplicative inverses (by long division). In a more advanced course one could verify some rules of differentiation like the product formula or the chain rule. All of this can be done by direct algebraic manipulation without concern for convergence or any mention of limits. The work is detailed and grubby, but routine. It is not, however, immediately applicable to the Calculus.

3.3.2 Example Consider the formal power series,

$$f = \langle 1, -2, 0, 0, 0, \ldots \rangle,$$

representing the function $f(x) = 1 - 2x$. By long division, one quickly sees

$$1/f = \langle 1, 2, 4, 8, \ldots \rangle$$

representing the function

$$g(x) = \sum_{k=0}^{\infty} 2^k x^k.$$

From $f * (1/f) = \langle 1, 0, 0, 0, \ldots \rangle$ one concludes

$$g(x) = \frac{1}{1-2x},$$

whence

$$\begin{aligned} g(1) &= 1+2+4+8+\ldots = \frac{1}{1-2\cdot 1} = -1 \\ g\left(\frac{1}{2}\right) &= 1+1+1+1+\ldots = \frac{1}{1-2\cdot\frac{1}{2}} = \frac{1}{0} \\ g\left(-\frac{1}{2}\right) &= 1-1+1+-1+\ldots = \frac{1}{1-2\cdot -\frac{1}{2}} = \frac{1}{2}, \end{aligned}$$

only the middle equation of which makes some sense in the Calculus.

This already condemns Lagrange's programme to failure. To found the Calculus on the theory of power series Algebra alone will not suffice; one must also discuss convergence. This necessity was further demonstrated by two counterexamples by Cauchy, and by numerous so-called pathological functions later produced by Bolzano, Riemann, Weierstrass, etc.

Cauchy's first pertinent example concerns the limitations of the algebraic treatment of the product of series and appeared in the *Cours d'analyse*[171] in 1821:

3.3.3 Example The Cauchy product of two convergent series can fail to converge, and thus can fail to equal the product of the two series. For, take both series to be

---

[171] Bradley and Sandifer, *op. cit.*, pp. 101–102.

$$\sum_{k=0}^{\infty} \frac{(-1)^k}{\sqrt{k+1}}. \tag{3.28}$$

The k-th term of the Cauchy product of this series with itself is given by

$$c_k = (-1)^k \left(\sum_{i=0}^{k} \frac{1}{\sqrt{i+1}\sqrt{k+1-i}} \right).$$

But, for $0 \leq x \leq k$, $(x+1)(k+1-x)$ is maximised at $x = k/2$. Thus,

$$\begin{aligned} \frac{1}{\sqrt{i+1}\sqrt{k+1-i}} &\geq \frac{1}{\sqrt{\frac{k}{2}+1}\sqrt{k+1-\frac{k}{2}}} \\ &\geq \frac{1}{\sqrt{\frac{k+2}{2}}\sqrt{\frac{k+2}{2}}} = \frac{2}{k+2}. \end{aligned}$$

Thus,

$$\begin{aligned} |c_k| &= \sum_{i=0}^{k} \frac{1}{\sqrt{i+1}\sqrt{k+1-i}} \\ &\geq \sum_{i=0}^{k} \frac{2}{k+2} = \frac{2(k+1)}{k+2} > 1, \text{ for } k > 1, \end{aligned}$$

and we see that the Cauchy product of (3.28) with itself oscillates.

This, of course, is not an example dealing with power series, but it applies to the power series

$$f(x) = \sum_{k=0}^{\infty} \frac{(-1)^k}{\sqrt{k+1}} x^k$$

at $x = 1$.

In the *Résumé des leçons* of 1823, at the end of the 38th lesson, he cites, without proof, the following example.

3.3.4 Example Let

$$f(x) = e^{-x^2} \text{ and } g(x) = e^{-x^2} + e^{-1/x^2}.$$

The functions f, g have identical Taylor expansions at $x = 0$, but they differ at all $x \neq 0$, and thus the expansion can only converge to one of them.[172]

[172] The proof consists of showing that the function

Although ultimately doomed to failure, Lagrange's programme was not a waste of time. It did produce things of lasting value — including the Mean Value Theorem — and we should take a brief look at it. Indeed, with the increased interest in the history of the mathematics of this period, it is high time that annotated English translations of Lagrange appeared. In the present work, I offer only a brief and somewhat simplified account taking us up to the proof of the Mean Value Theorem.

Lagrange gave a fairly broad definition of function and assumed every function expandable into a power series except at isolated singularities. Thus, if f is a function and x is a point, one could write

$$f(x+i) = f(x) + pi + qi^2 + ri^3 + \dots, \tag{3.29}$$

where $p, q, r, \dots$ were functions of x and thus independent of i. (Here we think of x as a constant and i as a variable.) The application of algebraic rules to such series was implicitly axiomatic; he gave no thought to justifying or even mentioning them. However, he was careful in other matters. His first task, for example, was to give an unconvincing proof that none of the terms could be of the form $ui^{m/n}$, where m/n was a fraction in lowest terms with $n > 1$. His argument was that if f were a proper single-valued function of x, then $f(x+i)$ would also be single-valued, but the term $ui^{m/n}$ would have n distinct values corresponding to the various n-th roots of i^m.[173]

His first real step[174] was to show successively that $p, q, r, \dots$ were unique. Because (3.29) contains no terms with fractional exponents, one can write

$$f(x+i) = f(x) + iP,$$

where P is a function of x and i and is not infinite when $i = 0$. It follows that

$$\begin{aligned} P &= \frac{f(x+i) - f(x)}{i} \\ &= p + qi + ri^2 + \dots \end{aligned}$$

and p is the value of P at $i = 0$. He notes that $P - p$ can be written as iQ and that Q is q when $i = 0$, etc. Although mathematical induction was known at the time, it was not Lagrange's style to present formal inductive arguments in these matters. He

---

(Footnote 172 continued)

$$h(x) = \begin{cases} e^{-1/x^2}, & x \neq 0 \\ 0, & x = 0 \end{cases}$$

and all of its derivatives vanish at $x = 0$, whence its Taylor series is $0 + 0x + 0x^2 + 0x^3 + \dots$ However, $h(x)$ is 0 only at the single point $x = 0$. The proof is not hard, but is a bit grubby. I refer the reader to Exercise 6.6 (pp. 184–185) of Smoryński, *Formalism*, *op. cit.*, for an outline of the proof.

[173] Grüson, *op. cit.*, §10, pp. 10–12.

[174] *Ibid.*, §§11–12, pp. 12–16.

established the first few cases of a result and then either asserted the general result or left that formality to the reader.

Establishing the uniqueness of $p, q, r, \ldots$ was not the point to these calculations; rather it was to determine $p, q, r, \ldots$ successively. This he did with the example of $f(x) = \sqrt{x}$. The first step is fairly trivial:

$$P(x, i) = \frac{\sqrt{x+i} - \sqrt{x}}{i} = \frac{(\sqrt{x+i} - \sqrt{x})(\sqrt{x+i} + \sqrt{x})}{i(\sqrt{x+i} + \sqrt{x})}$$
$$= \frac{x+i-x}{i(\sqrt{x+i} + \sqrt{x})} = \frac{1}{\sqrt{x+i} + \sqrt{x}},$$

whence

$$p(x) = P(x, 0) = \frac{1}{\sqrt{x+0} + \sqrt{x}} = \frac{1}{2\sqrt{x}}.$$

Finding q is a tiny bit more involved:

$$Q(x, i) = \frac{P - p}{i} = \frac{\frac{1}{\sqrt{x+i} + \sqrt{x}} - \frac{1}{2\sqrt{x}}}{i}$$
$$= \frac{2\sqrt{x} - \sqrt{x+i} - \sqrt{x}}{i(\sqrt{x+i} + \sqrt{x}) \cdot 2\sqrt{x}} = \frac{\sqrt{x} - \sqrt{x+i}}{i2\sqrt{x}(\sqrt{x+i} + \sqrt{x})}$$
$$= \frac{(\sqrt{x} - \sqrt{x+i})(\sqrt{x} + \sqrt{x+i})}{i2\sqrt{x}(\sqrt{x+i} + \sqrt{x})^2} = \frac{-1}{2\sqrt{x}(\sqrt{x+i} + \sqrt{x})^2},$$

whence

$$q(x) = Q(x, 0) = \frac{-1}{2\sqrt{x}(2\sqrt{x})^2} = \frac{-1}{8x\sqrt{x}}.$$

And finding r is an exercise:

3.3.5 Exercise Show

$$R = \frac{\sqrt{x+i} + 3\sqrt{x}}{8x\sqrt{x}(\sqrt{x+i} + \sqrt{x})^3},$$
$$r = \frac{1}{16x^2\sqrt{x}}.$$

The algebra can get a bit involved and better methods are needed.

3.3.6 Exercise Lagrange[175] simplifies the above computations as follows:

$$\sqrt{x+i} = \sqrt{x} + iP,$$

175 *Ibid.*, §13, pp. 16–18.

whence

$$x + i = x + 2\sqrt{x}\, iP + i^2P^2$$
$$i = 2\sqrt{x}\, iP + i^2P^2$$
$$1 = 2\sqrt{x}P + iP^2,$$

and

$$p = P(x, 0) = \frac{1}{2\sqrt{x}}.$$

Likewise, using

$$P = p + iQ,$$

one can solve for q. Do this and find r as well.

Having got this far, Lagrange now deserts pure algebra and introduces continuity considerations:

14.

But the greatest advantage of the method just shown is in that it shows how the functions p, q, r, &c. arise from the main function fx, and particularly because it demonstrates that the remainders iP, iQ, iR, &c., are quantities which must vanish when $i = 0$; from this we draw the important consequence that in the series

$$f(x) + pi + qi^2 + ri^3 + \&c.$$

which arises from the development of the function $f(x + i)$, one can always take i so small, that each member will be larger than the sum of all following terms; and the same holds for all smaller values of i.

For, since the remainders iP, iQ, iR, &c. are functions of i, which, in accordance with the nature of the development itself, vanish when $i = 0$, it follows that if one consider the curved line of which i will be the abscissa and one of the functions the ordinate, this curved line will cut the axes at the origin of the abscissas; and if only this point is not singular, which happens only for very special values of x as one easily sees with a little reflexion and through reasoning analogous to that of [section] number 10, so the course of this curved line will necessarily remain continuous from this point on; it will thus approach the axis more and more before it intersects, so that it comes closer to such a quantity which is smaller than each given quantity; one will thus always be able to find an abscissa i which belongs to an ordinate, which is smaller than a given quantity; and every smaller value of i corresponds too to a still smaller ordinate than the given quantity.

One can thus take i so small, without it being null, so that iP is smaller than $f(x)$, iQ smaller than p, and iR smaller than q, etc.; and therefore, i^2R will be smaller than iq; i^3R smaller than i^4q, &c. Thus too (number 11)

$$iP = ip + i^2q + i^3r + \&c.$$
$$i^2Q = i^2q + i^3r + \&c.$$
$$\text{and } i^3R = i^3r + \&c.$$

> so it follows, that one can always give i so small a value, that each term of the series $fx + ip + i^2q + i^3r + \&c.$ will be larger than the sum of all following terms; and then too the same condition will be fulfilled for any smaller value of i.
>
> This theorem should be regarded as one of the fundamental principles of the theory that we propose to develop: it is tacitly supposed in the differential calculus and in the calculus of fluxions, and it is here that one can say that one has the greatest hold on these calculi, especially in their application to geometrical and mechanical problems.[176]

Before the reader starts citing the obvious counterexamples, let me explain that in Lagrange's day "larger" and "smaller" referred to size, not order. And I imagine that he meant "nonzero term" when he referred to a "term" of the series. He certainly knew the sine series,

$$\sin x = x - \frac{x^3}{3!} + \frac{x^5}{5!} - \frac{x^7}{7!} + \ldots$$

and was not foolish enough to assert, for example, that

$$0 \cdot \frac{i^2}{2!} \text{ is greater than } -\frac{i^3}{3!} + \frac{i^5}{5!} - \frac{i^7}{7!} + \ldots$$

for small enough $i > 0$.

In modern terms Lagrange's argument would proceed as follows. Suppose

$$g(i) = a_0 + a_1 i + a_2 i^2 + \ldots \tag{3.30}$$

converges for some value i_0, and let $a_m \neq 0$. Ignoring the first m terms, we consider the tail

$$\begin{aligned} g_m(i) &= a_m i^m + a_{m+1} i^{m+1} + \ldots \\ &= a_m i^m + i^{m+1}(a_{m+1} + a_{m+2} i + \ldots) \\ &= i^m(a_m + iM(i)). \end{aligned}$$

By the continuity of $M(i)$ for $|i| < |i_0|$ and the fact that $iM(i) = 0$ at $i = 0$, it follows for any $\epsilon > 0$ we can find a $\delta > 0$ such that

$$|i| < \delta \Rightarrow |iM(i)| < \epsilon.$$

Simply choose $\epsilon = |a_m|$.

The proof is incomplete and today we would argue a bit differently, first proving that the convergence of the series (3.30) at i_0 entails the *absolute convergence* of the series for $|i| < |i_0|$. Indeed, one would show that this convergence is *uniform* for $|i| \leq \rho$ for any $\rho < |i_0|$, whence one would establish the continuity of $M(i)$ on $[-\rho, \rho]$. Precise definitions of these concepts — continuity, convergence, absolute

---

[176] *Ibid.*, pp. 18–20. Insofar as I am translating from a translation both here and below, I have been less scrupulous in preserving typographic conventions than usual.

convergence, and uniform convergence — would be given over the next few decades. We have already discussed continuity, and we will not be needing the other concepts below, so I will forego further discussion of these matters.

3.3.7 Exercise Our sketched replacement of Lagrange's argument shows that for

$$g(i) = a_0 + a_1 i + a_2 i^2 + \dots$$

and any m for which $a_m \neq 0$ there is an $i_0 > 0$ such that for all $|i| < i_0$,

$$\left|a_m i^m\right| > \left|a_{m+1} i^{m+1} + a_{m+2} i^{m+2} + \dots\right|.$$

Unless I have mistranslated the German or the German translator has mistranslated Lagrange's French, Lagrange asserts that i_0 can be found so that this inequality holds for all m. Find the Taylor expansion of

$$g(i) = \frac{1}{1 - i^2} + \frac{2i}{4 - i^2}$$

and show Lagrange's stronger assertion fails to hold for this g. [Hint: Compare $a_{2m+1} i^{2m+1}$ with the terms arising from the expansion of $1/(1 - i^2)$.]

The next important step in Lagrange's development was to define the derivative, which he called the *derived function* or the *first function* and which he denoted by f'. His definition was, given the expansion, purely algebraic and made no reference to limits: if

$$f(x + i) = f(x) + p(x)i + q(x)i^2 + r(x)i^3 + \dots,$$

then

$$f'(x) = p(x).$$

Working formally, without concern for issues of convergence, he then compared $f((x + o) + i)$ and $f(x + (i + o))$, and concluded successively

$$q(x) = \frac{p'(x)}{2},\ r(x) = \frac{q'(x)}{3},\ \dots$$

i.e.,

$$q(x) = \frac{f''(x)}{2},\ r(x) = \frac{f'''(x)}{3 \cdot 2},\ \dots,$$

whence

$$f(x + i) = f(x) + f'(x)i + \frac{f''(x)}{2!}i^2 + \frac{f'''(x)}{3!}i^3 + \dots$$

Following this Lagrange devotes a fair number of sections and pages performing routine tasks—establishing specific Taylor expansions, differentiating trigonometric functions, deriving the Chain Rule, etc.

In §39 he reaches L'Hôpital's Rule. Because of its customary association with the Mean Value Theorem, it is worth quoting him in full on the matter:

39.

On the occasion of the difficulty which we have just removed, we would like to demonstrate the theory of the method of finding the value of a fraction in the case where the numerator and denominator approach zero simultaneously.

Let $\frac{fx}{Fx}$ be such a fraction, where fx and Fx are functions of x, and where the assignment $x = a$ makes both zero simultaneously. One desires the value of this function when $x = a$.

One sets $y = \frac{fx}{Fx}$, and hence

$$yFx = fx.$$

If one takes $x = a$, this equation justifies itself independently of the value of x, which thus remains indeterminate; therefore it cannot serve for the determination of y in this condition of determinacy. Take however the first equation,[177]

$$\text{then one has } y'Fx + yF'x = f'x;$$

the assignment of $x = a$ lets the first term $y'Fx$ vanish; and the rest of the equation gives $y = \frac{f'x}{F'x}$. If it happens that the first functions $f'x$, $F'x$ approach zero through the same assignment, so one would find through the same principle, if one replaces fx, Fx by $f'x$, $F'x$ in the above cited equation, this new expression for y,

$$y = \frac{f''x}{F''x}, \text{ etc.}$$

One could also derive this directly from the first[178] equation if one considers that, as it justifies itself anew, it can no longer serve for the determination of y; that it consequently is necessary to go over to this second equation, which will be

$$y''Fx + 2y'F'x + yF''x = f''x.$$

Because the assignment of $x = a$ makes the functions Fx and $F'x$ equal to zero, the terms with y' and y'' will fall away and the remaining terms will give

$$y = \frac{f''x}{F''x},$$

as again above. One need not fear that the functions $fx, f'x, f''x$, &c. and $Fx, F'x, F''x$, &c. *ad infinitum* can simultaneously become zero through the assignment $x = a$, as it appears some geometers assume. For, because

$$f(x+i) = fx + if'x + \frac{i^2}{2}f''x \text{ \&c.}$$

one has, if $x = a$,

[177] I.e., differentiate once.

[178] I.e., derived.

$$f(x+i)=0$$

no matter what i is, which is impossible. It would be the same for $F(x+i)$. But it can happen that these functions through the given assignment of $x=a$ become infinite, which likewise would make the fractions

$$\frac{fx}{Fx}, \frac{f'x}{F'x}, \&c.$$

indeterminate: but the solution of this difficulty depends on the investigation of the second case of no. 34, with which we will now occupy ourselves.[179]

The above excerpt suffices to illustrate his approach and we will thus not look into the cases cited in §34 nor will we continue his discussion of the case in which the functions simultaneously become infinite. I have only two comments to make on Lagrange's form of L'Hôpital's Rule. First, he does not prove, as we do today, that if $\lim_{x\to a} f'(x)/F'(x)$ exists, then so does $\lim_{x\to a} f(x)/F(x)$; he *assumes* the limit exists. Second, his argument that the vanishing of a function and all its derivatives at a point cannot occur for a nonzero function is only valid if the function equals its Taylor expansion at some point other than a — as is *not* the case in Cauchy's Example 3.3.4 which showed conclusively the limitations of Lagrange's programme of founding the Calculus on Taylor series.

In §48 he offers something of a proof of the Strictly Increasing Function Theorem. The proof is not acceptable by today's standards, but can be salvaged by isolating the implicit assumptions.

We want to establish this general theorem: If a first function[180] of z, such as $f'z$, is always positive for all values of z, from $z=a$ to $z=b\,(b>a)$ the difference of the original functions, which correspond to those two values of z, namely $fb-fa$, is necessarily a positive quantity. We consider the function $f(z+i)$ of which the expansion is

$$fz+if'z+\frac{i^2}{2}f''z+\&c.$$

We have seen that we can always take the quantity i so small that each arbitrary term of this series is greater than the sum of all following terms (no. 14). Thus the term $if'z$ can be greater than the rest of the series; whence, if $f'z$ is a positive quantity, then one can take i positive and small enough that with the whole series

$$if'z+\frac{i^2}{2}f''z+\&c.$$

necessarily has a positive value; but this series is

$$=f(z+i)-fz;$$

consequently, if $f'z$ is a positive quantity, one can take for i a positive quantity which is small enough that the quantity

$$f(z+i)-fz$$

[179] Grüson, *op. cit.*, pp. 56–58. I haven't checked the French original, but the German text has a number of misprints which I assume were introduced in translation and have consequently corrected without notice.

[180] I.e., derived function.

is necessarily positive.

If we successively set in place of z the quantities

$$a,\ a+i,\ a+2i,\ \&c.,\ a+ni,$$

it will follow that one can take i positive and small enough so that all the quantities

$$f(a+i)-fa,\ f(a+2i)-f(a+i), f(a+3i)-f(a+2i),\\ \text{to } f[a+(n+1)i]-f(a+ni)$$

are necessarily positive if the quantities

$$f'a,\ f'(a+i),\ f'(a+2i),\ \&c. \text{ to } f'(a+ni)$$

are so. Thus it will also be in this case that the sum of the first quantities, that is, the quantity

$$f[a+(n+1)i]-fa,$$

is positive.

We now set

$$a+(n+1)i=b,$$

so one has

$$i=\frac{b-a}{n+1},$$

and one will conclude therefrom that the quantity $fb-fa$ is necessarily positive, if all the quantities

$$f'a,\ f'\left(a+\frac{b-a}{n+1}\right),\ f'\left(a+\frac{2(b-a)}{n+1}\right),\ f'\left(a+\frac{3(b-a)}{n+1}\right),\ \&c.$$

to

$$f'\left(a+\frac{n(b-a)}{n+1}\right)$$

are positive, however large one wishes to take n.

Therefore the quantity $fb-fa$ is all the more positive if f' is always a positive quantity, and one gives all possible values from $z=a$ to $z=b$, because the values

$$a,\ a+\frac{b-a}{n+1},\ a+\frac{2(b-a)}{n+1},\ \text{etc. } a+\frac{n(b-a)}{n+1}$$

will necessarily find themselves under these values, however large one wishes to take n.[181]

This proof, as I said, is incorrect. The problem is not its appeal to his false result of §14, as he only used the true consequence that the term $if'z$ is larger than the sum of all succeeding terms for sufficiently small i. The problem is that he assumes *uniformity* again — that the same smallness is sufficient for all $z \in [a,b]$. Now, this can be proven for Taylor series when $[a,b]$ lies in the interior of the interval of convergence, but Lagrange doesn't prove this as he seems to be unaware of the issue of uniformity.[182]

181 *Ibid.*, pp. 70–72.

182 Also, he doesn't have the tools.

Lagrange's proof is, nonetheless, not a complete loss: it is valid in some generality. One merely has to apply the technique of *proof-generated concept* to isolate and define those functions to which the proof applies. In applying this technique, I come up with the following tentative definition.

3.3.8 Definition Let $f : [a, b] \to \mathbb{R}$. f satisfies the *Lagrange condition* if we can write

$$f(z+i) = f(z) + if'(z) + i^2 Q(i, z),$$

where Q is bounded and f' is continuous on $[a, b]$.

3.3.9 Theorem (Lagrange's Strictly Increasing Function Theorem) *Let the function $f : [a, b] \to \mathbb{R}$ satisfy the Lagrange condition and suppose $f'(z) > 0$ for all $z \in [a, b]$. Then f is strictly increasing on $[a, b]$: for all $x, y \in [a, b]$,*

$$x < y \Rightarrow f(x) < f(y).$$

Proof. By the Extreme Value Theorem, f' assumes a minimum value m on $[a, b]$. By assumption, $|Q(i, z)|$ has an upper bound, say, M. For

$$|if'(z)| > \left|i^2 Q(i, z)\right|$$

it suffices that

$$|f'(z)| \geq m > |iM| \geq |iQ(i, z)|.$$

Hence, so long as $0 < i < m/M$, we have

$$f(z+i) - f(z) = if'(z) + i^2 Q(i, z) > 0$$

for all $z \in [a, b]$. The rest of Lagrange's proof proceeds as before. □

3.3.10 Exercise Recall the notion of uniform differentiability of Definition 2.3.12 in Chap. 2. Show that Lagrange's proof applies to f uniformly differentiable on $[a, b]$.

Today, of course, we prove the Strictly Increasing Function Theorem under the weaker assumption of differentiability, either by direct appeal to the Least Upper Bound Principle as on pages 131–136, above, or by appeal to the Mean Value Theorem as on page 139. Where most theoretical treatments of the Calculus derive the Strictly Increasing Function Theorem from the Mean Value Theorem, Lagrange's original proof of the latter theorem reverses the dependence.

The natural lemma to prove at this point is the Constant Function Theorem, the basis of the proof of which he has already presented in his discussion of L'Hôpital's Rule. It could be that I skipped over it in looking up his proof of the Mean Value Theorem, or that he reserves its presentation for the later discussion of the Integral Calculus, but I can report that he does not present it at this point in his exposition.

3.3.11 Theorem (Lagrange-Like Constant Function Theorem) *Let f equal its Taylor expansion in an interval I. Suppose $f'(x) = 0$ for all $x \in I$. Then f is constant.*

Proof. If $f'(x) = 0$ identically on I, then f' is constant and has zero derivative: $f''(x) = 0$ for all $x \in I$. But then $f'''(x) = 0$ for all $x \in I$. Etc.

Now, let $x, x + i \in I$, and observe

$$\begin{aligned} f(x+i) &= f(x) + if'(x) + \frac{i^2}{2}f''(x) + \dots \\ &= f(x) + i \cdot 0 + \frac{i^2}{2} \cdot 0 + \dots = f(x). \end{aligned}$$ □

Note that this proof requires the full force of the assumption that f equals its Taylor expansion.

We are now a tiny step away from proving the Mean Value Theorem. We can define $\int f(x)dx$ to be any anti-derivative F of f and define the definite integral by

$$\int_a^b f(x)dx = F(b) - F(a).$$

In particular,

$$\int_a^b f'(x)dx = f(b) - f(a).$$

But, by the Increasing Function Theorem, if M, N are the maximum and minimum values, respectively, of f' on $[a, b]$, it follows that

$$\int_a^b Ndx \leq \int_a^b f'(x)dx \leq \int_a^b Mdx.$$

Thus

$$N(b-a) \leq f(b) - f(a) \leq M(b-a),$$

i.e.,

$$N \leq \frac{f(b) - f(a)}{b-a} \leq M.$$

Applying the Intermediate Value Theorem to f' yields the existence of some c between where the extreme values M and N occur for which

$$f'(c) = \frac{f(b) - f(a)}{b-a}.$$

Lagrange's derivation is not quite as straightforward as this. He does not prove the Constant Function Theorem at this point but shows directly the Mean Value Inequality: for any antiderivative F of f',

$$N(b-a) \leq F(b) - F(a) \leq M(b-a), \tag{3.31}$$

and exhibits a specific F. Well, almost — he exhibits F as an antiderivative to a specific substitution instance of f'. The ensuing algebra is a masterful display that hides the motivation well. Moreover, he first proves a higher-order generalisation of (3.31) before first considering the simplest case.

The tail of the Taylor series,

$$f(x+z) = f(x) + zf'(x) + \frac{z^2}{2}f''(x) + \ldots,$$

assumes the form $z^m Z(z)$, where $m \in \{0, 1, 2, \ldots\}$ is the number of discarded terms. Z is itself a continuous and well-behaved function. Choose F so that $F'(z) = z^m Z(z)$ and let M, N be the maximum and minimum values of Z on an interval $[a, b]$ contained within the interval of convergence. As Lagrange will shortly be taking $a = 0, b = 1$, we assume outright that $0 \leq a < b$.

By choice of M and N, the quantities

$$M - Z(z) \text{ and } Z(z) - N$$

are nonnegative throughout $[a, b]$. For $z \geq 0$, it follows that

$$z^m(M - Z(z)) \text{ and } z^m(Z(z) - N)$$

are also nonnegative for all $z \in [a, b]$. It follows from the Increasing Function Theorem[183] that the antiderivatives

$$G(z) = \frac{z^{m+1}M}{m+1} - F(z) \text{ and } H(z) = F(z) - \frac{z^{m+1}N}{m+1}$$

are increasing functions on $[a, b]$. In particular,

$$G(b) \geq G(a),$$

[183] Lagrange is a bit sloppy in not distinguishing between $<$ and $\leq$ or between positive and nonnegative. He proved the Increasing Function Theorem for positive derivatives and strictly increasing functions and is now applying it for nonnegative derivatives and increasing functions. Perhaps it should be noted at this point that his proof of the Strictly Increasing Function Theorem can easily be modified to establish the result for non-negative derivatives and the weak inequality, but that the proof of Theorem 3.3.9 does not generalise.

i.e.,

$$\frac{b^{m+1}M}{m+1} - F(b) \geq \frac{a^{m+1}M}{m+1} - F(a),$$

i.e.,

$$F(b) \leq F(a) + \frac{M\left(b^{m+1} - a^{m+1}\right)}{m+1}.$$

Likewise,

$$F(a) + \frac{N\left(b^{m+1} - a^{m+1}\right)}{m+1} \leq F(b).$$

Put differently,

$$\frac{N\left(b^{m+1} - a^{m+1}\right)}{m+1} \leq F(b) - F(a) \leq \frac{M\left(b^{m+1} - a^{m+1}\right)}{m+1}. \tag{3.32}$$

One is tempted here to set $m = 0$, write

$$N \leq \frac{F(b) - F(a)}{b - a} \leq M$$

and apply the Intermediate Value Theorem to F and conclude

$$F'(c) = \frac{F(b) - F(a)}{b - a}$$

for some c in the interval. However, this yields the result for F as a function of z. Lagrange wants the result for f as a function of x and has to supply an additional argument. To this end, he considers the infinite series,

$$f(x+i) = f(x) + if'(x) + \frac{i^2}{2}f''(x) + \ldots,$$

and makes the substitutions $x - i$ for x and xz for i, obtaining

$$f(x) = f(x - xz) + xzf'(x - xz) + \frac{x^2z^2}{2}f''(x - xz) + \ldots$$

As z goes from 0 to 1, $x - xz$ will go from x to 0, thus the interval $[a, b] = [0, 1]$ will correspond to the interval $[0, x]$ or $[x, 0]$ in the Maclaurin expansion,

$$f(x) = f(0) + xf'(0) + \frac{x^2}{2}f''(0) + \ldots$$

Lagrange writes

$$f(x) = f(x - xz) + xP(x, z). \tag{3.33}$$

Thus

$$P = zf'(x - xz) + x\frac{z^2}{2}f''(x - xz) + x^2\frac{z^3}{6}f'''(x - xz) + \ldots$$

He now chooses F to be an antiderivative of P' with respect to the variable z, thinking of x as a constant. Differentiating P term by term,[184] he obtains

$$\begin{aligned} F'(z) = P'(z) =& f'(x - xz) + zf''(x - xz)(-x) + xzf''(x - xz) + \\ & x\frac{z^2}{2}f''(x - xz) + \frac{3x^2z^2}{6}f''(x - xz) + \ldots \\ =& f'(x - xz) + 0 + 0 + \ldots \\ =& f'(x - xz). \end{aligned} \tag{3.34}$$

Writing $F'(z) = z^m Z(z)$, we have $m = 0$ and $Z(z) = f'(x - xz)$, whence (3.32) reads

$$N(b - a) \leq F(b) - F(a) \leq M(b - a), \tag{3.35}$$

where M, N are the maximum and minimum values, respectively, of $f'(x - xz)$ for $z \in [0, 1]$.[185] We are taking $a = 0$, $b = 1$, and we might as well take $F = P$, translating (3.35) successively into

$$N \leq P(1) - P(0) \leq M$$

$$N \leq \left(f'(0) + \frac{x}{2}f''(0) + \frac{x^2}{6}f'''(0) + \ldots\right) - 0 \leq M$$

$$N \leq \frac{f(x) - f(0)}{x} \leq M, \text{ by (74).}$$

And we conclude there to be some $c \in [0, 1]$ such that

$$f'(x - xc) = \frac{f(x) - f(0)}{x},$$

i.e., there is some $u = x - xc$ between 0 and x such that

---

[184] One of the many procedures Lagrange did not rigorously justify: Later (1816) Bolzano thought he had proven every series of functions could be differentiated term by term, an error he continued to make in the 1830s. Credit for the result should probably go to Cauchy, whose *Résumé*, *op. cit.*, included Lesson 38 on convergence criteria, and the proof in Lesson 40 that one can perform integration term by term on a uniformly convergent series. I did not find an explicit statement that differentiation can be performed termwise on a power series in the *Résumé*, but the conclusion is immediate.

[185] Note that (3.35) is just the Mean Value Inequality established without appeal to the Constant Function Theorem, which itself follows from the Inequality on assumption that $N = M = 0$.

$$f(x) = f(0) + xf'(u).$$

Writing

$$f(x) = f(x - xz) + xzf'(x - xz) + x^2Q(x, z),$$

Lagrange similarly derives the Second Order Mean Value Theorem,

$$f(x) = f(0) + xf'(0) + \frac{x^2}{2}f''(u),$$

for some u between 0 and x. He also treats the Third Order Mean Value Theorem and asserts the result to hold for all higher orders. He then makes a new substitution to conclude

$$\begin{aligned} f(z + x) &= f(z) + xf'(z + u_1) \\ &= f(z) + xf'(z) + \frac{x^2}{2}f''(z + u_2) \\ &= f(z) + xf'(z) + \frac{x^2}{2}f''(z) + \frac{x^3}{6}f'''(x + u_3) \\ &\&c. \end{aligned}$$

for some $u_1, u_2, u_3, \ldots$ between 0 and x.[186]

I don't believe this proof would please the modern instructor or student much. The auxiliary function normally used to reduce the Classroom Mean Value Theorem to Rolle's Theorem, often criticised as an unmotivated trick, can, as we saw earlier, be motivated quite naturally. Perhaps I have simply not thought deeply enough about it, but the trick of expressing $f(x)$ in terms of the value of the function and all its derivatives at $x - xz$ seems inspired. After the computation of P' has been made, one can see that it is precisely this substitution that led to all the cancellations in (3.34) resulting in a single term for P', but other than it being the result of a search for some expression that will perform this trick, I fail to see how to come up with it.

Another thing is that the cancellations require the infinite series: Lagrange's proof of the Mean Value Theorem only works for real analytic functions and not for, say, continuously or m-times continuously differentiable functions.

This was not Lagrange's last word on the matter. In 1799 he lectured again on his theory, his new lectures, *Leçons sur le calcul des fonctions*, undergoing several printings in the ensuing years. I find this new work clearer and better organised, the new proof of the Mean Value Theorem more closely resembling the motivation behind his earlier proof cited on page 252, above. The big departure from this motivation is that he eschewed the use of the integral sign, which does not appear at all in *Théorie des fonctions analytique* and which I could only find in a supplement to the 1808

---

[186] Cf. Grüson, *op. cit.*, §49, pp. 72–74, for (3.32) and §§50–52, pp. 74–77, for the rest of the argument.

printing of the *Leçons*.[187] He preferred to refer to *primitive functions* and to deal with a specific primitive function taking the value 0 at the point around which the Taylor expansion is made (thus bypassing any need to appeal to the Second Fundamental Theorem of the Calculus and its dependence on the Constant Function Theorem).

In the *Leçons* Lagrange states more precisely the property of the derivative he will use in the proof, isolating it first as a Lemma. Grabiner has called it the *Lagrange property*:

3.3.12 Definition Let f be a function defined in a neighbourhood of a point x. f has the *Lagrange property at* x if one can write

$$f(x+i) = f(x) + if'(x) + iV,$$

where V goes to 0 with i.

This is slightly weaker than the Lagrange condition of Definition 3.3.8, which we read off the earlier proof. There it was assumed V took the form iQ with Q bounded, so that $\lim_{i\to 0} iQ = 0$. Here one replaced the explicit form of V by the conclusion drawn from it.

A little algebra yields

$$\frac{f(x+i)-f(x)}{i} - f'(x) = V.$$

But $V \to 0$ as $i \to 0$. Thus, for $\epsilon > 0$, there is a $\delta > 0$ such that

$$0 < |i| < \delta \Rightarrow |V| < \epsilon,$$

i.e.,

$$0 < |i| < \delta \Rightarrow \left|\frac{f(x+i)-f(x)}{i} - f'(x)\right| < \epsilon.$$

And we see that the Lagrange property is almost our modern definition of the derivative. The differences are that i. Lagrange, having defined the derivative already, assumes its existence prior to considering the limit, and ii. he has not explicitly defined any notion of limit.

The isolation of the Lagrange property almost allows one to state how Lagrange's new proof is superior to the old: it is more general. His first proof of the Mean Value Theorem was valid for functions equalling their Taylor expansions, i.e., real analytic functions; the new proof ostensibly carries over for functions for which the Lagrange property holds at all points of an interval. A careful reading of the proof shows that he assumes a uniform version of the Lagrange property — i.e., he assumes uniform differentiability — in proving the Strictly Increasing Function Theorem. And, in the

---

[187] Mind you, I could have overlooked such an occurrence, but I take that possibility as proof that any such occurrence was at best rare in his textbooks on the Calculus.

final step in applying the Intermediate Value Theorem to f', he uses the continuity of the derivative, itself a consequence of uniform differentiability. Thus, his new proof established the Mean Value Theorem for uniformly differentiable functions.

The new proof begins with Lagrange deriving the Strictly Increasing Function Theorem as before, but now appealing explicitly to the Lagrange property. Following this he sets out to trap

$$f(x+i) - f(x) - if'(x) - \frac{i^2}{2}f''(x) - \ldots - \frac{i^{\mu-1}}{(\mu-1)!}f^{(\mu-1)}(x)$$

between explicit bounds:

> Now here's how the principle in question applies to the determination of the limits of the development of $f(x+i)$:
>
> Let p and q first be values of $x+i$ which render the smallest and largest values of the derivative $f'(x+i)$ regarding x as given and i as varying from 0 to some given value of i. So $f'p$ will be the smallest value of $f'(x+i)$ and $f'q$ will be the largest, and therefore $f'(x+i) - f'p$ and $f'q - f'(x+i)$ will always be positive quantities.
>
> Looking at these two quantities as derived functions relative to the variable i, their primitive functions, taken so that they are zero when $i = 0$, will be, because x, p and q are constant,
>
> $$f(x+i) - fx - if'p, \text{ and } if'q - f(x+i) + fx.$$
>
> Thus, provided that $f'(x+i)$ is never infinite from $i = 0$ to the given value of i, which will be the case if $f'p$ and $f'q$ are not infinite quantities, we will have, by the previous rule, if i is positive,[188]
>
> $$f(x+i) - fx - if'p > 0, \text{ and } fx - f(x+i) + if'q > 0;$$
>
> from which one derives
>
> $$f(x+i) > fx + if'p, \text{ and } f(x+i) < fx + if'q.$$
>
> Now suppose that p and q are the values of $x+i$ that yield the smallest and the largest values of the second order derivative $f''(x+i)$ of the function, i varying from 0 to a given value, we have $f''p$ and $f''q$ for the smallest and largest values of $f''(x+i)$; consequently, $f''(x+i) - f''p$ and $f''q - f''(x+i)$ will always be positive quantities.
>
> Looking at these quantities as derived functions relative to the variable i, their primitive functions taken so that they are zero when $i = 0$ will be
>
> $$f'(x+i) - f'x - if''p, \text{ and } if''q - f'(x+i) + f'x.$$
>
> Thus, provided that f'' is never infinite throughout the range of i, which means that $f''p$ and $f''q$ will not be infinite, this means that these two quantities will, by the same principle, always be positive and finite, assuming i to be positive; viewing these as derived functions relative to i, their primitive functions, taken so that they are zero when $i = 0$, will be, because x, p and q are assumed constant,

---

188 Lagrange handles only the case in which i is positive, silently letting the reader conclude the result also to hold in the negative case. Moreover he is again being a bit sloppy in not distinguishing between positive and nonnegative.

$$f(x+i) - fx - if'x - \frac{i^2}{2}f''p$$

and

$$\frac{i^2}{2}f''q - f(x+i) + fx + if'x.$$

These new quantities will by the same principle also be positive; thus we will have

$$f(x+i) - fx - if'x - \frac{i^2}{2}f''p > 0;$$
$$fx - f(x+i) + if'x + \frac{i^2}{2}f''q > 0,$$

from which one derives

$$f(x+i) > fx + if'x + \frac{i^2}{2}f''p, \text{ and } f(x+i) < fx + ifx' + \frac{i^2}{2}f''q.$$

At[189] this point Lagrange repeats the argument for the next derivative. This passage repeats the arguments just given and thus takes fully as much space. He then[190] states without proof that for any i positive or negative, and any positive integer μ, the value $f(x+i)$ lies between the two quantities

$$f(x) + if'x + \frac{i^2}{2}f''x + \frac{i^3}{2.3}f'''x + \&c. + \frac{i^\mu}{2.3\ldots\mu}f^\mu p,$$
$$f(x) + if'x + \frac{i^2}{2}f''x + \frac{i^3}{2.3}f'''x + \&c. + \frac{i^\mu}{2.3\ldots\mu}f^\mu q,$$

where p and q are where the smallest and largest values, respectively, of $f^{(\mu)}$ occur.

Lagrange's immediate goal was not to prove the Mean Value Theorem, but to provide estimates for how well the first few terms of the Taylor expansion approximated the function being expanded. While today we view his work as providing a method of proving that the Taylor series converges to the given function in certain cases, we should remember that Lagrange assumed this convergence at the outset. He carried out the above as an applied rather than as a pure mathematician. Having done this work, he then worked through a few examples before coming to the Higher Order Mean Value Theorem:

> Since these limits correspond to the largest and the smallest value of $f^\mu i$, taking i from zero to the given value, it is clear that the exact value of the remainder of the development of function fi corresponds to an intermediate value of $f^\mu i$, which can be represented by $f^\mu j$, j being some quantity between zero and i. It follows that one can always represent in a finite way the development of any function fi, introducing an unknown quantity j less than i. One has thus the analytical theorem, remarkable for its simplicity,

---

[189] J.L. Lagrange, *Leçons sur le calcul des fonctions*, downloaded edition unknown, pp. 70–71. The term ifx' in the final line is obviously a misprint for $if'x$.

[190] *Ibid.*, p. 73.

$$fi = f + if' + \frac{i^2}{2}f'' + \frac{i^3}{2.3}f''' + \&c.$$
$$+ \frac{i^{\mu-1}}{2.3\ldots\mu-1}f^{\mu-1} + \frac{i^{\mu}}{2.3\ldots\mu}f^{\mu}j$$

where f, f', f'', &c., are the values of $fi, f'i, f''i$, &c., at $i = 0$, the exponent μ being arbitrary.[191]

As we have seen, Lagrange only gave a detailed proof of the Higher Order Mean Value Theorem for orders 1, 2, and 3. A simple proof for all orders is possible by induction, a method already well-established by Lagrange's day. One can only assume his failure to carry out such a proof was didactic. Proving the individual cases for orders 1, 2, and 3 would be sufficient to convince his students of the general truth of the Theorem. Moreover, the formal inductive proof is a matter of unenlightening bookkeeping — keeping track of various indices in sums — and would not aid in understanding at all.[192]

I haven't studied Lagrange sufficiently to pinpoint the location of the Mean Value Theorem for Integrals in his work, but he is generally credited with the result. It seems to be an unremarkable result in his context as he takes the integral to be defined by the antiderivative: The formula

$$f(b) - f(a) = f'(c)(b - a)$$

thus does double duty as both the Mean Value Theorem for derivatives (start with f and take its derivative f') and the Mean Value Theorem for Integrals (start with f' and consider f to be its antiderivative). Far more interesting is his generalisation of the Mean Value Theorem to functions of two or more variables.

Functions of several variables made an early appearance in the Calculus as everyone worked with differentials of all the variables in an equation. Lagrange's approach mirrored his earlier approach for the single variable case: He assumed every function of, say, two variables had a two-variable Taylor expansion:

$$f(x + i, y + o) = \sum_{n=0}^{\infty} \sum_{j+k=n} a_{jk}(x, y) i^j o^k.$$

Working purely formally without regard for issues of convergence it is easy to determine the coefficients a_{jk}:

$$a_{jk}(x, y) = \frac{1}{j!k!} \frac{\partial^{j+k}}{\partial x^j \partial y^k} f(x, y).$$

---

[191] *Ibid.*, p. 78.

[192] I refer the curious reader to Smoryński, *Formalism*, *op. cit.*, pp. 133–135 for the grubby details, slightly simplified.

As in the single variable case, Lagrange does some algebraic gymnastics, replacing x, y by $x - i, y - o$, respectively, and i, o by xz, yz, respectively, for $0 \le z \le 1$, to obtain

$$\begin{aligned} f(x,y) = f(x-xz, y-yz) + xz\frac{\partial}{\partial x}f(x-xz,y-yz) + yz\frac{\partial}{\partial y}f(x-xz,y-yz) + \\ \frac{x^2z^2}{2}\frac{\partial^2}{\partial x^2}f(x-xz,y-yz) + xyz^2\frac{\partial^2}{\partial x\partial y}f(x-xz,y-yz) + \\ \frac{y^2z^2}{2}\frac{\partial^2}{\partial y^2}f(x-xz,y-yz) + \dots \end{aligned} \quad (3.36)$$

Writing

$$f(x,y) = f(x-xz,y-yz) + P(z) \quad (3.37)$$

and differentiating with respect to z, one has

$$0 = f'(x-xz,y-yz) + P'(z),$$

i.e.,

$$\begin{aligned} P'(z) &= -f'(x-xz,y-yz) \\ &= -\left(-x\frac{\partial}{\partial x}f(x-xz,y-yz) - y\frac{\partial}{\partial y}f(x-xz,y-yz)\right), \end{aligned}$$

by the Chain Rule,

$$= x\frac{\partial}{\partial x}f(x-xz,y-yz) + y\frac{\partial}{\partial y}f(x-xz,y-yz). \quad (3.38)$$

In terms of z, (3.37) can be written

$$g(0) = g(z) + P(z), \text{ where } g(z) = f(x-xz,y-yz),$$

i.e.,

$$g(0) - g(z) = P(z). \quad (3.39)$$

But, by the Mean Value Theorem,

$$\begin{aligned} g(0) - g(z) &= -zg'(c), \text{ for some } 0 < c < z \\ &= zP'(c), \text{ by (80)} \\ &= z\left(x\frac{\partial}{\partial x}f(x-xc,y-yc) + y\frac{\partial}{\partial y}f(x-xc,y-yc)\right), \text{ by (79)}, \end{aligned}$$

and we have

$$f(x,y)-f(x-xz,y-yz)=xz\frac{\partial}{\partial x}f(x-xc,y-yc)+yz\frac{\partial}{\partial y}f(x-xc,y-yc).$$

Reintroducing i, o,

$$f(x,y)-f(x-i,y-o)=i\frac{\partial}{\partial x}f(x-xc,y-yc)+o\frac{\partial}{\partial y}f(x-xc,y-yc)$$

and replacing x, y by $x+i, y+o$,

$$f(x+i,y+o)-f(x,y)=i\frac{\partial}{\partial x}f(x+i-xc-ic,y+o-yc-oc)+$$
$$o\frac{\partial}{\partial y}f(x+i-xc-ic,y+o-yc-oc).$$

And, writing $xc=xz\cdot\frac{c}{z}=i\cdot\frac{c}{z}$, $yc=o\cdot\frac{c}{z}$, the arguments of the partials,

$$x+i-i\frac{c}{z}-ic,\ \ y+o-o\frac{c}{z}-oc,$$

are of the forms $x+\lambda i$ and $y+\lambda o$ for

$$\lambda=1-\frac{c}{z}-c=\frac{z-c-cz}{z}.$$

Hence we have, for some λ,

$$f(x+i,y+o)=f(x,y)+i\frac{\partial}{\partial x}f(x+\lambda i,y+\lambda o)+o\frac{\partial}{\partial y}f(x+\lambda i,y+\lambda o). \tag{3.40}$$

It is not algebraically obvious from the above, but because the points $x+\lambda i$, $y+\lambda o$ are on the line segments connecting $\langle x,y\rangle$ and $\langle x+i,y+o\rangle$, we have $0<\lambda<1$. I confess to find the treatment unnecessarily messy and complain of having had to fill in some of the details, when the result should actually be easy. Today we simply define

$$g(t)=f(x+it,y+ot)$$

and note

$$f(x+i,y+o)-f(x,y)=g(1)-g(0)=1\cdot g'(\lambda)$$

for some $0 < \lambda < 1$ by the Mean Value Theorem, and note further that

$$g'(t) = i\frac{\partial}{\partial x}f(x+it, y+ot) + o\frac{\partial}{\partial y}f(x+it, y+ot)$$

by the Chain Rule.

Lagrange emphasises that the same λ occurs in each partial derivative, but fails to note the requirement that the domain of f include the entire line segment connecting the points $\langle x, y\rangle$ and $\langle x+i, y+o\rangle$. And when he considers the second-order version of the theorem,

$$f(x+i, y+o) = \\ f(x,y) + i\frac{\partial}{\partial x}f(x,y) + o\frac{\partial}{\partial y}f(x,y) + \frac{i^2}{2}\frac{\partial^2}{\partial x^2}f(x+\lambda i, y+\lambda o) + \\ io\frac{\partial^2}{\partial x\partial y}f(x+\lambda i, y+\lambda o) + \frac{o^2}{2}\frac{\partial^2}{\partial y^2}f(x+\lambda i, y+\lambda o),$$

or indeed the Taylor expansion (3.36), he is unaware of the possibility that the mixed partials might differ:

$$\frac{\partial^2 f}{\partial x\partial y} \neq \frac{\partial^2 f}{\partial y\partial x}.$$

Equality of the two partials generally requires the continuity of each, a level of smoothness certainly satisfied when f is a real analytic function of two variables, with which Lagrange dealt. His proof is thus correct, but not as general or as complete as he believed.

3.4 Transition to Cauchy: Ampère

The second main event in the history of the Mean Value Theorem was Cauchy's treatment of the subject. Between Lagrange and Cauchy, however, there was Ampère. Best known as a physicist, André Marie Ampère (1775–1836) started his scientific career as a mathematician, later switching to chemistry, and finally settling down in physics. His mathematical work includes a paper on differentiation and the derivative entitled "Recherches sur quelques points de la théorie des fonctions dérivées qui conduisent à une nouvelle démonstration de la série de *Taylor*, et à l'expression finie des termes qu'on néglige lorsqu'on arrête cette série à un terme quelconque"[193] ["Investigation of some points of the theory of derived functions which lead to a new demonstration of the series of *Taylor*, and to the finite expression of the terms which are neglected when one stops this series at any term whatsoever"].

---

[193] *Journal de l'École Polytechnique* series 13, volume 6 (1806), pp. 148–181.

In her study of Cauchy, Grabiner describes Ampère's paper as follows:

> Unfortunately, Ampère's paper is confusing and poorly organized. On occasion it has been misread as an attempt to prove that every continuous function is differentiable. This misreading is due partly to Ampère himself, who wrote that a derivative "exists" when he meant that it was finite and nonzero, and partly to historians, among whom the prevailing view is that analysis prior to Cauchy lacked rigour and sophistication. For these reasons, Ampère's paper has been not only misinterpreted but neglected. What in fact does it say?[194]

The misinterpretation is natural enough as Ampère announces on the second page that he will prove the existence of the derivative. This proof must have been somewhat convincing as it appeared in textbooks for some time thereafter. However, the mistake or, viewed more positively, the result he actually proved is not what is important here about his paper. What is important for us is that Ampère, who was one of Cauchy's teachers, provides a bridge from Lagrange to Cauchy.

Ampère's main contribution is a pair of lemmas, variants of the Mean Value Inequality, derived from the Lagrange property of the derivative and his invoking of these lemmas to give a formal definition of the derivative. As Grabiner points out his exposition leaves a lot to be desired, and nothing is to be gained by following his presentation closely. Thus I describe rather than cite his work.

The first of these lemmas is the following combinatorial lemma:

3.4.1 Lemma (Discrete Mean Value Inequality) *Let* $f : [a,b] \to \mathbb{R}$ *and let*

$$a = a_0 < a_1 < \ldots < a_{n-1} < a_n = b.$$

Then, for some $0 \le i, j < n$,

$$\frac{f(a_{i+1}) - f(a_i)}{a_{i+1} - a_i} \le \frac{f(b) - f(a)}{b - a} \le \frac{f(a_{j+1}) - f(a_j)}{a_{j+1} - a_j}.$$

Proof. We can give a geometrically motivated proof as follows. Let

$$m = \frac{f(b) - f(a)}{b - a}, \quad m_i = \frac{f(a_{i+1}) - f(a_i)}{a_{i+1} - a_i}, \quad \text{for } i = 0, 1, \ldots, n-1.$$

If m does not lie somewhere in the midst of the m_i's, then either $m < m_i$ for all such i or $m > m_i$ for all such i. Suppose, for example, that the first of these holds. Then proceeding from $\langle a, f(a)\rangle$ to $\langle a_1, f(a_1)\rangle$, one stays above the line connecting $\langle a, f(a)\rangle$ to $\langle b, f(b)\rangle$ because the slope is greater. Proceeding from $\langle a_1, f(a_1)\rangle$ to $\langle a_2, f(a_2)\rangle$ takes one even farther above the given line, etc. At stage $n-1$ the last small secant places $\langle a_n, f(a_n)\rangle$ above $\langle b, f(b)\rangle$, clearly impossible since $a_n = b$.

More formally, one proves by induction that

$$f(a_k) > f(a) + m(a_k - a) \tag{3.41}$$

[194]Grabiner, *Origins*, *op. cit.*, p. 129.

by induction on k.

For the basis, note simply

$$f(a_1)=f(a_0)+m_1(a_1-a_0)=f(a)+m_1(a_1-a)>f(a)+m(a_1-a).$$

For the induction step, note that

$$\begin{aligned}f(a_{k+1})&=f(a_k)+m_k(a_{k+1}-a_k)\\&>\big(f(a)+m(a_k-a)\big)+m(a_{k+1}-a_k),\ \text{by induction hypothesis}\\&>f(a)+m(a_{k+1}-a).\end{aligned}$$

By induction, we conclude (3.41) holds for all $k \le n$, in particular,

$$f(a_n)>f(a)+m(a_n-a),$$

i.e.,

$$f(b)>f(a)+m(b-a),$$

i.e.,

$$\frac{f(b)-f(a)}{b-a}>m,$$

a contradiction. □

Ampère's proof is given with no geometric motivation cited.

Ampère's proof. Ampère proves this for $n = 2, 4$, and 8 in something like an inductive proof that the Lemma holds for n of the form 2^k.

The basis of an induction is given by $n = 2$. Let c be an intermediate point and observe, writing A, B, C for $f(a), f(b), f(c)$, respectively,

$$\begin{aligned}\frac{B-A}{b-a}-\frac{B-C}{b-c}&=\frac{Bb-Bc-Ab+Ac-(Bb-Ba-Cb+Ca)}{(b-a)(b-c)}\\&=\frac{-Ab+Ac+Ba-Bc-Ca+Cb}{(b-a)(b-c)}\end{aligned} \tag{3.42}$$

$$\begin{aligned}\frac{C-A}{c-a}-\frac{B-A}{b-a}&=\frac{Cb-Ca-Ab+Aa-(Bc-Ba-Ac+Aa)}{(b-a)(c-a)}\\&=\frac{-Ab+Ac+Ba-Bc-Ca+Cb}{(b-a)(c-a)}.\end{aligned} \tag{3.43}$$

The fractions (3.42) and (3.43) have equal numerators and positive denominators, whence they have the same sign. If this is positive,

$$\frac{C-A}{c-a}\ge\frac{B-A}{b-a}\ge\frac{B-C}{b-c},$$

i.e.,

$$\frac{f(b)-f(c)}{b-c} \leq \frac{f(b)-f(a)}{b-a} \leq \frac{f(c)-f(a)}{c-a};$$

while if the sign is negative the opposite inequalities hold,

$$\frac{f(c)-f(a)}{c-a} \leq \frac{f(b)-f(a)}{b-a} \leq \frac{f(b)-f(c)}{b-c}.$$

The induction step does not require one to double the number of intervals. Given

$$a = a_0 < a_1 < \ldots < a_{n-1} < a_n < a_{n+1} = b,$$

temporarily ignore a_n. By the induction hypothesis there are $0 \leq i, j < n$ such that

$$\frac{f(a^*_{i+1})-f(a_i)}{a^*_{i+1}-a_i} \leq \frac{f(b)-f(a)}{b-a} \leq \frac{f(a^*_{j+1})-f(a_j)}{a^*_{j+1}-a_j},$$

where

$$a^*_{k+1} = \begin{cases} a_{k+1}, & \text{if } k < n-1 \\ a_{n+1}, & \text{if } k = n-1. \end{cases}$$

If, say, $i = n-1$, apply the basis step to conclude either

$$\frac{f(a_{n+1})-f(a_n)}{a_{n+1}-a_n} \text{ or } \frac{f(a_n)-f(a_{n-1})}{a_n-a_{n-1}}$$

to be

$$\leq \frac{f(a_{n+1})-f(a_{n-1})}{a_{n+1}-a_{n-1}} \leq \frac{f(b)-f(a)}{b-a}.$$

And argue analogously for $j = n-1$. □

At this point Ampère notes that one can take equal intervals,

$$a_n - a_{n-1} = a_{n-1} - a_{n-2} = \ldots = a_1 - a_0 = i,$$

whence there are $x, y \in [a, b]$ such that

$$\frac{f(x+i)-f(x)}{i} \leq \frac{f(b)-f(a)}{b-a} \leq \frac{f(y+i)-f(y)}{i}.$$

He now says that if $i = 0$ the outer ratios are $f'(x)$ and $f'(y)$, respectively, interpreting this in accordance with the Lagrange property: Any ratio

$$\frac{f(z+i)-f(z)}{i}$$

is equal to $f'(z) + I$, where I is a function of z and i which vanishes with i. From this he will conclude the existence of $x, y \in [a, b]$ such that

$$f'(x) \leq \frac{f(b) - f(a)}{b - a} \leq f'(y). \tag{3.44}$$

Properly carried out, his argument requires the uniform differentiability of f on $[a, b]$: Suppose $f'(x)$ is never $\leq \big(f(b) - f(a)\big)/(b - a)$, i.e., assume it is always greater:

$$f'(x) - \frac{f(b) - f(a)}{b - a} > 0 \text{ for all } x \in [a, b]$$

Let ϵ be any positive number less than the minimum value of this difference and choose $\delta > 0$ small enough so that for all $0 < |i| < \delta$ and all $x \in [a, b]$,

$$-\epsilon < \frac{f(x + i) - f(x)}{i} - f'(x) < \epsilon,$$

and

$$\frac{f(x + i) - f(x)}{i} > f'(x) - \epsilon > \frac{f(b) - f(a)}{b - a}, \text{ by choice of } \epsilon$$

which contradicts Lemma 3.4.1. Likewise, for some $y \in [a, b]$,

$$\frac{f(b) - f(a)}{b - a} \leq f'(y).$$

Thus we have a version of the Mean Value Inequality:

3.4.2 Lemma (Mean Value Inequality) *Let $f : [a, b] \to \mathbb{R}$ be uniformly differentiable. There are $x, y \in [a, b]$ such that*

$$f'(x) \leq \frac{f(b) - f(a)}{b - a} \leq f'(y).$$

From this of course, we obtain the Mean Value Theorem for uniformly differentiable f by applying the Intermediate Value Theorem to f'.

3.4.3 Remark Ampère's proof can be thought of as a working out of the second intuitive explanation of the truth of the Mean Value Theorem cited in the beginning of the Preface. The idea was to consider the arc connecting two points A and B on a curve and, starting with the tangent line at A, moving the line along the arc from A to B rotating it as one goes in such a way that the line remains tangent to the curve at all points of contact. In the illustration of this (Fig. 1.3 on page 2) the moving tangent line starts at A with a slope steeper than that of the secant connecting A to B and finishes at B with a lesser slope. Assuming the right amount of smoothness of the curve, the secant itself being intermediate between the slopes of these two tangents, there must be a point C intermediate between A and B where the tangent had this slope.

Ampère's proof begins with a discrete version of this. Given f defined on $[a, b]$ and any partition $a = a_0 < a_1 < \ldots < a_{n-1} = b$, the slope of the secant connecting $\langle a, f(a)\rangle$ to $\langle b, f(b)\rangle$ lies between the minimum and maximum slopes of the individual secant lines connecting successive points $\langle a_i, f(a_i)\rangle$ and $\langle a_{i+1}, f(a_{i+1})\rangle$.

A passage to the limit, requiring the uniform differentiability of f, together with the continuity of f' (needed to obtain extreme values of f' and to allow the Intermediate Value Theorem to be invoked) yields the Mean Value Theorem for f.

After a decade and a half Cauchy would simplify the proof of the first lemma, and some time later Weierstrass would improve on the second lemma, reducing the uniform differentiability requirement to continuous differentiability. Finally, in 1974 Thomas Muirhead Flett (Sect. 3.10.2, below) was able to reduce the Mean Value Inequality to the Discrete Mean Value Inequality under the assumptions that f is continuous on $[a, b]$ and differentiable on (a, b), and that f' assumes minimum and maximum values on $[a, b]$.

Before citing the Mean Value Theorem or any other consequences, Ampère defined the derivative:

> We can draw from this a definition of the derived function $f'(x)$ which seems to me the most general and rigorous possible, and which leads immediately to geometrical and mechanical applications of the theory of derived functions.
>
> The derived function of $f(x)$ is a function of x such that $\frac{f(x+i)-f(x)}{i}$ is always between two of the values which this derived function takes from x to $x + i$, regardless of what x and i may be.[195]

Today we would present this more formally:

3.4.4 Definition Let $f, f' : I \to \mathbb{R}$ be given. f' is *a derivative* of f if, for all $x, x + i \in I$, there are y, z between x and $x + i$ such that

$$f'(y) \leq \frac{f(x+i) - f(x)}{i} \leq f'(z).$$

Here, I is any interval, open or closed; Ampère himself was none too specific about the domain of f.

Note that the formal definition refers to f' as *a* derivative while Ampère refers to it as *the* derivative. He does offer a brief argument for uniqueness prior to stating the definition, but it is not complete.

Grabiner criticises the definition as follows:

> All the rigorous nineteenth-century definitions of $f'(x)$ define it by the ratio $f(x + i) - f(x)/i$ and the inequalities that this ratio must satisfy; Ampère was thus the first to give such a definition. His definition has some major deficiencies, however. First, it defines $f'(x)$ at the point x in terms of its values on the whole interval; thus $f'(x)$ must exist on an entire interval

[195] Ampère, *op. cit.*, p. 156.

to be defined at a single point. This is much too restrictive (though not as restrictive as assuming that $f(x)$ has an entire Taylor series). Second, there is no reason to believe that any such $f'(x)$ exists at all. Third, it is not clear that $f'(x)$ is the only function that satisfies defining criterion (5.8),[196] though Ampère did try to prove that $f'(x)$ was unique.[197]

The first criticism is a bit Whiggish. While today we think of the derivative first as the derivative at a point and then as a function comprised of the derivatives at all the points of an interval, Ampère was following Lagrange's lead in thinking of the derived function *qua* function and using a characterisation of this function as the formal definition of the concept. Neither Ampère nor any of his contemporaries had ever met a function for which derivatives existed only at isolated points and there would have been no reason to consider such.

The second criticism makes no sense at all. There is no reason to assume $f'(x)$ exists under the modern definition either.

Grabiner's third criticism is cogent and is the reason I referred to *a* derivative rather than *the* derivative in Definition 3.4.4. Ampère did offer an argument for this just prior to giving his definition, but it is not convincing — and for good reason:

3.4.5 Exercise Let $f, f', g : I \to \mathbb{R}$ be given and let f', g be derivatives of f.

i. Show: If f', g are continuous, then $f' = g$.
ii. Show by example that uniqueness can fail if continuity of the derivatives is not assumed.

The real puzzle about the definition would seem to be why Ampère chooses a consequence of the Lagrange property rather than the property itself as the basis for his definition. The Lagrange property may be less intuitive than the expression of the derivative as the limit of the difference quotient, but Ampère's defining property is not intuitive at all. Sometimes, as with Carathéodory's definition of a measurable set, a non-intuitive definition is used because it simplifies the ensuing theoretical development. Ampère does tell us that his condition is a useful property in applications, but he does not use it in the routine development of the Differential Calculus. He does demonstrate the differentiability of the sine function by showing directly that the cosine satisfies the necessary condition, but he doesn't follow up with the rest of the routine work.

3.4.6 Exercise Consider the linearity of the derivative: Let $f, g : I \to \mathbb{R}$ be differentiable in Ampère's sense.

i. Show: For any constant c, cf is also differentiable in Ampère's sense and $(cf)'(x) = cf'(x)$.
ii. Can you show the Ampère differentiability of $f + g$?

The rest of Ampère's paper is a mass of algebraic calculation aimed at deriving Taylor's Theorem. Insofar as this involves the Lagrange form of the remainder, he is

---

[196] I.e., the second paragraph in the above quotation from Ampère.
[197] Grabiner, *Origins*, *op. cit.*, pp. 129–130.

offering a proof of the Higher Order Mean Value Theorem under weaker conditions than Lagrange's proof for real analytic functions and this work is relevant to our discussion. However, this material does not look to be rewarding reading and I, for one, will not even attempt it until someone else has translated it into English. Before leaving Ampère's paper and moving on to greener pastures, I have a few additional observations to make. The first, and simplest, is the remark that, as one of Cauchy's teachers, he was acknowledged by the latter in general terms in his books.

Second, we might pause to consider a simple verification of Ampère's claim of applicability:

3.4.7 Exercise Prove the Strictly Increasing Function Theorem, the Increasing Function Theorem, and the Constant Function Theorem for a function f possessing a derivative in the sense of Definition 3.4.4.

Third, Ampère's paper dealt primarily with functions of a single variable. It does, however, contain a brief addition defining the derivative of a function of two variables in a most unusual manner. While the emphasis of the present book is on the Classroom Mean Value Theorem and its rôle in the single variable Calculus course, I think a short digression on Ampère's treatment of functions of two variables is not unwarranted. At the very least it highlights his commitment to the Mean Value Inequality as the defining property of the derivative. Moreover, it provides a nice illustration of the nonlinearity of mathematical history as he veers even farther away from the clue to what we see today as the correct treatment of differentiation given by the Lagrange property. The reader who finds all of this unrewarding is invited to skip ahead to page 273 and the section on Cauchy.

The reader who chooses to continue with Ampère should read at least as far as Ampère's first displayed formula. After that he or she has the option of continuing or jumping ahead to the explanation in modern notation that immediately follows the passage.

NOTE *relative to functions of two variables.*

To apply to functions of two variables, as we said at the beginning of this paper, a new way to make rigorous and uniform the application of the theory of derived functions in geometry and mechanics, it is only necessary to see that if one takes the derivative function of $f(x, y)$, relative to one of the two variables, and then the derivative of the function resulting from this first operation, by varying the other variable, the function thus obtained, and which we represent by $f'^{,\prime}(x, y)$,[198] enjoys the property of always taking for values of x between X and x, and for values of y between y and Y, a larger value and a smaller value than

$$\frac{f(X, Y) - f(x, Y) - f(X, y) + f(x, y)}{(X - x)(Y - y)}.$$

For the demonstration, we first note that, regarding y as a constant in $f(x, y)$, one can apply to the quantity considered as a function of x alone, all that has been said of the functions of a single variable. Thus $f'^{,\prime}(x, y)$ will take for values of x from x up to X, a value greater and a value smaller than[199]

---

[198] He is using here a variant of Lagrange's notation for $\partial^2/\partial x\partial y$.

[199] Again, modifying notation of Lagrange, the missing first prime indicates that the derivative is only being taken with respect to the second variable: $\partial f/\partial y$.

$$\frac{f'_{\prime}(X,y) - f'_{\prime}(x,y)}{X-x},$$

y having the same value in any of the three functions. So if y is varied from y to Y, there will be among all the values we get for $f'_{\prime}{}'(x, y)$, making both x and y vary simultaneously, from x to X, and from y to Y, a larger and a smaller value than all those taken by

$$\frac{f'_{\prime}(X,y) - f'_{\prime}(x,y)}{X-x},$$

by making y always vary only between the same limits. But as X and x then must be considered as two constants, there will be, among all the values that quantity will take, a value that is larger and a value that is smaller than its primitive with respect to y taken from y to Y and divided by $Y - y$; to have this primitive, it is necessary that from those of its values that correspond to Y, i.e.,

$$\frac{f(X,Y) - f(x,Y)}{X-x},$$

subtract that corresponding to y, and which is

$$\frac{f(X,y) - f(x,y)}{X-x};$$

one will find by dividing the remainder by $Y - y$, the quantity

$$\frac{f(X,Y) - f(x,Y) - f(X,y) + f(x,y)}{(X-x)(Y-y)},$$

which will be between the largest and the smallest value which

$$\frac{f'_{\prime}(X,y) - f'_{\prime}(x,y)}{X-x}$$

is capable of having from y to Y; and since these two values are themselves between the largest and the smallest of $f'_{\prime}{}'(x, y)$, it follows that

$$\frac{f(X,Y) - f(x,Y) - f(X,y) + f(x,y)}{(X-x)(Y-y)}$$

will be between them.[200]

This is poorly expressed, so let me rephrase the argument using modern notation. We consider a function $f(x, y)$ defined on a rectangle $R = [x_0, X] \times [y_0, Y]$. If we now assume the continuity of the mixed second partial derivative $(\partial^2/\partial x\partial y)f$, this derivative will attain minimum and maximum values m and M, respectively, on R. By the Mean Value Inequality for $\partial f/\partial y$ we have

$$m \le \frac{\frac{\partial}{\partial y}f(X,y) - \frac{\partial}{\partial y}f(x_0,y)}{X - x_0} \le M.$$

[200]Ampère, *op. cit.*, pp. 178–179.

Then

$$\int_{y_0}^{Y} m\,dy \le \int_{y_0}^{Y} \frac{\frac{\partial}{\partial y}f(X,y) - \frac{\partial}{\partial y}f(x_0,y)}{X - x_0}dy \le \int_{y_0}^{Y} M\,dy,$$

i.e.,

$$(Y - y_0)m \le \left.\frac{f(X,y) - f(x_0,y)}{X - x_0}\right|_{y_0}^{Y} \le (Y - y_0)M$$

$$(Y - y_0)m \le \frac{f(X,Y) - f(x_0,Y) - \big(f(X,y_0) - f(x_0,y_0)\big)}{X - x_0} \le (Y - y_0)M,$$

and division by $Y - y_0$ yields

$$m \le \frac{f(X,Y) - f(x_0,Y) - f(X,y_0) + f(x_0,y_0)}{(X - x_0)(Y - y_0)} \le M,$$

as was to be shown: $(\partial^2/\partial x\partial y)f$ satisfies a two-variable mean value inequality which will make it the *derived function* of $f(x,y)$.

Like his contemporaries, Ampère is none too clear on the conditions f must satisfy for the above argument to work. Evidently, the continuity of $(\partial^2/\partial x\partial y)f$ suffices.

He now promises to prove that this is the only function satisfying the mean value inequality. Today we would say that it is the unique continuous function satisfying the inequality and we would prove this fairly directly using an ϵ-δ argument. Ampère's attempted proof is a bit less direct, confusingly presented, and incomplete. He begins by assuming f satisfies a possibly stronger condition than possessing continuous mixed second partial derivatives.

Let $\langle a, b\rangle$ be some reference point in the plane (e.g., it could be the origin $\langle 0, 0\rangle$ or even $\langle x_0, y_0\rangle$) and let $z = g(x,y)$ be a surface such that $f(x,y)$ is the volume of the solid trapped between the surface and the rectangle $[a,x] \times [b,y]$. Applying the Inclusion-Exclusion Principle to volumes, we see that the volume trapped between the surface $z = g(x,y)$ and the rectangle $R = [x_0, X] \times [y_0, Y]$ is

$$f(X,Y) - f(x_0,Y) - f(X,y_0) + f(x_0,y_0).$$

If we now let m and M denote the minimum and maximum values of g on R, this volume is trapped between

$$(X - x_0)(Y - y_0)m \text{ and } (X - x_0)(Y - y_0)M,$$

whence

$$m \le \frac{f(X,Y) - f(x_0,Y) - f(X,y_0) + f(x_0,y_0)}{(X - x_0)(Y - y_0)} \le M. \tag{3.45}$$

He now states, "whence we conclude the general formula of the evaluation of the value

$$z = f^{\prime,\prime}(x, y)".$$

I assume this refers to the calculation from

$$\int\int_R g(x, y) = \int_b^y \int_a^x g(x, y)\, dxdy,$$

of

$$\frac{\partial}{\partial x \partial y} f(x, y) = \frac{\partial}{\partial x}\left(\frac{\partial}{\partial y}\int_b^y \int_a^x g(x, y)\, dxdy\right)$$
$$= \frac{\partial}{\partial x}\int_a^x g(x, y)\, dx = g(x, y).$$

From this it follows that m and M are the minimum and maximum values, respectively, of $(\partial^2/\partial x \partial y)f$ on R and we have derived the Mean Value Inequality for $f(x, y)$ anew.

What we have not done is prove that any continuous function g satisfying (3.45) must equal f, only that any g defining a surface whose volumes are given by f must equal the mixed second partial derivatives of f. One needs yet to fill in the step taking one from (3.45) to f determining these volumes, which needs a theory of the integral as first developed later by Cauchy.

3.5 Cauchy and the Mean Value Theorem

Lagrange ushered in an era of foundational work despite the fact that his goal would seem to have been that of avoiding the key foundational issue — the nature of the limit. He provided some rigour in Analysis and even gave the first explicit statements of the Mean Value Theorem and the Mean Value Inequality, two results that would evolve into valuable tools of rigour. Ampère can be seen as a mechanic, fine-tuning the mechanism inherited from Lagrange rather than inventing anything new. Cauchy took up where they left off, extending the discussion to include continuity and integration. Cauchy did so much so well that he is generally credited as the originator of rigour in the Calculus.

Cauchy's first textbook on the Calculus, the *Cours d'analyse*, or *Analyse algébrique* as it is often called, developed the theories of continuous functions and series, and even provided important lemmas for the next course. The lectures for this course, published as the *Résumé des leçons*, or *Calcul infinitesimal* as many authors prefer, covered the Differential and Integral Calculi. Cauchy's treatment is much closer to our own, the key differences are the slightly stronger conditions in some of the definitions and the stronger hypotheses in theorems: his "continuity" is our "uniform continuity", his "differentiability" our "uniform differentiability", his definite integral is defined only for continuous functions, and his proofs of many theorems rely on these stronger hypotheses. Indeed, there are famous theorems of his that have long been incorrectly declared false because more recent mathematicians have read his statements of the results with the modern meanings of the terms.

The *Cours d'analyse* starts with a list of preliminaries beginning with an explanation of notation and ending with a discussion of general averages or *means*:

> We will finish these preliminaries by presenting several theorems on average quantities, the knowledge of which will be extremely useful in the remainder of this work. We call an *average* among several given quantities a new quantity between the smallest and the largest of those under consideration. From this definition it is clear that there are an infinity of averages among several unequal quantities, and that the average of several equal quantities is equal to their common value. Given this, we will easily establish, as one can see in Note II, the following propositions:
>
> **Theorem I.** — *Let $b, b', b'', \ldots$ denote n quantities of the same sign, and $a, a', a'', \ldots$ be the same number of arbitrary quantities. The fraction*
>
> $$\frac{a + a' + a'' + \ldots}{b + b' + b'' + \ldots}$$
>
> *is an average of the following quantities*
>
> $$\frac{a}{b}, \frac{a'}{b'}, \frac{a''}{b''}, \ldots$$

This[201] Theorem is a more general form of Ampère's combinatorial lemma, Lemma 3.4.1, and Cauchy defers its proof to the second of nine appendices he refers to as notes, the title of this one being "Note II — On formulas that result from the use of the signs $>$ or $<$, and on the averages among several quantities", where, following more elementary material, the Theorem in question is presented as Theorem XII:

> **Theorem XII.** — *Let $b, b', b'', \ldots$ be several quantities of the same sign, n in number, and let $a, a', a'', \ldots$ be any quantities, also n in number. Then we have*
>
> $$\frac{a + a' + a'' + \ldots}{b + b' + b'' + \ldots} = M\left(\frac{a}{b}, \frac{a'}{b'}, \frac{a''}{b''}, \ldots\right). \tag{17}$$ [202]

A few pages earlier Cauchy has introduced the notation $M(\ldots)$ for the average or mean of a finite list of quantities:

> Now we move on to theorems on averages. As we have already said (*Preliminaries*), we call an *average* among several given quantities a new quantity contained between the smallest and the largest of those under consideration. From this definition, the quantity h is an average between two quantities g and k, or among several quantities among which one of these values is the largest and the other is the smallest, if the two differences
>
> $$g - h \text{ and } h - k$$
>
> are of the same sign. Given this, if we use the notation
>
> $$M(a, a', a'', \ldots)$$

---

[201] Bradley and Sandifer, *op. cit.*, pp. 12–13.

[202] *Ibid.*, p. 298.

> for denoting an average among the quantities $a, a', a'', \ldots$, as we did in the *Preliminaries*, we establish the following propositions without trouble...[203]

He follows this with a number of theorems to the effect that certain expressions yield averages, the conclusion written in equational form. Following Theorem XII, are, for example, the arithmetic mean,

$$\frac{a + a' + a'' + \ldots}{n} = M(a, a', a'', \ldots), \tag{18}$$

and the geometric mean

$$\sqrt[n]{AA'A'' \cdots} = M(A, A', A'', \ldots). \tag{22}$$

As the expressions on the left in these two equations can differ, even for $a = A, a' = A', a'' = A'', \ldots$, we should not read an expression $f(a, a', a'', \ldots) = M(a, a', a'', \ldots)$ as an equation, but as a shorthand for "$f(a, a', a'', \ldots)$ is an average", or

$$k \leq f(a, a', a'', \ldots) \leq g,$$

for k the smallest and g the largest of the quantities.

In Note II, Cauchy develops a sort of mini-calculus for dealing with averages that he applied in both textbooks. For us, however, the interest in his general treatment of means is an indication of how far he had advanced beyond Lagrange's confusion of strong and weak inequality. Of specific interest here is his treatment of Theorem XII, both the statement of the result as compared to Ampère's and the simplicity and generality of the proof.

For convenience I will present Cauchy's proof using more modern notation. We let $b_0, b_1, \ldots, b_{n-1}, c_0, c_1, \ldots, c_{n-1}$ be $2n$ numbers, the c_i's all positive or all negative. Without loss of generality, we can assume the c_i's all positive. Let m, M be the minimum and maximum, respectively, of the ratios,

$$\frac{b_0}{c_0}, \frac{b_1}{c_1}, \ldots, \frac{b_{n-1}}{c_{n-1}}.$$

From the inequalities,

$$m \leq \frac{b_i}{c_i} \leq M,$$

we conclude

$$mc_i \leq b_i \leq Mc_i.$$

Summation yields

$$m(c_0 + c_1 + \ldots + c_{n-1}) \leq b_0 + b_1 + \ldots + b_{n-1} \leq M(c_0 + c_1 + \ldots + c_{n-1}),$$

[203] *Ibid.*, p. 296.

i.e.,

$$m \leq \frac{b_0 + b_1 + \ldots + b_{n-1}}{c_0 + c_1 + \ldots + c_{n-1}} \leq M,$$

i.e.,

$$\frac{b_0 + b_1 + \ldots + b_{n-1}}{c_0 + c_1 + \ldots + c_{n-1}} = M\left(\frac{b_0}{c_0}, \frac{b_1}{c_1}, \ldots, \frac{b_{n-1}}{c_{n-1}}\right),$$

to use Cauchy's notation.

Ampère's Lemma 3.4.1 follows for

$$a = a_0 < a_1 < \ldots < a_{n-1} < a_n = b$$

by choosing $b_i = f(a_{i+1}) - f(a_i)$ and $c_i = a_{i+1} - a_i$ for $i = 0, 1, \ldots, n-1$.

Cauchy's second text, the *Résumé des leçons*, is his course in 40 lessons and 2 appendices on the Differential and Integral Calculi. The book begins with two short chapters reviewing the concepts of limit and continuity, settling down to business in the third lesson defining the derivative and calculating the derivatives of the basic functions of the Calculus. The fourth lesson introduces differentials and the fifth covers the differentials of sums and products, as well as of complex-valued functions $f: \mathbb{R} \to \mathbb{C}$. The sixth lesson is a not particularly rigorous introduction to the applications of the derivative. It includes an imprecise treatment of the Strictly Increasing Function Theorem and its decreasing counterpart, the principle of finding the maxima and minima of a function by using its derivative, and a hand-waving argument for l'Hôpital's Rule.[204]

Lesson 7 is where things begin to get genuinely rigorous. Following some brief remarks on reducing the problem of finding the values of expressions of the form $\frac{\infty}{\infty}$, ∞^0, etc., to the problem for the form $\frac{0}{0}$ considered in Lesson 6, Cauchy comes to the Mean Value Theorem and the introduction of the now ubiquitous ϵ-δ notation:

> Let us now turn to a noteworthy[205] relation which exists between the derivative $f'(x)$ of an arbitrary function $f(x)$ and the ratio of finite differences $\dfrac{f(x+h) - f(x)}{h}$. If in this ratio we assign to x a particular value x_0, and if in addition we make $x_0 + h = X$, it takes the form $\dfrac{f(X) - f(x_0)}{X - x_0}$. That said, we establish the following proposition easily.
>
> Theorem. *If the function $f(x)$ is continuous between the limits $x = x_0$, $x = X$, and we denote by A the smallest and by B the largest of the values that the derived function $f'(x)$ receives in this interval, the ratio of finite differences*
>
> $$\frac{f(X) - f(x_0)}{X - x_0} \tag{4}$$
>
> *will necessarily lie between* A and B.

[204] Given on p. 175, above.

[205] Here Cauchy adds a footnote: "On this subject one can consult a memoir by Mr. Ampère inserted into the 13th series of the *Journal de l'École polytechnique*.".

Proof. Denote by δ, ε, two very small numbers, the first being selected in such a way that, for the numerical values[206] of i less than δ, and for any value of x between the limits x_0, X, the ratio

$$\frac{f(x+i)-f(x)}{i}$$

is always greater than $f'(x)-\varepsilon$, and less than $f'(x)+\varepsilon$. If, between the limits x_0, X we interpose $n-1$ new values of the variable x, namely,

$$x_1,\ x_2,\ \ldots\ldots\ x_{n-1},$$

so as to divide the difference $X-x_0$ into elements,

$$x_1-x_0,\ x_2-x_1,\ \ldots\ldots\ X-x_{n-1},$$

which are all of the same sign and have numerical values less than δ; the fractions

$$(5)\qquad \frac{f(x_1)-f(x_0)}{x_1-x_0},\ \frac{f(x_2)-f(x_1)}{x_2-x_1},\ \ldots,\ \frac{f(X)-f(x_{n-1})}{X-x_{n-1}},$$

lying, the first between the limits $f'(x_0)-\varepsilon, f'(x_0)+\varepsilon$, the second between the limits $f'(x_1)-\varepsilon, f'(x_1)+\varepsilon$ &c.... will all be greater than the quantity $A-\varepsilon$ and less than the quantity $B+\varepsilon$. Moreover, the fractions in (5) having denominators of the same sign, if the sum of their numerators is divided by the sum of their denominators, we get an *average* fraction, that is to say, lying between the smallest and the largest of those being considered [*see* l'Analyse algébrique, note II, 12th theorem]. The expression (4), with which this average coincides, will itself be enclosed between the limits $A-\varepsilon, B+\varepsilon$, and as this conclusion remains, no matter how small the number ε; we can say that the expression (4) will be between A and B.

Corollary. If the derived function $f'(x)$ is itself continuous between the limits $x=x_0, x=X$, passing from one limit to the other, this function will vary so as to remain always between the two values A and B, and to take successively all intermediate values. Thus any average quantity between A and B will be a value of $f'(x)$ corresponding to a value of x included between the limits x_0 and $X=x_0+h$, or, equivalently, to a value of x of the form

$$x_0+\theta h = x_0+\theta(X-x_0),$$

θ denoting a number less than unity. Applying this remark to the expression (4), we conclude that there exists between the limits 0 and 1 a specific value of θ verifying the equation

$$\frac{f(X)-f(x_0)}{X-x_0}=f'[x_0+\theta(X-x_0)],$$

or, what amounts to the same thing, the following

$$(6)\qquad \frac{f(x_0+h)-f(x_0)}{h}=f'(x_0+\theta h).$$

This last formula must persist, regardless of the value of x represented by x_0, provided that the function $f(x)$ and its derivative $f'(x)$ remain continuous between the extreme values $x=x_0, x=x_0+h$, we will have in general, under this condition,

$$(7)\qquad \frac{f(x+h)-f(x)}{h}=f'(x+\theta h),$$

then, writing Δx in place of h, we derive

$$(8)\qquad f(x+\Delta x)-f(x)=f'(x+\theta\Delta x).\Delta x.$$

[206] I.e., absolute values.

It is essential to observe that, in Eqs. (7) and (8), θ always denotes an unknown number, but less than unity.[207]

I have a couple of quick comments to make about Cauchy's proofs of the Theorem (the Mean Value Inequality) and its Corollary (the Mean Value Theorem). First, these results are simply fuller versions, more clearly presented, of Ampère's proofs. Cauchy's contribution up to this point is mainly in clarifying the conditions needed for the validity of the results—f must be "differentiable" on $[x_0, X]$ and f' "continuous" there. In modern terms he has established these results for f uniformly differentiable on $[x_0, X]$. He further explicitly requires f' to be uniformly continuous there, but this, as we saw in Lemma 2.3.13 in Chap. 2, is redundant.

A second comment is that his proof makes a clear appeal to the uniformity of the choice of δ for a given ε in placing the fractions of (5) close to the derivatives at $x_0, x_1, \ldots, x_{n-1}$. Cauchy's result should thus be read as follows:

3.5.1 Theorem (Classroom Mean Value Theorem; Cauchy Form) *Let $a < b$ and let $f : [a, b] \to \mathbb{R}$ be uniformly differentiable on $[a, b]$. There is a $c \in [a, b]$*[208] *such that*

$$\frac{f(b) - f(a)}{b - a} = f'(c).$$

One can replace "uniformly" here by "continuously", as first noted (to my knowledge) by Bolzano, who attempted to prove that a function continuously differentiable on a closed, bounded interval is also uniformly differentiable there.

As we already know from our earlier discussion, Cauchy's involvement with the Mean Value Theorem goes well beyond providing a better exposition of Ampère's attempt to improve Lagrange's result. Still to be found in the *Résumé des leçons* are the Cauchy Mean Value Theorem, the Higher Order Mean Value Theorem, the Mean Value Theorem for Integrals, and some applications of the Mean Value Theorem.

The first of these to be found in the book is an application of the Mean Value Theorem in the immediately following Lesson 8 on differentials of functions of several variables. The given application can still be found in textbooks today: Cauchy begins by defining partial derivatives in the usual way:

[207] Cauchy, *Résumé*, *op. cit.*, pp. 26–28; Cates, *op. cit.*, pp. 23–25.

[208] Cauchy says $c \in (a, b)$, but on reading him over carefully, I am not convinced he proves c not to be an endpoint of the interval. Consider, for example, the function $f(x) = \cos x$ with $x_0 = 0, X = 2\pi$. One has

$$\frac{f(X) - f(x_0)}{X - x_0} = \frac{1 - 1}{2\pi} = 0 = f'(0) = f'(2\pi).$$

There is another $c \in [0, 2\pi]$ for which $f'(c) = 0$, but he hasn't demonstrated this. We return to this point in the next section, pages 310–311, below.

$$\frac{\partial}{\partial x}f(x,y,z,\ldots) = \lim_{i\to 0}\frac{f(x+i,y,z,\ldots)-f(x,y,z)}{i}$$
$$\frac{\partial}{\partial y}f(x,y,z,\ldots) = \lim_{i\to 0}\frac{f(x,y+i,z,\ldots)-f(x,y,z)}{i}$$
$$\frac{\partial}{\partial z}f(x,y,z,\ldots) = \lim_{i\to 0}\frac{f(x,y,z+i,\ldots)-f(x,y,z)}{i}$$
$$\vdots$$

He does not use our modern notation, but writes instead $\varphi(x, y, z, \ldots)$ for $\partial f/\partial x$, $\chi(x, y, z, \ldots)$ for $\partial f/\partial y$, $\psi(x, y, z, \ldots)$ for $\partial f/\partial z$, and, eventually, for $u = f(x, y, z, \ldots)$,

$$\frac{d_x u}{dx} \text{ for } \varphi(x,y,z,\ldots),\quad \frac{d_y u}{dy} \text{ for } \chi(x,y,z,\ldots),\quad \frac{d_z u}{dz} \text{ for } \psi(x,y,z,\ldots),$$

and even

$$\frac{du}{dx} \text{ for } \frac{d_x u}{dx},\quad \frac{du}{dy} \text{ for } \frac{d_y u}{dy},\quad \frac{du}{dz} \text{ for } \frac{d_z u}{dz}.$$

He then defines what is essentially a directional derivative with respect to a not necessarily normalised vector $\vec{\beta} = \langle h, k, l, \ldots\rangle$ as follows

$$\frac{df}{d\vec{\beta}}(x,y,z,\ldots) = \lim_{\alpha\to 0}\frac{f(x+h\alpha, y+k\alpha, z+l\alpha,\ldots)-f(x,y,z,\ldots)}{\alpha}.$$

Writing $\Delta x = h\alpha$, $\Delta y = k\alpha$, $\Delta z = l\alpha, \ldots$, he applies the Mean Value Theorem successively:

$$f(x+\Delta x, y, z, \ldots) - f(x,y,z,\ldots) = \Delta x\frac{\partial f}{\partial x}(x+\theta_1\Delta x, y, z, \ldots)$$

$$f(x+\Delta x, y+\Delta y, z, \ldots) - f(x+\Delta x, y, z, \ldots) = \Delta y\frac{\partial f}{\partial y}(x+\Delta x, y+\theta_2\Delta y, z, \ldots)$$

$$f(x+\Delta x, y+\Delta y, z+\Delta z, \ldots) - f(x+\Delta x, y+\Delta y, z, \ldots) = \Delta z\frac{\partial f}{\partial z}(x+\Delta x, y+\Delta y, z+\theta_2\Delta z, \ldots)$$

$$\vdots$$

where $\theta_1, \theta_2, \theta_3, \ldots$ are in $[0, 1]$. Adding these up he gets

$$f(x + \Delta x, y + \Delta y, z + \Delta z, \ldots) - f(x, y, z) =$$
$$\Delta x \frac{\partial f}{\partial x}(x + \theta_1 \Delta x, y, z, \ldots) + \Delta y \frac{\partial f}{\partial y}(x + \Delta x, y + \theta_2 \Delta y, z, \ldots) +$$
$$\Delta z \frac{\partial f}{\partial z}(x + \Delta x, y + \Delta y, z + \theta_2 \Delta z, \ldots) + \ldots,$$

whence dividing by α and letting $\alpha \to 0$ he concludes[209]

$$\frac{df}{d\vec{\beta}} = \left(\lim_{\alpha \to 0} \frac{\Delta x}{\alpha}\right) \frac{\partial f}{\partial x}(x, y, z, \ldots) + \left(\lim_{\alpha \to 0} \frac{\Delta y}{\alpha}\right) \frac{\partial f}{\partial y}(x, y, z, \ldots) +$$
$$\left(\lim_{\alpha \to 0} \frac{\Delta z}{\alpha} \frac{\partial f}{\partial z}\right)(x, y, z, \ldots) + \ldots$$
$$= h \frac{\partial f}{\partial x}(x, y, z, \ldots) + k \frac{\partial f}{\partial y}(x, y, z, \ldots) + l \frac{\partial f}{\partial z}(x, y, z, \ldots) + \ldots$$

Following twenty lessons on the Differential Calculus, Cauchy takes up the Integral Calculus in Lesson 21 in which he defines the definite integral for continuous functions on closed, bounded intervals and proves that it exists. The proof makes essential use of the uniform continuity of the functions on these intervals, but, as Cauchy means "uniformly continuous" when he says "continuous", no mention of the uniformity is made. The key to his treatment of the existence proof, as well as his derivation of the Mean Value Theorem for Integrals, is his Theorem XII cited earlier, or, rather, one of its immediate corollaries[210]:

Corollary III. — If we denote by $\alpha, \alpha', \alpha'', \ldots$ new quantities which have the same sign, then by virtue of Eq. (17) we have

$$(20) \quad \begin{cases} \dfrac{\alpha a + \alpha' a' + \alpha'' a'' + \ldots}{\alpha b + \alpha' b' + \alpha'' b'' + \ldots} = M\left(\dfrac{\alpha a}{\alpha b}, \dfrac{\alpha' a'}{\alpha' b'}, \dfrac{\alpha'' a''}{\alpha'' b''}, \ldots\right) \\ \qquad = M\left(\dfrac{a}{b}, \dfrac{a'}{b'}, \dfrac{a''}{b''}\right). \end{cases}$$

The Mean Value Theorem for Integrals is our main concern, but its proof, briefly hinted at by Cauchy, harks back to the existence of the integral, so we should consider this latter proof.

Cauchy starts with a continuous function $f : [a, b] \to \mathbb{R}$. For any partition of the interval $[a, b]$:

$$X: \quad a = x_0 < x_1 < \ldots < x_{n-1} < x_n = b,$$

he defines the sum

$$S_X = (x_1 - x_0)f(x_0) + (x_2 - x_1)f(x_1) + \ldots + (x_n - x_{n-1})f(x_{n-1}). \qquad (3.46)$$

[209] By the continuity of the partial derivatives.

[210] Bradley and Sandifer, *op. cit.*, p. 300.

He then claims that, as the sizes $|x_{i+1} - x_i|$ of the individual subintervals of the partition get smaller and smaller, the values of the sums S_X tend to a limit he calls the *definite integral* of f on $[a, b]$ and which he denotes by

$$\int_a^b f(x)dx.$$

The first step is to apply his formula (20) to $a_i = f(x_i)$, $b_i = 1$ and $\alpha_i = x_{i+1} - x_i$:

$$\begin{aligned}\frac{\alpha a + \alpha' a' + \alpha'' a'' + \ldots}{\alpha b + \alpha' b' + \alpha'' b'' + \ldots} &= \frac{(x_1 - x_0)f(x_0) + \ldots + (x_n - x_{n-1})f(x_{n-1})}{(x_1 - x_0) + \ldots + (x_n - x_{n-1})} \\ &= \frac{S_X}{x_n - x_0} = \frac{S_X}{b-a} \\ &= M\big(f(x_0), f(x_1), \ldots, f(x_{n-1})\big),\end{aligned}$$

by Corollary III. Thus, if $m = \min\big\{f(x) \,\big|\, x \in [a, b]\big\}$, $M = \max\big\{f(x) \,\big|\, x \in [a, b]\big\}$,

$$m \leq \frac{S_X}{b-a} \leq M,$$

and by the Intermediate Value Theorem,

$$\frac{S_X}{b-a} = f\big(a + \theta(b-a)\big), \text{ for some } 0 \leq \theta \leq 1. \tag{3.47}$$

Now this argument applies to any partition of each of the subintervals $Y_i = [x_i, x_{i+1}]$ and the corresponding sum over that interval. Thus, if Y is any partition refining X by the inclusion of more points, one has

$$\begin{aligned}S_Y = (x_1 - x_0)f\big(x_0 + \theta_0(x_1 - x_0)\big) + (x_2 - x_1)f\big(x_1 + \theta_1(x_2 - x_1)\big) + \ldots \\ + (x_n - x_{n-1})f\big(x_{n-1} + \theta_{n-1}(x_n - x_{n-1})\big),\end{aligned} \tag{3.48}$$

where $\theta_0, \theta_1, \ldots, \theta_{n-1}$ lie between 0 and 1 are the numbers defining the mean values for the partitions of the subintervals $Y_0, Y_1, \ldots, Y_{n-1}$, respectively. But

$$\begin{aligned}S_Y - S_X = (x_1 - x_0)\left(f\big(x_0 + \theta_0(x_1 - x_0)\big) - f\big(x_0\big)\right) + \ldots \\ + (x_n - x_{n-1})\left(f\big(x_{n-1} + \theta_{n-1}(x_n - x_{n-1})\big) - f\big(x_{n-1}\big)\right),\end{aligned}$$

whence

$$\begin{aligned}\big|S_Y - S_X\big| \leq \left|(x_1 - x_0)\left(f\big(x_0 + \theta_0(x_1 - x_0)\big) - f\big(x_0\big)\right)\right| + \ldots \\ + \left|(x_n - x_{n-1})\left(f\big(x_{n-1} + \theta_{n-1}(x_n - x_{n-1})\big) - f\big(x_{n-1}\big)\right)\right|.\end{aligned}$$

But f is uniformly continuous, whence for any $\epsilon > 0$ there is some $\delta > 0$ such that for all $x, y \in [a, b]$,

$$|x - y| < \delta \Rightarrow \left|f(x) - f(y)\right| < \frac{\epsilon}{b - a}.$$

Assuming the subintervals of the original partition X all to have length $< \delta$, we have

$$\begin{aligned}\left|S_X - S_Y\right| &< (x_1 - x_0)\frac{\epsilon}{b-a} + \ldots + (x_n - x_{n-1})\frac{\epsilon}{b-a} \\ &< (x_n - x_0)\frac{\epsilon}{b-a} = (b-a)\frac{\epsilon}{b-a} = \epsilon.\end{aligned}$$

Calling the maximum length $|x_{i+1} - x_i|$ of a subinterval of a partition the *mesh* of the partition, we see that we have proven the following lemma:

3.5.2 Lemma *Let $f : [a, b] \to \mathbb{R}$ be continuous. For any $\epsilon > 0$ there is a $\delta > 0$ such that for any partition $X \subseteq [a, b]$ of mesh $< \delta$ and any refinement Y of the partition X (i.e., $X \subseteq Y$), we have $\left|S_X - S_Y\right| < \epsilon$.*

Next, Cauchy considers any two partitions X, Y of $[a, b]$.

3.5.3 Lemma *Let $f : [a, b] \to \mathbb{R}$ be continuous. For every $\epsilon > 0$ there is $\delta > 0$ such that for any partitions $X, Y \subseteq [a, b]$ of mesh $< \delta$, $\left|S_X - S_Y\right| < \epsilon$.*

The proof of this is very simple. Assume $\delta > 0$ small enough by Lemma 3.5.2 so that for any partition X of mesh $< \delta$ and any refinement $Z \supseteq X$, $\left|S_X - S_Z\right| < \epsilon/2$. Let X, Y be two partitions of $[a, b]$ of mesh $< \delta$. Let $Z \supseteq X \cup Y$ be any common refinement, and observe

$$\left|S_X - S_Y\right| = \left|S_X - S_Z + S_Z - S_Y\right| \leq \left|S_X - S_Z\right| + \left|S_Z - S_Y\right| < \frac{\epsilon}{2} + \frac{\epsilon}{2} = \epsilon.$$

All that remains to prove the existence of the integral is to specify some sequence of partitions $X_0, X_1, X_2, \ldots$ whose meshes tend to 0 and for which the sums $S_{X_0}, S_{X_1}, S_{X_2}, \ldots$ converge to a limit. Cauchy waves his hands at this point:

> So when the elements of the difference $X - x_0$ [211], become infinitely small, the mode of division has no more effect on the value of S than an insensible influence; and, if we decrease indefinitely the numerical values of these elements, increasing their number, the value of S will eventually be substantially constant, or in other words, it will eventually reach a limit that only depends on the form of the function $f(x)$, and extreme values x_0, X assigned to the variable X. this limit is called a *definite integral*.[212]

Today we would insist on a rigorous proof that the limit exists, probably giving it by exhibiting a sequence of partitions $X_0, X_1, X_2, \ldots$ for which the sequence of sums $S_{X_0}, S_{X_1}, S_{X_2}, \ldots$ is Cauchy convergent and then applying Lemma 3.5.3.

---

[211] I.e., the partition subintervals of $[x_0, X]$ = our $[a, b]$.

[212] Cauchy, *Résumé*, *op. cit.*, p. 83; Cates, *op. cit.*, p. 83.

3.5.4 Exercise Do this.

In the course of this proof, Cauchy isolates, among others, two formulæ he numbers (4) and (5). (4) is our (3.47), multiplied by $b - a$,

$$S_X = (b - a)f\big(a + \theta(b - a)\big);$$

and (5) is just (3.48). Lesson 22, "Formulas for the determination of exact or approximate values of definite integrals", uses the sums defining the integral to give direct estimates of some integrals and eventually mentions the Mean Value Theorem for Integrals in passing:

> An important point to make is that the forms under which the value of S presents itself in Eqs. (4) and (5) of the previous lesson, are also suitable for the integral (2).[213] Indeed, these equations, remaining valid, one and the other, while we subdivided the difference $X - x_0$ or the quantities $x_1 - x_0, x_2 - x_1, \ldots, X - x_{n-1}$ into infinitely small elements, will still be true in the limit, so that we will have
>
> (19) $$\int_{x_0}^{X} f(x)dx = (X - x_0)f\big[x_0 + \theta(X - x_0)\big]\ldots$$[214]

This is a bit loosely stated, but basically it says

$$S_X = (b - a)M\big(f(x_0), f(x_1), \ldots, f(x_{n-1})\big) = (b - a)M(m, M),$$

where $m = \min\big\{f(x)\,\big|\,x \in [a, b]\big\}$, $M = \max\big\{f(x)\,\big|\,x \in [a, b]\big\}$. Thus

$$\int_a^b f(x)dx = \lim_{\text{mesh}(X)\to 0} S_X = (b - a)M(m, M).$$

Thus, by the Intermediate Value Theorem,

$$\int_a^b f(x)dx = (b - a)f\big(a + \theta(b - a)\big) \text{ for some } 0 \le \theta \le 1.$$

If in Cauchy's Corollary III[215] one sets $b = b' = b'' = \ldots = 1$, one has

$$\frac{\alpha a + \alpha' a' + \alpha'' a'' + \ldots}{\alpha + \alpha' + \alpha'' + \ldots} = M(a, a', a'', \ldots),$$

from which follows, on multiplying by the single value $\alpha + \alpha' + \alpha'' + \ldots$,

$$\alpha a + \alpha' a' + \alpha'' a'' + \ldots = (\alpha + \alpha' + \alpha'' + \ldots)M(a, a', a'', \ldots).$$

---

[213] Formula (2) of Lesson 22 is just $\int_{x_0}^{X} f(x)dx$ — our $\int_a^b f(x)dx$.

[214] Cauchy, *Résumé*, *op. cit.*, p. 87; Cates, p. 87.

[215] P. 281, above.

Using this in Lesson 23 Cauchy generalises the Mean Value Theorem for Integrals to the following:

3.5.5 Theorem (Cauchy Mean Value Theorem for Integrals) *Let f, g be continuous on $[a, b]$. There is some $c \in [a, b]$ such that*

$$\int_a^b f(x)g(x)dx = f(c)\int_a^b g(x)dx.$$

Note that this reduces to the ordinary Mean Value Theorem for Integrals when g is identically 1, and the Theorem is indeed a generalisation of and not merely an analogue to the Mean Value Theorem for Integrals.

Cauchy's 26th Lesson proves the Fundamental Theorem of the Calculus. The first part is very simply and elegantly handled. For f continuous on $[x_0, X]$ he defines

$$\mathfrak{F}(x) = \int_{x_0}^{x} f(x)dx \tag{3.49}$$

and observes

$$\begin{aligned}\mathfrak{F}(x+\alpha) - \mathfrak{F}(x) &= \int_{x_0}^{x+\alpha} f(x)dx - \int_{x_0}^{x} f(x)dx \\ &= \int_{x}^{x+\alpha} f(x)dx = \alpha f(x+\theta\alpha),\ 0 \le \theta \le 1,\end{aligned} \tag{3.50}$$

by the Mean Value Theorem for Integrals. Dividing by α,

$$\frac{\mathfrak{F}(x+\alpha) - \mathfrak{F}(x)}{\alpha} = f(x+\theta\alpha) \to f(x) \text{ as } \alpha \to 0$$

by the continuity of f. It follows that $\mathfrak{F}'(x) = f(x)$ and from (3.50) that $\mathfrak{F}(x+\alpha) = \mathfrak{F}(x) + \mathfrak{F}'(x+\theta\alpha)\alpha$, which is the instance of the Mean Value Theorem for a function of the form (3.49).

Cauchy has almost given a proof of the Mean Value Theorem under the ostensibly weaker assumption of continuous differentiability. He has almost proven that, on a closed, bounded interval continuous differentiability implies uniform differentiability. For, the above proof of the differentiability of $\mathfrak{F}(x)$ readily adapts to yield the function's uniform differentiability: We have

$$\mathfrak{F}(x) - \mathfrak{F}(y) = (x-y)f\big(x+\theta(x-y)\big),\ 0 \le \theta \le 1, \tag{3.51}$$

for any $x, y \in [x_0, X]$ by the same argument used to establish (3.50). But, by the uniform continuity of f on $[x_0, X]$, for any $\epsilon > 0$ there is some $\delta > 0$ such that

$$|x-y| < \delta \Rightarrow |f(x) - f(y)| < \epsilon.$$

Note that

$$\left|y+\theta(x-y)-y\right|=\left|\theta(x-y)\right|=\theta|x-y|<\theta\delta\le\delta,\ \text{since } 0\le\theta\le 1.$$

Thus (3.51) yields

$$\left|\mathfrak{F}(x)-\mathfrak{F}(y)-(x-y)f(y)\right|=\left|(x-y)\big(f\big(y+\theta(x-y)\big)-f(y)\big)\right|<|x-y|\epsilon,$$

and

$$\left|\frac{\mathfrak{F}(x)-\mathfrak{F}(y)}{x-y}-f(y)\right|<\epsilon$$

for $0<|x-y|<\delta$.

To conclude that continuous differentiability on a closed, bounded interval entails uniform differentiability and that the Mean Value Theorem is valid under the assumption of continuous differentiability, one needs only to reduce the problem to the case for functions of the form (3.49). This is where the Second Fundamental Theorem of the Calculus comes in. If g is such that $g'(x)=f(x)$ on $[x_0,X]$, then

$$g(x)-g(x_0)=\int_{x_0}^{x}f(x)dx=\mathfrak{F}(x)-\mathfrak{F}(x_0),$$

i.e.,

$$g(x)=\mathfrak{F}(x)+g(x_0)-\mathfrak{F}(x_0)=\mathfrak{F}(x)+\text{constant}.$$

Both uniform differentiability of g and the Mean Value Theorem for g follow trivially from the corresponding results for $\mathfrak{F}$. And, of course, the Second Fundamental Theorem of the Calculus follows immediately from the First Fundamental Theorem via the Constant Function Theorem. Indeed, it is really the Constant Function Theorem more than the Second Theorem that is needed here and it is a proof of this which Cauchy immediately follows his proof of the First Fundamental Theorem with.

In Chap. 2[216] we saw how to prove the Strictly Increasing and Decreasing Function Theorems for differentiable functions without appeal to the Mean Value Theorem; in Exercise 2.3.25 in Chap. 2[217] we reduced the ordinary Increasing and Decreasing Function Theorems to their strict counterparts again without appeal to the Mean Value Theorem or uniform differentiability; and we then immediately derived the Constant Function Theorem from these results again without strong assumptions. Hence we now have an (admittedly ugly) proof that continuous differentiability implies uniform differentiability on $[x_0,X]$ and thus a proof of the Mean Value Theorem for continuously differentiable functions.

Directly after proving the First Fundamental Theorem of the Calculus, Cauchy proves the Constant Function Theorem by applying the Mean Value Theorem, his

---

[216] Pp. 131–134, above.

[217] P. 134, above.

proof of which relied on uniform differentiability. So he misses credit here for showing that continuous differentiability suffices for the Mean Value Theorem and that it implies uniform differentiability. We do, of course, know the modern proof of the Mean Value Theorem and can use it to derive the Constant Function Theorem quickly under more modest assumptions than Cauchy made. Combining this with Cauchy's proof of the First Fundamental Theorem of the Calculus we have a much simpler proof that continuous differentiability implies uniform differentiability on $[x_0, X]$ than that outlined above.[218]

Having said all of this about what Cauchy did and did not do, I should explain that this does not signal the end of our discussion of Cauchy. For, he had more to say about the Mean Value Theorem and its related results.

In Lessons 35 and 36 Cauchy offers a couple of proofs of a variant of the Higher Order Mean Value Theorem.

3.5.6 Theorem (Taylor's Theorem with the Cauchy Form for the Remainder) *Let f be n times uniformly differentiable on $[a, b]$. Then*

$$f(b) = f(a) + \frac{b-a}{1}f'(a) + \ldots + \frac{(b-a)^{n-1}}{(n-1)!}f^{n-1}(a) + \int_a^b \frac{(b-z)^{n-1}}{(n-1)!}f^{(n)}(z)dz.$$

If one applies the Cauchy Mean Value Theorem for Integrals (Theorem 3.5.5, above.) to the remainder term, one finds

$$\int_a^b \frac{(b-z)^{n-1}}{(n-1)!}f^{(n)}(z)dz = f^{(n)}(c)\int_a^b \frac{(b-z)^{n-1}}{(n-1)!}dz, \text{ some } c \in [a, b]$$
$$= f^{(n)}(c)\int_{b-a}^0 \frac{x^{n-1}}{(n-1)!}(-dx),$$

substituting $x = b - z$,

$$= f^{(n)}(c)\left[-\frac{x^n}{n!}\right]_{b-a}^0$$
$$= \frac{(b-a)^n}{n!}f^{(n)}(c).$$

Thus, Cauchy's result yields the Higher Order Mean Value Theorem, under the hypothesis of n-fold uniform differentiability and with c not necessarily lying in the interior of the interval.

I think I will omit the proofs of Theorem 3.5.6. Now that Cauchy has been translated into English, these proofs are readily accessible. I note that the proof in the 36th Lesson,[219] though it appears something of an unmotivated trick, is the simpler and more readable of the two.

---

[218] For an even simpler proof, see the proof of Lemma 3.6.10 on page 301, below.

[219] Cates, *op. cit.*, p. 151.

As to the importance of Cauchy's form of the remainder, I note simply that one of the applications of the Higher Order Mean Value Theorem is to prove that the Taylor series of a function converges to the function when this occurs. This is done by showing that the remainder term tends to 0 as n grows large without bound. Occasionally the Lagrange form of the remainder cannot be shown to tend to 0, while the Cauchy form can. This is the case, for example, for the binomial series, which, prior to Cauchy's moving the Mean Value Theorem into the position, occupied a central position in the foundations of analysis.

The *Résumé* contains, in addition to 40 lessons on the Differential and Integral Calculi, two appendices. He introduces them as follows:

> Since the printing of this book, I recognised that using a very simple formula one could return to the differential calculus the solutions of several problems that I had referred to the integral calculus. I will first give this formula; then I will indicate its main applications.[220]

The result was what is now called the Cauchy Mean Value Theorem and it, along with some immediate applications, occupies the first appendix called simply "Addition". The second appendix, more informatively titled "On the formulas of Taylor and Maclaurin", applies the result to Taylor's Theorem.

The proof of the Cauchy Mean Value Theorem borrows heavily from Lagrange, reducing the Theorem to the Strictly Increasing Function Theorem, a result Lagrange proved only under a stringent uniformity condition, Ampère sort of bypassed, and Cauchy himself gave an inadequate proof[221] of. However, the gaps in Cauchy's proof are easily filled in.

3.5.7 Lemma (Cauchy's Strictly Increasing Function Theorem) *Let the function $f : [a, b] \to \mathbb{R}$ be uniformly differentiable on $[a, b]$ and suppose $f'(x) > 0$ for all $x \in [a, b]$. Then f is strictly increasing on $[a, b]$: for all $x, y \in [a, b]$,*

$$x < y \Rightarrow f(x) < f(y).$$

Proof. f' is continuous on $[a, b]$ and therefore assumes a minimum value $m > 0$ somewhere on the interval $[a, b]$. By the uniform differentiability of f, there is a δ such that for all $x, y \in [a, b]$,

$$0 < |x - y| < \delta \Rightarrow \left| \frac{f(x) - f(y)}{x - y} - f'(y) \right| < \frac{m}{2},$$

[220]Cauchy, *Résumé*, *op. cit.*, p. 161; Cates, *op. cit.*, p. 170.

[221]Cauchy, *op. cit.*, p. 21; Cates, *op. cit.*, p. 18. Cauchy first attempts to show that f is locally increasing if $f'(x_0) > 0$ at a single point without appeal to the continuity of f'. Then, assuming f' always greater than 0, he attempts to conclude f is globally increasing by proceeding "by imperceptible steps" from x_0 to X. This latter step presupposes some uniformity.

i.e.,

$$-\frac{m}{2} < \frac{f(x)-f(y)}{x-y} - f'(y) < \frac{m}{2}$$
$$f'(y) - \frac{m}{2} < \frac{f(x)-f(y)}{x-y} < f'(y) + \frac{m}{2},$$

whence

$$0 < \frac{m}{2} = m - \frac{m}{2} \leq f'(y) - \frac{m}{2} < \frac{f(x)-f(y)}{x-y} = \frac{f(y)-f(x)}{y-x}.$$

Thus, if $x < y$ and $0 < |x-y| = y-x < \delta$,

$$0 < f(y) - f(x),$$

i.e., $f(x) < f(y)$.

Now let $x < y$ be any two elements of $[a, b]$, not necessarily within δ of each other. Choose $n > (y-x)/\delta$ so that $(y-x)/n < \delta$ and define

$$x_k = x + k\frac{y-x}{n}, \quad k = 0, 1, \ldots, n.$$

Then

$$x = x_0 < x_1 < \ldots < x_{n-1} < x_n = y$$

and

$$0 < x_{k+1} - x_k = \frac{y-x}{n} < \delta.$$

By what has already been proven,

$$f(x) = f(x_0) < f(x_1) < \ldots < f(x_{n-1}) < f(x_n) = f(y). \qquad \square$$

As in our earlier treatment, the Strictly Decreasing Function Theorem can either be given an analogous proof or reduced to Lemma 3.5.7 by considering $g(x) = -f(x)$. I also note that the Theorem generalises slightly to allow $f'(x)$ to equal 0 at either of the endpoints.

3.5.8 Exercise Let f be uniformly differentiable on $[a, b]$ and suppose $f'(x) > 0$ for all $x \in (a, b)$. By Lemma 3.5.7 f is increasing on $[a + (b-a)/n, b - (b-a)/n]$ for all $n > 2$. Use this to show f is strictly increasing on $[a, b]$.

With Lemma 3.5.7 Cauchy is able to prove the Cauchy Mean Value Theorem in the following form.

3.5.9 Theorem (Cauchy Mean Value Theorem) *Let f, F be uniformly differentiable on the interval $[x_0, X]$ and suppose $F'(x) \neq 0$ for all $x \in [x_0, X]$. Then, for some $\theta \in [0, 1]$,*

$$\frac{f(X) - f(x_0)}{F(X) - F(x_0)} = \frac{f'[x_0 + \theta(X - x_0)]}{F'[x_0 + \theta(X - x_0)]}.$$

Cauchy observes that this Theorem can be proven by an argument analogous to that he used to prove the Mean Value Theorem, but says the result can also be proven directly and proceeds to give the direct proof, which I paraphrase here.

Proof. By the continuity of F' on $[x_0, X]$ either $F'(x) > 0$ for all $x \in [x_0, X]$ or $F'(x) < 0$ for all $x \in [x_0, X]$. For, otherwise, one would have $x, y \in [x_0, X]$ such that $F'(x) < 0 < F'(y)$ and the Intermediate Value Theorem would yield the existence of some z between x and y such that $F'(z) = 0$, contrary to hypothesis.

For the sake of definiteness, we assume $F'(x) > 0$ for all $x \in [x_0, X]$.

Let $A = \min\left\{f'(x)/F'(x) \,\middle|\, x \in [x_0, X]\right\}$ and $B = \max\left\{f'(x)/F'(x) \,\middle|\, x \in [x_0, X]\right\}$. Then

$$\frac{f'(x)}{F'(x)} - A, \quad B - \frac{f'(x)}{F'(x)}$$

are nonnegative for all $x \in [x_0, X]$. Multiplying by $F'(x)$, we see that

$$F'(x)\left[\frac{f'(x)}{F'(x)} - A\right] = f'(x) - AF'(x),\; F'(x)\left[B - \frac{f'(x)}{F'(x)}\right] = BF'(x) - f'(x)$$

are again nonnegative for all $x \in [x_0, X]$.

But these expressions are derivatives of

$$f(x) - AF(x) \text{ and } BF(x) - f(x),$$

respectively, and we can apply the Increasing Function Theorem[222] to conclude that

$$f(x) - AF(x) \geq f(x_0) - AF(x_0), \quad BF(x) - f(x) \geq BF(x_0) - f(x_0) \tag{3.52}$$

for all $x \in [x_0, X]$. Rearranging terms of (3.52) we have

$$B\big(F(x) - F(x_0)\big) \geq f(x) - f(x_0) \geq A\big(F(x) - F(x_0)\big),$$

[222] Cauchy's Note II of the *Cours d'analyse* is one of the pioneering studies of inequalities. Still he is a bit cavalier in ignoring the differences between $<$ and $\leq$ and between positive and nonnegative. He should probably argue that, for all $\epsilon > 0$

$$f'(x) - AF'(x) - \epsilon > 0 \text{ and } BF'(x) - f'(x) - \epsilon > 0,$$

whence

$$f(x) - AF(x) - \epsilon x > f(x_0) - AF(x_0) - \epsilon x_0,$$

etc. One can now fix x, x_0 and let $\epsilon \to 0$ to conclude (3.52).

and, setting $x = X$,

$$A \leq \frac{f(X) - f(x_0)}{F(X) - F(x_0)} \leq B.$$

One can now apply the Intermediate Value Theorem to f'/F' to conclude the existence of some $c = x_0 + \theta(X - x_0)$, $0 \leq \theta \leq 1$, such that

$$\frac{f'(c)}{F'(c)} = \frac{f(X) - f(x_0)}{F(X) - f(x_0)}. \quad \square$$

In 1829 Cauchy published another calculus textbook, *Leçons sur le calcul différentiel*[223] [*Lessons on the differential calculus*]. In this work he presented a slight modification of the proof for the special case in which $f(x_0) = F(x_0) = 0$. The general case then follows by applying the special result to $g(x) = f(x) - f(x_0)$, $G(x) = F(x) - F(x_0)$. In effect, the new exposition proceeds by first proving L'Hôpital's Rule and then deriving the Cauchy Mean Value Theorem from it.

In 1840 François Napoléon Marie Moigno (1804–1884) published *Leçons de calcul différentiel et de calcul intégral*,[224] an exposition of Cauchy's work on the Calculus based on both published and unpublished work of the master. In this work he went back to the 1823 proof of the *Résumé*.

3.6 Bolzano and the Mean Value Theorem

The next most important figure after Cauchy in the early 19th century rigorisation of the Calculus was Bolzano, whom we have already encountered in Chap. 2. Like Cauchy, Bolzano was influenced by Lagrange and there is consequently some similarity in their works, but there are also differences. The most glaring difference is that Bolzano's definitions of continuity and differentiability were pointwise, while Cauchy's were uniform. But there were also differences in style. Bolzano's published papers on foundations covered special topics, while Cauchy's textbooks were systematic developments. Bolzano did attempt a systematic development in the early 1830s, after Cauchy's textbooks had appeared, but this work was never completed and not published until the 20th century, some partial accounts in the first half of the century and the full work in the latter half. Moreover, while Cauchy wrote for students and is easy to read, Bolzano's programme was philosophical, incomplete, imperfect, and difficult to read. And, as it was only published well after everything new in it had been rediscovered, it played no rôle in the historical development of

---

223 De Bure, Paris, 1829. A German translation appeared a few years later: C.H. Schnuse (trans.), *A.L. Cauchy's Vorlesungen über die Differentialrechnung mit Fourier's Auflösungsmethode der bestimmte Gleichungen verbunden* [*A.L. Cauchy's Lectures on the Differential Calculus Bound with Fourier's Method of Solution of Definite Equations*], G.C.E. Meyer, Braunschweig, 1836.

224 Bachelier, Paris, 1840.

the Calculus. It is, however, quite interesting and we should take a brief look at it, particularly as to his discussion of the Mean Value Theorem.

Bolzano's first important paper on the foundations of the Calculus was his 1816 paper on the Binomial Theorem, the paper in which he also gave the first rigorous ϵ-δ proofs as well as some of the first formal definitions of some fundamental concepts of the Calculus. In this paper, in the course of "proving" a false lemma used to prove the termwise differentiability of a Taylor series and thus the uniqueness of such a series, he included a short proof of the Mean Value Theorem for real analytic functions.[225] This proof is essentially the same as Lagrange's second proof and need not be discussed here. He would do better later, almost improving on Cauchy's work.

Bolzano's work on the foundations of the Calculus goes deeper than that of Cauchy or any of his contemporaries. In the 7th chapter on "Infinite Quantity Concepts" of his "Reine Zahlenlehre" ["Pure theory of numbers"], he takes a close look at the real numbers themselves and their relation to the rational numbers. Today we would be inclined to interpret this as a construction of the real number system out of the rationals, much like the later constructions by Weierstrass (date uncertain), Dedekind (1858), and others (*c.* 1872). His intent, however, appears to have been more descriptive than constructive. To him, real numbers, as used by mathematicians, were given by *infinite number expressions*, e.g., infinite sums and products. Such an expression was *measurable* if it could be placed arbitrarily accurately with respect to the rational numbers. He never quite got this system to work, but he did derive basic properties of the real number system, such as the existence of a decimal expansion of any measurable real number. Weierstrass's later treatment of real numbers as sums of infinite collections of rational numbers can be viewed as a successful simplification of Bolzano's approach. I like to think that, had Bolzano approached the problem from a mathematical perspective instead of a philosophical one, he would have realised that the only infinite number expressions he needed were sums and he would have come up with Weierstrass's construction.

With his conception of real numbers as measurable infinite number expressions, his often criticised proof of the convergence of Cauchy sequences given in his 1817 paper proving the Intermediate Value Theorem is clearly seen to be correct. In this paper and his paper on the Binomial Theorem published the previous year, he gave his pointwise definitions of continuity and differentiation and produced occasionally poorly organised or poorly expressed, but rigorous, ϵ-δ proofs — albeit not using these particular letters.

His systematic study of the Calculus, however, is in his unpublished "Functionenlehre" ["Theory of functions"] and its list of improvements to be incorporated. The work begins with 36 short sections[226] on pre-Calculus topics — the dependency

[225] Russ, *op. cit.*, pp. 182–183. A thoughtful discussion of the proof in context was given in: Otto Stolz, "B. Bolzano's Bedeutung in der Geschichte der Infinitesimalrechnung", *Mathematische Annalen* 18 (1881), pp. 255–279; here: pp. 264–265.

[226] Sections of the work range from single sentences to several pages.

of one variable on another and the Δ-notation. Then follow §§1/37–82/118[227] on continuity. Highlights include the definitions of continuity and uniform continuity with an example to illustrate that continuity does not imply uniform continuity,[228] the Extreme Value Theorem, the Intermediate Value Theorem, and the construction of his famous nowhere differentiable function, which here he only shows to fail to be monotone in every interval.

The rest of the "Functionenlehre", §§1/119–99/217, covers differentiation. §§1/119–11/129 concern the definition of the derivative and questions regarding its uniqueness. There follow §§12/130–16/135[229] on the relation between continuity and differentiability, the last of these sections proving that his continuous function which had failed to be monotone in any interval is in fact not differentiable at every point in an everywhere dense set of real numbers.[230]

§17/136 deserves mention here as he criticises Lagrange and Galois. His function proves false one of their basic tenets:

> *Note*. The last part of the above theorem contradicts to a certain extent those which Lagrange and so many others partly expressly claimed, partly only silently assumed, that every function, with at most the exception of a few isolated values of their variables, have a *derivative* in all remaining cases.[231]

Bolzano quickly adds that Lagrange's notion of function is much narrower than his, encompassing only those "which can be expressed by one of the seven signs $a + x$, $a - x$, ax, $\frac{a}{x}$, x^n, a^x, $\log x$, or through a combination of several of these",[232] and the claim is true of these. He points out that there are others who have a more general notion of function and still make the claim. One of these was Évariste Galois (1811–1832), whose name will always be remembered for his fundamental work in algebra and who, were he alive today, would cringe at the mention of his analytic note cited by Bolzano.

Galois's paper consists of two short notes, the first of which contains an attempted proof that every function is differentiable. Bolzano cites the note in full and then criticises the proof. As the proof depends on a pseudo-mean value theorem, I cannot resist doing the same:

---

227 The German edition begins the numbering of the three parts of the work anew with each part: the preliminaries, continuity, and differentiability. I thus cite section numbers in the format §German/English.

228 This is given by $f(x) = 1/(1 - x)$ near 1. He does prove that continuity implies uniform continuity on a closed, bounded interval, but not in the main body of the text; this is given in the unincorporated improvements.

229 §135 has no counterpart in the German edition. It refers to his earlier construction of his nowhere differentiable function.

230 When his manuscript was rediscovered in the early 20th century, Karel Rychlík and Vojtěch Jarník independently proved in 1922 that it was in fact not differentiable at any argument.

231 Bernard Bolzano (author), Karl Petr (ed.), and Karel Rychlík (ed.), *Bernard Bolzano's Schriften. Band 1 Functionenlehre*, Prague, 1930, p. 96; Russ, *op. cit.*, p. 508.

232 *Ibid.* Presumably he allows a to assume complex values so that the trigonometric functions can be generated.

Notes on some points of analysis;

By Mr. Galais [*sic*], student at the Ecole normale.

I.

Demontration[sic] *of a theorem of analysis.*

THEOREM. Let Fx and fx be any two given functions; we have, for any x and h,

$$\frac{\mathrm{F}(x+h)-\mathrm{F}x}{\mathrm{f}(x+h)-\mathrm{f}x}=\varphi(k),$$

φ being a determinate[233] function, and k a quantity intermediate between x and $x+h$.

Demonstration. Indeed, put

$$\frac{\mathrm{F}(x+h)-\mathrm{F}x}{\mathrm{f}(x+h)-\mathrm{f}x}=P;$$

from which one deduces

$$\mathrm{F}(x+h)-P\mathrm{f}(x+h)=\mathrm{F}x-P\mathrm{f}x,$$

whence one sees that the function $\mathrm{F}x-P\mathrm{f}x$ does not change when changing x and $x+h$; whence it follows that, unless it remains constant between these limits, which could take place only in special cases, this function has, between x and $x+h$, one or more *maxima* and *minima*. Let k be the value of x corresponding to one of them; evidently we have

$$k=\psi(P),$$

ψ being a determinate function; thus we must also have

$$P=\varphi(k),$$

φ being another equally determinate function; which demonstrates the theorem.

Thence we can conclude, as a corollary, that the quantity

$$\mathrm{Lim.}\frac{\mathrm{F}(x+h)-\mathrm{F}x}{\mathrm{f}(x+h)-\mathrm{f}x}=\varphi(x),$$

for $h=0$, is necessarily a function of x, which demonstrates *à priori*, the existence of the derived function.[234]

By the Mean Value Theorem, Galois's theorem holds when $f(x)=x$ by choosing $\varphi(x)=\mathrm{F}'(x)$; and by the Cauchy Mean Value Theorem, it holds more generally by choosing $\varphi(x)=\mathrm{F}'(x)/\mathrm{f}'(x)$. Thus the theorem is a sort of non-specific weak version of these theorems. Or, rather, it would be were the proof not completely bogus.

---

[233] I.e., well-defined.

[234] Galois, "Notes sur quelques points d'analysis", *Annales de Mathématiques pures et appliquées* 21 (1831/32), pp. 182–184; here: pp. 182–183. The French "démonstration" would normally be rendered mathematically as "proof", but I couldn't resist preserving the misspelling of the word in the section title as it pairs so perfectly with the misspelling of the author's name above it. A third error, the omission of the accent in the appearance of *démonstration* in the body of the paper, does not show up in translation. This is corrected in the slim volume of his collected works I found online.

Despite having already given a counterexample to the conclusion Galois drew from his theorem, Bolzano finds it necessary to criticise the proof, which he does in a more philosophical than mathematical manner, i.e., through words rather than counterexamples:

> For me this proof is not satisfying. Without doubt the equation $\frac{F(x+h)-Fx}{f(x+h)-fx}=P$ demands that one regards P as a number which depends not only on x and h, but rather on the nature of the functions which should be expressed by F and f. That now the entire expression $Fx - Pfx$ does not alter its value when x passes over to $x+h$ is right: and from this follows (if the continuity of the functions Fx and fx was assumed) mind you, that each expression must have one or more *maxima* or *minima* between x and $x+h$. However, if one denotes one of these by k, that it *evidently* must be a function of the number P, is not at all clear to me. Namely, in the expression $Fx - P.fx$ not only P occurs, but also the symbols F and f: so it would possibly be, yes it is the case, that k depends not only on the value of P but rather on the nature of the functions which are denoted by F and f. Should one wish perhaps to reply to me that the influences which the nature of the functions Fx and fx has on the determination of k, though undeniable, that one can however manage the same by the determination of k, if one allows k to depend only on P, because it itself already depends on F and f: so I answer it is no certain conclusion: "If k and P both depend on one and the same function Fx (or on two functions Fx and fx): so it must also be possible to determine k and P from each other". — So will, for example, the length of a line s through the abscissa x and the function for the ordinate $y = fx$ be determined: the same holds true of the area P which this line enclosed with its coordinates. However, can we likewise say that $s = \psi(P)$ or $P = \varphi(s)$?[235]

This quote offers a good example of why Bolzano is so difficult to read. He starts out strong, pointing out correctly the dependence of P on x and h, but then goes on to criticise the unmade assumption that P is independent of the choice of F and f. And the rest of his argument is irrelevant. Moreover, he misses the second clear objection to Galois's proof.

The dependence of P on x and h leads easily to a simple counterexample to the theorem. Let $\mathrm{F}(x) = \sin x$, $\mathrm{f}(x) = x$ on the interval $[0, 3\pi]$ and consider $x_0 = 0, x_1 = 2\pi, h = \pi$:

$$\frac{\mathrm{F}(x_0+h)-\mathrm{F}(x_0)}{\mathrm{f}(x_0+h)-\mathrm{f}(x_0)} = \frac{\sin(\pi)-\sin(0)}{\pi} = 0$$
$$\frac{\mathrm{F}(x_1+h)-\mathrm{F}(x_1)}{\mathrm{f}(x_1+h)-\mathrm{f}(x_1)} = \frac{\sin(3\pi)-\sin(2\pi)}{\pi} = 0.$$

$\psi(0)$ must lie between x_i and $x_i + h$ for each i, i.e., between 0 and π and between 2π and 3π, which is clearly impossible. As Bolzano says, k depends on x and h as well as on P.

Even granted the existence of φ for some particular F and $\mathrm{f}(x) = x$, the conclusion that F is differentiable at x cannot be drawn. Writing $\varphi(x) = \lim_{h\to 0}\varphi(h)$ assumes this limit exists, i.e., that the derivative exists. All Galois can conclude from the hypothesis offered by the theorem is that if F is differentiable, then the differentiability of F follows.

---

[235] *Functionenlehre, op. cit*, pp. 97 – 98; Russ, *op. cit.*, pp. 509–510.

3.6.1 Example Let

$$\mathrm{F}(x) = \begin{cases} x \sin \dfrac{1}{x}, & x > 0 \\ 0, & x = 0 \end{cases}$$

and $\mathrm{f}(x) = x$. By the Mean Value Theorem, for $h > 0$, there is a function φ, namely F', and a number k with $0 < k < h$ such that

$$\frac{\mathrm{F}(0+h) - \mathrm{F}(0)}{h} = \varphi(k).$$

But $\mathrm{F}'(0)$ does not exist.

3.6.2 Remark Using different notation we can reframe the beginning of Galois's argument as follows: Let $a < b$ be given and let p be the fixed number

$$p = \frac{F(b) - F(a)}{f(b) - f(a)}.$$

We have

$$F(b) - pf(b) = F(a) - pf(a),$$

whence $G(b) = G(a)$ for the function $G(x) = F(x) - pf(x)$. Unless G is constant, it assumes an extreme value at some $c \in (a, b)$. Had Galois assumed differentiability, he might have noticed

$$G'(c) = F'(c) - pf'(c) = 0,$$

i.e.,

$$p = \frac{F'(c)}{f'(c)}.$$

He would thus have proven the Cauchy Mean Value Theorem assuming only the continuity of F, f on $[a, b]$, their differentiability on (a, b), and the non-vanishing of f' on (a, b). Bolzano seems to have been so intent on criticising the latter part of Galois's proof that he too did not take the hint.

Note that adding to G the constant $F(a) - pf(a)$ results in

$$\phi(x) = F(x) - F(a) + p\big(f(x) - f(a)\big),$$

which is a constant multiple of the auxiliary function used in section 1, above, to prove the Cauchy Mean Value Theorem. And, for $f(x) = x$, it is the standard auxiliary function used to prove the Mean Value Theorem for F.

Taking leave of Galois and getting back on track, I note that the next few sections of the "Functionenlehre" concern the relation between one- and two-sided derivatives. We finally arrive at the outskirts of Bolzano's discussion of the Mean Value Theorem in §24/141, where he proves, again using $f(x) = 1/(1-x)$ in, say $(0, 1)$ or $(1, 2)$, that differentiability does not imply uniform differentiability. He does not mention it, but f' is continuous in these intervals, whence he shows that continuous differentiability in an open interval need not imply uniform differentiability there.

Two additional counterexamples are discussed in the next two sections before reaching the all important §27/144 in which he attempts to prove that continuous differentiability on a closed, bounded interval entails uniform differentiability. At least that seems to be what he attempts to prove. His statement of this result is itself a formidable challenge to the reader:

> *Theorem.* If a function Fx has a derivative $F'x$ in both directions for all values of its variable lying between a and $a+h$, but at least one [derivative] in the same direction as h for the value $x = a$, and one in the opposing direction for the value $x = a+h$: if additionally this derivative obeys the law of continuity for all these just named values of x, it follows: there must be a number e small enough for one to state that the increase Δx need not be taken smaller than e, whereby the difference $\frac{F(x+\Delta x) - Fx}{\Delta x} - F'x$ falls in absolute value below any given fraction, so long as both x and $(x + \Delta x)$ do not lie outside a and $a+h$.[236]

The awkwardness of his phrasing of each clause of the hypothesis and the conclusion does more-or-less achieve precision in the statement, but he could have achieved the same using phrases like "continuous on $[a, a+h]$", "differentiable on $[a, a+h]$", etc. And, to the modern reader, the negative formulations like "do not lie outside" and "need never be taken smaller than" do not improve readability or add nuance.[237] Or, perhaps, the latter phrase does: Does specifying "e small enough to assert that the increase Δx need never be taken smaller than e" actually mean that whatever follows holds for e and all smaller numbers Δx? Or, is that merely what one expects it to mean?

When one is confronted with an ambiguous statement of a result like that of Bolzano's lemma, one can usually resolve the ambiguity either by reading the proof or seeing how the result is applied. As first noted by Vojtěch Jarník,[238] these approaches resolve the question differently in the present case. A partial explanation of this is that the "Functionenlehre" was a work in progress, incomplete in the details of its execution.

In the proof he invokes the Bolzano–Weierstrass Theorem through an appeal to an earlier intended section that appears not to have been written yet when he was citing it.[239]

---

236 *Ibid.*, p. 106, p. 515, respectively.

237 Evidently, "x lies inside the interval $[a, b]$" means $x \in (a, b)$, while "x does not lie outside the interval $[a, b]$" means $x \notin (-\infty, a) \cup (b, \infty)$, i.e., $x \in [a, b]$.

238 Vojtěch Jarník, "Bolzano's 'Functionenlehre' ", in: Vojtěch Jarník, *Bolzano and the Foundations of Mathematical Analysis*, Society of Czechoslovak Mathematicians and Physicists, Prague, 1981, pp. 43–66; here: pp. 62–64.

239 "But we know from § that there must be a certain measurable number c...".

The Bolzano–Weierstrass Theorem has two equivalent formulations. The first says that if $c_0, c_1, c_2, \ldots$ is a sequence of real numbers taken from a closed, bounded interval $[a, b]$, there is a subsequence $c_{i_0}, c_{i_1}, c_{i_2}, \ldots$ which converges to a limit $c \in [a, b]$. The second says that any such sequence possesses a limit point $c \in [a, b]$, where the notion of limit point is defined as follows.

3.6.3 Definition A number c is a *limit point* of a sequence $c_0, c_1, c_2, \ldots$ if, for every $\epsilon > 0$ and every positive integer n_0, there is an $n > n_0$ such that $|c - c_n| < \epsilon$.

In other words, the sequence comes within ϵ of c infinitely often — for any $\epsilon > 0$, no matter how small. This is weaker than the requirement for c to be a limit of the sequence, which is that the sequence not only comes within ϵ of c, but eventually stays there. It follows that, if a sequence has a limit, the limit is a limit point. In the Calculus one shows that limits, when they exist, are unique. Likewise, if a sequence has a limit, it is the only limit point. For sequences not possessing limits, the situation varies.

3.6.4 Example Let $c_n = n$. This sequence has no limit points: For any c, $|c - c_n| = |c - n| > 1$ for all $n > |c| + 1$.

3.6.5 Example Let $c_n = 1/(n+1)$. This sequence has 0 as a limit, hence has only one limit point. To show that 0 is the limit, let $\epsilon > 0$ be given and choose $n > 1/\epsilon$. Then $\epsilon > 1/n > 1/(n+1)$ and

$$\left|0 - \frac{1}{n+1}\right| = \left|\frac{1}{n+1}\right| < \epsilon.$$

To show directly that 0 is the only limit point, choose n_0 such that for all $n > n_0$

$$\left|0 - \frac{1}{n+1}\right| < \frac{\epsilon}{2}$$

and note that, if c is a limit point of the sequence, there is an $n > n_0$ such that

$$\left|c - \frac{1}{n+1}\right| < \frac{\epsilon}{2}.$$

But then

$$\begin{aligned} |c| = |0 - c| &\leq \left|0 - \frac{1}{n+1} + \frac{1}{n+1} - c\right| \\ &\leq \left|0 - \frac{1}{n+1}\right| + \left|\frac{1}{n+1} - c\right| \leq \frac{\epsilon}{2} + \frac{\epsilon}{2} = \epsilon. \end{aligned}$$

As $\epsilon > 0$ was arbitrary, $c = 0$.

3.6.6 Example Let $\{r_0, r_1, \ldots, r_{k-1}\}$ be a finite set of real numbers and consider the periodic sequence $c_0, c_1, c_2, \ldots$ given by

$$r_0, r_1, \ldots, r_{k-1}, r_0, r_1, \ldots, r_{k-1}, r_0, r_1, \ldots$$

Each r_i occurs infinitely many times in the sequence and is thus a limit point. Any other number c is not a limit point because if $\epsilon = \min\{\,|r_i - c| \,:\, i = 0, 1, \ldots, k-1\}$, then $|c - c_n| > \epsilon/2$ for all n.

3.6.7 Example Let $c_0, c_1, c_2, \ldots$ be any enumeration of the rationals in $[0, 1]$ in which every rational number is repeated infinitely many times, e.g.,

$$0, \frac{0}{1}, \frac{1}{1}, \frac{0}{2}, \frac{1}{2}, \frac{2}{2}, \frac{0}{3}, \frac{1}{3}, \frac{2}{3}, \frac{3}{3}, \frac{0}{4}, \frac{1}{4}, \frac{2}{4}, \frac{3}{4}, \frac{4}{4}, \frac{0}{5} \ldots,$$

i.e.,

$$0, 0, 1, 0, \frac{1}{2}, 1, 0, \frac{1}{3}, \frac{2}{3}, 1, 0, \frac{1}{4}, \frac{1}{2}, \frac{3}{4}, 1, 0 \ldots$$

Every number in the interval is a limit point of the sequence.

Somewhat deeper is the following example.

3.6.8 Example Let $0 < \alpha < 1$ and consider the sequence $0\alpha, 1\alpha, 2\alpha, \ldots$ *modulo* 1, i.e., let c_n be the *fractional part* of $n\alpha$,

$$\{n\alpha\} = n\alpha - [n\alpha],$$

where $[x]$ denotes the greatest integer $\le x$. If α is rational, say $\alpha = p/q$, where p, q are relatively prime positive integers, then $c_0, c_1, c_2, \ldots$ is a periodic sequence of the form given in Example 3.6.6. If, on the other hand, α is irrational, every number in $[0, 1]$ is a limit point of the sequence. [The assertion for α rational is an easy exercise. The irrational case is more difficult and has some big names behind it: Peter Gustav Lejeune Dirichlet proved that 0 is a limit point of the sequence, Leopold Kronecker proved every $c \in [0, 1]$ to be a limit point, and Hermann Weyl proved that in a definite sense no limit point is approached more frequently than any other.[240]]

The most general existence theorem for limit points of sequences is the Bolzano–Weierstrass Theorem:

3.6.9 Lemma (Bolzano–Weierstrass Theorem) *Every sequence $c_0, c_1, c_2, \ldots$ of elements of $[a, b]$ has a limit point $c \in [a, b]$.*

I don't know how Bolzano intended to prove the result[241] in the missing section, but the following proof is fairly simple.

---

[240] Cf., e.g., Edmund Hlawka, *Theorie der Gleichverteilung*, Bibliographisches Institut, Mannheim, 1979, pp. 1–10.

[241] It has been reported that no one has found the proof in any of Bolzano's papers. He did, however, cite the result.

Proof. Let $a < b$ be given and $c_0, c_1, c_2, \ldots$ a sequence of points in $[a, b]$. We define two sequences $x_0, x_1, x_2, \ldots$ and $y_0, y_1, y_2, \ldots$ inductively as follows:

$$x_0 = a, \quad y_0 = b.$$

Note that infinitely many elements[242] of the sequence $c_0, c_1, c_2, \ldots$ lie inside $[x_0, y_0]$. Note also that the length of the interval $[x_0, y_0]$ is

$$y_0 - x_0 = b - a = \frac{b-a}{2^0}.$$

Suppose one has defined x_n, y_n such that

$$y_n - x_n = \frac{b-a}{2^n},$$

and infinitely many elements of the sequence $c_0, c_1, c_2, \ldots$ are in the interval $[x_n, y_n]$. Let $z = (x_n + y_n)/2$ be the midpoint of the interval $[x_n, y_n]$. At least one of the two subintervals $[x_n, z]$ and $[z, y_n]$ contains infinitely many elements of the given sequence. If $[x_n, z]$ contains them, set $x_{n+1} = x_n, y_{n+1} = z$; otherwise choose $x_{n+1} = z, y_{n+1} = y_n$. In either case

$$y_{n+1} - x_{n+1} = \frac{1}{2}(y_n - x_n) = \frac{1}{2} \cdot \frac{b-a}{2^n} = \frac{b-a}{2^{n+1}}$$

and $[x_{n+1}, y_{n+1}]$ contains infinitely many elements of the sequence $c_0, c_1, c_2, \ldots$

Note that

$$x_0 \le x_1 \le x_2 \le \ldots \le y_2 \le y_1 \le y_0,$$

whence the sequence $x_0, x_1, x_2, \ldots$ is bounded above. Let c be the least upper bound of this sequence. Since each y_n is an upper bound for the x's, it follows that $c \le y_n$ for each n. And, since c is an upper bound for the x's, we simultaneously have $x_n \le c$, i.e., $c \in [x_n, y_n]$ for all n.

Let now $\epsilon > 0$ be given and choose n_0 so large that

$$2^{n_0} > \frac{b-a}{\epsilon}, \quad \text{i.e.,} \quad \frac{b-a}{2^{n_0}} < \epsilon.$$

For $n > n_0$,

$$y_n - x_n = \frac{b-a}{2^n} < \frac{b-a}{2^{n_0}} < \epsilon.$$

But $[x_n, y_n]$ contains infinitely many elements of the sequence $c_0, c_1, c_2, \ldots$ Choose any of these, c_m, for $m > n_0$ and observe

---

[242] In fact, all of them.

$$c_m \in [x_n, y_n] \Rightarrow |c - c_m| < |y_n - x_n| < \epsilon.$$

Thus c is a limit point of the sequence $c_0, c_1, c_2, \ldots$ □

This brings us to Bolzano's lemma of §24/141. The result we expect to see proven is the following.

3.6.10 Lemma *Let $f : [a, b] \to \mathbb{R}$ be continuously differentiable on $[a, b]$. Then: f is uniformly differentiable there.*

This result is eminently true and easily proven by appeal to the Mean Value Theorem: Let $\epsilon > 0$ be given and note, for $x, x + h \in [a, b]$,

$$\frac{f(x+h) - f(x)}{h} - f'(x) = f'(c) - f'(x),$$

for some c between x and $x + h$. But f' is continuous on $[a, b]$, whence uniformly continuous there, and there is a $\delta > 0$ such that

$$|c - x| < \delta \Rightarrow |f'(c) - f'(x)| < \epsilon.$$

Thus, if $|h| < \delta$, one has

$$\left| \frac{f(x+h) - f(x)}{h} - f'(x) \right| < \epsilon.$$

There are two problems here. First, Bolzano intends to use the lemma to prove the Mean Value Theorem, and an appeal to this latter to prove the lemma would thus make his final proof circular. Second, at this point in his exposition Bolzano has not yet proven that continuity implies uniform continuity. He gets to this in §6 of the improvements he intended to incorporate in the never completed final version of "Functionenlehre". This result, which is even simpler than Lemma 3.6.10 is given a longer, more roundabout treatment than is his lemma. A correct proof in that case, however, is quite simple:

Proof that continuity implies uniform continuity on $[a, b]$. Let $f : [a, b] \to \mathbb{R}$ be continuous. If f were not uniformly continuous, there would be some ϵ_0 such that for all $\delta > 0$ there are $x, y \in [a, b]$ such that $|x - y| < \delta$ and yet $|f(x) - f(y)| > \epsilon_0$. Let, for each n, x_n be an x for which such a y exists for $\delta = 1/(n + 1)$. By the Bolzano–Weierstrass Theorem, the sequence $x_0, x_1, x_2, \ldots$ has a limit point c.

But f is continuous at c, so there is a $\delta > 0$ such that for all $y \in [a, b]$,

$$|y - c| < \delta \Rightarrow |f(y) - f(c)| < \frac{\epsilon_0}{2}.$$

By choice of c, there is some element x_n of the sequence with $n > 2/\delta$ within $\delta/2$ of c. Choose y such that

$$|x_n - y| < \frac{1}{n+1} \text{ and } |f(x_n) - f(y)| \geq \epsilon_0.$$

Now

$$|y - c| = |y - x_n + x_n - c| \leq |y - x_n| + |x_n - c| < \frac{1}{n+1} + \frac{\delta}{2} < \frac{\delta}{2} + \frac{\delta}{2} = \delta,$$

whence

$$|f(y) - f(c)| < \frac{\epsilon_0}{2}.$$

But

$$\begin{aligned} |f(x_n) - f(y)| &\leq |f(x_n) - f(c) + f(c) - f(y)| \\ &\leq |f(x_n) - f(c)| + |f(c) - f(y)| < \frac{\epsilon_0}{2} + \frac{\epsilon_0}{2} = \epsilon_0, \end{aligned}$$

a contradiction. □

Bolzano's exposition of this proof, which covers almost two and a half pages in the English translation,[243] is not as clear as this and the verdict is mixed on whether he proved the result or only nearly proved it. Regardless of the imperfection of his execution, he certainly had the right ideas and all it took to improve the presentation was more experience in giving ϵ-δ proofs. I am thus inclined to credit him with the result. His attempt to apply the same idea to prove his lemma in §27/144 was, however, less successful. The situation here is much more subtle, and his treatment is a good half page shorter[244] than that of the simpler case. And, it is not even clear exactly what it was he was trying to prove or what he has proven. What is clear is that he has not proven Lemma 3.6.10.

Jarník, one of Bolzano's 20th century rediscoverers, discusses the matter as follows:

> Thus the aim of §27[245] was, roughly speaking, … a proof of the following statement:
> *Theorem* A. If both $f(x)$ and $f'(x)$ are continuous in $[a, b]$ then for every $\varepsilon > 0$ there is $\delta > 0$ such that
>
> (2) $$\left|\frac{f(x+h) - f(x)}{h} - f'(x)\right| < \varepsilon$$
>
> provided $a \leq x \leq b$, $a \leq x + h \leq b$, $0 < |h| \leq \delta$.
>
> Bolzano actually applies this theorem in §28; however, the theorem presented instead in §27 is very complicated and rather vague; it is possible either to conclude that Bolzano desired to present Theorem A, or to interpret the theorem from §27 in the following way:
> *Theorem* B. If both $f(x)$ and $f'(x)$ are continuous in $[a, b]$ then for every $\varepsilon > 0$ there is $\delta > 0$

[243] Russ, *op. cit.*, pp. 575–577.

[244] It occupies most of pp. 515–516 of the English translation, *ibid.*

[245] In line with Rychlík's 1930 publication of the "Functionenlehre", Jarník begins the numbering of sections anew for the chapter on differentiation. Thus §144 of the English translation is §27 according to him and §145 becomes §28.

such that for every x ($a \le x \le b$) there exists at least one number h with $|h| \ge \delta$ such that

$$\left|\frac{f(x+h)-f(x)}{h} - f'(x)\right| < \varepsilon \ (a \le x+h \le b).$$

Theorem B is evidently weaker than Theorem A, for Theorem A guarantees the inequality (2) for all $|h| \le \delta$ while B does so evidently for a single h, $|h| \ge \delta$, which can moreover depend also on x. Bolzano's proof in §27 (provided we interpret some minor vaguenesses on Bolzano's behalf) demonstrates correctly Theorem B but not Theorem A. Definitely, Bolzano's proof of the Mean Value Theorem is not quite correct: if the theorem from §27 is interpreted as Theorem A, then the proof in §27 fails; if it is interpreted as Theorem B, then §28 is incorrect, for Bolzano uses Theorem A there.[246]

Like his proof of the Uniform Continuity Theorem, Bolzano's proof of his lemma proceeds by contradiction. The proof starts by assuming the lemma to be false. Thus to determine whether he has given an incorrect proof of Theorem A or a correct proof of Theorem B actually requires us to know what he is trying to prove! The vagueness of Bolzano's language makes this difficult if not impossible. As he needs Theorem A in the next section, let us assume he was trying to prove Theorem A, i.e., Lemma 3.6.10 and see where his proof fails.

Attempted proof of Lemma 3.6.10. Assume by way of contradiction that the result is false. Then there is some $\epsilon_0 > 0$ such that for all $\delta > 0$ and all h with $0 < |h| \le \delta$ there is some $x \in [a, b]$ with $x + h \in [a, b]$ such that

$$\left|\frac{f(x+h)-f(x)}{h} - f'(x)\right| \ge \epsilon_0. \tag{3.53}$$

For each n, set $h_n = (b-a)/(n+1)$ and taking h_n for δ, h let x_n satisfy (3.53), i.e.,

$$\left|\frac{f(x_n+h_n)-f(x_n)}{h_n} - f'(x_n)\right| \ge \epsilon_0.$$

This gives us a sequence $x_0, x_1, x_2, \ldots$ in $[a, b]$ to which we can apply the Bolzano–Weierstrass Theorem. Let $c \in [a, b]$ be a limit point of the sequence. f is differentiable at c, whence there is some $\delta > 0$ such that

$$0 < |x-c| < \delta \Rightarrow \left|\frac{f(x)-f(c)}{x-c} - f'(x)\right| < \frac{\epsilon_0}{2}.$$

Because c is a limit point of the sequence there will be infinitely many terms x_n of the sequence such that $|x_n - c| < \delta/2$, $|h_n| < \delta/2$. For such n,

$$|x_n - c| < \delta \ \text{ and } \ |x_n + h_n - c| < \delta.$$

Choose such an n, say m and fix it and h_m.

f and f' are continuous on $[a, b]$, whence uniformly so. Hence we can find $\delta_0 > 0$ so small that for $0 < |\eta| < \delta_0$,

[246] Jarník, *op. cit.*, pp. 62–63.

$$\left|f'(c+\eta)-f'(c)\right| < \frac{\epsilon_0}{6}, \tag{3.54}$$

$$\left|f(c+\eta)-f(c)\right| < \frac{\epsilon_0|h_m|}{6}, \tag{3.55}$$

$$\left|f(c+\eta+h_m)-f(c+h_m)\right| < \frac{\epsilon_0|h_m|}{6}. \tag{3.56}$$

Bolzano now observes that

$$\begin{aligned}
&\left|\frac{f(c+\eta+h_m)-f(c+\eta)}{h_m}-f'(c+\eta)-\left(\frac{f(c+h_m)-f(c)}{h_m}-f'(c)\right)\right| \\
&\quad = \left|\frac{f(c+\eta+h_m)-f(c+h_m)}{h_m}-\frac{f(c+\eta)-f(c)}{h_m}-\left(f'(c+\eta)-f'(c)\right)\right| \\
&\quad \leq \left|\frac{f(c+\eta+h_m)-f(c+h_m)}{h_m}\right|+\left|\frac{f(c+\eta)-f(c)}{h_m}\right|+\left|f'(c+\eta)-f'(c)\right|,
\end{aligned}$$

by (3.54)–(3.56),

$$< \frac{\epsilon_0}{6}+\frac{\epsilon_0}{6}+\frac{\epsilon_0}{6}=\frac{\epsilon_0}{2}.$$

But

$$\begin{aligned}
&\frac{f(c+\eta+h_m)-f(c+\eta)}{h_m}-f'(c+\eta)= \\
&\quad \frac{f(c+\eta+h_m)-f(c+\eta)}{h_m}-f'(c+\eta)- \\
&\quad\quad \left(\frac{f(c+h_m)-f(c)}{h_m}-f'(c)\right)+\left(\frac{f(c+h_m)-f(c)}{h_m}-f'(c)\right),
\end{aligned}$$

whence

$$\left|\frac{f(c+\eta+h_m)-f(c+\eta)}{h_m}-f'(c+\eta)\right| < \frac{\epsilon_0}{2}+\frac{\epsilon_0}{2}=\epsilon_0,$$

which would only be a contradiction if $c+\eta = x_m$. One can find x_n so close to c with $|h_n| < |\eta|$ and $x_n+\eta$ within δ_0 of c, thus yielding

$$\left|\frac{f(x_n+h_m)-f(x_n)}{h_m}-f'(x_n)\right| < \epsilon_0,$$

but n is not necessarily m and we haven't shown

$$\left|\frac{f(x_n+h_n)-f(x_n)}{h_n}-f'(x_n)\right| < \epsilon_0,$$

which would achieve the desired contradiction. The result is thus unproven by this argument. □

I leave it to the reader to check if the proof adapts to yield Theorem B. As B is insufficient in the application to the next step in deriving the Mean Value Theorem and as I imagine his goal was to prove Theorem A, I do not myself find Theorem B interesting.

Although the proof is incorrect, Bolzano has made an advance. Jarník continues his discussion with the remark that in spite of the fact that the proof is incorrect, "Bolzano's attempt in §27 deserves our respect; here we have one more occasion when Bolzano met with 'uniformity' and recognized that such a notion would be necessary".[247] Cauchy, as we have remarked, defined continuity, differentiability, and even series convergence as uniform concepts. But he was apparently unaware of the nonuniform concepts and was thus not explicit about this uniformity. Niels Henrik Abel had already misinterpreted Cauchy as giving pointwise definitions. Most likely Bolzano now did the same and regarded Cauchy's proof of the Mean Value Theorem as being in need of repair. Cauchy's statement of the Theorem asserted its validity for "differentiable functions with continuous derivatives". Not realising Cauchy meant, in our terms, "uniformly differentiable functions with continuous derivatives", Bolzano set out to prove this uniformity and then prove the Mean Value Theorem using this uniformity. His proof of this first step failed, but he can be credited with the conjecture.

It is of course true that continuous differentiability on a closed bounded interval implies uniform differentiability, and, as we noted back on page 286, this can be proven without appeal to the Mean Value Theorem by a rather ugly proof. We shall give a nicer proof in Sect. 3.10.2 on pages 372–373 after we have introduced the Heine–Borel Theorem.

The rest of Bolzano's proof of the Mean Value Theorem is now fairly straightforward. In §28/145 he proves a lemma which we may write as follows.

3.6.11 Lemma *Let f be continuously differentiable on $[a, b]$ and let $\epsilon > 0$ be given. For sufficiently large n and for $h = (b - a)/n$*

$$\left|f(b) - f(a) - h\big[f'(a) + f'(a+h) + \ldots + f'(a+(n-1)h)\big]\right| < \epsilon.$$

In words, he is saying that a particular Cauchy approximation to the integral of f' over the interval $[a, b]$ is close to $f(b) - f(a)$, i.e., he is stating a variant of the Fundamental Theorem of the Calculus.

Proof. By the supposedly established uniform differentiability of f, for any $\epsilon > 0$ there is a $\delta > 0$ such that for all $x, y \in [a, b]$,

$$0 < |x - y| < \delta \Rightarrow \left|\frac{f(x) - f(y)}{x - y} - f'(y)\right| < \frac{\epsilon}{b - a}.$$

[247] *Ibid.*, p. 63.

If $n > (b-a)/\delta$, one has $\delta > (b-a)/n = h$, whence

$$\left|\frac{f\big(a+(k+1)h\big)-f\big(a+kh\big)}{h} - f'\big(a+kh\big)\right| < \frac{\epsilon}{b-a}$$

for $k = 0, 1, \ldots, n-1$. Hence

$$\begin{aligned} f(b)-f(a) &= \sum_{k=0}^{n-1}\Big(f\big(a+(k+1)h\big)-f\big(a+kh\big)\Big) \\ &= h\left(\sum_{k=0}^{n-1}\frac{f\big(a+(k+1)h\big)-f\big(a+kh\big)}{h}\right) \\ &= h\sum_{k=0}^{n-1}\Big(f'(a+kh)+\epsilon_k\Big), \end{aligned}$$

where each ϵ_k has absolute value $< \epsilon/(b-a)$. Thus

$$\left|f(b)-f(a)-h\sum_{k=0}^{n-1}f'(a+kh)\right| < h\sum_{k=0}^{n-1}|\epsilon_k| < \frac{b-a}{n}\sum_{k=0}^{n-1}\frac{\epsilon}{b-a} = \epsilon. \qquad \square$$

If one assumes Cauchy's treatment of the integral, this lemma says

$$f(b)-f(a) = \int_a^b f'(x)dx. \tag{3.57}$$

But, for $m = \min\{f'(x) \,|\, x \in [a,b]\}$ and $M = \max\{f'(x) \,|\, x \in [a,b]\}$, one sees directly

$$\frac{b-a}{n}\sum_{k=0}^{n-1}m \le \frac{b-a}{n}\sum_{k=0}^{n-1}f'(a+kh) \le \frac{b-a}{n}\sum_{k=0}^{n-1}M,$$

whence

$$(b-a)m \le \int_a^b f'(x)dx \le (b-a)M,$$

i.e.,

$$m \le \frac{1}{b-a}\int_a^b f'(x)dx \le M.$$

Applying the Intermediate Value Theorem to f' yields the existence of c such that

$$f'(c) = \frac{1}{b-a}\int_a^b f'(x)dx,$$

i.e., the Mean Value Theorem for the Integral of f'. And from (3.57) we obtain the Mean Value Theorem for f:

$$f'(c) = \frac{1}{b-a}\big(f(b) - f(a)\big).$$

Bolzano's "Functionenlehre" does not include a chapter on integration and thus his derivation of the Mean Value Theorem from Lemma 3.6.11 is not quite this straightforward. Besides, Cauchy's proof of (3.57) relies on the Mean Value Theorem and such a proof would be circular. Bolzano begins §30/146 with a statement of the Mean Value Theorem:

> *Theorem*. If a function Fx has a derivative in both directions for all values of x lying between a and $a+h$, for $x = a$ at least one in the same direction as h, and for $x = a+h$ one in the direction opposite to h; if above this we know that this derivative follows the law of continuity for all such values of x: then there is always a number μ not outside 0 and 1, or (what amounts to the same) a number $a + \mu h$ not lying outside a and $a+h$ for which the equation
>
> $$F(a+h) = Fa + h.F'(a+\mu h)$$
>
> holds.[248]

A couple of things to note before considering the proof: First, since his proof that continuous differentiability implies uniform differentiability is not complete, his proof of this theorem requires the ostensibly stronger condition of uniform differentiability as a premise. Second, he is not placing $c = a + \mu h$ strictly between a and $a+h$, but allowing for the possibility that c be a or b. He will deal with that later.

Proof of Bolzano's Theorem of §31/146. I prefer using the modern notation of Lemma 3.6.11 to Bolzano's. Thus, write b for $a+h$, h for $(b-a)/n$, etc. By the Lemma

$$f(b) = f(a) + h\Big[f'(a) + f'(a+h) + \ldots + f'\big(a + (n-1)h\big)\Big] + r_n,$$

where the error term r_n can be made smaller than any given ϵ by choosing n sufficiently large.

Bolzano considers two cases.

Case 1. For infinitely many n one has

$$f'(a) = f'(a+h) = f'(x+2h) = \ldots = f'\big(a+(n-1)h\big).$$

---

[248] *Functionenlehre*, p. 111; Russ, *op. cit.*, p. 519.

In this case one has, for infinitely many n,

$$\begin{aligned} f(b) &= f(a) + h[nf'(a)] + r_n \\ &= f(a) + \frac{b-a}{n}[nf'(a)] + r_n \\ &= f(a) + (b-a)f'(a) + r_n, \end{aligned}$$

where $r_n \to 0$ as $n \to \infty$. Taking the limit,

$$f(b) = f(a) + (b-a)f'(a),$$

and we have

$$f(b) = f(a) + (b-a)f'\big(a + \mu(b-a)\big)$$

for $\mu = 0$.

Case 2. For all large n, the numbers

$$f'(a), f'(a+h), \ldots, f'\big(a + (n-1)h\big)$$

are not all the same. Then if p, q are where f' assumes its maximum and minimum values respectively, one has

$$nf'(q) < \sum_{k=0}^{n-1} f'\big(a + k(b-a)\big) < nf'(p)$$

and

$$f'(q) < \frac{1}{n}\sum_{k=0}^{n-1} f'\big(a + k(b-a)\big) < f'(p)$$

$$(b-a)f'(q) < \frac{b-a}{n}\sum_{k=0}^{n-1} f'\big(a + k(b-a)\big) < (b-a)f'(p).$$

Taking the limit of the sum,

$$(b-a)f'(q) \le f(b) - f(a) \le (b-a)f'(p),$$

whence

$$f'(q) \le \frac{f(b) - f(a)}{b-a} \le f'(p)$$

and the Intermediate Value Theorem yields some c between p, q (hence between a and b) such that

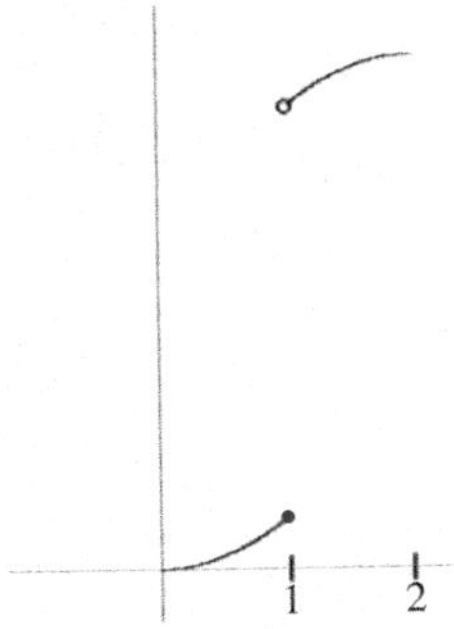

Fig. 3.17 Bolzano's counterexample

$$f'(c) = \frac{f(b) - f(a)}{b - a}.$$

And, of course, since $c \in [a, b]$, we have $c = a + \mu(b - a)$ for some $0 \leq \mu \leq 1$. □

Bolzano ends §30/146 with a numerical example and follows up in §31/147 with two examples to show the necessity of the premises of the theorem. The first of these examples illustrates how far from complete his study was. For he cites it as an example of the necessity of the continuity of the derivative.

3.6.12 Example Let

$$f(x) = \begin{cases} x^2, & 0 \leq x \leq 1 \\ 4x - x^2 + 6, & 1 < x \leq 2. \end{cases}$$

The graph consists of two disconnected pieces of parabolas, as in Fig. 3.17. Bolzano points out that f is not differentiable at $x = 1$, where only the one-sided derivative exists. He also notes that for $x \neq 1$,

$$f'(x) = \begin{cases} 2x, & 0 \leq x < 1 \\ 4 - 2x, & 1 < x \leq 2. \end{cases}$$

Now $f(2) = 10, f(0) = 0$, but $f(2) \neq f(0) + 2f'(c)$ for any c since this would require $f'(c) = (10 - 0)/2 = 5$, but

$$2x = 5 \Rightarrow x = \frac{5}{2} \notin [0, 2]$$
$$4 - 2x = 5 \Rightarrow x = -\frac{1}{2} \notin [0, 2].$$

In giving this example, Bolzano failed to observe that f is in fact discontinuous at $x = 1$ and he subsequently gave another similar counterexample to demonstrate the necessity of the continuity of f as a premise to the theorem.

We know today, of course, that the continuity of the derivative is not a necessary condition for the validity of the Mean Value Theorem. What must be assumed is the existence of the derivative at interior points.

3.6.13 Exercise Construct a continuous function $f : [a, b] \to \mathbb{R}$ which fails to have a derivative at some point in the open interval (a, b) and for which the Mean Value Theorem fails: There is no $c \in [a, b]$ such that $f(b) = f(a) + (b - a)f'(c)$.

In §31/148 Bolzano addresses an issue that earlier writers had ignored or whose treatments I have overlooked in my admittedly perfunctory examination of their works. This concerns the positioning of c in the interior of the interval and the relaxation of the differentiability requirement at the endpoints of the interval:

> *Theorem*: If a function Fx has a derivative from both directions for all values of x lying between a and $a + h$, which beyond this follows the law of continuity for the stated values of x; if further the function Fx also has continuity for both values $x = a$ and $x = a + h$ for the first at least in the same sense as h, [and] for the second at least in the opposite sense: then in this case too the equation of the preceding theorem still holds[249]:
>
> $$F(a + h) = Fa + h.F'(a + \mu h).$$

He neglects to mention here that μ lies strictly between 0 and 1 in the statement of the Theorem and in his proof says only that μ does not lie outside $[0, 1]$, i.e., that $0 \leq \mu \leq 1$. This would not necessarily follow from his proof, which attempts to reduce the result to the already established result on closed subintervals $[\alpha, \alpha + i]$ of $(a, a + h)$. His proof takes a limit of the derivative at an endpoint, which cannot in general be justified. A correct proof via reduction is, however, easily supplied and it places $a + \mu h$ squarely in the interior.

We begin by restating the result in modern language.

3.6.14 Theorem (Bolzano's Mean Value Conjecture) *Let $f : [a, b] \to \mathbb{R}$ be continuous and suppose f is continuously differentiable on (a, b). Then there is an element $c \in (a, b)$ such that $f(b) = f(a) + f'(c)(b - a)$.*

Geometrically, the idea of the proof is very simple. Consider the secant line

$$y = f(a) + \frac{f(b) - f(a)}{b - a}(x - a)$$

connecting the points $\langle a, f(a)\rangle$ and $\langle b, f(b)\rangle$. If the secant coincides with the curve, there is nothing to prove: $f'(x)$ equals $\big(f(b) - f(a)\big)/(b - a)$ for all $x \in [a, b]$ and every $x \in (a, b)$ can be taken to be the point c we are looking for.

If the secant does not coincide with the curve, there is some point $p \in (a, b)$ such that $\langle p, f(p)\rangle$ does not lie on the secant. Consider the line

$$y = f(p) + \frac{f(b) - f(a)}{b - a}(x - p)$$

---

[249] *Functionenlehre*, p. 114; Russ, *op. ci.*, p. 521.

parallel to the secant and passing through $\langle p, f(p)\rangle$. Either it has no other point of intersection with $y = f(x)$ and the line is tangent to the curve, whence

$$f'(p) = \frac{f(b) - f(a)}{b - a},$$

and p is the sought-after c, or the line passes through the curve at another point $\langle p^*, f(p^*)\rangle$, with $p^* \in (a, b)$. In this case apply the earlier result to f on the subinterval with p, p^* as endpoints: There is some $c \in [p, p^*]$ (if $p < p^*$) or $\in [p^*, p]$ (if $p^* < p$) such that

$$f'(c) = \frac{f(p^*) - f(p)}{p^* - p} = \frac{f(b) - f(a)}{b - a},$$

the last equation holding since parallel lines have equal slopes. But the closed interval $[p, p^*]$ or $[p^*, p]$ is contained in (a, b), whence $c \in (a, b)$.

A rigorous analytic proof merely expresses this algebraically.

Proof of Theorem 3.6.14. Let f, a, b be as stated and define

$$g(x) = f(a) + \frac{f(b) - f(a)}{b - a}(x - a).$$

If f and g coincide, then

$$f'(x) = g'(x) = \frac{f(b) - f(a)}{b - a}$$

for all $x \in [a, b]$ and c can be taken to be any of these.

If f and g do not coincide, there is some $p \in (a, b)$ such that $f(p) \neq g(p)$. Consider the function

$$h(x) = f(x) - g(x).$$

We have $h(a) = h(b) = 0$, $h(p) \neq 0$. Let $d = h(p)/2$. By the Intermediate Value Theorem, there are $p_1 \in (a, p)$, $p_2 \in (p, b)$ such that $h(p_1) = d = h(p_2)$. But h is evidently continuously differentiable on $[p_1, p_2]$ whence for some $c \in [p_1, p_2] \subseteq (a, b)$, one has

$$h'(c) = \frac{h(p_2) - h(p_1)}{p_2 - p_1} = \frac{d - d}{p_2 - p_1} = 0.$$

But

$$0 = h'(c) = f'(c) - g'(c) = f'(c) - \frac{f(b) - f(a)}{b - a} \Rightarrow f'(c) = \frac{f(b) - f(a)}{b - a}. \qquad \square$$

As we saw in the preceding sections, Lagrange was careless about inequalities and cannot be credited with proving c to lie strictly between a and b. Cauchy is also not very explicit about his claim and, in any event, his first proof doesn't really establish where c lies. Bolzano's Theorem 3.6.14 is quite explicit both about the position of c and the lack of necessity for the derivative to be defined at the endpoints of the interval. His overall proof of Theorem 3.6.14 is unfortunately incorrect on two counts. First, he left unproven the important lemma asserting continuous differentiability to imply uniform differentiability on closed, bounded intervals. This lemma is true, but the most obvious proof rests on a stronger version of the Mean Value Theorem first proven only years later. Second, despite the simplicity of the reduction of the Theorem to the case in which f' was differentiable at the endpoints, he failed to give a correct proof of this reduction. Hence we find Bolzano merely giving a variant of the proof of Cauchy's result for f uniformly differentiable on $[a, b]$ and merely *conjecturing* a more general result somewhat closer to the modern formulation.

Bolzano's work on the Mean Value Theorem was left unpublished until the 20th century and is thus not part of the story of its development. The obvious next question is: Who ultimately replaced him by rediscovering Lemma 3.6.10 and proving it, and who did the same for Theorem 3.6.14? Indeed, did anyone prove Theorem 3.6.14? Perhaps everyone assumed Cauchy had proven this because it agrees with his statement of the Mean Value Theorem, but nobody noticed Cauchy meant uniform differentiability when he wrote "differentiability". Bolzano saw the distinction but he had no impact and uniformity would only become widely recognised about the time the modern proof of the Mean Value Theorem was discovered some decades later.

3.7 More Textbooks but No Progress

The turn from the 19th to the 20th century saw the publication of a massive multivolume mathematical encyclopædia. The article on the Differential and Integral Calculi was penned by Aurel Voss (1845–1931).[250] The section thereof on the Mean Value Theorem is called "Der Mittelwertsatz nach Cauchy, Darboux und Weierstrass" ["The Mean Value Theorem after Cauchy, Darboux and Weierstrass"]. Voss begins by referring to the Mean Value Theorem as the "Fundamental Theorem of the Differential Calculus", thus testifying to the importance the result had assumed for the foundations of Analysis by the time the article was written. His historical remarks on the Mean Value Theorem are all too brief and but for some footnoted references to a couple of results of the Integral Calculus they do not include any work between Cauchy and the publication in 1868 by Joseph Alfred Serret (1819–1885) of Ossian Bonnet's (1819–1892) modern proof of the result. Following Serret's book

[250] Aurel Voss, "Differential- und Integralrechnung", in: H. Burkhardt, W. Wirtinger, and R. Fricke (eds.), *Encyklopädie der mathematischen Wissenschaften mit Einschluss ihrer Anwendungen. II. Analysis*, Verlag von B.G. Teubner, Leipzig, 1899–1916.

he cites a number of generalisations to functions of a complex variable, functions of several real variables, and higher derivatives, which he attributes to mathematicians like Darboux, Axel Harnack (1851–1888), Riemann, Hermann Amandus Schwarz (1843–1921), and Weierstrass. Most of these results lie beyond the scope of this book and I will not go into them here; an exception is the Linear Function Theorem of Schwarz, for which see page 345, below.

The decades that passed between Cauchy's and Serret's texts saw quite a number of mathematical publications and a fair amount of textbooks on the Calculus, many of which are available online and all of which a serious historian will have to read. The period also saw a gradual lessening of the French hegemony on Analysis as the Germans began their rise to mathematical power and as the British emerged from their self-imposed exile from the mainstream that had resulted from their insistence on staying true to Newton's fluxions. And other European nations were not idle either. Insofar as I imagine any published breakthrough[251] during the period would have been reported by Voss, I have decided not to play the serious historian but to take the easy way out and report only on a few select texts, first one by Augustus de Morgan (1806–1871).

English mathematics had lagged behind that on the continent as a result of the priority dispute over the invention of the Calculus. The British Royal Society's fallacious finding in a report written by Newton himself that Leibniz had plagiarised Newton resulted in the isolation of English analysts from the greater developments of the followers of Leibniz and the Bernoullis.[252] By the 1810s England was well behind and some undergraduates at Cambridge founded the Cambridge Analytical Society with the avowed goal of introducing Leibnizian Calculus into England. De Morgan was a beneficiary of their programme and carried it a bit further by incorporating some of the work of Cauchy into his expositions.

De Morgan was a prolific writer who wrote a number of textbooks. Most central to our current interest was *The Differential and Integral Calculus*, a work published in instalments between 1836 and 1842. In the *Dictionary of Scientific Biography* the book is described thus:

> In *The Differential and Integral Calculus* (1842) there is a good discussion of fundamental principles with a definition of the limit which is probably the first precise analytical formulation of Cauchy's somewhat intuitive concept. The same work contains a discussion of infinite series with an original rule to determine convergence precisely when simpler tests fail. De Morgan's rule, which is proved rigorously, is that if the series is given by
>
> $$\sum \frac{1}{\phi(n)},$$
>
> then if
>
> $$e = \lim_{n\to\infty} \frac{n\phi'(n)}{\phi(n)},$$

---

[251] As opposed to, e.g., the work of Bolzano.

[252] The classic account of this battle is given in: A. Rupert Hall, *Philosophers at War; The quarrel between Newton and Leibniz*, Cambridge University Press, Cambridge, 1980.

the series converges for $e > 1$ but diverges for $e \leq 1$.[253]

The convergence criterion is irrelevant to our present purposes, and the crediting to de Morgan of the first precise analytic formulation of continuity is a wild overstatement. De Morgan was influenced by Cauchy and fellow Cambridge scholar William Whewell who himself followed Newton in emphasisisng limits. Following a peliminary publication in 1835,[254] in 1836 de Morgan began publishing in 25 instalments his "most substantial educational volume" with the Society for the Diffusion of Useful Knowledge. Historian Ivor Grattan-Guinness summarises the importance of this work as follows:

> During these years Cauchy's approach gradually gained ascendency, to reach a position of dominance which (for better or worse) it has enjoyed ever since. De Morgan's book is a quite important source for this change in Britain, in what amounted to a second reform in the calculus; but it proceeded much more gradually than had the putsch of the Analytical Society.[255]

De Morgan announces his intention in the "Advertisement" prefaced to the work:

> The following Treatise will differ from most others, for better or worse, in several points. In the first place, it has been endeavoured to make the theory of *limits*, or *ultimate ratios*, by whichever name it may be called, the sole foundation of the science, without any aid whatsoever from the theory of series, or algebraical expansions. I am not aware that any work exists in which this has been avowedly attempted, and I have been the more encouraged to make the trial from observing that the objections to the theory of limits have usually been founded either upon the difficulty of the notion itself, or its *unalgebraical* character, and seldom or never upon anything not to be defined or not to be received in the conception of a limit, or not to be admitted in the usual consequences, when drawn independently of expansions, that is, of developments under assumed forms.[256]

This is a bit problematic. The latter remarks, which extend to further defend unalgebraical techniques, may well be an answer to Lagrange's attempt to use only algebraic manipulation and the expansion of functions into series. The opening line and his lack of awareness of any earlier work attempting to base the Calculus on limits would seem to imply that de Morgan hadn't read Cauchy — or, as Grattan-Guinness charitably suggests, de Morgan referred only to works in English.

Following a chapter of preliminaries, the real work of the book begins with Chap. II, "On the general theory of functional increments and differentiation". His style is not the now familiar Definition-Theorem-Proof presentation, but is far more informally explanatory. This can be attributed to his concern for pædagogical matters, as well as the fact that he was writing his treatise for the Society for the Diffusion

---

[253] John M. Dubbey, "De Morgan, Augustus", in: Charles Coulston Gillispie (ed.), *Dictionary of Scientific Biography*, vol. 4, Charles Scribner's Sons, New York, p. 35.

[254] Meant is a preliminary work: *The Elements of Algebra, Preliminary to the Differential Calculus and Fit for the Higher Classes of Schools*.

[255] Ivor Grattan-Guinness, "An eye for method: Augustus De Morgan and mathematical education", *Paradigm*, no. 9 (1992), pp. 1–7. I quote from an online single web page reproduction of the paper and can only say that this quote comes from near the end of the essay.

[256] Augustus de Morgan, *The Differential and Integral Calculus*, London, 1836–1842, p. 3.

of Useful Knowledge, which intended to put the knowledge out there for the working class. In any event, he devoted a page to explaining the notation ϕa when given an expression ϕx. There were two possible meanings. Usually, one would plug a into ϕx and simplify until a numerical value resulted. When this happened, the numerical value was ϕa. For example, if

$$\phi x = (1 - x)^{(1-x)},$$

and $a = 2$, one calculates

$$\phi 2 = (1 - 2)^{(1-2)} = (-1)^{-1} = \frac{1}{-1} = -1.$$

But it could also happen that ϕa resulted in an indeterminate form. For example,

$$\phi 1 = (1 - 1)^{(1-1)} = 0^0.$$

In this case, the value of ϕa would be determined, if possible, by the values ϕb for b near a, as the limit of these values:

$$\phi 1 = \lim_{b \to 1} \phi b = \lim_{b \to 1} (1 - b)^{(1-b)}.$$

This is not the best example he could have chosen, but today with our graphing calculators it is easy for any beginner to see that

$$\lim_{b \to 1^-} \phi b = 1, \quad \lim_{b \to 1^+} \phi b \text{ does not exist.}$$

He ignores the right-hand fact and concludes

> If we can prove, as we may hereafter do, that the preceding function also approaches without limit to 1 when x approaches without limit to 1, we may then abbreviate the preceding proposition into these words "when x is 1, $(1 - x)^{1-x}$ is also 1:" but we use the preceding sentence in no other signification. Therefore we have the following definition.
> DEFINITION.—The function is said to have the value A when x has the value a, either when the common arithmetical sense of these phrases applies, or when by making x sufficiently near to a, we can make the function as near as we please to A. In the first case A is simply called a value, or an ordinary value, of the function: in the second case A is called a *singular* value.
> *Postulate* 1.—If ϕa be an ordinary value of ϕx, then h can always be taken so small that no singular value shall lie between ϕa and $\phi(a + h)$, that is, no singular value shall correspond to any value of x between $x = a$ and $x = a + h$.
> The truth of this postulate is a matter of observation. We always find singular values separated by an infinite number of ordinary values.[257]

Today we do not consider Postulate 1 as a postulate, but as a restriction on the class of functions to be considered. It is a much milder restriction than Lagrange's

---

[257] *Ibid.*, pp. 44–45.

assumption of analyticity. It is true of the so-called *elementary functions* — those generated from $a, x, a+x, a-x, ax, a/x, e^x, \ln x, \sin x$, and $\cos x$ by composition, where a denotes an arbitrary constant. In general, however, the Postulate loses its meaning as it rests on the notion of *ordinary* value. The elementary functions come with rules of computation attached to them, but general functions do not. There is no explanation of what an ordinary value is for functions in general.

3.7.1 Exercise Explain why Postulate 1 holds for $f(x) = \sqrt{x}$ at $x = 0$.

> *Postulate* 2.—If ϕa be any finite value of ϕx, it is always possible to take h so small, that $\phi(a+h)$ shall be as near to ϕa as we please, and that ϕx shall remain finite from $x = a$ to $x = a + h$, and always lie between ϕa and $\phi(a+h)$ in magnitude.
> This again is a part of our experience of algebraical functions. It is generally assumed under the name of the *law of continuity*. The latter part of the postulate may be true of the whole extent of some functions: thus, however great h may be, x^2 perpetually increases between a^2 and $(a+h)^2$.[258]

What can I say? Our experience is already greater than his. We have $f(x) = x\sin(1/x)$ which oscillates infinitely often near $x = 0$ where $f(0) = 0$ is a singular value. And Bolzano had already discovered, albeit not published, an example of a continuous function for which this happened at all points a. Of course, Bolzano also noted that not everyone accepted as broad a definition of function as he did. De Morgan was clearly one of these. Perhaps if de Morgan had added the requirement that ϕa be an ordinary value his postulate would still have been true of the elementary functions. He continues his commentary on this postulate, perhaps addressing this new problem, but immediately moving on to other matters.

> It is possible to imagine a function which does not observe this law, but we cannot, without further consideration of singular values, find the means of expressing it algebraically. For instance, in the following figure [Fig. 3.18], the function represented by ABCDEF is *dis*continuous at B and D. But we have no means of expressing such a function in common algebra. We may call the law expressed in this postulate the law of *continuity of value*, to distinguish it from that of the next postulate; and we may say that functions, which do not obey this law, if any, are discontinuous in *value*.[259]

De Morgan's comments on this Postulate stop here; he offers no explanation of the bar the Postulate raises to unlimited local oscillation, which was not an unknown phenomenon. Cauchy had cited the functions

$$f(i) = i^3 \sin\frac{1}{i} \quad \text{and} \quad f(i) = i \sin\frac{1}{i},$$

which oscillate wildly around $i = 0$, in the "Addition" to the *Résumé*. And Bolzano, admittedly unpublished, was aware of a similar example. Moreover, in 1829, Dirichlet (1805–1859) had published his famous example of an everywhere discontinuous

---

[258] *Ibid.*, p. 45.

[259] *Ibid.*, pp. 45–46.

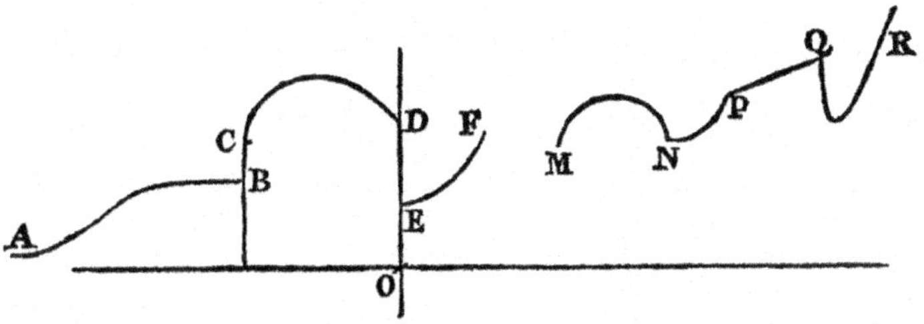

Fig. 3.18 De Morgan's illustration

function now called the *Dirichlet function* or even the *Dirichlet monster*[260]:

$$D(x) = \begin{cases} c, & x \text{ rational} \\ d, & x \text{ irrational}, \end{cases}$$

where $c \neq d$.[261] These are not the sorts of examples one easily forgets, and de Morgan's book owes enough to Cauchy that one assumes he has at least read this author and would be aware of such oscillation.

3.7.2 Exercise Let

$$f(x) = \begin{cases} x^3 \sin \dfrac{1}{x}, & x \neq 0 \\ 0, & x = 0. \end{cases}$$

i. Show that f is continuously differentiable on $[-1, 1]$.
ii. Show that f' is not differentiable at 0.
iii. Show that f satisfies de Morgan's Postulate 1, but, if he allows $f(0)$ to be a singular value, f does not satisfy Postulate 2.

> *Postulate* 3.—If any function follow one law for every value of x between $x = a$ and $x = a + h$, however small h may be, it follows the same law throughout: that is, the curves of no two algebraical functions can entirely coincide with each other, for any arc, however small. If ϕx be x^2 for every value of x between a and $a + h$, however small h may be, it is x^2 for every other value of x. This we may call the law of *continuity of form*, or *permanence* of form.
> Exceptions to this law may be represented, but cannot yet be algebraically formed. As in MNPQR, we may conceive a function which is represented by an arc of a circle joined to one of a parabola, which itself is joined to a part of a straight line, and so on. [Cf. Fig. 3.18.] Such a

[260] Functions which did not behave like the familiar old functions were often called "monsters". Poincaré referred to them as "teratological". Today one may still see them described as "pathological" or referred to as "counterexamples". Most functions, however, are not at all like the familiar functions encountered in the Calculus or pre-Calculus courses and "pathological monsters" are the rule, not the exception.

[261] Lejeune-Dirichlet, "Sur la convergence des séries trigonométriques servent à représenter une fonction arbitraire entre des limites-données", *Journal für die reine und angewandte Mathematik* 4 (1826), pp. 157–169; here: p. 169.

function would be called *discontinuous in form*, and though not now exhibited algebraically, may actually occur in practice.[262]

Postulate 3 is certainly strange to the modern eye and is susceptible of several interpretations. The most straightforward, suggested by his explanatory remark, is that definition by cases is not allowed. He is aware that one can define functions by cases, but wishes to rule them out. In a beginning course on the Calculus, one usually considers functions given by one defining formula on an interval. A function spliced together by successive defining formulæ on successive intervals is generally broken into pieces and each piece treated separately.

Less likely, he is preparing the way for an easy evaluation of certain limits. For example, if $f(x) = x^2$ and one considers

$$g(x) = \frac{f(x) - f(a)}{x - a} = \frac{x^2 - a^2}{x - a},$$

then $g(a) = 0/0$ is of an indeterminate form and one must take

$$g(a) = \lim_{x \to a} \frac{x^2 - a^2}{x - a}.$$

But, for $x \neq a$,

$$\frac{x^2 - a^2}{x - a} = \frac{(x + a)(x - a)}{x - a} = x + a.$$

By Postulate 3, since $g(x) = x + a$ on any interval not containing a, the two functions $g(x)$ and $h(x) = x + a$ must agree everywhere. Thus

$$g(a) = a + a = 2a.$$

But $g(a) = \lim_{x \to a}(x^2 - a^2)/(x - a)$, whence

$$\lim_{x \to a} \frac{x^2 - a^2}{x - a} = 2a.$$

Postulate 3 also touches on a key feature of *analytic* functions. An analytic function was initially any function expressed by an analytic formula and it was believed that if two such expressions defined the same function on one domain, they did so wherever they were both defined. And, indeed, this was true when the mode of expression was limited, but is not true for more general classes of functions. It is true, however, for those functions we refer to today as analytic functions, namely, those infinitely differentiable functions which equal their Taylor series locally around all points in their domains. This result is usually proven in a first year course in Complex Analysis and relies on a number of results about analytic functions in the complex domain.

---

[262] De Morgan, *op. cit.*, p. 46.

In Real Analysis, the result may be proven by simple appeals to the Mean Value Theorem and the Least Upper Bound Principle and, although the proof doesn't fit into our present discussion of de Morgan's book, it is relevant to the general discussion of my book and I digress to present it here.[263]

3.7.3 Definition A function $f : I \to \mathbb{R}$ defined on an open interval I is *real analytic* on I if for each $a \in I$ there is $\delta > 0$ such that $(a - \delta, a + \delta) \subseteq I$ and there are $c_0, c_1, c_2, \ldots$ such that for all $x \in (a - \delta, a + \delta)$

$$f(x) = \sum_{k=0}^{\infty} c_k (x - a)^k. \tag{3.58}$$

In other words, f is a real analytic function if f expands into a power series in some neighbourhood around any point in its domain.

We assume known that real analytic functions are continuous, differentiable, that the series (3.58) can be differentiated term-by-term,

$$f'(x) = \sum_{k=0}^{\infty} kc_k (x - a)^{k-1} = \sum_{k=1}^{\infty} kc_k (x - a)^{k-1},$$

and that

$$c_k = \frac{f^{(k)}(a)}{k!}.$$

3.7.4 Definition A point a is a *limit point* of a set X if for every $\epsilon > 0$ there is a point $b \in X$ with $b \neq a$ such that $|b - a| < \epsilon$.

A limit point of a set is like the limit point of a sequence except that, with sequences $a_0, a_1, a_2, \ldots$ there is no requirement that $a_k \neq a$. If a sequence $a_0, a_1, a_2, \ldots$ has no repetitions, then a is a limit point of the sequence just in case a is a limit point of the set $\{a_0, a_1, a_2, \ldots\}$. Conversely, if a is a limit point of a set X, one can define a sequence $a_0, a_1, a_2, \ldots$ of distinct elements of X such that (a is not only a limit point of the sequence, but, in fact) $a = \lim_{k\to\infty} a_k$. For, choose $a_0 \in X$ such that $a_0 \in X$ such that $a_0 \neq a$ and $|a_0 - a| < 1$. Given $a_0, a_1, a_2, \ldots, a_k$, choose $a_{k+1} \in X$ such that $a_{k+1} \neq a$ and

$$|a_{k+1} - a| < \min\left\{\frac{1}{k+2}, |a_0 - a|, |a_1 - a|, \ldots, |a_k - a|\right\}.$$

The elements of the sequence of such a_k's are distinct because each successive a_{k+1} is strictly closer to a than all the preceding members of the sequence. And the limit is a because

[263] Partially: here I assume the basics of the Theory of Taylor series and present only the proof of the unicity result. These basic results, which i never intended to prove in the present book, have wormed their way in below in Sect. 3.12.3.

$$|a_k - a| < \frac{1}{k+1} < \epsilon$$

for any $k > 1/\epsilon$.

3.7.5 Lemma *Let f be real analytic on an open interval I and suppose the set of zeroes of f, $Z = \{z \in I \,|\, f(z) = 0\}$ has a limit point in I. Then f is identically 0 on I.*

Proof. We first show that f is identically 0 in some neighbourhood of any limit point of Z. To this end, let $a \in I$ be a limit point of Z. Because $a \in I$ we can write

$$f(x) = \sum_{k=0}^{\infty} c_k(x-a)^k$$

for all x in some interval $(a - \delta, a + \delta) \subseteq I$.

Let $a_0, a_1, a_2, \ldots \in Z$ have a as its limit and assume as above that $|a_k - a| < 1/(k+1)$. By the continuity of f, $f(a) = \lim_{k\to\infty} f(a_k) = \lim_{k\to\infty} 0 = 0$. Thus $c_0 = f(a) = 0$.

We now apply the Mean Value Theorem infinitely many times to find $b_0, b_1, b_2, \ldots$, each b_k lying between a_k and a_{k+1}, such that

$$f'(b_k) = \frac{f(a_{k+1}) - f(a_k)}{a_{k+1} - a_k} = \frac{0-0}{a_{k+1} - a_k} = 0.$$

But

$$\begin{aligned}|b_k - a| &= |b_k - a_k + a_k - a| \le |b_k - a_k| + |a_k - a| \\ &\le |a_{k+1} - a_k| + |a_k - a| = |a_{k+1} - a + a - a_k| + |a_k - a| \\ &\le |a_{k+1} - a| + |a - a_k| + |a_k - a| < \frac{1}{k+2} + \frac{1}{k+1} + \frac{1}{k+1} \\ &< \frac{3}{k+1} < \epsilon\end{aligned}$$

for $k > 3/\epsilon$. Thus $b_k \to a$ and f' satisfies the same conditions as f. Thus $c_1 = f'(a) = 0$.

Likewise $c_2 = c_3 = c_4 = \ldots = 0$. Thus, in $(a - \delta, a + \delta)$,

$$f(x) = \sum_{k=0}^{\infty} 0(x-a)^k = 0.$$

Thus, if a is a limit point of Z, f is identically 0 in some neighbourhood of a.

To see that f is identically 0 in I, let $c \in I$ be such that $f(c) \neq 0$. Consider, without loss of generality, the case $c > a$, where a is a limit point of Z in I. Let

$$X = \{x \in I \,|\, \forall y \in I (a \le y < x \Rightarrow f(y) = 0)\}.$$

X is bounded above by c, whence it has a least upper bound, say, b. But $(a, b) \subseteq Z$ and b is a limit point of (a, b), whence it is a limit point of Z and f is identically 0 on some interval $(b - \delta_1, b + \delta_1) \subseteq I$ with $\delta_1 > 0$ This puts $b + \delta_1$ into X, contrary to b's being an upper bound on X. □

3.7.6 Theorem *Let $f, g : I \to \mathbb{R}$ be real analytic functions on an open interval I. Suppose $f(x) = g(x)$ on some nonempty open subinterval $J \subseteq I$. Then f is identically equal to g on all of I.*

Proof. Apply the Lemma to $h(x) = f(x) - g(x)$: Every point of $J \subseteq I$ is a limit point of Z. □

De Morgan's postulates are true of the real analytic functions, but these functions do not suffice for the Calculus and its applications. Indeed, the so-called elementary functions studied in the standard course in the Calculus do not satisfy all of his postulates. Unbeknownst to anybody, Bolzano had already produced an everywhere oscillating function violating Postulate 2. And Postulate 3 not only fails for functions like that of de Morgan's curve MNPQR, but for functions given by expressions of analysis — i.e., as Cauchy first showed in a paper of 1844,[264] the very notion of discontinuity of form made little sense:

$$|x| = \begin{cases} x, & x \geq 0 \\ -x, & x < 0 \end{cases}$$

is ostensibly discontinuous in form, but

$$|x| = \sqrt{x^2} = \frac{2}{\pi} \int_0^\infty \frac{x^2}{t^2 + x^2} dt$$

is clearly continuous in form. Indeed, Alfred Pringsheim (1850–1941) would later observe that the Dirichlet monster, which would seem to be everywhere discontinuous in form, is also continuous in form[265]:

$$D(x) = (c - d) \lim_{n \to \infty} \lim_{m \to \infty} (\cos n!\pi x)^m + d.$$

When statements like these postulates are unclear, mathematics affords us another method of understanding them: we can look to see how they are used. De Morgan does this by immediately presenting his version of what I suppose is Ampère's argument

---

[264] Augustin Louis Cauchy, "Mémoire sur les fonctions continues", *Compte Rendus* 18 (1844), pp. 116–130; here: pp. 116–117.

[265] Alfred Pringsheim, "Grundlagen der allgemeinen Funktionenlehre", in: H. Burkhardt, W. Wirtinger, and R. Fricke (eds.), *Encyklopädie der mathematischen Wissenschaften mit Einschluss ihrer Anwendungen. II. Analysis*, Verlag von B.G. Teubner, Leipzig, 1899–1916; here: p. 7. The parentheses around the cosine are missing in the original. Peano had given a similar but slightly more complicated formula earlier in 1884—cf. Kennedy, *Works of Peano*, *op. cit.*, p. 44.

that all functions are differentiable except at isolated points.[266] His proof, which I find unconvincing, begins with the words,

> Let ϕx be a function, continuous in form and value, which we always mean unless when the contrary is expressed.[267]

Thus, he is saying in effect, "Let ϕx be a function for which Postulates 2 and 3 hold". By the end of his proof he also invokes Postulate 1. Following the details of the proof, which I will not repeat here, he concludes

> ϕx being any function of x, and a and $a+h$ any consecutive values of x, where h may be given as small as we please, there must be finite limits to the fraction $\dfrac{\phi(x+\theta)-\phi x}{\theta}$, in which θ diminishes without limit, for some values of x between $x=a$ and $x=a+h$. The limit of $\dfrac{\phi(x+\theta)-\phi x}{\theta}$ is called the *differential coefficient* of ϕx with respect to x, and the theorem just proved is as follows:—Every function either has a finite differential coefficient when x has the specific value a, or when it has a value $a+k$ where k may be as small as we please.[268]

Personally, I find this a bit hard to read. I think the last statement says that either $f'(c)$ exists, or, for all sufficiently small $k \neq 0, f'(x)$ exists — which means the points a where $f'(a)$ fails to exist are isolated from each other. We could try to verify this by reading the proof to see what is established, but the proof must be incorrect as Bolzano's function shows. De Morgan himself acknowledges that his proof may not be convincing:

> There are points in the preceding demonstration which lie open to certain objections, depending upon the way in which the terms of the postulates are understood. The student may, if he pleases, consider it only as giving a very high degree of probability to the fact stated, since we shall presently demonstrate of all classes of functions separately, that the preceding fraction has a finite limit for all values of x, with the exception of a limited and assignable number of values for each function.[269]

The rest of the chapter presents the usual rules for differentiating elementary functions. This is done in intuitive, none-too-rigorous fashion that makes me wonder if Dubbey were not describing some other work in his comment from the *Dictionary of Scientific Biography* cited above. De Morgan does, however, end the chapter on a note promising something approaching rigour later:

> We are now to proceed to the application of this calculus to algebra. We must call the attention of the student to the fact that we have not assumed any algebraical development into an infinite series, directly or indirectly. He may therefore dismiss from his mind entirely (until further proof shall be offered) all such developments and their consequences. The assumption

---

[266]Recall my refusal to read any more of Ampère than I had to: The fact that there is disagreement among scholars who do know French of whether or not he attempted to prove such a result convinces me of the futility of my looking into the matter. Whatever Ampère attempted to prove, it is clear that de Morgan was trying to prove all functions to be differentiable — somewhere.

[267]De Morgan, *op. cit.*, p. 46.

[268]*Ibid.*, pp. 47–48.

[269]*Ibid.*, p. 48.

which is usually made in algebraical works for the establishment of such developments, is that certain functions of x, $(a+x)^{\frac{m}{n}}$ for example, can be expanded in a series of whole powers of x of the form

$$\mathrm{A} + \mathrm{B}x + \mathrm{C}x^2 + \mathrm{E}x^3 + \&\mathrm{c}.$$

where A, B, C, &c. are not functions of x. Of this no legitimate proof was ever given depending entirely on algebra. Nor is the assumption universally true. That we may make use of infinite series, we shall find; but it should be matter of proof, not of assumption. By rejecting infinite series we are unable as yet to complete the differentiation of a^x. We have only found it to be $ca^x \log a$, and *have assumed* that c is 1 when $\log a$ is the Naperian logarithm. This assumption, which is excusable while we are only inquiring into what will be its consequences if it be true, must be abandoned in all applications until we can produce a proof of it.[270]

Up to this point the greatest debt owed to Cauchy by de Morgan was his uncited application of Cauchy's Theorem XII[271] in his Ampèrean proof of the existence of derivatives. The next chapter, "On algebraical development", though still not mentioning Cauchy relies more heavily on the *Résume*. First, however, is an unattributed reference[272] to the Lagrange property:

Assuming $u = \phi x$, we have shown how to find another function $\phi' x$, which has the property, that $\dfrac{\phi(x+\Delta x) - \phi x}{\Delta x}$ may be made as near as we please to $\phi' x$, by taking Δx sufficiently small. Let the first of these differ from the second by P, which is therefore a function of x and Δx, having this property, that whatever x may be, it diminishes without limit with Δx. There may be special exceptions in each particular function. For instance, if $u = \log(x-a)$, $\dfrac{du}{dx} = \dfrac{1}{x-a}$, which is finite for every value of x except only $x = a$. These cases, observe, we except for the present; that they must be finite in number, or, if infinite in number, belonging only to a particular class of values, separated by intervals in which no such thing takes place, appears as follows. The only cases in which we can conceive them to happen, are those in which such a value is first assigned to x as makes a numerator or a denominator, or an exponent, one or any of them, nothing or infinite. Now, in all known functions, the values of x which satisfy such a condition are separated by intervals of *finitude*, and there is no function which is nothing or infinite for every value of x between a and $a+b$ (for any value of b however small) in all the functions of algebra. If there be such, we have notified in the postulates at the head of Chap. II. that they do not form a part of what we have called the Differential and Integral Calculus, but their consideration forms a science by itself. This condition is expressed or implied in every treatise on the subject.[273]

De Morgan now turns to a proof of the Mean Value Theorem. That such is the first result of the Chapter testifies once more to the importance the result had assumed for the Differential Calculus. His proof follows familiar lines, but is not quite the same in some details as any of the proofs offered by Lagrange, Ampère, or Cauchy. Moreover, the conditions on ϕ guaranteeing the validity of his proof are not explicitly

---

[270] *Ibid.*, p. 65.

[271] Cf. p. 274, above.

[272] I emphasise the lack of attribution because de Morgan himself criticised the French for their failure to cite their sources. Two possible justifications for his own failure here are: i. he was laying no claim to originality in the work as regards to results, and ii. the names of Lagrange, Ampère, and Cauchy would have meant nothing to his intended audience.

[273] De Morgan, *op. cit.*, pp. 65–66.

stated, as they are by the three Frenchmen. It is safe to say, however, that he has in mind the elementary functions one encounters in the Calculus course. For it is these for which he has proven via exhibiting the rules of differentiation that the derivatives exist and are also elementary.

Aside from his not carefully stating his assumptions and some careless proofreading, he lays the proof out rather nicely. He begins spelling out one assumption:

> Let there be two limits a and $a+h$, such that neither for them nor between them, are there any singular values of ϕx.[274]

He also assumes $\phi' x$ to have no singular values, an assumption he "reminds" us of later. Following the example of $\log x$ having no singular values at or between 2 and 3, he gets down to business:

> We now have P,[275] a *comminuent*[276] with Δx, whatever the value of x may be, between a and $a+h$. Consequently, P and Δx will still remain comminuent, even though, while Δx diminishes, x should vary in any manner between a and $a+h$. Thus, for instance, Δx and $x\Delta x$ are comminuents, even though, while Δx diminishes without limit, x increase from a to $a+h$. Let us suppose Δx to be the nth part of h, so that Δx diminishes without limit as n increases without limit. Let P, which is a function of x and Δx, be denoted by $f(x, \Delta x)$, and we then have
> $$\frac{\phi(x+\Delta x)-\phi x}{\Delta x}=\phi' x+f(x,\Delta x);$$

I interrupt him mid-sentence to point out that f is clearly the difference of two elementary functions of two variables, hence an elementary function of two variables. For each fixed x it is a continuous function of Δx and for each $\Delta x \neq 0$ it is continuous as a function of x. He is later going to assume that f is a continuous function of both variables, which is not clear considering that f has each $\langle x, 0\rangle$ giving a singular value. The uniform differentiability of ϕ guarantees this simultaneous continuity in both variables.

De Morgan continues, setting up a list of fractions to which to apply Cauchy's Theorem XII:

> now substitute successively $x+\Delta x$ for x until we come to have $\phi(x+n\Delta x)$ or $\phi(x+h)$ in the numerator, which will give the following set of equations (n in number):—

---

[274] *Ibid.*, p. 66.

[275] The difference between $\dfrac{\phi(x+h)-\phi x}{h}$ and $\phi' x$.

[276] De Morgan inserts the explanatory footnote:

> To avoid the tedious repetition of "a quantity which diminishes without limit when Δx diminishes without limit", I have coined this word. If ever the constant recurrence of a long phrase justified a new word, here is a case. There are sufficient analogies for the derivation, or at any rate we must not want words because Cicero did not know the Differential Calculus. Hence we add to our dictionary as follows:—To *comminute* two quantities, is to suppose them to diminish without limit together: *comminution*, the corresponding substantive; *comminuents*, quantities which diminish without limit together. To *comminute* has been used in the sense of to *pulverize*, and is therefore recognised English.

The terminology did not catch on, at least according to my spell-checker.

$$\frac{\phi(x+\Delta x)-\phi x}{\Delta x}=\phi' x+f(x,\Delta x)$$
$$\frac{\phi(x+2\Delta x)-\phi(x+\Delta x)}{\Delta x}=\phi'(x+\Delta x)+f(x+\Delta x,\Delta x)$$
$$\frac{\phi(x+3\Delta x)-\phi(x+2\Delta x)}{\Delta x}=\phi'(x+2\Delta x)+f(x+2\Delta x,\Delta x)$$
$$\cdot\quad\cdot\quad\cdot\quad\cdot\quad\cdot\quad\cdot\quad\cdot\quad\cdot\quad\cdot$$
$$\cdot\quad\cdot\quad\cdot\quad\cdot\quad\cdot\quad\cdot\quad\cdot\quad\cdot\quad\cdot$$
$$\frac{\phi(x+\overline{n-1}\Delta x)-\phi(x+\overline{n-2}\Delta x)}{\Delta x}=\phi'(x+\overline{n-2}\Delta x)+f(x+\overline{n-2}\Delta x,\Delta x)$$
$$\frac{\phi(x+\bar{n}\Delta x)-\phi(x+\overline{n-1}\Delta x)}{\Delta x}=\phi'(x+\overline{n-1}\Delta x)+f(x+\overline{n-1}\Delta x,\Delta x).$$

Form the fraction which has the sum of the numerators of the preceding for its numerator, and the sum of the denominators for its denominator, it being clear that all the denominators have the same sign. This gives

$$\frac{\phi(x+\Delta x)-\phi x+\phi(x+2\Delta x)-\phi(x+\Delta x)+..+\phi(x+n\Delta x)-\phi(x+\overline{n-1}\Delta x)}{n\Delta x}$$

or $\dfrac{\phi(x+n\Delta x)-\phi x}{n\Delta x}$ or $\dfrac{\phi(x+h)-\phi x}{h}$,
which must therefore lie between the greatest and least of the preceding fractions,[277] or of their equivalents, all contained under the formula

$$\phi'(x+k\Delta x)+f(x+k\Delta x,\Delta x).$$

Now let the first value of x be a, and let C and c be the values of x which give $\phi' x$ the greatest and least possible values it can have between $x=a$ and $x=a+h$. (We have supposed that $\phi' x$ does not become infinite between these limits.) And let C′ and K′ be the values of x and k which give $f(x+k\Delta x,\Delta x)$ the greatest value it can have between the limits, and c' and k' those which give it the least. Then still more do we know that[278]

$$\frac{\phi(a+h)-\phi a}{h}\text{ lies between }\phi\mathrm{C}+f(\mathrm{C}'+\mathrm{K}'\Delta x,\Delta x)$$
$$\text{and }\phi c+f(c'+k'\Delta x,\Delta x),$$

in which the two functions marked f are, as we have shown, comminuents with Δx. Now, if a quantity always lie between two others, it must lie between their limits: for if not, let it be ever so little greater than the greater limit, then we can bring the greater quantity nearer to that limit than the one we have supposed to be always intermediate. Or, in illustration, suppose P and Q to be moving points which perpetually approach the limits A and B: if X (a fixed point) must always lie between the two, P and Q, it must lie between A and B; for if not, let it be at X, then by the notion of a limit, Q may be brought nearer to B than X, or X does not always lie between A and B; which is a contradiction. The limits of the preceding, when n

277 He is here invoking Cauchy's Theorem XII.

278 Typo! The terms ϕC and ϕc on the right should be ϕ'C and $\phi' c$, respectively.

Fig. 3.19 De Morgan's illustration

increases or Δx diminishes, are ϕC and ϕc[279]: whence we have the following THEOREM:— If ϕx be a function which is finite and without singular values from $x = a$ to $x = a + h$ inclusive, and if the differential coefficient be the same, and if C and c be the values of x which make $\phi' x$ greatest and least between these limits, then it follows that[280]

$$\frac{\phi(a+h) - \phi a}{h} \text{ lies between } \phi\text{C and } \phi c.$$

COROLLARY.—Since, by the law of continuity of value, a function does not pass from its greatest to its least without passing through *every* intermediate value, and since $\dfrac{\phi(a+h) - \phi a}{h}$ is *an* intermediate value of ϕx[281] between ϕC[282] and ϕc[283] and since $a + \theta h$ where θ lies between 0 and 1, is, by properly assuming θ, a representative of any value which falls between a and $a + h$, and consequently between C and c, it follows that

$$\frac{\phi(a+h) - \phi a}{h} = \phi'(a + \theta h)$$

is true for some positive value of θ less than unity[.][284]

De Morgan follows this with a numerical example and then a proof of the Cauchy Mean Value Theorem. This is the proof alluded to but not given in the "Addition" to the *Résume* (Fig. 3.19):

Let there now be two functions ϕx and ψx, the second of which has the property of always increasing or always decreasing, from $x = a$ to $x = a + h$, in other respects fulfilling the conditions of continuity in the same manner as ϕx.

Let $$\frac{\psi(x + \Delta x) - \psi x}{\Delta x} = \psi' x + f_1(x, \Delta x),$$
whence $f_1(x, \Delta x)$ is comminuent with Δx. We have then, as before, a series of equations of the form

$$\frac{\dfrac{\phi(x + k\Delta x) - \phi(x + \overline{k-1}\Delta x)}{\Delta x}}{\dfrac{\psi(x + k\Delta x) - \psi(x + \overline{k-1}\Delta x)}{\Delta x}} = \frac{\phi'(x + \overline{k-1}\Delta x) + f(x + \overline{k-1}\Delta x, \Delta x)}{\psi'(x + \overline{k-1}\Delta x) + f_1(x + \overline{k-1}\Delta x, \Delta x)}$$

or

$$\frac{\phi(x + k\Delta x) - \phi(x + \overline{k-1}\Delta x)}{\psi(x + k\Delta x) - \psi(x + \overline{k-1}\Delta x)} = \frac{\phi'(x + \overline{k-1}\Delta x) + f(x + \overline{k-1}\Delta x, \Delta x)}{\psi'(x + \overline{k-1}\Delta x) + f_1(x + \overline{k-1}\Delta x, \Delta x)},$$

from which, by summing the numerators and denominators of the first sides, which gives $\dfrac{\phi(a+h) - \phi a}{\psi(a+h) - \psi a}$ if the first value of x be a, and if $n\Delta x = h$; by observing that the denominators

[279] Again, replace ϕ by ϕ'.

[280] Again, replace ϕC and ϕc by ϕ'C and $\phi' c$.

[281] Again, replace ϕ by ϕ'.

[282] Again, replace ϕ by ϕ'.

[283] Again, replace ϕ by ϕ'.

[284] De Morgan, *op. cit.*, pp. 66–67.

are all of one sign by the supposition either of continual increase or decrease in ψx from $x = a$ to $x = a + h$; we find the preceding fraction to lie between the greatest and least values of the fractions on the second side of the set, and therefore (using the preceding reasoning) between

$$\frac{\phi' C}{\psi' C} \text{ and } \frac{\phi' c}{\psi' c} \text{ the greatest and least values of } \frac{\phi' x}{\psi' x},$$

from $x = a$ to $x = a + h$. And this must as before correspond to some value of $\frac{\phi' x}{\psi' x}$ for a value of x lying between $x = a$ and $x = a + h$. Let it be $x = a + \theta h$ as before, and we have the following THEOREM:—
If ϕx and ψx be continuous in value from $x = a$ to $x = a + h$, and if in addition $\phi' x$ and $\psi' x$ be the same, and if also ψx always increases or always decreases from $x = a$ to $x = a + h$, then

$$\frac{\phi(a+h) - \phi a}{\psi(a+h) - \psi a} = \frac{\phi'(a+\theta h)}{\psi'(a+\theta h)} \quad \theta < 1.$$

COROLLARY.—If the two functions be such that $\phi a = 0$ and $\psi a = 0$ without any discontinuity or singularity of value, we then have[285]

$$\frac{\phi(a+h)}{\psi(a+h)} = \frac{\phi'(a+\theta h)}{\psi'(a+\theta h)} \quad \theta < 1.$$

Given that the first ratio is singular at $h = 0$, the Corollary is a version of L'Hôpital's Rule:

$$\frac{\phi a}{\psi a} = \lim_{h \to 0} \frac{\phi(a+h)}{\psi(a+h)} = \lim_{\theta h \to 0} \frac{\phi'(a+\theta h)}{\psi'(a+\theta h)} = \frac{\phi' a}{\psi' a}.$$

De Morgan next proceeds to discuss higher order instances of L'Hôpital's Rule and finally comes to Taylor's Theorem

If we were at once to proceed with the consequences of this theorem, the student would not be well able to see why so apparently cumbrous an apparatus of proof is necessary to obtain what is called Taylor's Theorem: we shall therefore make what is often given as a proof precede what we consider as really a proof.
THEOREM. If it be allowable to suppose that $\phi(x + h)$ can be expanded in a series of whole powers of h, of the form

$$f^0 \text{ of } x + \begin{pmatrix}\text{another}\\ f^0.\text{ of } x\end{pmatrix} \times h + \begin{pmatrix}\text{a third}\\ f^0.\text{ of } x\end{pmatrix} \times h^2 + \begin{pmatrix}\text{a fourth}\\ f^0.\text{ of } x\end{pmatrix} \times h^3 + \&\text{c.}$$

then that series must be the following, and no other[286]:

$$\phi x + \phi' x.h + \phi'' x.\frac{h^2}{2} + \phi''' x \frac{h^2}{2.3} + \phi^{\text{iv}} x \frac{h^4}{2.3.4} + \&\text{c.}$$

[285] *Ibid.*, pp. 68–69. In this passage in the original the Greek letters are inconsistently typeset in both slanted and upright positions; I have taken the liberty of unifying these.
[286] *Ibid.*, pp. 69–70.

De Morgan now proceeds to give the heuristic argument, pointing out the "doubtful assumption[s]", first that the series can be differentiated term-by-term and second that when $h = 0$ the series is reduced to its first term. In a footnote he adds, "Observe that we do not say these assumptions are *untrue*, but not self-evident, and therefore not to be assumed without proof".[287] In fact, both assumptions are true, the latter by the definition of convergence, and the former by deeper reasoning. That a power series can be differentiated term-by-term had long been accepted as self-evident before Cauchy finally proved it to be the case. Thus, the proof given by De Morgan is not really incorrect; it just contains a gap. And it doesn't prove Taylor's Theorem that $\phi(a+h)$ can be expanded into an infinite series, merely that, *if* it can, the series must have the form exhibited. De Morgan follows his presentation of the proof with the remark that, "We shall treat the preceding process as nothing more than rendering it highly probable that $\phi(a+h)$ and $\phi a + \phi' a.h + \phi'' a\frac{h^2}{2}$ + &c. have relations which are worth inquiring into".[288]

He has at least demonstrated that Taylor's Theorem needs a stronger argument and he now proceeds to apply the Higher Order L'Hôpital's Rule to conclude Taylor's Theorem with the Lagrange Form for the Remainder, i.e., the Higher Order Mean Value Theorem. Following this, before considering any examples, he offers a summary:

> ...we have
>
> $$f(a+h) = fa + f'ah + f''a\frac{h^2}{2} + \ldots + f^{(n)}a\frac{h^n}{2.3\ldots n} + f^{(n+1)}(a+\theta h)\frac{h^{n+1}}{2.3\ldots n+1},$$
>
> subject only to the condition that no one of the set $fa, f'a \ldots$, up to $f^{(n)}a$ is infinite. We may carry this series (if no diff. co.[289] become infinite) as far as we please: it will afterwards remain to be pointed out *what are the cases in which we may legitimately suppose it carried ad infinitum*. Whatever these cases may be, in them we have
>
> $$f(a+h) = fa + f'a.h + f''a.\frac{h^2}{2} + f'''a.\frac{h^3}{2.3} + \text{\&c. } ad\ infin.$$
>
> which is TAYLOR'S THEOREM; and we see that we may stop at any term, and give an expression for the value of the rest, beginning at that term, by writing $a + \theta h$ instead of a in the term we stop at, and expunging all that come after, the value of this accession lying in its having been proved that θ is less than 1. This is LAGRANGE'S THEOREM ON THE LIMITS OF TAYLOR'S SERIES.[290]

The first example he considers is the Binomial Theorem — or what he calls the Binomial Theorem. The Finite Binomial Theorem familiar from elementary mathematics courses gives a formula for the expansion of a binomial raised to a positive

---

[287] *Ibid.*, p. 70.

[288] *Ibid.*, p. 71.

[289] I.e., differential coefficient = derivative.

[290] De Morgan, *op. cit.*, p. 73.

integral power:

$$(a+b)^n = \binom{n}{0}a^n + \binom{n}{1}a^{n-1}b + \binom{n}{2}a^{n-2}b^2 + \ldots + \binom{n}{n-1}ab^{n-1} + \binom{n}{n}b^n.$$

Newton generalised this to arbitrary rational exponents and an infinite series:

$$(a+b)^q = a^q\left(1+\frac{b}{a}\right)^q = a^q\sum_{k=0}^{\infty}\frac{q(q-1)\cdots(q-k+1)}{k!}\left(\frac{b}{a}\right)^k,$$

the series converging so long as $|b| < |a|$. Newton's Binomial Theorem became one of the cornerstones of a Calculus that relied heavily on power series until Cauchy moved the Mean Value Theorem front and centre, and there were numerous failed attempts in the 18th century to provide the Theorem with a rigorous proof. Such a proof was eventually given (for arbitrary real exponents) by Bolzano (1816), and again by Cauchy (1821). Cauchy's proof, however, relied on uniform interpretations of basic concepts and Abel, interpreting these concepts pointwise, gave a new, very nearly correct, proof (1826)[291] which led slowly to the recognition of the distinction between the pointwise and uniform convergence of series. De Morgan betrays no knowledge of any of this:

> We can now demonstrate the binomial theorem: for if $\phi x = x^n$ we have $\phi' x = nx^{n-1}$, $\phi'' x = n(n-1)$[292] and therefore $\phi a = a^n$, $\phi' a = na^{n-1}$, &c. This gives[293]
>
> $$\begin{aligned}(a+h)^n = a^n + na^{n-1}h + n(n-1)a^{n-2}\frac{h^2}{2} + n(n-1)(n-2)a^{n-3}\frac{h^3}{2.3} +\\ + \ldots\ldots + n.(n-1)\ldots\ldots\ldots(n-p)a^{n-p-1}\frac{h^{p+1}}{2.3\ldots.p+1}\\ + n(n-1)\ldots.(n-p-1)(a+\theta h)^{n-p-2}\frac{h^{n+2}}{2.3\ldots.p+2}.\end{aligned}$$
>
> or
>
> $$\begin{aligned}(a+h)^n &= a^n + n(a+\theta h)^{n-1}h\\ &= a^n + na^{n-1}h + n\frac{n-1}{2}(a+\theta h)^{n-2}h^2\\ &= a^n + na^{n-1}h + n\frac{n-1}{2}a^{n-2}h^2 + n\frac{n-1}{2}\frac{n-2}{3}(a+\theta h)^{n-3}h^3, \text{\&c.},\end{aligned}$$
>
> where, however, it must be observed that though θ is less than unity in every one of these cases, it is not the same in all.[294]

---

[291] Cf. Smoryński, *Treatise*, *op. cit.* or Henrik Kragh Sørensen, "Exceptions and counterexamples: Understanding Abel's comment on *Cauchy's Theorem*", *Historia Mathematica* 32 (2005), pp. 435–480, for fuller accounts of the story.

[292] The latter term is missing the factor x^{n-2}.

[293] The last exponent should, of course, be $p+2$.

[294] De Morgan, *op. cit.*, pp. 73–74.

There is no mention here of the infinite series! The error term,

$$\binom{n}{k}(a+\theta h)^{n-k}h^k,$$

cannot easily be shown to approach 0 as k gets large without bound.[295] Indeed, this is the classical example of a series for which one needs the Cauchy form of the remainder to prove convergence for all h with $|h| < |a|$.

We see in the above cited passages that de Morgan has acquired some of the *results* and understanding of Cauchy. His treatment of the Binomial Theorem suggests this acquisition was incomplete at the time of writing. Moreover, we do not see evidence that he has acquired Cauchy's *rigour*, for it is simply not here. On the other hand, we do not see that he has not acquired this rigour; for, this is not a treatise aimed at his fellow mathematicians, but a tract undertaken for the Society for the Diffusion of Useful Knowledge for the benefit of the occupants of the lower levels of the British Beehive. Our modern textbooks also lack rigour, stating without proof a number of fundamental results — the Intermediate Value Theorem, The Extreme Value Theorem, and the Existence of the Definite Integral. And the reason is that the modern Calculus textbooks are not written only for mathematics majors, but for a broader spectrum of students, including engineering students who are often indifferent to the beauty of mathematical reasoning and only care to learn technique. So if I criticise de Morgan now for the shortcomings of his presentation, I cannot say whether it is due to culture lag, indicative of how rigour was not catching on, or if it was due to a conscious pædagogical decision to present what de Morgan thought would be an acceptable amount of rigour to his intended audience.

So, what are the shortcomings of de Morgan's exposition? First, for wanting to base his development of the Calculus on the notion of limit, he has been remarkably quiet on the nature of limit. I would say he is no clearer than was Newton. For limit in general, Cauchy was also intuitive rather than formal, but he did define continuity and convergence of series, albeit not too clearly: Cauchy interpreted these definitions uniformly; everyone after him read the definitions as pointwise. De Morgan simply did not get this far. Indeed, his treatment of continuity was entirely intuitive, assuming as obvious the intermediate and extreme value properties — two properties Bolzano had recognised as in need of proof and had supplied proofs of. Cauchy too saw the need in the former case and gave a proof of the Intermediate Value Theorem. De Morgan did not even acknowledge that this was a Theorem and not an obvious property.

Lagrange's assumption that every function was real analytic was false, but can be read as a restriction of the class of functions under consideration. Lagrange had cheated and dragged in limit and continuity considerations, but otherwise based his treatment on a well-defined set of functions. Ampère tried and Cauchy succeeded in defining the class of uniformly differentiable functions and proved theorems about

---

[295] n is rational, not necessarily integral, so the binomial coefficient $\binom{n}{k}$ will not always be 0 for all sufficiently large k.

them using only what was inherent in the definitions. De Morgan tried with his three postulates to define the class of functions he wanted to restrict the Calculus to. He actually did not define a class of functions, but via Postulate 3, he implicitly defined a property of such classes to which his treatment, e.g., the proof of the existence of derivatives, applied. But whenever necessary he made new assumptions, sometimes without mention. To turn his proof of the Mean Value Theorem into a rigorous one, for example, one has to read the proof carefully and identify what is being assumed — I make it out to be uniform differentiability.

And, of course, de Morgan does not justify his claim that $\theta < 1$, by which, presumably, he means $0 < \theta < 1$ or $0 \leq \theta < 1$. Of those mathematicians considered here, only Bolzano recognised the problem and tried to prove θ could indeed be chosen to satisfy the strict inequality. But this is a relatively simple technical problem, easily solved, as we saw in the preceding section. It does not require a clarification of concepts as does the whole range of pointwise/uniform concepts.

Two years before the last instalment of de Morgan's text was published, Moigno's book "drawn from the methods and published and unpublished works of Mr. A.-L. Cauchy"[296] appeared. The fifth lesson concerns the Mean Value Theorem and its applications. It starts with paragraph 21 relating the sign of the derivative and the growth of the function, then moves to paragraph 22 and Cauchy's proof of the Cauchy Mean Value Theorem:

> **21**. Let Δx, Δy, the simultaneous increases in variables x and $y = \mathrm{F}(x)$, the ratio $\frac{\Delta y}{\Delta x}$ having for its limit the derivative y' will eventually have the sign of the limit, when Δx is small enough, and will therefore be positive if the derivative is positive, negative if the derivative is negative. In the first case, the infinitely small differences, Δy, Δx, being of the same sign, the function y will increase or decrease simultaneously with the variable x; in the second case, the infinitely small differences are of opposite signs, the function y will grow if the variable x decreases and will decrease if the variable y increases.
>
> *Corollary* 1st. Imagine that the function $y = \mathrm{F}(x)$ is continuous between two given limits x_0, X, and that we make the variable x increase by insensible degrees from the first limit to the second. The function cannot cease increasing to decrease or decreasing to increase unless the derivative $\mathrm{F}'(x)$ changes from positive to negative or from negative to positive. It is essential to note that in this passage the derived function will become zero if it does not cease to be continuous, and infinite if, without ceasing always to be real, it is discontinuous.
>
> *Corollary* 2nd. Suppose that the function $y = \mathrm{F}(x)$ vanishes for the particular value x_0, and is continuous in the vicinity of this value. We will have
>
> $$\mathrm{F}(x_0 + \Delta x) = \Delta x \mathrm{F}'(x_0) + \varepsilon_0 :$$
>
> so assuming that the value $x_0 + \Delta x = x$ differs very little from x_0,
>
> $$\mathrm{F}(x_0 + \Delta x) = F(x) \text{ will be positive if } \mathrm{F}'(x_0) > 0,$$
> $$\mathrm{F}(x_0 + \Delta x) = F(x) \text{ will be negative if } \mathrm{F}'(x_0) < 0.$$
>
> **22**. Let $\mathrm{F}(x)$ and $f(x)$ be two real functions of x which are continuous along with their derivatives within the limits x and $x + h$; also assume that the derivative of the second

[296] Abbé Moigno, *Leçons de calcul différential et de calcul intégral, rédigées d'après méthodes et les ouvrages publiés ou inédits de M. A.-L. Cauchy*, Bachelier, Paris, 1840.

function, $f'(x)$, does not change sign between the limits in question, or that between these limits the function $f(x)$ is always ascending or always descending, the ratio of the two differences

$$\mathrm{F}(x+h)-\mathrm{F}(x), \quad f(x+h)-f(x),$$

will be equal to one of the values which is taken between the limits x and $x+h$ by the ratio of the derivatives $\mathrm{F}'(x), f'(x)$, that is to say, for some θ_1 less than unity, we will have

$$\frac{\mathrm{F}(x+h)-\mathrm{F}(x)}{f(x+h)-f(x)}=\frac{\mathrm{F}'(x+\theta_1 h)}{f'(x+\theta_1 h)}.$$

Demonstration. Let A be the smallest and B the largest of the values which can be taken by the ratio $\frac{\mathrm{F}'(x)}{f'(x)}$ between the limits x and $x+h$; the two differences

$$\frac{\mathrm{F}'(x)}{f'(x)}-A, \quad \frac{\mathrm{F}'(x)}{f'(x)}-B,$$

will have opposite signs; it will be the same for these other two

$$\mathrm{F}'(x)-Af'(x), \quad \mathrm{F}'(x)-Bf'(x);$$

since $f'(x)$ is constantly of the same sign: but these two last differences are the derivatives of the two functions

$$\mathrm{F}(x)-Af(x), \quad \mathrm{F}(x)-Bf(x);$$

one of these functions will thus be increasing and the other decreasing, and consequently, if from what they become each of these two functions is subtracted from what they were, the differences obtained,

$$\mathrm{F}(x+h)-\mathrm{F}(x)-A[f(x+h)-f(x)],$$
$$\mathrm{F}(x+h)-\mathrm{F}(x)-B[f(x+h)-f(x)],$$

will one be positive and the other negative; and because $f(x+h)-f(x)$ is by hypothesis a quantity always positive or always negative, the two differences

$$\frac{\mathrm{F}(x+h)-\mathrm{F}(x)}{f(x+h)-f(x)}-A, \quad \frac{\mathrm{F}(x+h)-\mathrm{F}(x)}{f(x+h)-f(x)}-B,$$

will again necessarily be of opposite signs, and consequently the ratio $\frac{\mathrm{F}(x+h)-\mathrm{F}(x)}{f(x+h)-f(x)}$ greater than A, less than B, will be between the largest and the smallest values of the ratio $\frac{\mathrm{F}'(x)}{f'(x)}$. Moreover, if, as we have assumed, the derived functions are themselves continuous, while x will pass from the value x to the value $x+h$, the ratio $\frac{\mathrm{F}'(x)}{f'(x)}$ will pass through all intermediate values between A and B; but $\frac{\mathrm{F}(x+h)-\mathrm{F}(x)}{f(x+h)-f(x)}$ is one of these intermediate values; there thus exists a value of x of the form $x+\theta_1 h$ to satisfy the equation $\frac{\mathrm{F}(x+h)-\mathrm{F}(x)}{f(x+h)-f(x)}=\frac{\mathrm{F}'(x+\theta_1 h)}{f'(x+\theta_1 h)}$, which was to be proved.

Corollary 1st. Assume the equation which precedes,

$$\frac{\mathrm{F}(x_0+h)-\mathrm{F}(x_0)}{f(x_0+h)-f(x_0)}=\frac{\mathrm{F}'(x_0+\theta_1 h)}{f'(x_0+\theta_1 h)},$$

and if the two functions $F(x)$ and $f(x)$ vanish for $x = x_0$,[297]

$$\frac{F(x_0+h)}{f(x_0+h)} = \frac{F'(x_0+\theta_1 h)}{f'(x_0+\theta_1 h)}.$$

This last corollary, of course, leads to L'Hôpital's Rule and the Higher Order Mean Value Theorem.

Two points are to be made here. First, the treatment in paragraph 21, as I remarked back in Chap. 2 (Remark 2.3.22), is inadequate. This was common in the period. Indeed, the first rigorous treatment that I am aware of is in Weierstrass's unpublished lectures of 1861 on the Differential Calculus.

The second point is that the proofs of the Mean Value Theorem had now become standardised. For f continuously differentiable on $[a, b]$, one first proved

$$\min\left\{f'(x) \mid x \in [a,b]\right\} \le \frac{f(b)-f(a)}{b-a} \le \max\left\{f'(x) \mid x \in [a,b]\right\}$$

either by appeal to uniform differentiability to place $\dfrac{f(b)-f(a)}{b-a}$ being a mean between the greatest and lowest values of quotients $\dfrac{f(x_i)-f(x_{i-1})}{x_i - x_{i-1}}$ for a partition $a = x_0 < x_1 < \ldots < x_{n-1} < x_n = b$ of $[a, b]$, each quotient being within a fixed ϵ of the corresponding derivative $f'(x_{i-1})$, or by appeal to the Increasing Function Theorem as just done — and the only satisfactory proof of this latter theorem we have seen so far in the history has relied on uniform differentiability, a concept only Bolzano seems to have been aware of.

I did check another textbook of the period, namely the English language *A Treatise on the Differential Calculus and the Elements of the Integral Calculus with Numerous Examples* published in 1852 by Isaac Todhunter, a former student of de Morgan. The presentation is overall clearer than de Morgan's, but not yet fully rigorous. His treatment of the Mean Value Theorem follows Moigno, whose textbook was a standard until the 1870s.

All of this was soon to change, beginning with Weierstrass.

3.8 Weierstrass, Bonnet, Serret, and the Mean Value Theorem

The main contributors to rigour in the 19th century were Bolzano, Cauchy, and Weierstrass. Bolzano started with proofs of two basic theorems, the Binomial Theorem and the Intermediate Value Theorem, based on precise definitions of limit, continuity, differentiation, etc., in the 1810s. Cauchy took over in the 1820s, providing rigour, giving new proofs, and publishing the first reasonably rigorous textbooks.

---

297 *Ibid.*, pp. 33–36.

Bolzano produced an unfinished work on the foundations of the Calculus by the 1830s, but it was flawed as well as unfinished and it went unpublished until 1930, nearly a century later. The man who largely completed the task of rigorising the Calculus was Weierstrass, who lectured fairly regularly every two years on analytic functions, starting 1863/64 until 1884/85, wherein he presented his basic concepts of the Calculus. He did not, however, publish this material, leaving it in the classroom. Fortunately, he had students who were not shy about their master's accomplishments — as we saw in discussing Heine on uniform continuity in Chap. 2. Thus, we have a good knowledge of Weierstrass's contributions, if not always a good chronology of them.

Today we might find the main foundational tasks of this period to be the exposing of the nature of the real numbers themselves, the recognition of the pointwise/uniform distinction, and the exposition, particularly in textbooks, of these foundations.

The first work on the foundation of the real number system was Bolzano's theory of *measurable infinite number expressions*, a description of the real numbers as those infinite expressions built up from rational numbers that could be placed arbitrarily accurately among the rationals. The technical problems at this level of generality proved too great for him and his programme failed. At some indefinite later date Weierstrass developed a theory of real numbers in what more-or-less amounts to a modest version of Bolzano's approach: he constructed the real numbers as infinite sums of rational numbers.[298] However, he did not publish his theory, which only appeared in print in the publications of his students Ernst Kossak (1872), Salvatore Pincherle (1880), Otto Biermann (1887), and Gösta Mittag-Leffler (1920). In 1858 Dedekind had worked out his alternative theory of Dedekind cuts which reminds one of the old Eudoxian theory of proportion found in Euclid's *Elements*. But he too delayed publication until 1872, just as Méray,[299] Heine, and Cantor were publishing their constructions of the real numbers as Cauchy convergent sequences of rational numbers.

The remaining task for the foundations of the Calculus was thus the sorting out of the pointwise/uniform distinction. Bolzano was the earliest to have clearly understood the distinction at least as regards continuity and differentiability. He showed by example that the pointwise and uniform notions do not coincide in general, and

[298] Mittag-Leffler (G. Mittag-Leffler, "Die Zahl: Einleitung zur Theorie der analytischen Funktionen", *Tohoku Mathematics Journal* 17 (1920), pp. 157–209; here: pp. 206–207) reports that Weierstrass had the basic conception as early as 1841/42, lectured on it in 1859/60 and in the summer of 1860, and probably already lectured on it in the summer of 1857. These lectures do not seem to have been preserved. Reinhard Bölling informs me that the earliest of Weierstrass's lectures on the subject that he has seen are from Schwarz's notes from the winter 1863/64 semester. He also points to a tantalising hint from Schwarz's notes of the lectures of the 1861 summer semester indicating that Weierstrass had at least the basic conception by that date, and to a comment from Georg Hettner's notes of the lectures of the 1874 summer semester to the effect that Weierstrass had developed his theory of real numbers in response to the lack of understanding of his presentation in earlier years of various theorems depending on properties of the real numbers. This makes Mittag-Leffler's claim of an early development unlikely.

[299] Actually, Méray's first publication of the construction was in a paper of 1869, but his book of 1872 garnered more attention.

that, on closed bounded sets pointwise continuity implied uniform continuity. And he attempted to prove that continuous differentiability implied uniform differentiability on such intervals. He seems not to have considered the problem of uniform convergence, although the proof of the continuity of his nowhere differentiable function, being the limit of a uniformly convergent sequence of continuous functions, depends on establishing the uniformity of the convergence.

Dirichlet lectured on uniform continuity in 1854,[300] but also did not publish on it. We thus owe the concept to Weierstrass, via its first publication in 1872 by Heine. I don't know when uniform differentiability was first recognised in print. It may well have been Jordan's acknowledgement in 1884 of having used it in his proof of the Mean Value Theorem in his textbook.[301] Its main use seems to be in proving the Mean Value Theorem, which Cauchy had claimed valid for "differentiable functions", meaning uniformly differentiable functions, with continuous derivatives; and mathematicians, overlooking the uniformity assumption implicit in the proof, simply assumed the validity of the theorem for continuously differentiable functions.

If I cannot report anything definite about the history of uniform differentiability, I can say that the history of uniform convergence has been studied in some depth. The story begins with Cauchy's proof that the sum of a convergent series of continuous functions is itself continuous. Abel was the first to draw attention to the problem. Reading "convergent" as "pointwise convergent", he announced in a famous footnote to his paper on the Binomial Theorem that Cauchy's theorem seemed to "suffer exceptions".[302] He did not see wherein lay the error. Indeed, in the proof of Theorem V of that paper, he makes exactly the error of assuming uniformity where it does not hold.

If one takes Cauchy's references to infinitesimals and infinitely large integers at all seriously, one sees that his disputed proofs are indeed correct. Conversion of his definitions of continuity, differentiability, and series convergence into equivalent standard terms making no reference to such extended reals does not result in our standard notions of continuity, differentiability, and convergence, but in the standard uniform notions. Nonetheless, the judgment has come down against Cauchy on some of his results, the "continuity" of the sum of a "convergent" series of "continuous" functions being one of these. It wasn't until after the successful introduction of Nonstandard Analysis by Curt Schmieden (1905–1991) and Detlef Laugwitz (1932–2000) in the late 1950s and Abraham Robinson (1918–1974) in 1960 that mathematicians took a second look at Cauchy's "false proofs" and recognised their validity in their intended settings.[303]

---

[300] Russ, *op. cit.*, p. 350.

[301] More on this in the next section.

[302] Niels Henrik Abel, "Untersuchungen über die Reihe: $1 + \frac{m}{1}x + \frac{m.(m-1)}{1.2}.x^2 + \frac{m.(m-1).(m-2)}{1.2.3}.x^3 + \ldots..$ u.s.w.", *Journal für die reine und angewandte Mathematik* 1 (1826), pp. 311–339.

[303] Of particular interest here are works of Detlef Spalt and Laugwitz. Cf. Smoryński, *Formalism, op. cit.*, Chap. III, Sect. 6, Subsection 4, for references and a summary in English of their work.

From the modern point of view, reading the terms pointwise, Cauchy's result is false. Indeed, in 1837 Dirichlet, in studying the pointwise convergence of Fourier series, showed that many discontinuous functions were the limits of convergent series of continuous functions. The apparent discrepancy between Cauchy's proof and this plethora of counterexamples was first explained in print by Dirichlet's former student Philipp Seidel (1821–1896) in 1847,[304] and again, independently, by George Gabriel Stokes (1819–1903) in 1848.[305] They came up with two slightly different conditions — *arbitrarily slow convergence* and *infinitely slow convergence*, respectively — under which it was possible for the sum of a convergent series of continuous functions to fail to be continuous. A third concept, our modern *uniform convergence* related to, but distinct from, the negations of Seidel's and Stokes's concepts had, in fact, already been introduced by Christoph Gudermann (1798–1852) in 1838 and it would be used by Weierstrass as early as 1841 and finally emphasised by him in the early 1860s. Cauchy himself got into the act in 1853, offering a better statement of his result — still stated in terms of infinitesimals and infinitely large integers.

The whole story of uniform convergence is not really germane to our discussion, other than to indicate that by 1872 all the concepts and tools for the rigorous treatment of the Calculus of functions of a single real variable were in place. Thus, I have given only this most abbreviated discussion of uniform convergence and refer the interested reader to the literature for more.[306]

Thus we have the development of the foundations of the Calculus in brief, in broad terms not mentioning the Mean Value Theorem — the central point of interest in the present book. What did Weierstrass have to say about or contribute to the development of this Theorem?

In the summer semester of 1861 Weierstrass lectured on the Differential Calculus at one of the technical schools in Berlin. Notes were taken by Schwarz, who eventually deposited a typewritten copy[307] in the library of the Humboldt University in Berlin. A short preface by Schwarz begins

> The booklet before you contains an abridged elaboration of the lectures on Differential Equations, which Herr Professor Weierstrass held at the Royal Industrial Institute of Berlin in the *summer semester* 1861. Since these lectures were the first lectures on Differential Equations, which I had the good fortune to have heard, and since the elaboration itself had to be completed before the close of the summer semester of 1861, so the elaboration is plagued

---

[304] P.L. Seidel, "Note über eine Eigenschaft der Reihen welche discontinuirliche Funktionen darstellen", *Abhandlungen bayerische Akademie der Wissenschaften* 5 (1847), pp. 381–394.

[305] G.G. Stokes, "On the critical values of the sums of periodic series", *Transactions of the Cambridge Philosophical Society* 8 (1849), pp. 533–583.

[306] A good philosophical discussion of the issues involved can be found in Appendix I of Imre Lakatos, *Proofs and Refutations; The Logic of Mathematical Discovery*, Cambridge University Press, Cambridge, 1976. A mathematically superior and more up-to-date discussion is in Chap. 5, Sect. 4 (pp. 202–208), of Bottazzini, *op. cit.* An older, more technical, account worth a look is G.H. Hardy, "Sir George Stokes and the concept of uniform convergence", *Proceedings of the Cambridge Philosophical Society* 19 (1918), pp. 148–156.

[307] Karl Weierstrass and Hermann Amandus Schwarz, *Differential Rechnung, nach einer Vorlesung des Herrn Professor Weierstrass in Sommersemester 1861*, Hdschr. Koll. N 37.

with all the weaknesses and incompletenesses which are only so easily explainable by a first such work.[308]

The rest of the preface will be mentioned after presenting the main material. For now I note only that it more-or-less shows that the manuscript was not deposited in the library before the end of February 1870, for it mentions a result that can be dated almost exactly to the 24th or 25th of February of that year.

The notes do not include fundamental results about the real numbers, limits, or continuity other than giving a proof of the Intermediate Value Theorem on pages 3–4, but begin with the basics of differentiation. This missing material could have been presented in an earlier course, or could have been omitted by Schwarz in abridging the work as he may only have needed to make available to his students the material relevant to whatever course (presumably Differential Equations) he was teaching at the time he made his abridgement.

So far as I know the full version of the notes has never been published. The abridged version is available through a hard to read, somewhat blurry and low-contrast scan from the Humboldt University in Berlin. For those who read French or German, there is a long paper by Pierre Dugac discussing the lectures in some detail.[309] Excerpts from the introductory portions of the lecture notes, translated into English from Dugac's paper can be found in Calinger's source book.[310] Here I present in English translation the proof given by Weierstrass as told by Schwarz of the Mean Value Theorem and the lemmas leading up to it. I have kept some of the inconsistencies of style and the rough edges of the unpolished typescript to give the reader some taste for the original document, but have made a few typographical changes such as using italics instead of underlining for emphasis and italicising variables and functions for the sake of readability. I have also broken some of the longer run-on sentences up into more readily digestible chunks, but have not tampered with the paragraph structure. All in all, the work is quite readable and so as not to disrupt the flow I have opted to reserve my comments until the end.

Investigation of the Course of a Function[311]

A function of first order of a single variable quantity varies continuously with its argument and, indeed, proportionally to the variation of the argument. If the coefficient of the variable quantity is positive, it happens that the variation of the function is in the *same direction* as he variation of the argument; if on the other hand this is negative, it happens that the variation [of the function] is in the *opposite* direction. In both cases the function will either *steadily* be in the same direction or steadily in the opposite direction of the arguments' variation, but will never encounter a *change* [in direction]. Many complex relations connect conjoined functions; the following theorems concern themselves with one of these. It will

[308] *Ibid.*, unnumbered first page.

[309] P. Dugac, "Eléments d'analyse de Karl Weierstraß", *Archive for History of Exact Science* 10 (1973), pp. 41–176. Dugac discusses several of Weierstrass's contributions to the foundations of analysis. In addition to the main text with background and summary in French, he includes excerpts in German from Weierstrass's works. This includes the abridged text of the 1861 summer lectures, but not the material concerning the Mean Value Theorem.

[310] Ronald Calinger, *Classics of Mathematics*, Moore Publishing Company, Oak Park (Illinois), 1982.

[311] Weierstrass and Schwarz, *op. cit.*, pp. 20–26.

be assumed that x is a continuous variable quantity between two limits, and $f(x)$ and $f'(x)$ are two continuous, single-valued functions of the same.

I. Theorem.

If, for a certain value x_0 of x the derivative of the function $f(x)$ is not null, then there are in the vicinity of x_0 always values of x for which $f(x)$ is larger than $f(x_0)$ and also such for which $f(x)$ is smaller than $f(x_0)$.
Let $f'(x_0)$ not be null, x lie in the vicinity of x_0, and let $x - x_0 = h, x = x_0 + h$. $f(x_0 + h) - f(x_0)$ then consists of two parts $f'(x_0).h + hh_1$, where h_1 is generally an entirely unknown function of h, which simultaneously becomes infinitely small with h; $f(x_0 + h) - f(x_0) = h\big(f(x_0) + h_1\big)$.[312] Now, if $f'(x_0)$ is not null, h_1 can by decreasing h be made smaller than any given quantity however small. One determines thus for h a bound δ such that for all values of h which do not exceed δ in absolute value, h_1 will have absolute value smaller than $f'(x_0)$, so $f'(x_0) + h_1$ will have the same sign as $f'(x_0)$ and also $h\big(f'(x_0) + h_1\big)$ will have the same sign as $hf'(x_0)$. From this it is obvious that for opposite values of h also $f(x_0 + h) - f(x_0)$ must have opposite signs. It must be, if one gives h opposite values within the bounds, $f(x_0 + h)$ will be smaller the one time and larger the other time.

II. Theorem. If for two distinct values x_1 and x_2 of the argument $f(x_1)$ is equal to $f(x_2)$, then there necessarily exists between x_1 and x_2 a value x_0 for which the first derivative of $f(x)$ will equal null.[313]

One imagines given between x_1 and x_2 a value x_3 for which $f(x_3)$ is different from $f(x_1)$, — such a value must exist, for otherwise $f(x)$ would be a constant and not a function of x, — so two cases are possible, either $f(x_3) > f(x_1)$ thus too $> f(x_2)$, or $f(x_3) < f(x_1)$. Let $f(x_3) > f(x_1)$, so that among all the different values which the function can take, as x assumes all values from x_1 to x_2, there must be one which is not smaller than all of the rest: Let x_0 be this value, so $f'(x_0)$ must $= 0$, because, if this were not the case, the preceding theorem would yield in the vicinity of x_0 a value for which $f(x) > f(x_0)$ and this would be in contradiction with the assumption that $f(x_0)$ is not smaller than any of the values which $f(x)$ can assume for all the values of x between x_1 and x_2. — Now let $f(x_3) < f(x_1)$, so there must necessarily be among all the values of $f(x)$, for x between the limits x_1 and x_2, a value $f(x_0)$ which is not greater than all the rest; for this value x_0 of the argument one can show as in the first case that $f'(x_0)$ must equal 0.
III. Now the *largest* and the *smallest* values must be determined which $f(x)$, a continuous function of x defined between the limits x_1 and x_2, can take on between these limits if for all these values of x the first derivative of $f(x)$ is positive. Neither the greatest nor the smallest value of $f(x)$ can lie between $f(x_1)$ and $f(x_2)$; because, assuming the value x_3 where the largest or smallest value occurs there, between x_1 and x_2, by the assumption that $f'(x_3)$ does not equal null, there will thus be by the first theorem of this section values in the vicinity of x_3 for which $f(x)$ must be larger and others for which $f(x)$ is smaller than $f(x_3)$. Thus $f(x_3)$ can neither be the largest nor the smallest value. One of the two values $f(x_1)$ and $f(x_2)$ is thus necessarily the largest, the other the smallest. Of the two values x_1 and x_2 one is necessarily the larger, the other the smaller; it will be shown, that at the larger value of the argument the larger value of the function belongs. Let $x_2 > x_1$ and x_3 any arbitrary value between x_1 and x_2, thus in the case before us greater than x_1 and smaller than x_2. Now $f(x_3)$ can neither be equal to $f(x_1)$ nor $f(x_2)$, because otherwise by the second theorem $f'(x)$ will equal null for some value x_3 of x lying between x_3 and x_1 or x_2, which contradicts the assumption. $f(x_3)$ must thus necessarily lie between $f(x_1)$ and $f(x_2)$. Now for values of x in the neighbourhood of x_3, by which $x - x_3 = h$ does not exceed a certain bound, it will be that $f(x_3 + h) - f(x_3) = h.f'(x_3) + h(h)$. In consequence of the definition of the function $f'(x)$,

---

[312]There is a typo here: the right-hand side of the equation should be $h\big(f'(x_0) + h_1\big)$.

[313]This is the earliest statement and proof of Rolle's Theorem I am aware of.

(h) is a quantity that becomes infinitely small simultaneously with h. One now determines as in the first theorem of this section such a δ for h that the values of $x_3 + h$ lie between x_1 and x_3, and that for all values of h whose absolute value does not exceed δ, (h) has absolute value less than $f'(x_3)$, so $f(x_3 + h) - f(x_3)$ will be with $hf'(x_3)$ and since $f'(x)$ is assumed positive for all values x lying between x_1 and x_2 thus too for $x_3, f(x_0 + h)$ will be greater than $f(x_3)$. If h is negative, $f(x_3 + h)$ will be smaller than $f(x_3)$. However, if h is positive $x_3 + h$ lies between x_2 and x_3; if h is negative $x_3 + h$ lies between x_3 and x_1. For all values of x between x_3 and x_2 the function assumes a greatest and a least value for each of the two values x_3 and x_2. If now, however, $f(x_3 + h)$, where h is positive and $x_3 + h$ lies between x_3 and x_2, is larger than $f(x_3)$ as shown above, so $f(x_3)$ must be the smallest and therefore $f(x_2)$ the largest value which the function can take for x between x_3 and x_2; thus $f(x_2) > f(x_3)$. Analogously it can be shown that $f(x_1) < f(x_3)$, that thus $f(x_1) < f(x_3) < f(x_2)$, that thus to the larger value of the argument the larger value of the function belongs. One considers thus the different values of a function for all values of the argument between two limits between which it itself as well as its derivative is continuous, and the latter only assuming positive values, so to a larger value of the argument a larger value of the function also belongs.
The same theorem also holds if the first derivative is such that it is negative for all values of the argument between the distinct limits, while the case that it will equal null one or more times is in no way ruled out. For instance, let $f'(x) = 0$ at x_3 for one value lying between x_1 and x_2, so the derivative for values of x lying between x_1 and x_3 is positive and one has by the preceding, if $x_1 < x_3 < x_2$ then $f(x_3) > f(x_1)$; the derivative however is also positive between x_2 and x_3, thus $f(x_2) > f(x_3)$, and it follows that $f(x_2) > f(x_1)$. The same will hold for the case that $f'(x)$ equals null several times. From this follows thus the theorem:
If the derivative of a function never becomes negative so long as x remains in the interval x_1 and x_2, and if $x_2 > x_1$, then also $f(x_2) > f(x_1)$, i.e., the larger function belongs to the larger argument.
IV. If on the other hand under the same assumption the derivative is never positive so long as x finds itself in the interval x_1 to x_2, then $f(x_1) > f(x_2)$, so the larger value of the function belongs to the smaller of the argument.
If $f'(x)$, the derivative of $f(x)$, never becomes positive, so $-f'(x)$, the derivative of $-f(x)$, will never become negative; one now applies to $-f(x)$ the above theorem, so

$$-f(x_2) > -f(x_1) \text{ or } f(x_1) > f(x_2).$$

From this follows the following.
If the derivative of a function is negative for no value of the argument within an interval, then the value of the function is steadily increasing as the argument grows steadily from the lower to the upper limit of the interval.
If on the other hand the derivative of the function becomes positive for no value within the interval, then the value of the function steadily decreases as the argument increases from the lower to the upper limit of the interval.
To investigate the course of a function within an interval within which the derivative of the same never changes sign the two preceding theorems suffice.
On the other hand if the derivative changes its sign one or more times in the interval $a \ldots b$, so the values of the argument for which this occurs may be denoted in a series $x_1\, x_2\, x_3\, x_4 \ldots$ The derivative will then have equal signs in the intervals $ax_1, x_2x_3, x_4x_5, ..$ on the one side and $x_1x_2, x_3x_4, \ldots$ on the other. Now if the derivative in the first interval is positive, so will be $f'(x)$ while x continuously grows from a to b, so long as x finds itself in the 1st, 3rd, 5th,...$2n + 1^{\text{th}}$ intervals; and continuously decreases if x moves in the 2nd, 4th, 6th,...$2n$th intervals. The opposite happens if the derivative is negative in the first interval.

Main Theorem

If a function $f(x)$ with its first derivative $f'(x)$ is continuous for all values of its argument within the limits x_1 and x_2, there must be among all the values which the derivative can assume, so long as x remains in the indicated interval, a largest and a smallest — let the first be A and the second B. Then $A - f'(x)$ is a function which can never be negative.[314] These functions are now the derivatives of the following

$$A(x - x_1) - \big(f(x) - f(x_1)\big) \text{ and } B(x - x_1) - \big(f(x) - f(x_1)\big),$$

where x_1 is taken to be a constant and may be the smaller of the values x_1 and x_2. If one now applies to these functions the former theorems, one obtains $A(x - x_1) - \big(f(x) - f(x_1)\big)$ as steadily increasing, $B(x - x_1) - \big(f(x) - f(x_1)\big)$ as steadily decreasing, while x goes continuously from the smallest value x_1 to the largest x_2. The first function attains its smallest, the second its largest for $x = x_1$ and indeed both values equal null; if one now gives x the value x_2, so $A(x_2 - x_1) - \big(f(x_2) - f(x_1)\big) > 0$ and $B(x_2 - x_1) - \big(f(x_2) - f(x_1)\big) < 0$. From this it follows that

$$f(x_2) - f(x_1) < A(x_2 - x_1)$$
$$f(x_2) - f(x_1) > B(x_2 - x_1).$$

Then therefore $\dfrac{f(x_2) - f(x_1)}{x_2 - x_1}$ always lies between the bounds A and B, the largest and smallest values which $f'(x)$ can assume, so long as x finds itself in the interval from x_1 to x_2. The quotient $\dfrac{f(x_2) - f(x_1)}{x_2 - x_1}$ is always a mean value between the largest and smallest values, which the derivative $f'(x)$ can assume between the limits x_1 and x_2 of the argument.
Now let α be that value of the argument for which the derivative is the largest and β that value for which it is the smallest; one thus has $f'(\alpha) = A, f'(\beta) = B$, so there must obviously be between α and β a value γ for which $f'(\gamma)$ equals any given value between A and B. Let γ be chosen so that $f'(\gamma) = \dfrac{f(x_2) - f(x_1)}{x_2 - x_1}$, so there is, if one denotes a mean value between x_1 and x_2 by $(x_1 \ldots\ldots x_2)$ or by $x_1 + \varepsilon(x_2 - x_1)$, where ε is a proper positive fraction, a mean value between α and β, thus too between x_1 and x_2. So it must always be possible to satisfy the equation

$$\frac{f(x_2) - f(x_1)}{x_2 - x_1} = f'(x_1 \ldots\ldots x_2) \text{ or } f(x_2) - f(x_1) = (x_2 - x_1)f'(x_1 \ldots\ldots x_2)$$
$$\text{or } f(x_2) - f(x_1) = (x_2 - x_1)f'\big(x_1 + \varepsilon(x_2 - x_1)\big).$$

Writing now h for $x_2 - x_1$, thus for $x_2 = x_1 + h$, one has

$$f(x_1 + h) - f(x_1) = h.f'(x_1 + \varepsilon h)$$
$$f(x_1 + h) - f(x_1) = h.f'(x_1) + h\big(f'(x_1 + \varepsilon h) - f'(x_1)\big).$$

The formula shows clearly the decomposition of $f(x_1 + h) - f(x_1)$ into the differential[315] and the part becoming infinitely small in relation to h, where $f'(x)$ is assumed to be a continuous function. If now a second derivative of $f(x)$ is continuous between the limits x_1 and x_2 of the argument, so one can apply the same theorem to the difference $f'(x_1 + \varepsilon h) - f'(x_1)$ and one has

$$f'(x_1 + \varepsilon h) - f'(x_1) = \varepsilon h f''(x_1 + \varepsilon' \varepsilon h),$$

where ε' is again a proper positive fraction. If thus in a special case $f'(x_1)$ is equal to null and one writes ε for $\varepsilon'\varepsilon$, then

---

[314] Schwarz omits a clause: $B - f'(x)$ is a function which can never be positive.

[315] A literal translation would be "differential change", which is synonymous in the notes with the differential.

$$\begin{aligned} f(x_1+h)-f(x_1) &= hf'(x_1+\varepsilon h) \\ &= h.(f'(x_1+\varepsilon h)-f'(x_1)) = \varepsilon h^2 f''(x_1+\varepsilon h). \end{aligned}$$

This is a marvellous document, a bit unpolished, but important for the advances Weierstrass exhibits over his predecessors, as well as for his occasional shortcoming as viewed with the modern eye. His goal is to prove the Mean Value Theorem for *continuously differentiable* functions, the result as by then commonly stated, but never actually proved. He plans to prove this by appealing to the continuity of f' on an interval $[x_1, x_2]$ to conclude the maximum and minimum values A and B, respectively, of f' on this interval to exist, to demonstrate that

$$B \leq \frac{f(x_2)-f(x_1)}{x_2-x_1} \leq A,$$

and finally to appeal again to the continuity of f' to apply the Intermediate Value Theorem and conclude the existence of some $\gamma \in [x_1, x_2]$ such that

$$f'(\gamma) = \frac{f(x_2)-f(x_1)}{x_2-x_1}.$$

To this end he lays out the assumption right at the beginning that f and f' are both continuous on $[x_1, x_2]$. A modern expositor would not make this assumption at this point because the continuity of f' is not used until after the key lemmas are proven.

Theorem I is the familiar theorem stating that a function is increasing (decreasing) at a point where the derivative is positive (negative, respectively). Although he proves this explicitly, he states the result in a slightly less informative manner—if $f'(x_0) \neq 0$ there are points in any neighbourhood of x_0 where $f(x)$ is greater than $f(x_0)$ as well as points where $f(x)$ is less than $f(x_0)$. Not mentioning in the statement that, for, say, $f'(x_0) > 0$, the former x's are to the right of x_0 and the latter to the left, will complicate slightly the proof in paragraph III, where he will repeat the argument. Putting aside this expositional weakness, one must sing the praise of Weierstrass for this Theorem, not because he saw the relation between the growth of the function and the sign of the derivative — everybody saw this —, but because he saw what the hypothesis did not imply and that more was needed to prove the Strictly Increasing Function Theorem.

To prove the Strictly Increasing Function Theorem, Weierstrass first proved Rolle's Theorem (Theorem II) — ostensibly under the assumption of continuous differentiability on the closed interval, but his proof works for mere differentiability on the open interval. Today we would dismiss the case of the constant function differently, acknowledging that it does require proof by sneering at its triviality rather than declaring a constant function not to be a function. This raises a question for the serious student of history: did Weierstrass really mean this? Or was it Schwarz's attempt to fill in a gap in the lecture as delivered? Schwarz was a student and might not have had as broad a notion of function as Weierstrass must have had. In any event, other than giving the "wrong" reason for dismissing the trivial case, the proof is quite perfect.

In the course of proving Rolle's Theorem Weierstrass proves that if f assumes a maximum or a minimum at $x_0 \in (x_1, x_2)$ and $f'(x_0)$ exists, then $f'(x_0) = 0$. This, of course, was known already under vaguely understood conditions, as we saw in the first chapter when discussing Fermat et alia and as we saw earlier in this chapter in discussing Bhāskara II. Weierstrass, unaware that Bolzano had already proven the Extreme Value Theorem in his unpublished work, had given his own proof, and thus combined the two in proving Rolle's Theorem.

Paragraph III is a long, nearly unbroken paragraph devoted to proving the Strictly Increasing Function Theorem. The proof is perfectly correct if rather more involved than one might expect. The exposition could have been improved by breaking the argument up into more easily digestible chunks, or possibly by drawing a picture. Weierstrass is, however, the most commonly cited example of an algebraical/logical as opposed to a geometrical/intuitive thinker. One will not find many pictures in his works. Bölling explains

> We find drawings in "Mitschriften" of W's lectures so that we can assume that in fact W used them. In my opinion, in the later lectures the pictures disappear more and more. Even in the "Ausarbeitung" (i.e. carefully written after the lecture) of W's 1874 lecture "Introduction to the theory of analytic functions" by Hettner we find only a few drawings. Not a single picture even for the Bolzano-W Theorem. W accepted geometric representation as an illustration to make the contents clearer, but never as a part of a rigorous proof. He prefers purely arithmetic reasoning. E.g., when he introduced complex numbers, he defined these objects first arithmetically and, only after that, W gave the interpretation by means of points in Gauss's complex plane.[316]

In proving this result along his lines, I would argue that, if $f'(x) > 0$ for all $x \in (x_1, x_2)$, then the function must be strictly increasing or strictly decreasing. For, if not, one of three things occurs:

i. Two values of x take on the same y-value: $f(a) = f(b)$. This yields $c \in (a, b)$ with $f'(c) = 0$, contrary to hypothesis.

ii. There are $a < b < c$ in $[x_1, x_2]$ such that $f(a) < f(b)$ and $f(c) < f(b)$. For any y between $\max\{f(a), f(c)\}$ and $f(b)$ there are, by the Intermediate Value Theorem, $d \in (a, b)$, $e \in (b, c)$ such that $f(d) = y = f(e)$ (see Fig. 3.20), thus yielding an instance of case i, which we have just ruled out.

iii. There are $a < b < c$ in $[x_1, x_2]$ such that $f(a) > f(b)$ and $f(c) > f(b)$. This is symmetrical to case ii and is similarly ruled out.

Thus there is no change of direction, which can be determined by applying Theorem I to any interior point $x_0 \in (x_1, x_2)$. Weierstrass does not simply apply what has already been proven, but repeats the reasoning of paragraph I.

---

[316]Private correspondence. "Mitschriften", literally "with-writings", means "accompanying writings", i.e., lecture notes.

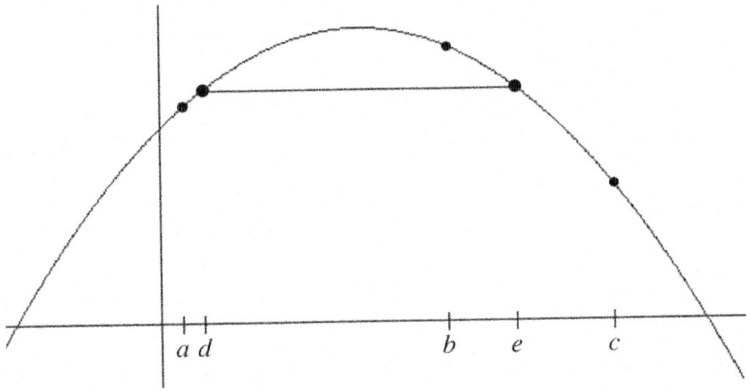

Fig. 3.20 Illustration of case ii

The end of paragraph III and the beginning of IV overlap in purpose, which is to give Weierstrass's proof that the function is still strictly increasing (decreasing) provided only that $f'(x)$ is never negative (positive, respectively). This, of course, is true when $f'(x)$ takes on the value 0 only finitely often or if between any two values at which $f'(x) = 0$ a value of x for which $f'(x) \neq 0$ can be found, i.e., if f' is not identically 0 on any subinterval. (*Exercise*. Use the Strictly Increasing Function Theorem and the Increasing Function Theorem to prove this.)

Weierstrass finishes paragraph IV with the usual first-year Calculus approach to functions: Break the interval up into alternating subintervals over which the function is strictly increasing and decreasing. This cannot always be done, at least not as straightforwardly as Weierstrass indicates.

3.8.1 Example Let f be any antiderivative to $x \sin(1/x)$ on any interval $[-\epsilon, \epsilon]$ for $\epsilon > 0$. f' changes sign infinitely often in $(-\epsilon, \epsilon)$ and there is no interval $[a, x_1]$ in which f is monotone for $a = 0$ and $0 < x_1 < \epsilon$.

Weierstrass, who would a decade later shock the world with an example of a function that was nowhere differentiable because it oscillated infinitely often in every neighbourhood, seems not yet to have been aware of such oscillation.

Weierstrass's proof of the Mean Value Theorem contains a slight, but easily correctible error: He wants to go from $A - f'(x) \geq 0$ and $f'(x) - B \geq 0$ for all $x \in [x_1, x_2]$ to

$$A(x - x_1) - \big(f(x) - f(x_1)\big) > 0 \text{ and } B(x - x_1) - \big(f(x) - f(x_1)\big) < 0,$$

where $A = \max\left\{f'(x) \,\middle|\, x \in [x_1, x_2]\right\}$, $B = \min\left\{f'(x) \,\middle|\, x \in [x_1, x_2]\right\}$. The correction is simply to replace A and B by $A + \epsilon$ and $B - \epsilon$, respectively, for ϵ an arbitrarily small positive number to conclude

$$B - \epsilon < \frac{f(x_2) - f(x_1)}{x_2 - x_1} < A + \epsilon$$

for all $\epsilon > 0$, whence

$$B \leq \frac{f(x_2) - f(x_1)}{x_2 - x_1} \leq A.$$

One then gets γ somewhere between the points where the extreme values occur for which

$$f'(\gamma) = \frac{f(x_2) - f(x_1)}{x_2 - x_1}.$$

Placing γ *strictly* between x_1 and x_2 requires an extra argument. We have seen this same oversight in Cauchy.

Weierstrass continues from here, giving a higher order Mean Value Theorem in the case in which the first n derivatives vanish at a point, concludes the Cauchy Mean Value Theorem, and lays the groundwork in general for deriving Taylor's Theorem with the Lagrange Form of the Remainder before finishing the section. The next section concerns finding maxima and minima.

What Weierstrass has missed, which would fit in nicely with the discussion of the relation between the growth of a function and the sign of the derivative, is the Constant Function Theorem. Schwarz remarks on this and remedies the situation in his prefatory comments:

> The proof for the theorem: "A continuous function, whose first derivative within a given interval of the argument is null everywhere, reduces itself to a constant" is not contained in the work before you. Let $f(x)$ be the function, $a \leqq x \overline{\overline{<}} b$ the interval, k a small positive quantity different from null whose smallness is not bound by any restriction. One considers
>
> $$\varphi(x) = \big(f(x) - f(a)\big) \pm k(x - a)$$
>
> so, since $\varphi'(x) = \pm k$, the Main Theorem is applicable to *this* function, whence, however small k may be, the relation
>
> $$-k(x - a) < f(x) - f(a) < k(x - a)$$
>
> holds. It follows that
>
> $$f(x) - f(a) = 0, \ \text{i.e.}\ f(x) = f(a).$$

The argument is more carefully laid out in a letter written by Schwarz to Cantor on 25 February 1870:

> Let k be a small positive quantity and consider the functions
>
> $$F(x) - F(a) - k(x - a) \ \text{and} \ F(x) - F(a) + k(x - a).$$
>
> The derivatives of these functions are respectively
>
> $$-k \ \text{and} \ +k;$$
>
> For $x = a$ both functions are equal to 0; by Theorem III and its corollary: ($f'(x)$ negative),
>
> $$F(b) - F(a) - k(b - a) \ \text{is negative, while}$$
> $$F(b) - F(a) + k(b - a) \ \text{is positive.}$$

> The difference $F(b) - F(a)$ in any case lies between $k(b-a)$ and $-k(b-a)$. Now however one can choose the quantity k arbitrarily small, whence the difference $F(b) - F(a)$, whose value is entirely independent of the value of k, exactly equals 0. One can now however replace b by each value of x between a and b and likewise conclude consequently $F(x) = F(a)$, i.e., a constant.
> The preceding proof appears to me to be completely rigorous; it is the foundation of the Differential and integral Calculus.[317]

This proof is preceded in the letter by brief versions of paragraphs I to III of Weierstrass's 1861 lectures, but begins with the words

> Even though I have written in detail yesterday, this is no obstacle to my writing to you again today; for I have an announcement to make to you which will certainly interest you, that for several hours I am in possession of what seems to me a rigorous proof of the theorem:
>
> If for each individual value of x for $a \leq x \leq b$,
>
> $$\lim \frac{F(x+h) - f(x)}{h} = 0$$
>
> for $\lim h = 0$, then $F(x)$ is a constant.[318]

This, of course, dates the depositing of the lectures in the library in Berlin to some time after 25 February 1870.

On the 6th of April of that year, just over a month after receiving this letter, Cantor submitted a paper[319] on the uniqueness of the expansion of a function into a trigonometric series. The proof made use of a generalisation Schwarz had proved of the Constant Function Theorem:

3.8.2 Theorem (Linear Function Theorem) *Let f be continuous on $[a, b]$ and suppose, for all $x \in (a, b)$,*

$$\lim_{h \to 0} \frac{f(x+h) - 2f(x) + f(x-h)}{h^2} = 0.$$

Then, f is a linear function:

$$f(x) = f(a) + \frac{f(b) - f(a)}{b-a}(x-a).$$

Proof. Let k be an arbitrarily small positive number and let ε be 1 or -1, and define

---

[317] Herbert Meschkowski, *Denkweise großer Mathematiker*, Friedr. Vieweg & Sohn, Braunschweig, 1961,p. 78. There is an English translation of Meschkowski's book: Herbert Meschkowski (John Dyer-Bennet, trans.), *Ways of Thought of Great Mathematicians: An Approach to the History of Mathematics*, Holden-Day, Inc., San Francisco, 1964.

[318] Meschkowski, *op. cit.*, p. 79.

[319] G. Cantor, "Beweis, dass eine für jeden reellen Werth von x durch eine trigonometrische Reihe gegebene Function $f(x)$ sich nur auf eine einzige Weise in dieser Form darstellen lässt", *Journal für die reine und angewandte Mathematik* 72 (1870), pp. 139–142.

$$\varphi(x) = \varepsilon\big[f(x) - f(a) - \frac{x-a}{b-a}\big(f(b) - f(a)\big)\big] - \frac{k}{2}(x-a)(b-x).$$

Note that φ is actually a function of ε and k as well as of x, but we consider these values fixed for now.

A little calculation shows

$$\varphi(x+h) - 2\varphi(x) + \varphi(x-h) = \varepsilon\big[f(x+h) - 2f(x) + f(x-h)\big] + kh^2,$$

whence

$$\lim_{h \to 0} \frac{\varphi(x+h) - 2\varphi(x) + \varphi(x-h)}{h^2} = k > 0.$$

One also calculates $\varphi(a) = \varphi(b) = 0$.

We claim that $\varphi(x)$ is not positive for any $x \in (a, b)$. For, if it were, φ would have a maximum value at some $c \in (a, b)$. But then for h small enough so that $c \pm h \in (a, b)$ one has

$$\varphi(c+h) - \varphi(c) \leq 0, \quad \varphi(c-h) - \varphi(c) \leq 0,$$

whence their sum is also nonpositive:

$$\varphi(c+h) - 2\varphi(c) + \varphi(c-h) \leq 0,$$

though for small h this sum should have the same positive sign as k.

It follows that, for any $\varepsilon = \pm 1$ and any $k > 0$, $\varphi(x) \leq 0$ for all $x \in (a, b)$ (and, by continuity, for $x = a, b$ as well):

$$f(x) - f(a) - \frac{x-a}{b-a}\big(f(b) - f(a)\big) - \frac{k}{2}(x-a)(b-x) \leq 0$$
$$-\left[f(x) - f(a) - \frac{x-a}{b-a}\big(f(b) - f(a)\big)\right] - \frac{k}{2}(x-a)(b-x) \leq 0,$$

i.e.,

$$-\frac{k}{2}(x-a)(b-x) \leq f(x) - f(a) - \frac{x-a}{b-a}\big(f(b) - f(a)\big) \leq \frac{k}{2}(x-a)(b-x).$$

This last holds for all $k > 0$, whence it holds when $k \to 0$,

$$0 \leq f(x) - f(a) - \frac{x-a}{b-a}\big(f(b) - f(a)\big) \leq 0,$$

i.e.,

$$f(x) = f(a) + \frac{f(b) - f(a)}{b-a}(x-a)$$

is linear. □

This is a nice little argument, but subtle. One must keep k, ε fixed until the right moment, then note that it works for both values of ε, and only then take the limit as $k \to 0$.

Cantor added a footnote about the use of the Extreme Value Theorem:

> This proof is based essentially on the frequently occurring and proven theorem from the lectures of Herr *Weierstrass*:
> "A given function $\varphi(x)$ of a real variable, continuous in an interval $(a \ldots b)$ (including the limits), *reaches* the maximum g of the values which it can assume for at least one value x_0 of the variable, so that $\varphi(x_0) = g$."
> *Ossian Bonnet* has introduced a similar proof, also resting on this [theorem], of the Fundamental Theorem of the Differential Calculus; it can be found in "*Cours de calcul différentiel et intégral*, by *J. A. Serret*, Paris, 1868" in the first volume, pp. 17 – 19.[320]

Bonnet's proof is similar also in the form of the auxiliary function used, and like Weierstrass's proof of Rolle's Theorem it uses the fact that the derivative, if it exists, is 0 at a local extremum.

Cantor and Schwarz both corresponded with Weierstrass, so it is surprising that, after Schwarz had produced his auxiliary function reducing the Linear Function Theorem to the Extreme Value Theorem and Cantor cited Bonnet's use of a similar auxiliary function in reducing the Mean Value Theorem to the Extreme Value Theorem, Weierstrass still proved the Mean Value Theorem only for continuously differentiable functions four years after the publication of Cantor's paper in his 1874 lectures on the theory of functions.[321]

Bonnet's proof, as Cantor noted, was published in Serret's book two years before Schwarz obtained his result. It might be noted that Serret's book does not begin to match the standards of rigour of Cauchy, Bolzano, or Weierstrass, and Serret is occasionally led into error.[322] His exposition of the proof of the Mean Value Theorem does not contain any false lemmas, but it does rest on unproven or inadequately proven results. It is given early on in his book, in the second chapter on "Differentiation of functions of one independent variable". The chapter begins with two short sections (12 and 13) on continuity and differentiability, proceeds in a third section (14) to prove the Mean Value Theorem, devotes the next two sections (15 and 16) to the Constant Function and Increasing Function Theorems, finally finishing the subject in a section (17) with the Cauchy Mean Value Theorem. Mathematically, the only novelty for us is the fact that Rolle's Theorem is not singled out in this exposition. For this, the brevity of Sects. 12 and 13, and the historical importance of Sect. 14, I present these sections here:

Of continuity.

12. A function $f(x)$ of the variable x is called *continuous* for values of x between two limits x_0 and X, where, for all values of x, the absolute value of the difference

$$f(x+h) - f(x)$$

[320] *Ibid.*, p. 141.

[321] I owe this observation to Reinhard Bölling.

[322] Cf. Peano's criticisms cited on pp. 355 and 358, below.

decreases indefinitely with h, or is infinitely small at the same time as h.
If the function $f(x)$ becomes infinite for a value of x between x_0 and X, it does not satisfy the preceding definition of continuity; one says then that it becomes *discontinuous* passing through infinity.

Of derivatives.

13. The function $f(x)$ being assumed continuous for the values of x between x_0 and X, the corresponding increments

$$h \text{ and } f(x+h) - f(x)$$

are simultaneously infinitely small, as we have said. The limit of the ratio

$$\frac{f(x+h)-f(x)}{h}$$

of these increments is in general a definite quantity independent of the sign of h; it depends on the value attributed to x and, consequently, it is a function of this variable. It was given the name of *derivative* of the function $f(x)$, and we represent it, following Lagrange, by the notation $f'(x)$; thus we will have

$$\frac{f(x+h)-f(x)}{h} = f'(x) + \varepsilon, \tag{1}$$

or

$$f(x+h) - f(x) = hf'(x) + h\varepsilon, \tag{2}$$

ε designating a quantity infinitely small simultaneously with h
It may happen that, for some particular values of x, the limit of the ratio

$$\frac{f(x+h)-f(x)}{h}$$

depends on the sign that is attributed to h in making this quantity tend to zero; in this case the derivative of the function ceases to be determinate.
From the foregoing, if we take h to be the principal infinitesimal, the increment

$$f(x+h) - f(x)$$

will be infinitely small of the first order, unless the derivative $f'(x)$ is zero or infinite. We shall see later that this circumstance can arise only for specific values assigned to x. When $f'(x)$ is zero, the increment of the function is an infinitesimal of order greater than 1; on the other hand this order is less than 1 when $f'(x)$ becomes infinite.
14. The simple notion of the derivative leads to several important proposals that we will establish.
THEOREM I. —*Let $f(x)$ be a function of x which remains continuous for the values of x between two given limits, and which, for these values, has a definite derivative $f'(x)$. If x_0 and* X *denote two values of x between these same limits, we have*

$$\frac{f(\mathrm{X}) - f(x_0)}{\mathrm{X} - x_0} = f'(x_1),$$

x_1 being a value between x_0 and X.
Indeed, the ratio

$$\frac{f(\mathrm{X}) - f(x_0)}{\mathrm{X} - x_0}$$

has, by hypothesis, a finite value, and if we denote this value by A, we have
(1) $[f(\mathrm{X}) - \mathrm{AX}] - [f(x_0) - \mathrm{A}x_0] = 0.$
Denote by $\varphi(x)$ the function of x defined by the formula
(2) $\varphi(x) = [f(x) - \mathrm{A}x] - [f(x_0) - \mathrm{A}x_0],$
we have, because of the equality (1),

$$\varphi(x_0) = 0, \quad \varphi(\mathrm{X}) = 0,$$

so that $\varphi(x)$ vanishes for $x = x_0$ and for $x = \mathrm{X}$. Suppose, to fix ideas, $\mathrm{X} > x_0$ and make x grow from x_0 to X; the function $\varphi(x)$ is initially zero. If we accept that it is not constantly zero, for the values of x between x_0 and X, it will begin to grow by taking positive values, or to decrease by taking negative values, either from $x = x_0$ or from a value of x between x_0 and X. If the values in question are positive, since $\varphi(x)$ is continuous and it must vanish for $x = \mathrm{X}$, it is clear that there will be a value x_1 between x_0 and X such that

$$\varphi(x_1)$$

will be greater than or at most equal to the neighbouring values

$$\varphi(x - h), \quad \varphi(x + h),$$

h being a quantity as small as we please. If the function $\varphi(x)$, ceasing to be zero, takes negative values, the same reasoning proves that there exists a value x_1 between x_0 and X such that

$$\varphi(x_1)$$

will be less than or at most equal to the neighbouring values

$$\varphi(x - h), \quad \varphi(x + h).$$

Thus, in either case, the value of x_1 will be such that the differences

$$\varphi(x_1 - h) - \varphi(x_1), \quad \varphi(x_1 + h) - \varphi(x_1)$$

will be of the same sign, and, therefore, the ratios
(3) $$\frac{\varphi(x_1 - h) - \varphi(x_1)}{-h}, \quad \frac{\varphi(x_1 + h) - \varphi(x_1)}{h}$$
will be of contrary signs.
Note that we do not exclude the case in which one of the previous ratios is reduced to zero, which requires that the function $\varphi(x)$ retains the same value for values of x included in a finite interval. In particular, if the function $\varphi(x)$ is constantly zero for the values of x between x_0 and X, the ratios (3) are one and the other zero.
The ratios (3) tend to the same limit when h tends to zero, because we assume that the function $f(x)$ has a definite derivative, and the same takes place, accordingly with respect to $\varphi(x)$; moreover, these ratios have opposite signs, so the limit is zero. Thus we have

$$\lim \frac{\varphi(x_1 + h) - \varphi(x_1)}{h} = 0,$$

or, because of the Eq. (2),

$$\lim \left[\frac{f(x_1 + h) - f(x_1)}{h} - \mathrm{A}\right] = 0,$$

that is to say

$$\mathrm{A} = \lim \frac{f(x_1 + h) - f(x_1)}{h} = f'(x_1).$$

We thus have

$$\frac{f(X) - f(x_0)}{X - x_0} = f'(x_1),$$

or

(4) $$f(X) - f(x_0) = (X - x_0)f'(x_1),$$

as was announced.

We have assumed $X > x_0$, but as the preceding formula does not change by the permutation of the letters x_0, X, it is obviously independent of this hypothesis.

If we make

$$X = x_0 + h,$$

the quantity x_1, between x_0 and $x_0 + h$, can be represented by $x_0 + \theta h$, θ being a quantity between 0 and 1; we can thus write

(5) $$f(x_0 + h) - f(x_0) = hf'(x_0 + \theta h).$$

REMARK. — The proof which precedes is due to Mr. Ossian Bonnet. It should be noted that it does not suppose in any way the continuity of the derivative $f'(x)$; it only requires that the derivative exists and has a definite value.[323]

3.8.3 Remark The similarity of Bonnet's formula (1) to Galois's near proof of the Cauchy Mean Value Theorem (pp. 294–297) is striking. Had Bonnet read Galois's note and realised what Galois had missed? Or, did he, which is equally likely, discover the argument independently? Unfortunately, we do not have Bonnet's own account of the discovery but only Serret's report of the proof.

With Serret's book we pretty much have the modern exposition of the Mean Value Theorem. His book does not prove the Extreme Value Theorem and thus his proof of the Mean Value Theorem is inadequate, but the reduction of the Mean Value Theorem to this unproven result differs from the standard classroom proof mainly in that Serret does not explicitly flag Rolle's Theorem for the reader's attention as he proves it in passing. Our history of the Mean Value Theorem thus almost ends here. What was now needed was a good accessible exposition of the theory of the Calculus, one discussing the foundations of the real number line *à la* Weierstrass, Dedekind, or any of the others that would appear in print in 1872, which applied this discussion to prove those theorems — Intermediate Value Theorem, Extreme Value Theorem — that most mathematicians had been applying without proof, which clarified the pointwise/uniform distinction also in print by 1872, and which rigorously treated the Mean Value Theorem and integration. This was finally accomplished in 1878 with the publication in Pisa of Ulisse Dini's (1845–1918) textbook *Fondamenti per la teorica della funzioni di variabili reali* [*Foundations for a theory of functions of a real variable*]. This book was the first textbook written with modern rigour. Volkert describes it thus:

> In Dini's textbook we find all the themes and concepts which are characteristic of the then "modern" Analysis: construction of the real numbers from the rationals (in the Italian edition of 1878 with cuts,[324] in the German of 1892 with the aid of Cauchy sequences ("because it enjoys now, at least in Germany, a greater following")), Weierstraß's theorem on the

[323] J.A. Serret, *Cours de calcul différentiel et intégral. Tome Premier. Calcul différentiel*, Gauthier-Villars, Paris, 1868, pp. 15–19.

[324] I.e., via Dedekind's construction.

assumption of the maximum,[325] difference between pointwise and uniform properties, etc. Dini's textbook remained for a long time a leader in its field. Later the new edition of Jordan's "Cours d'analyse" arrived as a competitor.[326]

There would be other rigorous texts to follow, in various languages, for example, Axel Harnack (1881) in German, but Dini's was the most popular until it was supplanted by later editions of Jordan's 1882 French language textbook. Peano would consider Dini's book and its proof of the Mean Value Theorem to be perfect. But not even a perfect proof exposited in a perfect textbook takes immediate effect, as we shall now see.

3.9 Peano and the Mean Value Theorem

Jordan's *Cours d'analyse* had a flaw which Peano brought to the attention of the mathematical world in the correspondence of the *Nouvelles annales de mathématiques* in 1884. The ensuing exchange affords us a nice example of mathematical culture lag. The definitive treatment of the Mean Value Theorem was out there for all to see, but not everyone saw it:

Extract of a letter of Mr. Dr. J. Peano.[327] —In his *Cours d'Analyse de l'École Polytechnique*, p. 21, Mr. Jordan gives a not very rigorous demonstration of the following theorem.
"Let $y = f(x)$ be a function of x whose derivative is finite and definite for each x varying in a certain interval.[328]
"Let a and $a+h$ be two values of x taken from this interval. We have

$$f(a+h) - f(a) = \mu h,$$

μ designating a quantity intermediate between the greatest and the least value of $f'(x)$ in the interval from a to $a+h$."
In fact, the author says, give to x a series of values $a_1, a_2, \ldots, a_{n-1}$ intermediate between a and $a+h$; put

---

[325] I.e., the Extreme Value Theorem.

[326] Volkert, Klaus Volkert, *Geschichte der Analysis*, Bibliographisches Institut & F.A. Brockhaus AG, Zürich, 1988, p. 225.

[327] "J" for "Joseph", the French equivalent of Giuseppe Peano's first name.

[328] The conditions of a derivative being finite and definite are usually subsumed under the existence condition today. But one can also allow infinite derivatives, which would occur at vertical tangents as in Fig. 2.2.43 on page 142, where

$$\lim_{h\to 0} \frac{f(x+h)-f(x)}{h} = +\infty.$$

This is distinguished from, e.g., a cusp like that of Fig. 1.6 on page 43 in which

$$\lim_{h\to 0+} \frac{f(x+h)-f(x)}{h} = +\infty, \text{ but } \lim_{h\to 0-} \frac{f(x+h)-f(x)}{h} = -\infty.$$

The word "definite" is usually replaced today by "well-defined", i.e., asserting both existence and uniqueness. Here, in particular, the left- and right-sided derivatives are assumed to agree.

$$f(a_r) - f(a_{r-1}) = (a_r - a_{r-1})[f'(a_{r-1}) + \varepsilon_r].$$

Suppose, maintaining the intermediate values $a_1, \ldots, a_{n-1}$ to be indefinitely multiplied (and approaching more closely).[329] The quantities $\varepsilon_1, \varepsilon_2, \ldots$ all tend to zero, because ε_r is the difference between $\dfrac{f(a_r) - f(a_{r-1})}{a_r - a_{r-1}}$ and its limit $f'(a_{r-1})$.

This assertion is not justified; for

$$f'(a_{r-1}) = \lim \frac{f(a_r) - f(a_{r-1})}{a_r - a_{r-1}}$$

when we suppose a_{r-1} is fixed, and the variable a_r approaches indefinitely to a_{r-1}; but we cannot affirm this when a_r and a_{r-1} vary simultaneously, if we do not suppose that the derivative is continuous.

Thus, for example, put

$$y = f(x) = x^2 \sin \frac{1}{x},$$

with

$$f(0) = 0;$$

its derivative

$$f'(x) = 2x \sin \frac{1}{x} - \cos \frac{1}{x}$$

for $x \gtrless 0$, and $f'(0) = 0$, remains finite and definite, but discontinuous.

Let

$$a = 0, \ h > 0;$$

set

$$a_1 = \frac{1}{2n\pi}, \ a_2 = \frac{1}{(2n+1)\pi},$$

$a_3, a_4, \ldots$ whatever.

We have

$$\varepsilon_2 = \frac{f(a_2) - f(a_1)}{a_2 - a_1} - f'(a_1):$$

but

$$f(a_1) = 0, \ f(a_2) = 0, \ f'(a_1) = -1;$$

thus

$$\varepsilon_2 = 1,$$

and its limit is not zero.

Almost the same error is committed by Mr. Hoüel (*Cours de Calcul infinitésimal*, vol. 1, p. 145). I add lastly that one can demonstrate very easily the formula

$$f(x_0 + h) - f(x_o) = hf'(x_0 + \theta h),$$

without supposing the continuity of the derivative.

Extract of a letter of Mr. C. Jordan.— I have nothing to respond to the criticism of Mr. Dr. Peano, which is perfectly justified. I have to admit implicitly into my demonstration that $\dfrac{f(x+h) - f(x)}{h}$ tends *uniformly* to $f'(x)$ in the interval from a to b. This is one of the

[329] I.e., keep taking finer partitions of the interval $[a, a+h]$.

points over which I propose moreover to correct in my third volume.
Mr. Peano says that it is easy to demonstrate the formula

$$f(x+h) - f(x) = hf'(x+\theta h),$$

without supposing the continuity of the derivative. I would be pleased by his communicating to me his argument, because I do not know of any which seems to me entirely satisfactory.[330]

As we see, Jordan had given the Ampère-Cauchy proof, but without carefully noting the conditions necessary for the proof. Almost a decade and a half had passed between the publication of Serret's book and the appearance of Jordan's, and either he was unaware of the existence of the superior proof or he simply did not trust it.[331] Peano notes that the proof given by Jordan does not hold if f' is not continuous and gives Darboux's function as a counterexample. And Jordan correctly points out that the proof depends on uniform differentiability and promises to correct the oversight in the next volume of his work. This reads like a rejoinder to Peano, who seemed to suggest that continuous differentiability suffices. Peano, as we shall see, was well aware that continuous differentiability on a closed, bounded interval implies uniform differentiability, but, as we saw in our discussion of Bolzano, this is not an obvious fact easily proven without presupposing the Mean Value Theorem itself.

Peano did not respond in print right away to Jordan's request. First, a third party responded. This was a Belgian mathematician Louis Philippe Gilbert, whose letter appeared in a subsequent issue of the same journal:

Mr. Editor,
Allow me a few words of response to criticism of Mr. Dr Peano, which Mr. Jordan would have no difficulty answering himself, if he had not probably seen a more subtle difficulty behind it.
I observe first that it is not necessary that the ε tend to zero for *each* mode of division of the interval h into indefinitely decreasing parts δ; only that it takes place for *one* mode of division, and the theorem in question will be shown. Mr. Peano supposes, in his criticism and in his example, that the quantities a_r are not *fixed* values of the variable x. However, nothing prevents us from conceiving decreasing the intervals between consecutive values of x, while assuming them fixed and inserting new values of x between them, which will remain fixed in turn, between them new values equally fixed, and so on indefinitely.[332] The δ intervals, forever subdivided, can be decreased below any given quantity, and each inserted value of x remaining fixed, the ratio

$$\frac{f(x+\delta) - f(x)}{\delta} = \varphi(x, \delta)$$

can not tend, for each of them, to a different limit for $f'(x)$. Unless therefore that, for *each* mode of division of the interval h into indefinitely decreasing δ parts, the difference

$$\varphi(x, \delta) - f'(x)$$

remains above a fixed limit for a finite or indefinitely increasing number of values of x, when all the δ intervals tend simultaneously to zero, the demonstration can continue forever in the same manner.

---

[330] *Nouvelles annales de mathématiques*, 3rd series, vol. 3 (1884), pp. 45–47.

[331] The proof given by Serret was not above criticism. Cf. p. 358, below.

[332] Gilbert inserts a footnote: "This is indeed, judging by his words, the thought of Jordan.".

Fig. 3.21 Gilbert's "counterexample"

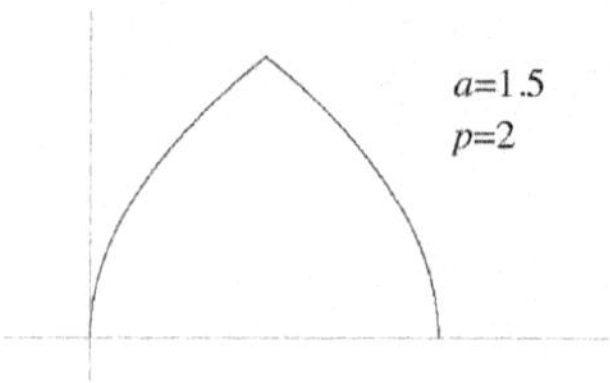

This is not the case, we see without difficulty, for the function

$$x^2 \sin \frac{1}{x};$$

so the theorem challenged by him is perfectly applicable. In making $a_1 = \frac{1}{2n\pi}$ and $a_2 = \frac{1}{(2n+1)\pi}$ and therefore making a_1 and a_2 tend simultaneously to zero, Mr. Peano arbitrarily introduced an unnecessary requirement. The demonstration can not be done *in this way*, that's all.

Mr. Peano believes it is easy to prove the formula

$$f(x+h) - f(x) = hf'(x - \theta h),$$

without assuming the continuity of the derivative. Mr. Jordan asked, not without malice, to see this demonstration, which is impossible, since the theorem is inexact.[333]

Suppose a function $f(x)$ equal to $\sqrt{2px}$ from $x = 0$ to $x = a$, and to $\sqrt{2p(2a-x)}$ from $x = a$ to $x = 2a$. [See Fig. 3.21.] This function is continuous, but its derivative ceases to be for $x = a$, where it passes from the value $\sqrt{\frac{p}{2a}}$ to the value $-\sqrt{\frac{p}{2a}}$.

We obviously have, h being $< a$,

$$f(a+h) - f(a-h) = \sqrt{2p(a-h)} - \sqrt{2p(a-h)} = 0:$$

yet there is no value of x between $a - h$ and $a + h$ for which $f'(x)$ reduces to zero.

Note that the theorem of Mr. Jordan remains true, however, in this case, since zero is between the values

$$\sqrt{\frac{p}{2(a-h)}} \text{ and } -\sqrt{\frac{p}{2(a-h)}}$$

of $f'(x)$ which correspond to $a - h$ and $a + h$. And yet here Mr. Peano could renew his objection, since $f(a+\delta) - f(a-\delta)$ does not have the limit $f'(a)$ when δ tends to zero.

PH. GILBERT,

Professor at the University of Louvain

It was now time for Peano to respond, which he did in a subsequent issue[334]:

[333]The French *inexact* seems to be used here to mean imprecise, perhaps in the sense of admitting counterexamples.

[334]*Nouvelles annales de mathématiques*, 3rd series, vol. 3 (1884), pp. 153–155.

Sir,
Permit me to respond to the letter of Mr. Gilbert. His comments add nothing to the rigour of the demonstration of Mr. Jordan. It assumes fixed the values successively interpolated in the interval considered; but, in my example, one may well consider them fixed, and the reasoning will still remain, if the first two retain the forms

$$\frac{1}{(2n+1)\pi} \text{ and } \frac{1}{2n\pi}.$$

Finally there is always a system of quantities $\varepsilon_1, \varepsilon_2, \ldots$ each of which has a limit of zero, but the number increases indefinitely; and when that happens, we cannot conclude in general that their maximum also tends to zero.
Mr. Gilbert says that the theorem will be demonstrated if it is proven that for *one* mode of division, the ε will have zero as the limit. If we mean by these words that, for a mode of division, the maximum of the ε have zero as a limit, the proposition is just; but as this does not happen for *every* mode of division, the resulting theorem is demonstrated only when Mr. Gilbert has found this particular mode of division for which the above condition is satisfied.
And I say this without malice, because this mode exists, but I leave it to Mr. Gilbert to find; and to settle the matter, I suggest he prove this theorem, which he uses:
If $f(x)$ has a definite and finite derivative $f'(x)$ for each value of x belonging to a finite interval (a, b), an arbitrarily small quantity ε being fixed, one can always divide the interval (a, b) with the points

$$x_0 = a, x_1, x_2, \ldots, x_{n-1}, x_n = b,$$

in such fashion that each of the differences

$$\frac{f(a_{r+1}) - f(a_r)}{a_{r+1} - a_r} - f'(a_r) \quad (r = 0, 1, \ldots, n-1)$$

will be, in absolute value, less than ε.
I said in my first letter, one easily proves the formula

$$f(x+h) - f(x) = hf'(x + \theta h),$$

without assuming the continuity of the derivative, but only its existence (that is to say the existence of a definite and finite derivative for all values of the variable in the interval in question). I learned this demonstration from Mr. Genocchi, when I was a student; it is due to Ossian Bonnet, and is found in the *Cours de Calcul* of Mr. Serret (2nd edition, p. 17); but there are some small imperfections, which may explain why Mr. Jordan has doubts about its rigour. But it is found also perfectly rigorous in:
DINI, *Fondamenti per la teorica delle funzioni di variabili reali*, p. 75; Pisa, 1878.
HARNACK, *Differential- und Integralrechnung*, p. 64; Leipzig, 1881.
PASCH, *Einleitung in die Diff.- und Integralrechnung*, p. 83; Leipzig, 1882, etc.[335]
The example cited by Mr. Gilbert, to prove that the theorem is inexact, does not satisfy the conditions of the theorem. Indeed, the function of Mr. Gilbert has, for $x = a$, what we call a left derivative and a right derivative (*rückwärts und vorwärts genommene Differential-Quotienten*),[336] and does not have a definite ordinary derivative.
Finally, I would add, in response to Mr. Jordan, that if $\dfrac{f(x+h) - f(x)}{h}$ tends uniformly to $f'(x)$, this derivative is continuous, and vice versa.

---

[335] Peano adds the footnote: "I sent this demonstration to Mr. Jordan, it is there some time; but you will find it attached.".

[336] *Backwards and forwards taken differential quotients.*

I have the honour to be, etc. G. PEANO.
Here is the demonstration of the formula

$$f(x+h) - f(x) = hf'(x+\theta h).$$

THEOREM I (of Rolle). — *If $f(x)$ has a definite and finite derivative $f'(x)$ for each of the values of x belonging to the interval (a, b), and if $f(a) = 0, f(b) = 0$, one has, for a certain value x_1 of x contained in the interior of the same interval,*

$$f'(x_1) = 0.$$

Indeed, $f(x)$, having a derivative, is continuous; and, varying x between a and b, $f(x)$ assumes its largest and its smallest value. If these extreme values are both null, the function will be constantly null, and we also have

$$f'(x) = 0.$$

If they are not both null, let x_1 be the value of x for which $f(x)$, not being null becomes maximum or minimum. The value x_1 is interior to the interval (a, b), for $f(a) = 0$, and $f(b) = 0$; and for $x = x_1$ the function is neither increasing nor decreasing; thus $f'(x_1)$ is neither negative nor positive; and, since we have supposed that it is definite and finite, it will be null.

THEOREM II. — *If $f(x)$ has a definite and finite derivative $f'(x)$ for each of the values of x belonging to the interval (a, b), one has*

$$f(b) - f(a) = (b-a)f'(x_1),$$

where x_1 is a value of x contained between a and b.
Indeed, applying the preceding theorem to the function

$$\mathrm{F}(x) = f(x) - f(a) - \frac{x-a}{b-a}[f(b) - f(a)],$$

for which

$$\mathrm{F}(a) = 0, \ \mathrm{F}(b) = 0, \ \mathrm{F}'(x) = f'(x) - \frac{f(b) - f(a)}{b-a},$$

we have

$$\mathrm{F}'(x_1) = 0$$

or

$$f'(x_1) = \frac{f(b) - f(a)}{b-a}. \qquad \text{q.e.d.}$$

One sees that one does not assume the continuity of the derivative, but only its existence. We can ignore its existence for the values $x = a$ and $x = b$, but assuming $f(x)$ continuous for these values; the theorem still remains true if the derivative becomes infinite, but of definite sign.
This demonstration is from Mr. Ossian Bonnet.

G. P.[337]

Peano offers quite a lot in this letter. Beginning with Lagrange, after some improvement by Ampère, and a perfecting by Cauchy, one standard proof of the Mean Value Theorem proceeded in two steps. First one noted that the ratio

$$\frac{f(b) - f(a)}{b-a}$$

[337] *Nouvelles annales de mathématiques*, 3rd series, vol. 3 (1884), pp. 252–256.

was an average of ratios[338]

$$\frac{f(a_{r+1}) - f(a_r)}{a_{r+1} - a_r} \tag{3.59}$$

for any partition $a = a_0 < a_1 < \ldots < a_{n-1} < a_n = b$. One then *implicitly* appealed to the *uniform* differentiability of f to place these latter ratios between $f'(a_r) - \varepsilon$ and $f'(a_r) + \varepsilon$ for ε arbitrarily small. Letting m and M be the minimum and maximum values of $f'(x)$ on $[a, b]$, respectively, this yielded the Mean Value Inequality

$$m \leq \frac{f(b) - f(a)}{b - a} \leq M.$$

One then appealed to the continuity of f' and the Intermediate Value Theorem for continuous functions (or, since 1875, one could appeal to Darboux's Intermediate Value Theorem for Derivatives) to conclude the existence of some $c \in [a, b]$ such that

$$f'(c) = \frac{f(b) - f(a)}{b - a}.$$

Gilbert said, and Peano now affirmed, that, provided m, M existed, one did not need uniformity in establishing the Mean Value Inequality: for any $\varepsilon > 0$ one can construct a partition locating each ratio (3.59) within ε of $f'(a_r)$ assuming only the continuity of f on $[a, b]$ and its differentiability on (a, b). He did not choose to present a proof, but instead challenged Gilbert to do so. The reason why is unclear. He did not say, "Gilbert claims this and I call upon him to prove it because I do not know how to do this", but asserted the result to be true. My suspicion is that he regarded the proof as over-elaborate as compared to the Bonnet proof he tacked onto the end of his letter, and he didn't think including the details would add anything of value to his letter.

He then turns to what has since become the standard classroom proof of the Mean Value Theorem, stating that he learned it from Angelo Genocchi (1817–1889), but that it is due to Bonnet and is to be found in Serret's text. He hints at some slight reservation about Serret's book and points to the texts of Dini, Harnack, and Pasch for perfectly rigorous proofs.

The letter finishes with two brief points before he gives the Bonnet proof in an attachment. The first dismisses Gilbert's counterexample with the simple remark that the function cited by Gilbert is not differentiable at a, though it does have one-sided derivatives there. Thus, it is not a counterexample. The second of these closing points also seems slightly dismissive. Peano answers Jordan's emphasis that his error was in assuming uniform differentiability where Peano refers to continuous differentiability — an emphasis given "not without malice"? — was misguided criticism: the two notions agree. Peano neglects to say "obviously", but given that he offers no hint of a proof, I take that as implied. And I would fault him for this, as the proof that continuity implies uniform differentiability is only trivial after the Mean Value Theorem has

338"Average" in the sense of Cauchy; cf. pp. 274–275.

been proven — as we saw in discussing Bolzano — and this implication is only true in closed, bounded intervals.

While this exchange was taking place, Peano was putting the finishing touches on his edition of Genocchi's lectures on the Calculus. Peano had met with the director of the Fratelli Bocca publishers in 1883 and the director had broached the subject of publishing Genocchi's lectures. Genocchi had been injured in an accident and Peano, who had been substituting for him in teaching the course, offered to write the lectures up, submitting them to him for his approval and publishing them under Genocchi's name. This was done and in September of 1884 the book appeared,[339] the title reading

ANGELO GENOCCHI

CALCOLO DIFFERENZIALE

E

PRINCIPII DI CALCOLO INTEGRALE

PUBBLICATO CON AGGIUNTE

DAL

D.r GIUSEPPE PEANO,

i.e., "Angelo Genocchi, *Differential Calculus and Principles of the Integral Calculus, Published by the Assistance of Dr. Giuseppe Peano*". The book consists of two parts, first a long list of annotations made by Peano to the lectures, and then the lectures themselves following the lines of Genocchi's course. Paragraphs 43–46 form a section titled "Theorems on the derivative", with 43 presenting Lemma 2.3.18 in Chap. 2 and its negative counterpart on a function's being increasing at a point, 44 presenting Rolle's Theorem, 45 the Mean Value Theorem and the Cauchy Mean Value Theorem, and 46 the Constant Function Theorem and the uniqueness (up to an additive constant) of the anti-derivative. The proofs in paragraphs 44 and 45 yielding the Mean Value Theorem are essentially the same as that given in Peano's second letter to the *Nouvelles annales de mathématiques* and need not be repeated here.

What is of particular interest is what he has to say in the annotations:

N. **44-45**.

The demonstration given here of the fundamental formula of the Calculus is attributed to OSSIAN- - BONNET. Cf. SERRET, *Calcul*, etc., N. 14.[340] As exhibited by SERRET, it lends itself to some objections. The words "il faudra qu'elle (the function) commence à croître

---

[339]For more on the writing and reception of this book, I refer the reader to Kennedy, *Life of Peano*, *op. cit.* pp. 11–19.

[340]The "etc." obviously refers to the rest of the publication data for Serret's book, which is given in an annotation for an earlier paragraph. "N. 14" refers to Sect. 14 in Serret's book wherein Bonnet's proof is to be found.

en prenant des valeurs positives ou à decroître..."[341] express an inexact concept, because a function can for a special value of the variable be neither increasing, decreasing, nor constant, as is for example the case with the function[342] $x \sin \frac{1}{x}$ for $x = 0$.

The demonstration given by JORDAN, *Analysis*, etc., supposes that $\frac{f(x+h)-f(x)}{h}$ converges uniformly to $f'(x)$ for all values of x contained in the interval (a, b), which requires the continuity of the derivative. (Cf. the Exercises 9 and 14.)[343] On this proposition see my note published in the *Nouvelle Annales*, 1884, page 45. See there also pages 153 and 252. It suffices, for the validity of Theorem N. 45, that the function $f(x)$ have a derivative at each value of x in the interval (a, b) and $[f]$ be continuous at the endpoints a and b; this derivative can be infinitely large for any value of x provided only that it has a definite sign there. Cf. DINI, *Fondamenti*, etc., pp. 69 and following.[344]

* *

A more general formula like that in N. 45 is the following. If $f(x), \varphi(x), \psi(x)$ possess derivatives for all values of x belonging to the interval (a, b), then for some value of x_1 between a and b

$$\begin{vmatrix} f'(x_1) & \varphi'(x_1) & \psi'(x_1) \\ f(a) & \varphi(a) & \psi(a) \\ f(b) & \varphi(b) & \psi(b) \end{vmatrix} = 0.$$

Setting $\psi(x) = 1$, one obtains the second formula,[345] and moreover [setting] $\varphi(x) = x$ [one obtains] the first.[346, 347]

3.9.1 Exercise Prove Peano's generalisation of the Mean Value Theorem by considering the determinant

$$\begin{vmatrix} f(x_1) & \varphi(x_1) & \psi(x_1) \\ f(a) & \varphi(a) & \psi(a) \\ f(b) & \varphi(b) & \psi(b) \end{vmatrix}$$

and differentiating its expansion across the top row.[348]

---

[341] Basically, Serret, *op. cit.*, p. 16, claims that if $f'(c) > 0$, then f is not only increasing *at* c, but in some neighbourhood of c. This, as we saw with Exercise 2.3.20 in Chap. 2, is not always the case.

[342] A typo; the function should be $x^2 \sin \frac{1}{x}$. $x \sin \frac{1}{x}$ is not differentiable at 0 and is not a counterexample. Cf. his remarks on Jordan, p. 355, above.

[343] This reference to Jordan is absent in the German translation of 1899. By the time this translation appeared, Jordan had incorporated the Bonnet proof into a new edition of his text.

[344] In the German translation reference is given to page 90 of the German translation of Dini's text.

[345] I.e., the Cauchy Mean Value Theorem.

[346] I.e., the Mean Value Theorem.

[347] Angelo Genocchi, *Calcolo differenziale*, Fratelli Bocca, Torino, 1884, pp. xiv–xv. A German translation by G. Bohlmann and A. Schepp, appeared in 1899: Angelo Genocchi, *Differentialrechnung und Grundzüge der Integralrechnung, herausgegeben von Giuseppe Peano*, Verlag von B.G. Teubner, Leipzig, 1899. In this edition, the annotations occur after the main text, more in line with traditional end-notes, and the passage cited occurs on pp. 316–317.

[348] Peano generalises this further on page xxii of the *Calcolo* to $n+1$ functions replacing f, φ, ψ and n points replacing a, b. See Kennedy, *op. cit.*, p. 45, for an English translation.

3.9.2 Exercise Peano follows his statement of this last result with a generalisation of Rolle's Theorem: Let f be continuous on $[a, b]$ and differentiable on (a, b). If $f(a) = f(b) = 0$, then for any real number k, there is some $c \in (a, b)$ such that $f'(c) = kf(c)$.
i. Prove this.
ii. For $f(x) = x^2 - 1$, $a = -1$, $b = 1$, and any value of k, find an explicit $c \in [a, b]$ satisfying $f'(c) = kf(c)$.
[Hint. i. Consider $g(x) = f(x)e^{-kx}$.]

3.10 Gilbert Revisited

3.10.1 Gilbert's Final Words

As a mathematician and scientist, Gilbert was not deemed important enough to be given an entry in the *Dictionary of Scientific Biography* when it was published in the 1970s, nor was he included the following decade when two supplementary volumes were published. One can find some information on him at the MacTutor web site, and Jean Mawhin discusses him in an article[349] covering, among other things, the correspondence translated in the preceding section. Mawhin sums up Gilbert as follows:

> More famous in mechanics for his *barogyroscope*, a mechanical device showing Earth rotation, Gilbert, in analysis, seems to be better known for his polemic than for his contributions.[350]

And he cites a devastating critique found in a letter from Charles Hermite to Genocchi:

> To use a French expression, Mr. Gilbert is an awkward customer[351]...Mr. Picard expresses to me on the subject of the letter of Gilbert to the editor, p. 153,[352] an opinion which I completely share...He judges the communication of Mr. Gilbert archstupid; it consequently does not merit your consideration.[353]

Gilbert wrote another letter to the editor in response to Peano's challenge. At first sight, one is tempted to dismiss the letter with the same contempt Picard and Hermite dismissed his earlier letter. Peano himself evidently thought it unnecessary to respond once again to Gilbert as Gilbert's letter ends the correspondence. Mawhin, however, finds some redeeming value in Gilbert's contribution — not, unfortunately, with respect to the Mean Value Theorem. It is, nonetheless, worth considering here.

---

[349] Jean Mawhin, "Some contributions of Peano to analysis in the light of the work of Belgian mathematicians", in: Fulvia Skof (ed.), *Giuseppe Peano between Mathematics and Logic*, Springer-Verlag Italia, Milan, 2011.

[350] *Ibid.*, p. 14.

[351] *Mauvais coucheur.*

[352] Cited on page 353, above.

[353] Mawhin, *op. cit.*, p. 14.

Letter of Mr. Ph. Gilbert. Professor at the University of Louvain.

I will, if you permit, continue with Mr. Dr Peano the examination of an important point of analysis.
My observations (this volume, p. 153[354]) were not intended to "add anything to the rigour of the demonstration of Mr. Jordan", but to prove that Mr. Peano's objection against this demonstration was not enough, and to clarify the defective point of the demonstration.
The objection (this volume, p. 46[355]) was that "if $f'(a_{r-1})$ is the limit of $\dfrac{f(a_r)-f(a_{r-1})}{a_r - a_{r-1}}$ when we assume a_{r-1} fixed and a_r approaching indefinitely to a_{r-1}, one can not affirm (in general) when a_r and a_{r-1} vary at the same time["]: that which is verified about the example

$$f(x) = x^2 \sin \frac{1}{x}$$

with

$$a_1 = \frac{1}{(2n+1)\pi}, \quad a_2 = \frac{1}{2n\pi}.$$

This objection fails, as I pointed out (this volume, p. 153,[356] if we decrease the δ intervals by the successive interposition of new *fixed* values of x, as is allowed, and as we can in particular for the alleged example.[357]
The fault of the demonstration by Mr. Jordan (and others before him) is thus not there. It is, as I said (p. 154), in that "when one makes all the δ intervals tend simultaneously to zero", by successively interposing new values of x, we can assume that the difference

$$\frac{f(x+\delta)-f(x)}{\delta} - f'(x)$$

always remains above a fixed limit for a finite or indefinitely increasing number of x values, precisely because we always introduced new [ones]. This is the "most subtle difficulty" of which I spoke, and of which Mr. Jordan no doubt spoke in saying that his argument assumed that the ratio of increments tends *uniformly* to $f'(x)$.
For that which regards the formula

$$f(x+h) - f(x) = hf'(x+\theta h),$$

I confess to having misinterpreted the terms of Peano: "one shows very easily *without assuming the continuity of the derivative*". I understood by this that he extended to all possible discontinuities of the derivative, whereas he assumed, for each value of x, [a derived value that is] *finite, definite* and *equal for both directions*, that which returns the strong theorem obtained by Mr. O. Bonnet, and restricts significantly the scope of the formula. It is for this reason that I opposed him the kinds of discontinuity that occur most often in the derivative, for which the above equation does not apply, while the theorem stated by Mr. Jordan remains. I am therefore inclined to believe this latter theorem more general, and it would be desirable for it to be established rigorously in all its generality.

---

[354]Cited on page 360, above.

[355]Cited on page 351, above.

[356]Cited on page 353, above.

[357]Gilbert adds a footnote here: "In my example, said Mr. Peano, one can still assume them fixed and the reasoning still remains, if the first two retain the form $\dfrac{1}{(2n+1)\pi}, \dfrac{1}{2n\pi}$." Without quibbling over a point of no importance, I would note that the interval between two consecutive values cannot decrease below any given number, *that which is essentially assumed by Mr. Jordan*, unless n becomes infinite, that is to say that a_1 and a_2 tend simultaneously to zero.

In this regard, the proposition that Mr. Peano has given proves to be of no use because it assumes precisely this restriction that the derivative $f'(x)$ be *finite* and *unique* for each value of x in the interval (a, b), a restriction which I wish to rule out if possible. Indeed, Mr. Peano is aware that when the derivative $f'(x)$ has a *unique* value at each point, albeit possibly infinite at some points, one can demonstrate rigorously, *without making use of the proposition that he states*, that the ratio $\frac{f(b)-f(a)}{b-a}$ is between the smallest and largest value of $f'(x)$ in the interval (a, b) (Mr. Jordan's theorem), and that this property remains even for some cases where the function $f(x)$ itself is discontinuous. The proposition in question can therefore not serve for my purpose, which is why I have not tried much to improve the demonstration that I give below; but as the theorem itself provides some interest, I hope that Mr. Peano will want to publish his demonstration, which will probably be better.

If $f(x)$ has a finite and definite[358] *derivative for all values of x belonging to a finite interval (a, b), and if we fix a quantitiy ε as small as we want, it is always possible to divide the interval (a, b) into a finite number of values of x:*

$$a, a_1, a_2, \ldots a_{n-1}, b,$$

in such a fashion that each of the differences

$$\frac{f(a_{r+1})-f(a_r)}{a_{r+1}-a_r}-f'(a_r) \quad (r=0,1,2,\ldots,n-1)$$

is, in absolute value, less than ε.

Suppose $a < b$. By the hypothesis, it is always possible to find a specific and positive quantity δ_1 such that one has in absolute value

$$\frac{f(a+\theta\delta_1)-f(a)}{\theta\delta_1}-f'(a)<\varepsilon,$$

θ denoting, *in general*, an arbitrary quantity > 0 and equal to or less than unity. Let δ_1 be as large as possible, and set $a+\delta_1=a_1$. We can likewise find a sequence of specific quantities $\delta_2, \delta_3, \ldots$ such that we always have

$$\left\|\left[\frac{f(a_1+\theta\delta_2)-f(a_1)}{\theta\delta_2}-f'(a_1)\right]\right\|<\varepsilon,\ a_1+\delta_2=a_2;$$

$$\left\|\left[\frac{f(a_2+\theta\delta_3)-f(a_2)}{\theta\delta_3}-f'(a_2)\right]\right\|<\varepsilon,\ a_2+\delta_3=a_3,$$

and, in general,

$$(1) \qquad \left\|\left[\frac{f(a_r+\theta\delta_{r+1})-f(a_r)}{\theta\delta_{r+1}}-f'(a_r)\right]\right\|<\varepsilon.$$

The quantities $a, a_1, a_2, \ldots, a_r, \ldots$ forming an ever increasing sequence, only two hypotheses are possible: 1st either the intervals $\delta_1, \delta_2, \ldots, \delta_r, \ldots$ will never become smaller than a fixed quantity δ, and in this case, for a finite value $n-1$ of the number r, we have[359]

$$a_{n-1}<b,\quad a_{n-1}+\delta_n=b,$$

---

[358] Gilbert adds a footnote: The same in both directions.

[359] Presumably, he intends $a_{n-1}+\delta_n \geq b$.

from which, attributing to θ a suitable value equal to or less than 1, $a_{n-1} + \theta\,\delta_n = b$. In this case, according to relation (1), the quantities

$$a, a_1, a_2, \ldots, a_{n-1}, b$$

will be the values of x which satisfy the required condition.
2nd or the successive values $a_1, a_2, \ldots, a_r, \ldots$, while constantly growing, cannot reach the value b, thus requiring that the intervals $\delta_1, \delta_2, \ldots, \delta_r, \ldots$ eventually become smaller than any given quantity. Then the increasing quantities $a_1, a_2, \ldots, a_r, \ldots$converge to a fixed limit c less than or at most equal to b, of the sort that, however small the positive quantity σ, between $c - \sigma$ and c there will be an infinity of quantities $a_p, a_{p+1}, \ldots$ or an infinite number of intervals δ within the interval $(c - \sigma, c)$.
But, since c is between a and b, the derivative has for $x = c$ a unique and definite value $f'(c)$. It is therefore always possible to assign a finite interval σ such that one has (θ always having the same meaning as that above)

$$\left|\left[\frac{f(c-\theta\sigma) - f(c)}{-\theta\sigma} - f'(c)\right]\right| < \frac{\varepsilon}{2}$$

or, equivalently, denoting by η a quantity which depends on θ, but which remains between -1 and $+1$,

(2) $$\frac{f(c) - f(c-\theta\sigma)}{\theta\sigma} = f'(c) + \eta\frac{\varepsilon}{2}.$$

On the other hand, by the hypotheses on $f(x)$ and the theorem of Mr. Bonnet, we have

(3) $$\frac{f(c) - f(c-\theta\sigma)}{\theta\sigma} = f'(\xi),$$

ξ designating a certain value of x, such that

$$c - \theta\sigma < \xi < c;$$

Thus if we combine the Eqs. (2) and (3), we have

(4) $$f'(\xi) = f'(c) + \eta\frac{\varepsilon}{2}.$$

Thus there necessarily exists, in the interval $(c - \sigma, c)$ as defined above, at least one value ξ of x, such that the derivative $f'(\xi)$ will differ from $f'(c)$ by a quantity smaller in absolute value than $\frac{\varepsilon}{2}$.
However, according to a remark made above, this value ξ will coincide with one of the quantities $a_p, a_{p+1}, \ldots$, or will be between two of them, a_r and a_{r+1}; it can therefore be represented by $a_r + \theta\,\delta_{r+1}$, so that one will have, in any case, according to (1),

(5) $$\left|\left[\frac{f(\xi) - f(a_r)}{\xi - a_r} - f'(a_r)\right]\right| < \varepsilon.$$

Let us now apply to this value ξ of x the relation (2), making, in the latter, $c - \theta\sigma = \xi$; we have

$$\frac{f(c) - f(\xi)}{c - \xi} = f'(c) + \eta'\frac{\varepsilon}{2},$$

where $\eta' - \eta$ being smaller than 2,

(6) $$\left|\left[\frac{f(c) - f(\xi)}{c - \xi} - f'(\xi)\right]\right| < \varepsilon.$$

Inequalities (1), (5), (6) show that one can pass *effectively* from the value a to the value c by a *finite* number of values of x

$$a, a_1, a_2, \ldots, a_r, \xi, c,$$

such that two consecutive values x' and x'' always verify the relation

$$\frac{f(x'') - f(x')}{x'' - x'} - f'(x') < \varepsilon$$

in absolute value.

By reasoning over the interval (c, b) as we have reasoned over the interval (a, b), we eventually establish that we can always go from a to b by a finite number of intervals δ, which satisfy the requirement contained in the statement of the theorem.

One can easily see what changes would be required if the demonstration had $c = b$.[360]

The proof could have been more clearly written. And it could have been a little more complete in showing why the process stops after finitely many steps. But a completely rigorous presentation of his proof is not what matters here. At first sight the undertaking is as archstupid as Gilbert's first letter: He uses the Mean Value Theorem to prove a lemma which would make Jordan's proof of the Mean Value Theorem valid — but circular, proving that the Mean Value Theorem follows from itself. In discussing this letter, however, Mawhin states that the result is not inconsequential. As provided by Gilbert it may not be of use in proving the Mean Value Theorem, but i. there are proofs of the result that do not depend on the Mean Value Theorem,[361] and ii. the lemma has application elsewhere, in modern integration theory .[362]

3.10.2 *Thomas Flett and the Vindication of Gilbert*

So far as I know, the first successful proof of the Mean Value Theorem using the Peano–Gilbert result without assuming uniform differentiability is due to Thomas Muirhead Flett (1923–1976), who proved this result without appealing to the Mean Value Theorem. I have not seen Flett's proof, but Mawhin presents a proof[363] of a variant of the result that is sufficiently strong to yield the Mean Value Theorem for f differentiable on $[a, b]$ provided that f' assumes maximum and minimum values on this interval. Mawhin's lemma is the following.

3.10.1 Lemma *Let f be differentiable on $[a, b]$. For any $\epsilon > 0$, there is a partition $a = a_0 < a_1 < a_2 < \ldots < a_{n-1} < a_n = b$ and points $x_0, x_1, x_2, \ldots, x_{n-1}$ such that, for each i, $x_i \in (a_i, a_{i+1})$ and*

$$\left|f(a_{i+1}) - f(a_i) - (a_{i+1} - a_i)f'(x_i)\right| < \epsilon(a_{i+1} - a_i).$$

Mawhin tells us that this proof is an application of the Heine–Borel Theorem, and he presents a simpler proof using a technical refinement of the Heine–Borel Theorem. The Heine–Borel Theorem, which is a lemma extracted by Émile Borel

[360] *Nouvelles annales de mathématiques*, 3rd series, vol. 3 (1884), pp. 475–482.

[361] Mawhin, *op. cit.*, p. 18, 27–28.

[362] *Ibid.*, p. 28.

[363] Mawhin, *op. cit.*, pp. 18, 27–28.

from Heine's proof of the Uniform Continuity Theorem, also follows easily from the Least Upper Bound Principle. The Lemma requires a couple of simple definitions.

3.10.2 Definition Let X be a set of real numbers and $\mathcal{O}$ a collection of open intervals (u, v). $\mathcal{O}$ is a *cover* of X if X is contained in the union of $\mathcal{O}$,

$$X \subseteq \bigcup \mathcal{O} = \{x \mid \exists uv(x \in (u, v) \in \mathcal{O})\}.$$

3.10.3 Definition Let $\mathcal{O}$ be a cover of a set X. A subset $\mathcal{O}_0 \subseteq \mathcal{O}$ is a *subcover* of $\mathcal{O}$ if $\mathcal{O}_0$ is itself a cover of X. $\mathcal{O}_0$ is a *finite subcover* if it contains only finitely many intervals.

3.10.4 Lemma (Heine–Borel Theorem) *Let $a < b$. Every cover $\mathcal{O}$ of $[a, b]$ has a finite subcover.*

Proof. The usual continuous inductive proof using the Least Upper Bound Principle goes through easily. One defines

$$X = \{x \in [a, b] \mid \text{some finite } \mathcal{O}_0 \subseteq \mathcal{O} \text{ covers } [a, x]\}.$$

X is non-empty since $a \in X$. Moreover X is bounded above by b.

If $x \in X$ and $x \neq b$, then for some $\delta > 0, x + \delta \in X$. For, given $\mathcal{O}_0$ covering $[a, x]$, there is some $(u, v) \in \mathcal{O}_0$ with $x \in (u, v)$. If we take

$$\delta = \frac{1}{2} \min \{b - x, v - x\},$$

then $[x, x + \delta] \subseteq (u, v)$ and we see $[a, x + \delta]$ is covered by $\mathcal{O}_0$.

Thus, let c be the least upper bound of X and choose $(u, v) \in \mathcal{O}$ such that $c \in (u, v)$. By an argument analogous to that just given, there is some $\delta > 0$ such that $c - \delta \in (u, v)$ and $c - \delta \in [a, c]$. By the choice of c, there is some $x \in X$ with $c - \delta < x$. Choose a finite $\mathcal{O}_0 \subseteq \mathcal{O}$ that covers $[a, x]$ and note that

$$[a, c] \subseteq [a, x] \cup [c - \delta, c] \subseteq \bigcup \mathcal{O}_0 \cup (u, v),$$

whence $\mathcal{O}_0 \cup \{(u, v)\}$ is a finite cover of $[a, c]$, i.e., $c \in X$. But X can only have b as a maximum, whence $c = b$ and $\mathcal{O}$ contains a finite subcover of $[a, b]$. □

The truth of the Heine–Borel Theorem is not as intuitively clear as that of the Least Upper Bound Principle. However, once one has proven it, its application can be easier than the direct inductive proofs applying the Least Upper Bound Principle. For example:

Proof of the Uniform Continuity Theorem. Let f be continuous on $[a, b]$. Let $\epsilon > 0$ and for each $x \in [a, b]$, choose $\delta_x > 0$ so that for all $y \in [a, b]$,

$$|x - y| < \delta_x \Rightarrow |f(x) - f(y)| < \frac{\epsilon}{2}.$$

Let $\mathcal{O} = \{(x - \delta_x/2, x + \delta_x/2) \mid x \in [a, b]\}$ and notice that $\mathcal{O}$ is a cover of $[a, b]$. Let $\mathcal{O}_0$ be a finite subcover,

$$\{(x_0 - \delta_{x_0}/2, x_0 + \delta_{x_0}/2), (x_1 - \delta_{x_1}/2, x_1 + \delta_{x_1}/2), \ldots,$$
$$(x_{n-1} - \delta_{x_{n-1}}/2, x_{n-1} + \delta_{x_{n-1}}/2)\}$$

and let $\delta = \min\{\delta_{x_0}, \delta_{x_1}, \ldots, \delta_{x_{n-1}}\}$. Let $y, z \in [a, b]$ and assume $|y - z| < \delta/2$. Choose x_i such that $y \in (x_i - \delta_{x_i}/2, x_i + \delta_{x_i}/2)$. Then

$$|y - x_i| < \delta_{x_i}/2,$$

and

$$|z - x_i| = |z - y + y - x_i| < |z - y| + |y - x_i| < \frac{\delta}{2} + \frac{\delta_{x_i}}{2} \leq \delta_{x_i},$$

since $\delta \leq \delta_{x_i}$. But

$$\begin{aligned} |y - x_i| < \delta_{x_i} \ \& \ |z - x_i| < \delta_{x_i} &\Rightarrow \left|f(y) - f(x_i)\right| < \frac{\epsilon}{2} \ \& \ \left|f(z) - f(x_i)\right| < \frac{\epsilon}{2} \\ &\Rightarrow \left|f(y) - f(x_i) + f(x_i) - f(z)\right| < \frac{\epsilon}{2} + \frac{\epsilon}{2} \\ &\Rightarrow \left|f(y) - f(z)\right| < \epsilon \end{aligned}$$

□

Like the original cover $\mathcal{O}$, a finite subcover $\mathcal{O}_0$ may be highly redundant, containing intervals contained in other intervals or contained in the union of other intervals. To apply the Heine–Borel Theorem to the Peano–Gilbert problem, we will need to avoid this by using a minimal cover:

3.10.5 Definition Let $[a, b]$ be given. A cover $\mathcal{O}$ of $[a, b]$ is *minimal* if no proper subset $\mathcal{O}' \subset \mathcal{O}$ also covers $[a, b]$.

3.10.6 Lemma (Extended Heine–Borel Theorem) *Let $a < b$. Every cover $\mathcal{O}$ of $[a, b]$ has a minimal finite subcover.*

Proof. Let

$$\mathcal{X} = \{\mathcal{O}' \subseteq \mathcal{O} \mid \mathcal{O}' \text{ covers } [a, b] \text{ and is finite}\}.$$

The set of cardinalities of elements of $\mathcal{X}$ is a nonempty set of positive integers and thus has a least element n. Let $\mathcal{O}_0 \in \mathcal{X}$ have exactly n elements. $\mathcal{O}_0$ is minimal since any proper subset covering $[a, b]$ would belong to $\mathcal{X}$ and have cardinality $< n$, contrary to the definition of n. □

3.10.7 Lemma *Let $\mathcal{O}$ be a minimal finite cover of $[a, b]$ of cardinality n. There are $a_0, a_1, \ldots, a_{n-1}, b_0, b_1, \ldots, b_{n-1}$ such that $\mathcal{O} = \{(a_0, b_0), (a_1, b_1), \ldots, (a_{n-1}, b_{n-1})\}$ and*

$$a_0 < a \leq a_1 < b_0 \leq a_2 < b_1 \leq \ldots \leq a_{n-1} < b_{n-2} \leq b < b_{n-1}, \tag{3.60}$$

i.e.,
i. $a \in (a_0, b_0)$
ii. $b \in (a_{n-1}, b_{n-1})$
iii. $b_i \in (a_{i+1}, b_{i+1})$ *for* $i = 0, 1, \ldots, n-2$
iv. $(a_i, b_i) \cap (a_{i+2}, b_{i+2}) = \emptyset$ *for* $i = 0, 1, \ldots, n-3$
v. $\bigcup_{i \le n-1} (a_i, b_i) = (a_0, b_{n-1})$.

Before proving this, let me quickly note that the cover

$$\mathcal{O} = \left\{ \left(\frac{-1}{2}, \frac{1}{2}\right), \left(\frac{1}{4}, \frac{3}{4}\right), \left(\frac{1}{2}, \frac{3}{2}\right), \left(\frac{3}{4}, \frac{7}{4}\right), \left(\frac{3}{2}, \frac{5}{2}\right) \right\}$$

of $[0, 2]$ shows that we cannot always replace the weak inequalities in the chain of inequalities of the Lemma by strict ones.

Proof. We proceed inductively.

For the basis step, we observe that a belongs to a unique element of $\mathcal{O}$. For, if

$$a \in (c, d) \in \mathcal{O} \text{ and } a \in (e, f) \in \mathcal{O},$$

one has $d < f$ or $f \le d$. If $d < f$, then

$$[a, b] \cap [c, d) = [a, d) \subseteq [a, f) = [a, b] \cap [e, f),$$

and dropping (c, d) from $\mathcal{O}$ results in a smaller cover of $[a, b]$, contrary to $\mathcal{O}$'s minimality. Likewise, if $f \le d$, the interval (e, f) is redundant. Thus a belongs to only one element, say, $a \in (a_0, b_0) \in \mathcal{O}$.

For the induction step, assume $a_0, a_1, \ldots, a_k, b_0, b_1, \ldots, b_k$ satisfying (3.60) up to k have been found, i.e., condition i holds and conditions iii, iv, and v hold for n replaced by $k-1$.

If $k < n-1$, $b \notin [a_0, b_k)$ because otherwise $\{(a_0, b_0), \ldots, (a_k, b_k)\}$ is a cover of $[a, b]$, contrary to the minimality of $\mathcal{O}$. Thus $b_k \le b$. As with a, b_k must belong to a unique element of $\mathcal{O}$: If $b_k \in (c, d) \in \mathcal{O}$ and $b_k \in (e, f) \in \mathcal{O}$, then one of $(a_k, b_k) \cup (c, d)$ and $(a_k, b_k) \cup (e, f)$ contains the other and one of (c, d), (e, f) can be dropped from $\mathcal{O}$ by the minimality of $\mathcal{O}$: one of (c, d), (e, f) is contained in the union of two other intervals in $\mathcal{O}$. Thus let (a_{k+1}, b_{k+1}) be the unique element of $\mathcal{O}$ to which b_k belongs. Condition iii is automatically satisfied for $i = k$.

To see that iv holds for $i = k-1$, note that we have
$(a_{k-1}, b_{k-1}) \cap (a_{k+1}, b_{k+1}) \neq \emptyset \Rightarrow$

$$(a_{k-1}, b_{k-1}) \cup (a_k, b_k) = (a_{k-1}, b_k) \subseteq (a_{k-1}, b_{k+1}) = (a_{k-1}, b_{k-1}) \cup (a_{k+1}, b_{k+1})$$

and (a_k, b_k) is redundant, contrary to the minimality of $\mathcal{O}$.

Finally, to see that v holds for $k+1$, observe that

$$\begin{aligned}\bigcup_{i\leq k+1}(a_i,b_i) &= \bigcup_{i\leq k}(a_i,b_i)\cup(a_{k+1},b_{k+1})\\ &= (a_0,b_k)\cup(a_{k+1},b_{k+1}),\ \text{by induction hypothesis}\\ &= (a_0,b_{k+1})\end{aligned}$$

since $a_0 < a_{k-1} < b_k < b_{k+1}$.

The induction step can be iterated until $k = n-1$. For, then all elements of $\mathcal{O}$ have been listed among the intervals $(a_0,b_0),\ldots,(a_k,b_k)$. Since $b\in\bigcup_{i\leq k}(a_i,b_i)$ and $b\notin\bigcup_{i\leq k-1}(a_i,b_i)$, it follows that $b\in(a_k,b_k)$. Thus $i-v$ are true and the proof is complete. □

In proving Lemma 3.10.1, we begin with a technical lemma offering a precise statement that, if $f'(x)$ exists and $y\leq x\leq z$ with $y\neq z$ and $|z-y|$ very small, the ratio $\big(f(z)-f(y)\big)/(z-y)$ is close to $f'(x)$. Geometrically this says that the slopes of the secant lines connecting $\langle y,f(y)\rangle$ and $\langle z,f(z)\rangle$ approach the slope of the tangent line through $\langle x,f(x)\rangle$ so long as y,z approach x from opposite directions simultaneously. That is, as in Newton's maligned proof of the product formula (pp. 116–117, above), one does not need to anchor one of y,z at x so long as they straddle x. (To conclude the same to hold when y,z approach x from only one direction, one must assume the continuity of f'.)

3.10.8 Lemma *Let f be differentiable on $[a,b]$ and let $\delta_0>0$. For every $x\in[a,b]$ and every $\epsilon>0$ there is a $\delta>0$ with $\delta<\delta_0$ such that for all $y,z\in[a,b]$,*

$$x-\delta<y\leq x\leq z<x+\delta\Rightarrow\big|f(z)-f(y)-(z-y)f'(x)\big|\leq\epsilon(z-y),$$

with equality holding only when $y=z$.

Proof. By the differentiability of x, for any $\epsilon>0$ there is a $\delta>0$, which we may take to be less than the given δ_0, such that for all $w\in[a,b]$, if $0<|w-x|<\delta$,

$$\left|\frac{f(w)-f(x)}{w-x}-f'(x)\right|<\epsilon,$$

i.e.,

$$\big|f(w)-f(x)-(w-x)f'(x)\big|<\epsilon|w-x|. \tag{3.61}$$

Note that, if we replace "$<$" by "$\leq$", (3.61) remains true when $w=x$ and this is the only value of $w\in(x-\delta,x+\delta)$ making the two sides of the inequality equal.

Let now $x-\delta<y\leq x\leq z<x+\delta$ with $y\neq z$, thus at most one of y,z equal to x. We have

$$\begin{aligned}\big|f(z)-f(x)-(z-x)f'(x)\big| &\leq \epsilon(z-x)\\ \big|f(x)-f(y)-(x-y)f'(x)\big| &\leq \epsilon(x-y),\end{aligned}$$

with equality holding at most once. Then

$$\begin{aligned}&\left|f(z)-f(y)-(z-y)f'(x)\right|\\&\qquad\leq\left|f(z)-f(x)-(z-x)f'(x)+f(x)-f(y)-(x-y)f'(x)\right|\\&\qquad<\epsilon(z-x)+\epsilon(x-y)=\epsilon(z-y).\end{aligned}$$ □

Proof of Lemma 3.10.1. For each $x\in[a,b]$ choose δ_x in accordance with Lemma 3.10.8 and define $D(x)=(x-\delta_x,x+\delta_x)$. The set

$$\mathcal{O}=\left\{D(x)\,\middle|\,x\in[a,b]\right\}$$

covers $[a,b]$ and thus has a minimal finite subcover $\mathcal{O}_0=\{D(x_0),D(x_1),\ldots,D(x_{n-1})\}$. By Lemma 3.10.7, we may assume $x_0<x_1<x_2<\ldots<x_{n-1}$,

$$D(x_i)\cap D(x_{i+1})\neq\emptyset,\quad\text{and}\quad D(x_i)\cap D(x_{i+2})=\emptyset.$$

Choose $a_{i+1}\in D(x_i)\cap D(x_{i+1})$ with $x_i<a_i<x_{i+1}$ for $i=0,1,\ldots,n-2$. Extend the list by setting $a_0=a$, $a_n=b$. We have

$$a=a_0<x_0<a_1<x_1<\ldots<a_{n-1}<x_{n-1}<a_n=b,$$

and, for $i\leq n-1$, $a_i,a_{i+1}\in D(x_i)$.

By Lemma 3.10.8,

$$\begin{aligned}a_i,a_{i+1}\in D(x_i)&\Rightarrow x-\delta_{x_i}<a_i<x_i<a_{i+1}<x_i+\delta_{x_i}\\&\Rightarrow\left|f(a_{i+a})-f(a_i)-(a_{i+1}-a_i)f'(x_i)\right|<\epsilon(a_{i+1}-a_i).\end{aligned}$$

This completes the proof of Lemma 3.10.1. □

We can now establish the Mean Value Theorem in the case where f' exists on all of $[a,b]$ and assumes maximum and minimum values — in particular for f' continuous on $[a,b]$. For,

$$\begin{aligned}&\left|\sum\left(f(a_{i+1})-f(a_i)-(a_{i+1}-a_i)f'(x_i)\right)\right|\\&\qquad\qquad\leq\sum\left|f(a_{i+1})-f(a_i)-(a_{i+1}-a_i)f'(x_i)\right|\\&\qquad\qquad<\sum\epsilon(a_{i+1}-a_i)=\epsilon(b-a).\end{aligned}$$

Thus

$$-\epsilon(b-a)<\sum\left(f(a_{i+1}))-f(a_i)\right)-\sum(a_{i+1}-a_i)f'(x_i)<\epsilon(b-a),$$

i.e.,

$$-\epsilon(b-a) < f(b) - f(a) - \sum(a_{i+1} - a_i)f'(x_i) < \epsilon(b-a), \tag{3.62}$$

whence

$$\sum(a_{i+1} - a_i)f'(x_i) - \epsilon(b-a) < f(b) - f(a) < \sum(a_{i+1} - a_i)f'(x_i) + \epsilon(b-a).$$

If we let m, M denote the minimum and maximum values, respectively, of $f'(x)$ on $[a, b]$, we have

$$(b-a)m = \sum(a_{i+1} - a_i)m \le \sum(a_{i+1} - a_i)f'(x_i)$$
$$\sum(a_{i+1} - a_i)f'(x_i) \le \sum(a_{i+1} - a_i)M = (b-a)M.$$

Thus

$$(b-a)m - \epsilon(b-a) < f(b) - f(a) < (b-a)M + \epsilon(b-a)$$
$$m - \epsilon < \frac{f(b) - f(a)}{b-a} < M + \epsilon.$$

This holds for all ϵ, whence

$$m \le \frac{f(b) - f(a)}{b-a} \le M.$$

One now wants to apply the Intermediate Value Theorem to f' to conclude the existence of $c \in [a, b]$ such that

$$f'(c) = \frac{f(b) - f(a)}{b-a}.$$

To this end, we can now assume f continuously differentiable on $[a, b]$, thus proving the Mean Value Theorem along Jordan's lines under the assumption of continuous differentiability that Peano had declared Jordan's proof to depend on. Jordan, it will be recalled, noted after Peano's criticism that his proof required uniform differentiability. The two notions, as Bolzano had tried unsuccessfully to prove and as Peano noted without proof in the correspondence, are equivalent on $[a, b]$.

So what does this say about Gilbert's attempt to vindicate Jordan? At first sight it seems just an increase in complication, making Gilbert's letter as "archstupid" as it appears on first impression. But if one looks at it, the Flett-Mawhin elaboration of Gilbert's idea does yield a bit more:

(1) For f differentiable on $[a, b]$ and f' bounded there, with $m \le f'(x) \le M$ for all $x \in [a, b]$, it yields a Mean Value Inequality:

$$m \leq \frac{f(b) - f(a)}{b - a} \leq M.$$

(2) For f differentiable on $[a, b]$ and f' assuming a maximum and a minimum on $[a, b]$, one can appeal to Darboux's Intermediate Value Theorem for Derivatives to conclude the existence of $c \in [a, b]$ such that

$$f'(c) = \frac{f(b) - f(a)}{b - a}.$$

These offer greater generality than continuous differentiability, holding for example for Darboux's function

$$f(x) = \begin{cases} x^2 \sin\left(\frac{1}{x}\right), & x \neq 0 \\ 0, & x = 0, \end{cases}$$

which is differentiable with extreme derivatives in any closed bounded interval, but is not continuously differentiable at 0.

On the other hand, these results are still not as general as the Bonnet result, not applying for example to the related function

$$f(x) = \begin{cases} x^2 \sin\left(\frac{1}{x^2}\right), & x \neq 0 \\ 0, & x = 0, \end{cases}$$

which is also differentiable, but whose derivatives are unbounded in any interval containing 0. (*Exercise.*)

As we saw in discussing Bolzano (pp. 310–311, above) and Weierstrass (pp. 342–343, above), the latter with respect to the Strictly Increasing Function Theorem, an extra argument can now be applied to generalise the result further. Note that the points x_i at which the derivatives are taken are all in the interior (a, b) of $[a, b]$, whence the only reason for assuming f' differentiable at the endpoints was to guarantee the existence of minimum and maximum values. Thus, (1) should read

(1′) For f continuous on $[a, b]$, differentiable on (a, b), and f' bounded on (a, b), say, $m \leq f'(x) \leq M$ for all $x \in (a, b)$, the Mean Value Inequality holds:

$$m \leq \frac{f(b) - f(a)}{b - a} \leq M.$$

Ad hoc arguments can now be applied to conclude the Mean Value Theorem. First, let $\frac{f(b) - f(a)}{b - a}$ assume one of the extreme values.

Suppose

$$\frac{f(b) - f(a)}{b - a} = M.$$

Let x be any element of (a, b). Note that (1′) yields

$$\frac{f(x)-f(a)}{x-a} \le \operatorname{lub}\left\{f'(y) \mid y \in (a,x)\right\} \le \operatorname{lub}\left\{f'(y) \mid x \in (a,b)\right\} = M$$
$$\frac{f(b)-f(x)}{b-x} \le \operatorname{lub}\left\{f'(y) \mid y \in (x,b)\right\} \le \operatorname{lub}\left\{f'(y) \mid x \in (a,b)\right\} = M,$$

where lub X is the least upper bound of X. But by Ampère's Discrete Mean Value Inequality (Lemma 3.4.1, page 265, above), we have

$$M = \frac{f(b)-f(a)}{b-a} \le \frac{f(x)-f(a)}{x-a} \quad \text{or} \quad M = \frac{f(b)-f(a)}{b-a} \le \frac{f(b)-f(x)}{b-x}.$$

Assume the first. Then
$$\frac{f(x)-f(a)}{x-a} = \frac{f(b)-f(a)}{b-a},$$

whence
$$f(x) = f(a) + \frac{f(b)-f(a)}{b-a}(x-a).$$

This holds for all $x \in (a, b)$. Thus, for such x,

$$f'(x) = \frac{f(b)-f(a)}{b-a}.$$

The other case is handled similarly, and the entire argument can be repeated for $\dfrac{f(b)-f(a)}{b-a} = m.$

If $\dfrac{f(b)-f(a)}{b-a}$ does not equal either m or M, there are $d_1, d_2 \in (a, b)$ such that

$$m < f'(d_1) < \frac{f(b)-f(a)}{b-a} < f'(d_2) < M.$$

Darboux's Intermediate Value Theorem for Derivatives yields c between d_1 and d_2, whence in (a, b), such that
$$f'(c) = \frac{f(b)-f(a)}{b-a}.$$

This same argument applies if f' is unbounded above and below on (a, b).

I leave to the reader to work out the details in case $f'(x)$ is bounded above but not below, or bounded below but not above. Once this is done, we have an overly complicated proof of the Bonnet form of the Mean Value Theorem for f continuous on $[a, b]$ and differentiable on (a, b).

It must be admitted, however, that the addenda to the proof relied in some cases on Darboux's result, the proof of which depends on the theorem that $f'(x) = 0$ at any extreme value, which very quickly yields Rolle's Theorem and the Mean Value Theorem, whence the application of Darboux's result to proving a version of the Mean Value Theorem under stricter conditions than the simple proofs may seem inappropriate. To this I can only remind the reader that this immediacy is logical, not historical: the modern proof of the Mean Value Theorem was a long time coming — Galois and Bolzano overlooking the last step in a proof *à la* that of Rolle's Theorem in the 1830s, Weierstrass proving Rolle's Theorem in 1861 but not recognising the reduction of the Mean Value Theorem to it, and finally Bonnet proving the Mean Value Theorem in 1868 — and that this whole exchange of letters was due to ignorance of this proof. People knew the Ampère-Cauchy approach and the approach *à la* Jordan using Riemann sums, essentially yielding the Mean Value Theorem for continuously differentiable functions as a corollary to the Fundamental Theorem of the Calculus.

Finally, I note that we can also apply the Extended Heine–Borel Theorem to the vindication of Bolzano by using these lemmas to prove Bolzano's conjecture that continuous differentiability implies uniform differentiability on a closed bounded interval without appeal to the Mean Value Theorem itself.

Proof that continuous differentiability implies uniform differentiability on closed bounded intervals (Lemma 3.6.10). Let f be continuously differentiable on $[a, b]$. Then f' is uniformly continuous on that interval. Let $\epsilon > 0$ and choose $\delta > 0$ such that, for all $x, y \in [a, b]$,

$$|x - y| < \delta \Rightarrow \left|f'(x) - f'(y)\right| < \frac{\epsilon}{2}. \tag{3.63}$$

Let $|x - y| < \delta$ be given. Suppose, for the sake of definiteness, that $x < y$ and consider the interval $[x, y]$. f is differentiable on $[x, y]$, whence for each $z \in [x, y]$ there is a $\delta_z > 0$ such that, for all $w \in [x, y]$,

$$|w - z| < \delta_z \Rightarrow \left|\frac{f(w) - f(z)}{w - z} - f'(z)\right| < \frac{\epsilon}{2}. \tag{3.64}$$

Define

$$D(z) = \left\{w \,\middle|\, z - \frac{\delta_z}{2} < w < z + \frac{\delta_z}{2}\right\}$$

and

$$\mathcal{O} = \left\{D(z) \,\middle|\, z \in [x, y]\right\}.$$

$\mathcal{O}$ is an open cover of $[x, y]$, whence the Extended Heine–Borel Theorem yields a minimal finite subcover given by elements $z_0 < z_1 < \ldots < z_{n-1}$ satisfying

$$z_0 - \frac{\delta_{z_0}}{2} < x \leq z_1 - \frac{\delta_{z_1}}{2} < z_0 + \frac{\delta_{z_0}}{2} \leq \ldots$$

$$< z_{m-1} - \frac{\delta_{z_{m-1}}}{2} < z_{m-2} + \frac{\delta_{z_{m-2}}}{2} \leq y < z_{m-1} + \frac{\delta_{z_{m-1}}}{2}.$$

Now, for any pair z_k, z_{k+1}, we have

$$\begin{aligned} z_{k+1} - z_k &< \frac{\delta_{z_k}}{2} + \frac{\delta_{z_{k+1}}}{2} \\ &< \max\{\delta_{z_k}, \delta_{z_{k+1}}\}, \end{aligned}$$

whence, for z_k^* equalling one of z_k, z_{k+1}, (3.64) yields

$$\left| \frac{f(z_{k+1} - f(z_k)}{z_{k+1} - z_k} - f'(z_k^*) \right| < \frac{\epsilon}{2},$$

i.e.,

$$f'(z_k^*) - \frac{\epsilon}{2} < \frac{f(z_{k+1}) - f(z_k)}{z_{k+1} - z_k} < f'(z_k^*) + \frac{\epsilon}{2}. \tag{3.65}$$

By Ampère's Discrete Mean Value Inequality (Lemma 3.4.1), there are $0 \leq i, j < n$ such that

$$\frac{f(z_{i+1}) - f(z_i)}{z_{i+1} - z_i} \leq \frac{f(y) - f(x)}{y - x} \leq \frac{f(z_{j+1}) - f(z_j)}{z_{j+1} - z_j}. \tag{3.66}$$

By (3.65),

$$\begin{aligned} f'(z_i^*) - \frac{\epsilon}{2} &< \frac{f(z_{i+1}) - f(z_i)}{z_{i+1} - z_i} < f'(z_i^*) + \frac{\epsilon}{2} \\ f'(z_j^*) - \frac{\epsilon}{2} &< \frac{f(z_{j+1}) - f(z_j)}{z_{j+1} - z_j} < f'(z_j^*) + \frac{\epsilon}{2}, \end{aligned}$$

whence (3.66) yields

$$f'(z_i^*) - \frac{\epsilon}{2} < \frac{f(y) - f(x)}{y - x} < f'(z_j^*) + \frac{\epsilon}{2}. \tag{3.67}$$

But now we can apply (3.63) to conclude

$$f'(x) - \frac{\epsilon}{2} < f'(z_i^*) < f'(x) + \frac{\epsilon}{2}, \quad f'(x) - \frac{\epsilon}{2} < f'(z_j^*) < f'(x) + \frac{\epsilon}{2},$$

which, combined with (3.67), yields

$$f'(x) - \epsilon < \frac{f(y) - f(x)}{y - x} < f'(x) + \epsilon,$$

i.e.,

$$\left|\frac{f(y)-f(x)}{y-x}-f'(x)\right|<\epsilon. \qquad \square$$

Once this has been proven, one can, with Bolzano, derive the Mean Value Theorem for continuously differentiable functions simply by repeating any proof of the Theorem for uniformly differentiable functions.

3.10.3 *Generalisations of the Mean Value Theorem*

In his second letter Gilbert attempted to clarify his reason for having introduced the counterexample that Peano shot down:

> I confess to having misinterpreted the terms of Peano: "one shows very easily *without assuming the continuity of the derivative*". I understood by this that he extended to all possible discontinuities of the derivative, whereas he assumed, for each value of x, [a derived value that is] *finite, definite* and *equal for both directions*, that which returns the strong theorem obtained by Mr. O. Bonnet, and restricts significantly the scope of the formula. It is for this reason that I opposed him the kinds of discontinuity that occur most often in the derivative, for which the above equation does not apply, while the theorem stated by Mr. Jordan remains. I am therefore inclined to believe this latter theorem more general, and it would be desirable for it to be established rigorously in all its generality.
> *In this regard*, the proposition that Mr. Peano has given proves to be of no use because it assumes precisely this restriction that the derivative $f'(x)$ be *finite* and *unique* for each value of x in the interval (a, b), a restriction which I wish to rule out if possible. Indeed, Mr. Peano is aware that when the derivative $f'(x)$ has a *unique* value at each point, albeit possibly infinite at some points, one can demonstrate rigorously, *without making use of the proposition that he states*, that the ratio $\frac{f(b)-f(a)}{b-a}$ is between the smallest and largest value of $f'(x)$ in the interval (a, b) (Mr. Jordan's theorem), and that this property remains even for some cases where the function $f(x)$ itself is discontinuous.[364]

He was absolutely right. The Mean Value Inequality does not require all derivatives to be two-sided, and the finiteness of the derivative can be weakened in the Mean Value Theorem itself.

Let us first consider the case of the occasional infinite derivative. In the present Chap. I have ignored the case of a function continuous on a closed interval $[a, b]$ and only failing to have a derivative at some point in the interior (a, b) in that the limit

$$\lim_{h\to 0}\frac{f(x+h)-f(x)}{h}$$

may be infinite at some point $x \in (a, b)$. One can mean this in the weak sense that, as h gets smaller and smaller, the absolute value of the difference quotient gets large without bound, but is not necessarily always of the same sign for h sufficiently small. This includes the case of a cusp like that of

[364] Above, pp. 361–362.

$$f(x) = \left|x^{1/3}\right| \text{ at } x = 0$$

where one has

$$\lim_{h\to 0^-} \frac{f(h) - f(0)}{h} = -\infty \text{ and } \lim_{h\to 0^+} \frac{f(h) - f(0)}{h} = +\infty.$$

For any value of $x > 0$ one has

$$f(-x) = \left|-x^{1/3}\right| = \left|x^{1/3}\right| = f(x),$$

but there is no point $c \in (-x, x)$ such that

$$f'(c) = \frac{f(x) - f(-x)}{x - (-x)} = 0,$$

whence the Mean Value Theorem fails for this function.

On the other hand, if the left- and right-difference quotients have infinite limits of the same sign wherever the limits are infinite, the Mean Value Theorem will again hold:

3.10.9 Theorem (Full Mean Value Theorem) *Let $f : [a, b] \to \mathbb{R}$ be continuous and suppose for each $x \in (a, b)$ one of the following holds:*

i. $f'(x) = \lim_{h\to 0} \frac{f(x+h) - f(x)}{h}$ *exists and is finite*

ii. $\lim_{h\to 0} \frac{f(x+h) - f(x)}{h} = +\infty$

iii. $\lim_{h\to 0} \frac{f(x+h) - f(x)}{h} = -\infty.$

Then: there is an element $c \in (a, b)$ such that

$$f'(c) = \frac{f(b) - f(a)}{b - a}.$$

We have already seen in Chap. 2 that this holds for curves $y = f(x)$ that have smooth parametrisations. For example consider the curve $y = f(x) = x^{1/3}$ of Fig. 2.2.43 on page 142, above. For this function $f'(0) = +\infty$ and the Classroom Mean Value Theorem does not apply. However, on any interval $[a, b]$, we can simply give the parametrisation

$$\left.\begin{aligned} x(t) &= t^3 \\ y(t) &= t \end{aligned}\right\} \text{ on } \left[a^{1/3}, b^{1/3}\right]$$

and conclude there to be some $d \in \left[a^{1/3}, b^{1/3}\right]$ such that the tangent to the curve at $\langle x(d), y(d)\rangle$ is parallel to the line connecting

$$\left\langle x\left(a^{1/3}\right), y\left(a^{1/3}\right)\right\rangle = \left\langle a, a^{1/3}\right\rangle = \langle a, f(a)\rangle$$

to

$$\left\langle x\left(b^{1/3}\right), y\left(b^{1/3}\right)\right\rangle = \left\langle b, b^{1/3}\right\rangle = \langle b, f(b)\rangle.$$

As the line connecting these points is not vertical, the slope at

$$\langle x(d), y(d)\rangle = \left\langle d^3, d\right\rangle = \left\langle d^3, f\left(d^3\right)\right\rangle$$

is finite and equal to $f'\left(d^3\right)$. Thus, for $c = d^3$, one has $a < c < b$ and

$$f'(c) = \frac{f(b) - f(a)}{b - a}.$$

This argument is a bit *ad hoc*, requiring us to find a smooth parametrisation of the curve $y = f(x)$. The Weierstrass–Bonnet proof handles this case without the extra fuss.

Proof of Theorem 3.10.9. Define

$$g(x) = f(x) - f(a) - \frac{x - a}{b - a}\big(f(b) - f(a)\big).$$

As usual, $g(a) = 0 = g(b)$. Let an extreme value of g occur at $c \in (a, b)$. A little algebra yields

$$\frac{g(c+h) - g(c)}{h} = \frac{f(c+h) - f(c)}{h} - \frac{f(b) - f(a)}{b - a}. \tag{3.68}$$

Obviously, we have

$$\lim_{h\to 0} \frac{g(c+h) - g(c)}{h} \begin{Bmatrix} \text{is finite} \\ +\infty \\ -\infty \end{Bmatrix} \text{ iff } \lim_{h\to 0} \frac{f(c+h) - f(c)}{h} \begin{Bmatrix} \text{is finite} \\ +\infty \\ -\infty \end{Bmatrix}.$$

If $g'(c) \neq 0$, then $g'(c)$ is either positive (finite or infinite) or negative (finite or infinite).

Assume first that $g'(c) = +\infty$. Then, for h sufficiently small,

$$\frac{g(c+h) - g(c)}{h} > 1. \tag{3.69}$$

If $h > 0$, this means $g(c+h) - g(c) > 0$ and, $g(c)$ being an extremum, $g(c)$ is thus a minimum. But for $h < 0$, (3.69) tells us $g(c+h) - g(c)$ is negative, whence, $g(c)$ being an extremum, it must be a maximum. This contradiction tells us $g'(c) \neq +\infty$.

Likewise $g'(c) \neq -\infty$. And, for $g'(c)$ finite, Lemma 2.3.17 (page 131, above) yields the contradiction.

Thus, $g'(c) = 0$ and Eq. (3.68) yields

$$0 = g'(c) = \lim_{h \to 0} \frac{f(c+h) - f(c)}{h} - \frac{f(b) - f(a)}{b-a} = f'(c) - \frac{f(b) - f(a)}{b-a},$$

and we conclude

$$f'(c) = \frac{f(b) - f(a)}{b-a}.$$

□

I am not sure whom to credit this slight generalisation of the Mean Value Theorem to. I do know that a complete proof can be found in Dini's *Fondamenti* (paragraph 71, pp. 69–71) of 1878.

The other weakening of the assumption on f I wish to consider is the case in which the derivative itself does not necessarily exist, but the one-sided derivatives exist at all points in the interior of the given interval. The result in this case is, of course, not that a point can be found with the desired derivative, nor that one of the one-sided derivatives assumes the desired value. What happens, however, is that there is a point at which the one sided derivatives straddle the desired value.

Up till now I have taken what mathematicians call a *naïve* approach to single sided derivatives, assuming their meaning clear enough, especially in reference to the derivatives at the endpoints of an interval, so that a formal definition was not needed. And up till now we have made sparse mention of one-sided derivatives — on page 140 with reference to points of the extended cycloid where tangents do not exist; on page 297 where Bolzano's statement of a theorem made reference to the existence of limits in one direction at the endpoints of the interval in question, and the same again on pages 307 and ?? in his statements of the Mean Value Theorem; and the subject popped up in our discussions of Peano and Gilbert (e.g., footnote 327 on page 351, Peano's response to Gilbert on page 355, and Gilbert's response on page 361 to Peano's response). But now one-sided derivatives are the topic of discussion and I should give at least their formal definition.

3.10.10 Definitions Let f be a function on some interval I and let $a \in I$. We say f has a *limit L as x approaches a from the left* for a not the left endpoint of I, written

$$\lim_{x \to a-} f(x) = L,$$

if, for any $\epsilon > 0$ there is some $\delta > 0$ such that for all $x \in I$,

$$a - \delta < x < a \Rightarrow |f(x) - L| < \epsilon.$$

Similarly, we say f has a *limit L as x approaches a from the right* for a not the right endpoint of I, written

$$\lim_{x \to a+} f(x) = L,$$

if, for any $\epsilon > 0$ there is some $\delta > 0$ such that for all $x \in I$,

$$a < x < a + \delta \Rightarrow \left|f(x) - L\right| < \epsilon.$$

A limit from the left or from the right is called a *one-sided limit*. If both these limits exist and are equal, their common value is the ordinary limit, which one may call a *two-sided limit* for emphasis.

When there is no danger of confusion, one often writes $f(a-0)$ and $f(a+0)$ for the limits from the left and right, respectively.

3.10.11 Definitions A number L is the *left derivative* or the *derivative from the left of f at a* if

$$L = \lim_{h \to 0-} \frac{f(a+h) - f(a)}{h}.$$

L is the *right derivative* or the *derivative from the right of f at a* if

$$L = \lim_{h \to 0+} \frac{f(a+h) - f(a)}{h}.$$

Such derivatives are called *one-sided derivatives* and one occasionally refers to an ordinary derivative as a *two-sided derivative*.

Again, one may write $f'(a \pm 0)$ to denote the appropriate one-sided derivative, despite the multiple ambiguity of the expression.[365]

One-sided derivatives had been around for a while by the time Gilbert and Peano crossed paths, as we saw in discussing Bolzano. Gilbert's counterexample for Peano,

$$f(x) = \begin{cases} \sqrt{2px}, & 0 \le x \le a \\ \sqrt{2p(2a-x)}, & a < x \le 2a, \end{cases}$$

shows that the Mean Value Theorem fails for f in any interval $[a-h, a+h]$ for $h < a$:

$$f(x+h) - f(x-h) = \sqrt{2p(a-h)} - \sqrt{2p(a-h)} = 0,$$

but there is no value of $x \in (a-h, a+h)$ for which $f'(x) = 0$ or for which even one of $f'(x+0)$ or $f'(x-0)$ equals 0. But he is right about Jordan's inequality holding:

$$\frac{f(a+h) - f(a-h)}{2h} \tag{3.70}$$

lies between the maximum and minimum values,

$$\sqrt{\frac{p}{2(a-h)}}, \quad -\sqrt{\frac{p}{2(a-h)}}$$

[365] $f'(a+0)$ could mean the derivative of f at $a+0 = a$, the limit of $f'(x)$ as x approaches a from the right, or as the right derivative of f at a. The context should always make clear the sense in which the notation is to be understood.

of $f'(x)$ for $x \in [a - h, a + h]$. In fact, the ratio (3.70) lies between the left and right derivatives of f at a: for all $0 < h < a$,

$$-\sqrt{\frac{p}{2a}} \leq \frac{f(a+h) - f(a-h)}{2h} \leq \sqrt{\frac{p}{2a}},$$

i.e.,

$$f'(a+0) \leq \frac{f(a+h) - f(a-h)}{2h} \leq f'(a-0).$$

This generalises quite easily by applying the Weierstrass–Bonnet arguments.

3.10.12 Theorem (Rolle's Theorem for One-Sided Derivatives) *Let f be continuous on $[a, b]$ and suppose both one-sided derivatives exist for all $x \in (a, b)$. If $f(a) = f(b)$, there is some $c \in (a, b)$ such that 0 lies between $f'(c - 0)$ and $f'(c + 0)$.*

Proof. If f is constant, $f'(c) = 0$ for all $c \in (a, b)$ and we are done.

If f is not constant, since $f(a) = f(b)$, at least one of its extrema occurs in the interior. Let this happen at c, and consider

$$\frac{f(c+h) - f(c)}{h}.$$

If f assumes a maximum at c, the numerator is ≤ 0 for $h > 0$, whence the ratio and its limit as $h \to 0$ is ≤ 0, while for $h < 0$ the ratio and its limit as $h \to 0$ is ≥ 0, i.e.,

$$f'(c-0) \geq 0 \geq f'(c+0).$$

If f assumes a minimum at c, one similarly concludes

$$f'(c-0) \leq 0 \leq f'(c+0).$$

□

3.10.13 Theorem (Mean Value Theorem for One-Sided Derivatives) *Let f be continuous on $[a, b]$ and suppose both one-sided derivatives exist for all $x \in (a, b)$. Then there is some $c \in (a, b)$ such that*

$$\frac{f(b) - f(a)}{b - a}$$

lies between $f'(c + 0)$ and $f'(c - 0)$.

Proof. Exercise. □

I'm not sure I'm justified in naming these theorems as I did. Because they are inequalities and not equations, they are not quite analogous to Rolle's Theorem and the Mean Value Theorem. But they are not quite analogous to the Mean Value Inequality either.

I confess once again my ignorance of the authorship of these results. My interest in the results of this subsection has primarily been to further absolve Gilbert of his supposed archstupidity, and I did not pay much attention to such generalisations in looking into the history of the Mean Value Theorem. Bolzano mentioned one-sided derivatives, but I did not find a generalisation of the Mean Value Theorem to them in his work. Dini famously considered generalisations involving weaker notions of one-sided derivatives and others have proceeded from there. None of this, however, tends to be included in the first year Calculus course and it thus lies beyond the scope of the present book.

3.11 Acker and the Mean Value Theorem

In 1996, Felipe Acker published a new proof of the Mean Value Theorem.[366] The paper in which he published this proof was titled "The missing link", referring to a new multivariable generalisation of the Mean Value Theorem as the missing link "in the chain leading from the Fundamental Theorem of Calculus to Stokes' Theorem" presented therein. The paper's new proof of the Mean Value Theorem for single variable functions is also a missing link, connecting the Ampère-Cauchy inequality and Bonnet's form of the Mean Value Theorem.

In describing this proof, the author admits: "Everyone can see that this proof is not simpler than the usual one, and I do not pretend otherwise. The difference is that I am able to generalize it to higher dimensions."[367] This generalisation is not relevant to the purposes of the present book, but the proof itself is of some interest and I propose to consider it here.

The proof divides into two parts. First, one finds a sequence of nested subintervals $[a, b] = [a_0, b_0] \supset [a_1, b_1] \supset [a_2, b_2] \supset \ldots$ such that $\lim_{n\to\infty} a_n = \lim_{n\to\infty} b_n$ and at each stage

$$\frac{f(b_n)-f(a_n)}{b_n - a_n} = \frac{f(b)-f(a)}{b-a}. \tag{3.71}$$

The second half of the proof consists of showing

$$\lim_{n\to\infty} \frac{f(b_n)-f(a_n)}{b_n - a_n} = f'(c),$$

where c is the common limit of the two sequences $a_0, a_1, a_2, \ldots$ and $b_0, b_1, b_2, \ldots$

366 Felipe Acker, "The missing link", *Mathematical Intelligencer* 18, No. 3 (1996), pp. 4–9. The paper was reprinted in: Robin Wilson and Jeremy Gray (eds.), *Mathematical Conversations: Selections from the Mathematical Intelligencer*, Springer-Verlag, New York, Inc., 2001.

367 P. 211 of the anthologised version of the paper cited in the preceding footnote.

We begin the first half of the proof by setting $a_0 = a$, $b_0 = b$ and noting that (3.71) is trivially satisfied.

For the induction step, let a_n, b_n be given, assume (3.71) for n, and assume

$$b_n - a_n = \frac{b-a}{3^n}.$$

Let $h = (b_n - a_n)/3$ and consider the partition $\alpha_0 < \alpha_1 < \alpha_2 < \alpha_3$ given by $\alpha_i = a_n + hi$ for $i = 0, 1, 2, 3$. Also define

$$\overline{m} = \frac{f(b) - f(a)}{b-a} \quad \text{and} \quad m_i = \frac{f(\alpha_{i+1}) - f(\alpha_i)}{h} \quad \text{for } i = 0, 1, 2.$$

We need a small combinatorial lemma that is basically a special case of the Ampère-Cauchy inequality:

3.11.1 Lemma *One of the following holds:*
i. $m_1 = m_2 = m_3 = \overline{m}$
ii. for some $i, j \in \{1, 2, 3\}$, $m_i < \overline{m} < m_j$.

Proof. The telescoping sum,

$$\begin{aligned} f(b_n) - f(a_n) &= f(\alpha_3) - f(\alpha_0) \\ &= f(\alpha_3) - f(\alpha_2) + f(\alpha_2) - f(\alpha_1) + f(\alpha_1) - f(\alpha_0), \end{aligned}$$

yields, on dividing by $h = (b_n - a_n)/3$ and reversing the order of the terms on the right,

$$3\overline{m} = m_1 + m_2 + m_3, \tag{3.72}$$

If $m_1 = m_2 = m_3$, then (3.72) yields $3\overline{m} = 3m_1$ and $\overline{m} = m_1 = m_2 = m_3$. Otherwise, apply the Ampère-Cauchy inequality to conclude

$$\min\{m_1, m_2, m_3\} \le \overline{m} \le \max\{m_1, m_2, m_3\}.$$

If $\overline{m} = \min\{m_1, m_2, m_3\}$, then $\overline{m} \neq \max\{m_1, m_2, m_3\}$ and, letting, say, m_3 be the maximum,

$$3\overline{m} = \overline{m} + \overline{m} + \overline{m} \le m_1 + m_2 + \overline{m} < m_1 + m_2 + m_3,$$

contrary to (3.72). Thus $\overline{m}$ is not the minimum. Likewise $\overline{m}$ is not the maximum and

$$\min\{m_1, m_2, m_3\} < \overline{m} < \max\{m_1, m_2, m_3\}.$$ □

Continuing the first part of the proof of the Mean Value Theorem, we consider the first of the two cases given by the Lemma: $m_1 = m_2 = m_3 = \overline{m}$. In this case choose the endpoints of the middle interval, $a_{n+1} = a_n + h$, $b_{n+1} = a_n + 2h$. Note that

$$\frac{f(b_{n+1}) - f(a_{n+1})}{b_{n+1} - a_{n+1}} = m_2 = \overline{m}$$

by assumption. Also, note that $[a_{n+1}, b_{n+1}] \subset (a_n, b_n)$ and

$$b_{n+1} - a_{n+1} = h = \frac{b_n - a_n}{3} = \frac{b - a}{3^{n+1}}.$$

In the remaining case, let $m_i < \overline{m} < m_j$ and define the function $m(x)$ on $[a, b - h]$ by

$$m(x) = \frac{f(x + h) - f(x)}{h}.$$

Because $h = (b_n - a_n)/3$ is a constant and f is continuous, m is continuous. But, for the new values of $\alpha_0, \alpha_1, \alpha_2, \alpha_3$ partitioning $[a_n, b_n]$,

$$m(\alpha_i) = m_i < \overline{m} < m_j = m(\alpha_j),$$

for $i, j \leq 2$ as described. By the Intermediate Value Theorem, there is some $\overline{x}$ in (α_i, α_j) or (α_j, α_i) according as $\alpha_i < \alpha_j$ or $\alpha_j < \alpha_i$ such that $m(\overline{x}) = \overline{m}$.

If we choose $a_{n+1} = \overline{x}$, $b_{n+1} = \overline{x} + h$, we have

$$\frac{f(b_{n+1}) - f(a_{n+1})}{b_{n+1} - a_{n+1}} = \frac{f(\overline{x} + h) - f(\overline{x})}{h} = m(\overline{x}) = \overline{m}$$

by choice of $\overline{x}$. We also have

$$b_{n+1} - a_{n+1} = h = \frac{b_n - a_n}{3} = \frac{b - a}{3^{n+1}},$$

and

$$a_n = \alpha_0 \leq \min\{\alpha_i, \alpha_j\} < \overline{x} < \max\{\alpha_i, \alpha_j\} \leq \alpha_2 < \alpha_3 = b_n,$$

which implies $\overline{x} + h < \alpha_2 + h = \alpha_3$. Thus $[a_{n+1}, b_{n+1}] \subset (a_n, b_n)$.

The sequence $a_0, a_1, a_2, \ldots$ is strictly increasing with any b_n serving as an upper bound, whence it converges to some limit c. Similarly, the sequence $b_0, b_1, b_2, \ldots$ is strictly decreasing with any a_n serving as a lower bound, whence it converges to some limit d. But $b_n - a_n = (b - a)/3^n$, whence the standard $\epsilon/3$ argument,

$$\begin{aligned} |d - c| &\leq |d - b_n| + |b_n - a_n| + |a_n - c| \\ &< \frac{\epsilon}{3} + \frac{\epsilon}{3} + \frac{\epsilon}{3} = \epsilon, \text{ for large enough } n, \end{aligned}$$

shows $d = c$, i.e., the two sequences share a common limit.

The next step in the proof is to show that

$$\frac{f(b)-f(a)}{b-a}=\overline{m}=\lim_{n\to\infty}\frac{f(b_n)-f(a_n)}{b_n-a_n}=f'(c).$$

Basically, this means we must show

$$\lim_{x\neq y\to c}\frac{f(x)-f(y)}{x-y}=f'(c), \text{ for } y<c<x.$$

The reader should already have proven a special case of this in Exercise 2.3.8 on page 116, above. The general case is not too difficult. By the Ampère-Cauchy inequality,

$$\frac{f(x)-f(y)}{x-y} \text{ lies between } \frac{f(x)-f(c)}{x-c} \text{ and } \frac{f(c)-f(y)}{c-y}, \tag{3.73}$$

so

$$\frac{f(x)-f(y)}{x-y}-f'(c) \text{ lies between } \frac{f(x)-f(c)}{x-c}-f'(c) \text{ and } \frac{f(c)-f(y)}{c-y}-f'(c).$$

Thus, if K is the maximum of

$$\left|\frac{f(x)-f(c)}{x-c}-f'(c)\right| \text{ and } \left|\frac{f(c)-f(y)}{c-y}-f'(c)\right|,$$

one has

$$-K<\frac{f(x)-f(y)}{x-y}-f'(c)<K.$$

Now, let $\epsilon>0$ be given and choose $\delta>0$ so small that

$$0<|z-c|<\delta \Rightarrow \left|\frac{f(z)-f(c)}{z-c}-f'(c)\right|<\epsilon.$$

Choosing $0<|x-c|, |c-y|<\delta$, we have $K<\epsilon$ and

$$-\epsilon<-K<\frac{f(x)-f(y)}{x-y}-f'(c)<K<\epsilon,$$

i.e.,

$$\left|\frac{f(x)-f(y)}{x-y}-f'(c)\right|<\epsilon,$$

i.e.,

$$\lim_{x,y\to c} \frac{f(x)-f(y)}{x-y} = f'(c).$$

Giving x, y the successive values of b_n, a_n, we see that

$$f'(c) = \lim_{n\to\infty} \frac{f(b_n)-f(a_n)}{b_n-a_n} = \lim_{n\to\infty} \overline{m} = \overline{m} = \frac{f(b)-f(a)}{b-a},$$

and the Mean Value Theorem is proven.

In his exposition of this result, Acker omitted the details of the second half of this proof. In a later paper,[368] however, he notes that,

$$\frac{f(x)-f(y)}{x-y} = \frac{x-c}{x-y}\cdot\frac{f(x)-f(c)}{x-c} + \frac{c-y}{x-y}\cdot\frac{f(c)-f(y)}{c-y},$$

i.e.,

$$\frac{f(x)-f(y)}{x-y} = \lambda\cdot\frac{f(x)-f(c)}{x-c} + (1-\lambda)\cdot\frac{f(c)-f(y)}{c-y}$$

for some $0 < \lambda < 1$, i.e.,

$$\frac{f(x)-f(y)}{x-y} \text{ lies between } \frac{f(x)-f(c)}{x-c} \text{ and } \frac{f(c)-f(y)}{c-y},$$

i.e., (3.73) holds. He omits the rest of the argument, which can be finished as above.

I think the proof may be too involved for the introductory course in the Calculus, but I find it interesting nonetheless. Aside from Acker's generalisation of the proof to the higher dimensional case where its mathematical value is evident, it is of historical interest in that it uses nothing that wasn't rigorously known to Weierstrass and, though a bit less straightforward than the Bonnet proof which actually suggests itself once one recognises the geometric equivalence of the Mean Value Theorem and Rolle's Theorem, it is not unmotivated.

Speaking of geometry, I should also note that in this second paper Acker carries out the proof for smooth parametrically defined curves, thereby deriving the Cauchy Mean Value Theorem as well.

I might also note an historical curiosity: Recall that Bonnet and Schwarz used similar auxiliary functions to prove the Mean Value Theorem (Bonnet) and the Constant Function Theorem (Schwarz). A variant of the nested interval argument used here had previously been used by Leon W. Cohen[369] to prove the Constant Function Theorem.

---

368 Felipe Acker, "Um Teorema bem conhecido", *Matemática Universitária* no. 37, 2004, pp. 1–8.

369 Leon W. Cohen, "On being mean to the Mean Value Theorem", *The American Mathematical Monthly* 74 (1967), pp. 581–582.

3.12 Loose Ends

There is, of course, much more to the story of the Mean Value Theorem than we have covered above. Two topics of particular relevance that the interested reader might want to look further into are the Mean Value Theorem for Integrals and the Mean Value Theorem for Multivariable Functions. I cannot imagine coverage of these topics in a reasonably small number of pages, the present book is already longer than I had anticipated, and there is still the final chapter on the rôle of the Mean Value Theorem in the introductory Calculus course — the debate over which provided the initial stimulus for writing this book. Hence I shall not be discussing these topics myself. In this final section of this chapter I shall merely tie up a few loose ends left dangling in earlier sections.

3.12.1 Flett's Theorem

As I mentioned, I have not seen Flett's own proof of the Peano–Gilbert result, but have, instead followed Mawhin's presentation. I have however, seen a different paper[370] of Flett's which contains an interesting corollary to Rolle's Theorem that has attained a certain degree of popularity and has become known as Flett's Theorem.

3.12.1 Theorem (Flett's Theorem) *If f is differentiable in $[a, b]$ and $f'(a) = f'(b)$, then there is a point $\xi \in (a, b)$ such that*

$$f'(\xi) = \frac{f(\xi) - f(a)}{\xi - a}.$$

Flett offers a geometric interpretation of this result. If $y = f(x)$ is a differentiable curve over an interval $[a, b]$ and if the non-vertical tangents at $\langle a, f(a)\rangle$ and $\langle b, f(b)\rangle$ are parallel, then there is some $\xi \in (a, b)$ such that the secant line connecting $\langle a, f(a)\rangle$ and $\langle \xi, f(\xi)\rangle$ is tangent to the curve at $\langle \xi, f(\xi)\rangle$. Figure 3.22, illustrates the situation. Despite giving the geometric interpretation, Flett fails to use it to motivate his introduction of an auxiliary function in his proof. The motivation for the proof can be found in Valerio's work cited earlier.[371] Given a point outside a region and a closed curve forming the boundary of the region, he found a tangent connecting the point to the curve by rotating a line around the point until it touched the region. We can imagine doing the same here.

Proof of Theorem 3.12.1. Flett's proof is mildly complicated.

---

[370] T.M. Flett, "A mean value theorem", *The Mathematical Gazette* 42 (1958), pp. 38–39.

[371] Passing through $\langle a, f(a)\rangle$, Valerio's rotating line is determined by the slope. Where Valerio starts with a line not intercepting the curve and rotating it until it meets the curve, Flett more-or-less starts from the opposite direction, starting with the tangent at $\langle a, f(a)\rangle$.

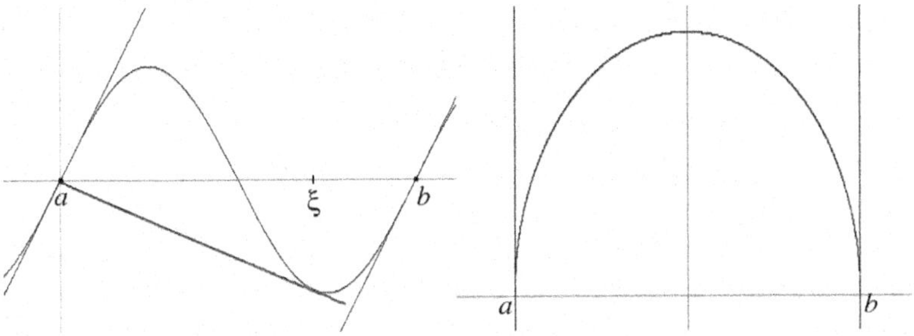

FLETT'S THEOREM   FAILURE OF FLETT'S THEOREM

Fig. 3.22 Flett's theorem

First, by considering $g(x) = f(x) - xf'(a)$, he reduces the result to the case in which the common value of $f'(a)$ and $f'(b)$ is 0. Under this assumption, he defines, for $x \in (a, b]$,

$$\psi(x) = \frac{f(x) - f(a)}{x - a}$$

to be the slope of the secant connecting $\langle a, f(a)\rangle$ to $\langle x, f(x)\rangle$. By the differentiability of f at a, we can extend ψ continuously to $[a, b]$ by setting $\psi(a) = f'(a)$.

Noting that

$$\begin{aligned}\psi'(x) &= \frac{(x-a)f'(x) - \big(f(x) - f(a)\big) \cdot 1}{(x-a)^2} \\ &= \frac{f'(x)}{x-a} - \frac{f(x) - f(a)}{(x-a)^2},\end{aligned}$$

we see that it suffices to find $\xi \in (a, b)$ such that $\psi'(\xi) = 0$. For

$$\begin{aligned}\psi'(\xi) = 0 &\Rightarrow f'(\xi) - \frac{f(\xi) - f(a)}{\xi - a} = 0 \\ &\Rightarrow f'(\xi) = \frac{f(\xi) - f(a)}{\xi - a}.\end{aligned}$$

There are three cases to consider, depending on wheree $\psi(b)$ sits relative to $\psi(a)$.

If $\psi(b) = 0 = f'(a) = \psi(a)$, then Rolle's Theorem immediately yields the desired ξ.

If $\psi(b) > 0$, then

$$\begin{aligned}\psi'(b) &= \frac{f'(b)}{b-a} - \frac{f(b) - f(a)}{(b-a)^2} = 0 - \frac{f(b) - f(a)}{(b-a)^2} \\ &= -\frac{\psi(b)}{b-a} < 0.\end{aligned}$$

Thus ψ is decreasing at b and there is some $x < b$ such that $\psi(x) > \psi(b)$. Thus $\psi(x) > \psi(b) > 0 = \psi(a)$, and by the Intermediate Value Theorem there is some $y \in (a, x)$ such that $\psi(y) = \psi(b)$. Rolle's Theorem now yields $\xi \in (y, b)$ such that $\psi'(\xi) = 0$.

If $\psi(b) < 0$, a similar argument applies. □

I find Flett's proof a little too complicated and initially simplified it somewhat only spotting my error after typesetting the argument. Thus I reverted to Flett's original proof and typeset it. Since then I have worked out a marginally simpler proof that bypasses the appeal to Rolle's Theorem and depends, *à la* Rolle's Theorem, directly on the Extreme Value Theorem and the vanishing of the derivative at a local extremum.

Alternate proof of Theorem 3.12.1. As before define

$$\psi(x) = \frac{f(x) - f(a)}{x - a}$$

for $x \in (a, b]$ to be the slope of the secant connecting $\langle a, f(a)\rangle$ to $\langle x, f(x)\rangle$ and extend ψ to include a in its domain by taking $\psi(a) = f'(a)$. As before,

$$\begin{aligned}\psi'(x) &= \frac{f'(x)}{x-a} - \frac{f(x)-f(a)}{(x-a)^2} \quad \text{for } x \in (a, b]\\ &= \frac{1}{x-a}\left[f'(x) - \frac{f(x)-f(a)}{x-a}\right].\end{aligned}$$

Note that we do not assume $f'(a) = 0$ as we did before.

If ψ is monotone increasing, then it is increasing at b. But

$$\begin{aligned}\psi'(b) &= \frac{1}{b-a}\left[f'(b) - \frac{f(b)-f(a)}{b-a}\right]\\ &= \frac{1}{b-a}\big[f'(a) - \psi(b)\big], \text{ since } f'(b) = f'(a)\\ &= \frac{1}{b-a}\big[\psi(a) - \psi(b)\big] = \frac{\psi(a)-\psi(b)}{b-a} \le 0, \end{aligned} \tag{3.74}$$

with equality only holding if ψ is constant on $[a, b]$, in which case, for any $\xi \in (a, b)$, one would have $\psi'(\xi) = 0$, i.e.,

$$\frac{1}{\xi - a}\left[f'(\xi) - \frac{f(\xi)-f(a)}{\xi - a}\right] = 0,$$

i.e.,

$$f'(\xi) = \frac{f(\xi)-f(a)}{\xi - a}.$$

If ψ is not constant on $[a, b]$, then the inequality in (3.74) is strict, $\psi'(b) < 0$, and ψ is not increasing at b.

Likewise, the assumption that ψ is monotone decreasing on $[a, b]$ leads to a contradiction should the function not be constant.

We now invoke a little lemma:

3.12.2 Lemma *Let g be continuous on $[a, b]$ and suppose g is not monotone. Then g has a local extremum in (a, b).*

It follows that ψ has a local extremum at some point $\xi \in (a, b)$. But $\psi'(\xi) = 0$ implies

$$f'(\xi) = \frac{f(\xi) - f(a)}{\xi - a} \qquad \square$$

I am tempted to declare the Lemma obvious or as something that was proven when discussing maxima and minima well before any form of the Mean Value Theorem was mentioned in one's class, but the fact is that one probably doesn't prove the result in an introductory course in the Calculus, and the proof must be given here — thus rendering questionable my words "marginally simpler" in describing the above proof.

Proof of Lemma 3.12.2. If g is not monotone on $[a, b]$, then there are $x < y, z < w$ in $[a, b]$ such that $g(x) < g(y)$ and $g(z) > g(w)$. There are 6 possible configurations of x, y, z, w:

$$\begin{array}{ll} x < y < z < w & x < z < y < w \\ z < x < w < y & z < w < x < y \\ x < z < w < y & z < x < y < w. \end{array}$$

In the first two cases, the values $g(x)$ and $g(w)$ at the endpoints are not maxima on $[x, w]$ since $g(x) < g(y)$ and $g(w) < g(z)$. Hence the maximum on $[x, w]$ occurs in the interior $(x, w) \subseteq (a, b)$.

In the second two cases, the values $g(z)$ and $g(y)$ at the endpoints are not minima, whence the minimum on $[z, y]$ occurs in the interior $(x, y) \subseteq (a, b)$.

If $x < z < w < y$ there are two subcases. If $g(x) < g(z)$, then $g(x)$ and $g(w)$ are not maxima on $[x, w]$ and the maximum occurs in $(x, w) \subseteq (x, y) \subseteq (a, b)$. If $g(x) \geq g(z)$, we have $g(y) > g(x) \geq g(z) > g(w)$ and the values $g(z), g(y)$ at the endpoints of $[z, y]$ are not minimal, whence the minimum on $[z, y]$ occurs in $(z, y) \subseteq (x, y) \subseteq (a, b)$.

The final configuration is handled similarly. $\square$

What we see is that Flett's Theorem can be proven by direct reduction to the Extreme Value Theorem and that, modulo some basic results on local extrema of continuous functions, by a marginally simpler proof than Flett's. As such, one might ask if it is more than just a curiosity — can we use Flett's Theorem to prove further results?

3.12.3 Corollary (Constant Function Theorem) *Let f be continuous on $[a, b]$, differentiable on (a, b), and suppose $f'(x) = 0$ for all $x \in (a, b)$. Then f is constant.*

Proof. By continuity of f, it suffices to show f constant on (a, b), which holds if f is constant on all closed subintervals $[\alpha, \beta]$. Thus we can assume, without loss of generality that f is differentiable on $[a, b]$.

Let $\epsilon > 0$ and choose, by the uniform continuity of f, a sufficiently small $\delta > 0$ such that, for all $x, y \in [a, b]$,

$$|x - y| < \delta \Rightarrow \left|f(x) - f(y)\right| < \epsilon.$$

By Flett's Theorem, there is some $\xi_0 \in (a, a + \delta)$ such that

$$f'(\xi_0) = \frac{f(\xi_0) - f(a)}{\xi_0 - a}.$$

But $f'(\xi_0) = 0$, whence $f(\xi_0) = f(a)$. Likewise, there is $\xi_1 \in (\xi_0, \xi_0 + \delta)$ such that $f(\xi_1) = f(\xi_0) = f(a)$. Etc.

Let X be the set of all $x \in [a, b]$ such that there is a sequence

$$a = a_0 < a_1 < \ldots < a_{k-1} < a_k = x$$

such that $|a_{i+1} - a_i| < \delta$ and $f(a_i) = f(a)$. $X \subseteq [a, b]$, whence it is bounded above by b. Let c be the least upper bound of X.

If $c \notin X$, there is a sequence $x_0 < x_1 < x_2 < \ldots$ of elements of X converging to c. We can construct such a sequence easily enough: Let $x_0 \in X$ be arbitrary. Having defined $x_0 < x_1 < \ldots < x_k$, all in X, note that

$$x_n + \frac{1}{2}(c - x_n) < x_n + c - x_n = c,$$

whence there is some $x \in X$ with

$$x_n < x_n + \frac{1}{2}(c - x_n) < x.$$

Let x_{n+1} be such an x. Then $x_n < x_{n+1}$ and

$$c - x_{n+1} < \frac{1}{2}(c - x_n) < \frac{1}{4}(c - x_{n-1}) < \ldots < \frac{1}{2^n}(c - x_0)$$

and $x_n \to c$. f is continuous, whence

$$f(c) = \lim_{n\to\infty} f(x_n) = \lim_{n\to\infty} f(a) = f(a).$$

But then $c \in X$ after all. For, if we choose $x_0 > c - \delta$ we have some sequence

$$a = a_0 < a_1 < a_2 < \ldots < a_{n-1} < a_n = x_0 < c$$

with all points listed mapped to $f(a)$ by f, and each difference being less than δ.

Now, if $c \neq b$, there is some $\xi \in (c, \min\{c + \delta, b\})$ such that $0 = f'(\xi) = \dfrac{f(\xi) - f(c)}{\xi - c}$, i.e., $f(\xi) = f(c) = f(a)$. Thus $c < \xi \in X$, a contradiction.

Thus $b \in X$ and there is a sequence

$$a = a_0 < a_1 < a_2 < \ldots < a_{n-1} < a_n = b$$

such that for all i, $|a_{i+1} - a_i| < \delta$. But, for any $x \in [a, b]$, either x is some a_i, whence $f(x) = f(a)$, or for some i, $|x - a_i| < \delta$, and by the choice of δ, this means

$$\left|f(x) - f(a)\right| = \left|f(x) - f(a_i)\right| < \epsilon.$$

This holds for all $\epsilon > 0$, whence

$$\left|f(x) - f(a)\right| = 0, \text{ i.e., } f(x) = f(a),$$

and f is constant. □

I don't think this proof is going to win any support for Flett's Theorem as a replacement for the Mean Value Theorem in the introductory course in the Calculus. Not only is it much more complicated than the Serret-Schwarz proof by appeal to Rolle's Theorem, but it relies on the Least Upper Bound Principle and the Uniform Continuity Theorem, the former of which is hardly mentioned if at all in such a course and the latter nearly universally ignored, at least in American textbooks. Nonetheless, Flett's Theorem is somewhat interesting and it is nice to see it has at least one significant consequence.[372]

3.12.2 *Finding the Mean Value*

The Mean Value Theorem, like the Intermediate Value Theorem and the Extreme Value Theorem, is an abstract existence theorem. It is not completely abstract, as it does tell one in general terms where to look to find $c \in (a, b)$ such that

$$f'(c) = \frac{f(b) - f(a)}{b - a}, \tag{3.75}$$

[372] For more, consult: O. Hutník and J. Molnárová, "Flett's mean value theorem: a survey", 2013, preprint.

namely, look for that c for which the point $\langle c, f(c)\rangle$ is at maximal distance from the line segment connecting $\langle a, f(a)\rangle$ to $\langle b, f(b)\rangle$. Geometrically, this is a task admitting a simple mechanical solution. One takes a collapsible parallelogram, positions one side on the segment, and expands it until the side opposite reaches the exit point. At this point the distance is maximised, the tangent at that point being determined by the parallel side.

Analytically, the problem is a bit more complicated. One wants as exact a numerical solution as the application at hand merits, and one wants it found as efficiently as possible. The proof doesn't yield this. The proofs of the Intermediate and Extreme Value Theorems do not yield computationally efficient procedures either, but the Calculus has devised many techniques to correct for this. And, computationally, the Mean Value Theorem is just a special case of a problem students are drilled in. Finding c satisfying (3.75) is an extremum problem: determine where

$$g(x) = f(x) - f(a) - (x-a)\frac{f(b) - f(a)}{b-a}$$

has an extreme value. I.e., solve $g'(x) = 0$. This reduces to solving (3.75), i.e., solving an equation $h(c) = 0$ for some function h. If h is reasonably well-behaved, the Newton-Raphson Method, usually called Newton's Method, for example, will very quickly determine c to a high level of accuracy.

In some special cases c can be found directly.

3.12.4 Example Let $f(x) = Ax^2 + Bx + C$ with $A \neq 0$ be a quadratic polynomial, $a < b$, and

$$m = \frac{f(b) - f(a)}{b-a}.$$

To solve (3.75), one simply differentiates and sets up the linear equation:

$$2Ax + B = m,$$

the solution to which is

$$c = \frac{m-B}{2A}.$$

Expressing $f(b), f(a)$, and thus m in terms of A, B, C, this means

$$\begin{aligned} c &= \left(\frac{Ab^2 + Bb + C - Aa^2 - Ba - C}{b-a} - B\right) \Big/ 2A \\ &= \left(\frac{A(b^2 - a^2)}{b-a} + \frac{B(b-a)}{b-a} - B\right) \Big/ 2A \\ &= \frac{A(b+a)}{2A} = \frac{b+a}{2}. \end{aligned}$$

Thus, c is the midpoint between a and b.

3.12.5 Example Let $f(x) = Ax^3 + Bx^2 + Cx + D$ with $A \neq 0$ be a cubic polynomial, $a < b$ and

$$m = \frac{f(b) - f(a)}{b - a}.$$

Then the solution c to (3.75),

$$3Ac^2 + 2Bc + C = m,$$

satisfies

$$c = \frac{-B \pm \sqrt{B^2 - 3A(C - m)}}{3A}.$$

Both the cubic and quartic equations can also be solved by radicals, so the Examples can be generalised to polynomials of degrees 4 and 5 as well.

Generally, the problem of actually finding c has not been of great importance in the history of the Calculus. This is because in applications one either requires only knowledge that c exists or bounds on $\frac{f(b) - f(a)}{b - a}$, in which case the Mean Value Inequality is all that is needed. Thus, Calculus textbooks do not traditionally cover the problem, giving at most something like Example 3.12.4 or 3.12.5 as a concrete example illustrating the existence of c in some special cases. There are, however, more general cases where c can be found or approximated, and increased attention has been paid to the problem in recent decades.

One straightforward approach is to fix a, and assume c is an analytic function of b — or, to simplify notation, of $h = b - a$. That is, to solve

$$f'(y) = \frac{f(b) - f(a)}{b - a} = \frac{f(a + h) - f(a)}{h}, \tag{3.76}$$

one writes

$$G(x, y) = \begin{cases} f'(y) - \dfrac{f(a + x) - f(a)}{x}, & x \neq 0 \\ f'(y) - f'(a), & x = 0, \end{cases} \tag{3.77}$$

and finds $g(x)$ such that $G\big(x, g(x)\big) = 0$.

In Multivariable Calculus, variants of the Implicit Function Theorem cite two conditions guaranteeing this can be done. The first is a condition specifying the level of smoothness of the function G, and the second is the assumption that $\partial G/\partial y$ is not 0. In the Calculus course, G is assumed to have continuous partial derivatives and $\partial G/\partial y$ is assumed not 0 in a neighbourhood of a given point $\langle x_0, y_0 \rangle$, and one concludes that in some neighbourhood of x_0 g exists and is continuously differentiable.

If we assume f is analytic, then, in fact, g will also be analytic and we can try to solve successively for the coefficients of the Taylor expansion of g. The condition that the partial derivative fails to equal 0 is just the condition that g be well-defined, i.e., that there exist exactly one solution to (3.76).

This latter fact is easily seen as follows. We know by the Mean Value Theorem that $G(x, y) = 0$ has at least one solution $c \in (a, b)$ for $0 < x < h = b - a$, and one solution $c = a$ for $x = 0$. If there were two such values c, c', then $f'(c) = f'(c')$ and, for some d between c and c' one would have $f''(d) = 0$. But a quick look at (3.77) shows $\partial G/\partial y(z, d) = f''(d)$ for all z. Thus, if we assume

$$f''(y) = \partial G/\partial y(x, y) \neq 0$$

for any $0 \leq x \leq h, a \leq y \leq b$, then there is, for each $x \in [0, h]$, a unique $c \in (a, b)$ satisfying $f'(c) = \dfrac{f(a+x) - f(a)}{x}$, i.e., g is well-defined.

I will not consider here the rôle of the smoothness of f in guaranteeing the appropriate level of smoothness of g. Suffice it to say that g is analytic if f is and once this version of the Implicit Function Theorem was proven, the method we are about to use was fully justified. The technique is much older, going back to Newton's method of *reverting* series, i.e., of determining the power series of functions inverse to those whose power series are known. Today, students learn the technique in a course on Differential Equations, an application nearly as old as Newton's reversion. It could have been applied at any time after Lagrange first proved the Mean Value Theorem for analytic functions. I have no idea who first applied the method to the problem at hand. The earliest I've seen is in texts of 1885 and 1886.[373] The proof that this method works depends on the appropriate form of the Implicit Function Theorem, which has its own complex history.[374]

But I digress.

Getting down to specifics, assume f'' is never 0 on $[a, b]$ and we can write f as a Taylor series around a,

$$f(a+x) = f(a) + \sum_{i=1}^{\infty} \frac{f^{(i)}(a)}{i!} x^i = \sum_{i=0}^{\infty} a_i x^i.$$

---

[373] Laurent, *op. cit.* and Edward, *op. cit.* Cf. footnotes 127 and 130 on pages 213 and 215 above, respectively, for the first of these references.

[374] For details of this history, cf. the following items: Giovanni Mingari Scarpello and Daniele Ritelli, "A historical outline of the theorem of implicit functions", *Divulgaciones Matemáticas* 10 No. 2 (2002), pp. 171–180; Steven G. Krantz and Harold R. Parks, *The Implicit Function Theorem; History, Theory, and Applications*, Birkhäuser, Boston, 2013; and Giuseppe Iurato, "On the role played by the work of Ulisse Dini on implicit function theory in the modern differential geometry foundations: the case of the structure of a differentiable manifold, 1", to appear. The last of these papers is more specialised than the other two sources, but it does contain a short history of the Implicit Function Theorem. Dini is generally credited with the first proof of the familiar form of the Theorem in his textbook of 1878. As for the analytic case, which is the relevant one here, credit might go to Weierstrass, as the result in this case is an immediate consequence of his Preparation Lemma which he first published in a paper of 1886 wherein he states in a footnote that he had lectured on the result repeatedly since 1860. Peter Ullrich informs us that notes taken by Wilhelm Killing from Weierstrass's lectures in 1868 include the analytic version of the Implicit Function Theorem in the complex case.

Assuming the value of c corresponding to x is given by $g(x)$, expand $g(x) - a$ into its own Taylor series,

$$g(x) - a = \sum_{j=0}^{\infty} b_j x^j.$$

But $g(0) = a$ because a is the unique solution to $G(0, y) = 0$ and simultaneously $g(0) = a - b_0$, whence $b_0 = 0$. Now

$$f'(a + x) = \sum_{i=0}^{\infty} i a_i x^{i-1} = \sum_{i=1}^{\infty} i a_i x^{i-1},$$

whence

$$f'(g(x)) = f'\left(a + \sum_{j=1}^{\infty} b_j x^j\right) = \sum_{i=1}^{\infty} i a_i \left(\sum_{j=1}^{\infty} b_j x^j\right)^{i-1}.$$

Setting $f'(g(x))$ equal to $\dfrac{f(a+x) - f(a)}{x}$ yields

$$\sum_{i=1}^{\infty} i a_i \left(\sum_{j=1}^{\infty} b_j x^j\right)^{i-1} = \frac{1}{x} \sum_{i=1}^{\infty} a_i x^i = \sum_{i=1}^{\infty} a_i x^{i-1}. \tag{3.78}$$

The idea now is to expand and collect the terms on the left side of (3.78) and equate the coefficients of the resulting terms to those of the far right side of the equation. The left expands to

$$\begin{aligned} &1 \cdot a_1 \left(b_1 x + b_2 x^2 + \ldots\right)^0 + 2a_2 \left(b_1 x + b_2 x^2 + \ldots\right)^1 + \\ &3a_3 \left(b_1 x + b_2 x^2 + \ldots\right)^2 + 4a_4 \left(b_1 x + b_2 x^2 + \ldots\right)^3 + \ldots, \end{aligned} \tag{3.79}$$

the first term of which reduces to a_1.

If one now expands the other terms and collects like terms of the resulting expansion, one will find the constant terms to be

$$a_1 = f'(a).$$

The sums in all the terms in (3.79) all begin with the term $b_1 x$. This reduces all the remaining calculations to finite operations and allows the recursive determination of the coefficients $b_1, b_2, b_3, \ldots$

The coefficient of the linear term in the expansion of (3.79) comes only from the term $2a_2 \left(b_1 x + b_2 x^2 + \ldots\right)$ and is $2a_2 b_1$. The coefficient of the linear term of the right-hand side of (3.78) being a_2, this yields

$$2a_2b_1 = a_2, \text{ i.e., } b_1 = \frac{a_2}{2a_2} = \frac{1}{2}.$$

[Note again the reliance on the assumption that $a_2 = f''(a)$ is nonzero.]

The coefficient of the quadratic term of the expansion of (3.79) uses the second and third terms of the unexpanded expression and is

$$2a_2b_2 + 3a_3b_1^2 = a_3.$$

Plugging in the known value of b_1 yields successively

$$\begin{aligned} 2a_2b_2 + \frac{3a_3}{4} &= a_3 \\ 2a_2b_2 &= \frac{a_3}{4} \\ b_2 &= \frac{a_3}{8a_2}. \end{aligned}$$

The coefficient of x^3 in the expansion will be

$$2a_2b_3 + 3a_3\left(b_1b_2 + b_2b_1\right) + 4a_4b_1^3 = a_4,$$

which can be solved for b_3 to yield

$$b_3 = \frac{4a_2a_4 - 3a_3^2}{16a_2^2}.$$

And the coefficient of x^4 will be

$$2a_2b_4 + 3a_3\left(b_3b_1 + b_2^2 + b_1b_3\right) + 4a_4\left(b_2b_1^2 + b_1b_2b_1 + b_1^2b_2\right) + 5a_5b_1^4 = a_5,$$

which I leave to the more daring reader to solve for b_4.

The bookkeeping at this point is getting a bit messy, but it is clear that $b_0 = 0$, $b_1 = 1/2$, and each b_{n+1} to follow can be expressed in terms of $b_1, b_2, \ldots, b_n$ and $a_2, a_3, \ldots, a_{n+1}$ and a recursion is *theoretically* possible. However, except for finding the first few coefficients $b_1, b_2, \ldots$, the method is not feasible. Note that, in expanding (3.79), before collecting like terms, there will be 2^{n-1} summands in the coefficient of x^n. Collecting like terms — e.g., $b_1^2b_2x^4$, $b_1b_2b_1x^4$, $b_2b_1^2x^4$ into the single $3b_1^2b_2x^4$ — will reduce the number of summands, but before plugging in the known values of b_j for $j < n$, the number of distinct terms like b_1^4, $3b_1^2b_2$, b_2^2, $2b_1b_3$ comprising the coefficient of x^4, still grows exponentially. There will be exponentially many multiplications, and the summation of an exponential number of terms. This is certainly not a task one wants to carry out by hand.

For small h, the first few b_j's should suffice to give a decent approximation to c — as was the case in Sect. 2.2.2.3 (pages 213–214, above). And, for h small enough,

$b_1 = 1/2$ will suffice, as we saw in Exercises 3.2.6 and 3.2.7 on the pages cited. This result, which required only that f be twice continuously differentiable and $f''(x) \neq 0$ on $[a, a+h]$ has been generalised to approximate the mean values arising in applications of the Higher Order Mean Value Theorem. I refer the curious reader to the following papers:

B. Jacobson, "On the mean value theorem for integrals", *The American Mathematical Monthly* 89 (1982), pp. 301–302.

R. Mera, "On the determination of the intermediate point in Taylor's theorem, *The American Mathematical Monthly* 99 (1992), pp. 56–58.

Ulrich Abel, "On the Lagrange Remainder of the Taylor Formula", *The American Mathematical Monthly* 110 (2003), pp. 627–633.

Emil C. Popa, "Continuity properties relative to the intermediate point in a mean value theorem", *General Mathematics* 12 No. 3 (2004), pp. 53–59.

Rick Kreminski, "Taylor's Theorem: The elusive c is not so elusive", *The College Mathematics Journal* 41 (2010), pp. 186–192.

There are a couple of additional results I'd like to present in this subsection. The first does not concern itself directly with finding c, but is instead the converse of Example 3.12.4: If f is such that c is always found midway between two arguments of f, then f is in fact a quadratic polynomial. Assuming f twice differentiable, Rudolf Rothe[375] proved the following:

3.12.6 Theorem *Let f be differentiable in an interval (a, b) and suppose, for all $x, y \in (a, b)$ with $x \neq y$,*

$$f'\left(\frac{x+y}{2}\right) = \frac{f(x) - f(y)}{x - y}. \tag{3.80}$$

Then f is a polynomial of degree at most 2: There are A, B, C such that $f(x) = Ax^2 + Bx + C$.

Proof. This proof is simplest when one assumes f is differentiable on all of $\mathbb{R}$. I will first prove the result under this assumption and then discuss the modifications needed to prove the result for f defined on a bounded open interval.

The first step, which is fully general, is to note that f' is continuous on the interval (a, b): Let $x \in (a, b)$ and choose h small enough so that $x + h \in (a, b)$. Observe

$$f'(x+h) = f'\left(\frac{x+2h+x}{2}\right) = \frac{f(x+2h) - f(x)}{2h} \to f'(x)$$

since f is differentiable at x.

The next step is to prove a lemma that is relatively easy when the domain of f is all of $\mathbb{R}$.

375 Rudolf Rothe, "Zum Mittelwertsatze der Differentialrechnung", *Mathematische Zeitschrift* 9 (1921), pp. 300–325.

3.12.7 Lemma *For f' differentiable satisfying* (3.80) *for all x, y, with $x \neq y$, one has, for all x, y,*

$$f'\left(\frac{x+y}{2}\right)=\frac{f'(x)+f'(y)}{2}.$$

Proof. This is trivial when $x = y$, so assume $x \neq y$.
Let $h = (y - x)/2$ and $u = x - h$. We have

$$x = u + h, \quad y = u + 3h.$$

Observe

$$\begin{aligned} f'\left(\frac{x+y}{2}\right) &= f'\left(\frac{u+h+u+3h}{2}\right) = f'\left(\frac{u+4h+u}{2}\right) \\ &= \frac{f(u+4h)-f(u)}{4h} \\ &= \frac{1}{2}\cdot\frac{f(u+4h)-f(u+2h)}{2h}+\frac{1}{2}\cdot\frac{f(u+2h)-f(u)}{2h} \\ &= \frac{1}{2}f'(u+3h)+\frac{1}{2}f'(u+h) \\ &= \frac{f'(y)}{2}+\frac{f'(x)}{2}=\frac{f'(x)+f'(y)}{2}. \end{aligned}$$

□

Continuing the proof of Theorem 3.12.6, we now look at $g(x) = f'(x)$ and note that g is continuous and satisfies

$$g\left(\frac{x+y}{2}\right)=\frac{g(x)+g(y)}{2}.$$

This is a fairly simple functional equation which ought to be easy to solve and, indeed, it is. In fact, the reader may already have seen something like this, namely the definition of a convex function: A function h defined on an interval I is *convex* if, for all $x, y \in I$ and any $\lambda \in [0, 1]$,

$$h\big(\lambda x+(1-\lambda)y\big) \leq \lambda h(x)+(1-\lambda)h(y). \tag{3.81}$$

h is *weakly convex* if (3.81) holds for $\lambda = 1/2$. A useful fact about functions is that a function h is convex iff it is continuous and weakly convex. The familiar proof from the literature[376] adapts easily to yield an analogous result with equality instead of inequality:

---

[376] A popular exposition is: Emil Artin, *Einführung in die Theorie der Gammafunktion*, Verlag von B.G. Teubner, Leipzig, 1931. English translation: Emil Artin (Michael Butler, trans.), *The Gamma Function*, Holt, Rinehart and Winston, New York, 1964. Cf. pp. 5–6 in the English edition.

3.12.8 Lemma *Let g be defined on an interval I. Suppose g is continuous and, for all $x, y \in I$,*

$$g\left(\frac{x+y}{2}\right) = \frac{g(x)+g(y)}{2}.$$

Then, for any $\lambda \in [0, 1]$,

$$g\big(\lambda x + (1-\lambda)y\big) = \lambda g(x) + (1-\lambda)g(y).$$

Proof. We first show that, for $n > 1$ and $x_0, x_1, \ldots, x_{n-1} \in I$,

$$g\left(\frac{x_0 + x_1 + \ldots + x_{n-1}}{n}\right) = \frac{g(x_0) + g(x_1) + \ldots + g(x_{n-1})}{n}. \tag{3.82}$$

This is a slightly tricky induction. The basis holds for $n = 2$ by assumption.

We first show that, if (3.82) holds for some value of n, then it holds for $2n$:

$$g\left(\frac{x_0 + \ldots + x_{2n-1}}{2n}\right) = g\left(\frac{\frac{x_0 + \ldots + x_{n-1}}{n} + \frac{x_n + \ldots + x_{2n-1}}{n}}{2}\right)$$
$$= \frac{1}{2}\left[g\left(\frac{x_0 + \ldots + x_{n-1}}{n}\right) + g\left(\frac{x_n + \ldots + x_{2n-1}}{n}\right)\right],$$

by the basic assumption of the Lemma,

$$= \frac{1}{2}\left[\frac{g(x_0) + \ldots + g(x_{n-1})}{n} + \frac{g(x_n) + \ldots + g(x_{2n-1})}{n}\right],$$

by the assumption that (3.82) holds for n,

$$= \frac{g(x_0) + \ldots + g(x_{n-1})}{2n}.$$

Next we show that if (3.82) holds for $n+1$, it also holds n. Let $x_0, x_1, \ldots, x_{n-1}$ be given and define

$$x_n = \frac{x_0 + \ldots + x_{n-1}}{n}.$$

Observe

$$g(x_n) = g\left(\frac{n+1}{n+1}x_n\right) = g\left(\frac{nx_n + x_n}{n+1}\right)$$
$$= g\left(\frac{x_0 + \ldots + x_{n-1} + x_n}{n+1}\right) = \frac{g(x_0) + \ldots + g(x_{n-1}) + g(x_n)}{n+1}.$$

Multiplying by $n+1$

$$(n+1)g(x_n) = g(x_0) + \ldots + g(x_{n-1}) + g(x_n).$$

Thus

$$ng(x_n) = g(x_0) + \ldots + g(x_{n-1})$$

and division by n yields

$$g\left(\frac{x_0 + \ldots + x_{n-1}}{n}\right) = g(x_n) = \frac{g(x_0) + \ldots + g(x_{n-1})}{n}.$$

One can now begin the induction in earnest. We know (3.82) holds for $n = 2$ by our initial assumption. By our first observation it holds for 4, and by the second it holds for 3. Thus it holds for 6 and 8, whence for 5 and 7. But this means it holds for 10, 12, 14, and 16, whence for 9, 11, 13, 15. Etc.

For $\lambda \in [0, 1]$ a positive rational number, write $\lambda = k/n$, with k, n nonnegative integers, and let $x_0 = x_1 = \ldots = x_{k-1} = x$, $x_k = x_{k+1} = \ldots = x_{n-1} = y$. Then

$$\begin{aligned}
g\big(\lambda x + (1-\lambda)y\big) &= g\left(\frac{k}{n}x + \frac{n-k}{n}y\right) \\
&= g\left(\frac{x + x + \ldots + x + y + y + \ldots + y}{n}\right) \\
&= \frac{g(x) + g(x) + \ldots + g(x) + g(y) + g(y) + \ldots + g(y)}{n} \\
&= \frac{kg(x) + (n-k)g(y)}{n} = \lambda g(x) + (1-\lambda)g(y).
\end{aligned}$$

Thus we have proven the Lemma for rational values of λ. If λ is not rational, we can write $\lambda = \lim_{k\to\infty} \lambda_k$ for a sequence $\lambda_0, \lambda_1, \ldots$ of rational numbers. Note that for any u, v

$$\lambda u + (1-\lambda)v = \lim_{k\to\infty} \big(\lambda_k u + (1-\lambda_k)v\big).$$

But g is continuous, whence

$$\begin{aligned}
g\big(\lambda x + (1-\lambda)y\big) &= g\big(\lim_{k\to\infty} (\lambda_k x + (1-\lambda_k)y)\big) \\
&= \lim_{k\to\infty} g\big(\lambda_k x + (1-\lambda_k)y\big) \\
&= \lim_{k\to\infty} \big(\lambda_k g(x) + (1-\lambda_k)g(y)\big) \\
&= \lambda g(x) + (1-\lambda)g(y).
\end{aligned}$$

□

The Lemma entails the linearity of g: Let $a_0 < b_0$ be in the domain of g and let $a_0 < x < b_0$. Then

$$x = a_0 + \frac{x - a_0}{b_0 - a_0}(b_0 - a_0),$$

and, writing $\lambda = (x - a_0)/(b_0 - a_0)$, this yields

$$x = a_0 + \lambda(b_0 - a_0) = \lambda b_0 + (1 - \lambda)a_0.$$

Thus

$$\begin{aligned} g(x) &= \lambda g(b_0) + (1 - \lambda)g(a_0) \\ &= \frac{x - a_0}{b_0 - a_0}g(b_0) + \left(1 - \frac{x - a_0}{b_0 - a_0}\right) g(a_0) \\ &= D_0 x + E_0 \end{aligned}$$

after collecting like terms. Now, D_0, E_0 do not depend on the choice of interval (a_0, b_0). For, if x lies in two such intervals (a_0, b_0) and (a_1, b_1) yielding two pairs of coefficients, D_0, E_0 and D_1, E_1, respectively, then the two functions

$$g(x) = D_0 x + E_0 = D_1 x + E_1$$

agree on the intersection of the intervals, i.e., at infinitely many points; and it follows that $D_0 = D_1$ and $E_0 = E_1$.

A simple integration yields the Theorem. Since $g = f'$, we have

$$f(x) = \frac{D_0}{2}x^2 + E_0 x + C = Ax^2 + Bx + C$$

for $A = D_0/2$, $B = E_0$, and C some constant.

All that remains is to discuss the modifications necessary when we do not assume the domain of f to be all of $\mathbb{R}$. In this case, x and y can be so far apart that $u = x - h$, h equalling half the distance between x and y, is outside the domain of f. What one has to do in this case is to establish Lemma 3.12.7 for x, y sufficiently close to each other, prove f' locally linear, and piece the linear bits together into a single linear function defined on the domain of f. I leave the details to the more ambitious reader. □

The details of the proof are a little grubby. I came across Theorem 3.12.6 in a pleasant exercise book called *The Calculus Integral*.[377] In working this exercise, Lemma 3.12.7 is quickly found and familiarity with the proof that continuous weakly convex functions are convex yields the Theorem without much additional thought. As I do not assume familiarity with convex functions or the proof that continuous weakly convex functions are convex, I included that proof in the form of Lemma 3.12.8. It is this proof that forms the bulk of the work and makes the proof as grubby

[377] Brian S. Thomson, *The Calculus Integral*, www.ClassicalRealAnalysis.com, 2010, Exercise 141, p. 29. Devoted mostly to the integral, this book has a nice collection of exercises dealing with the Mean Value Theorem.

as it is. Knowledge can work against one. Just as Weierstrass did not initially see that the Mean Value Theorem reduced quickly to Rolle's Theorem because, as I surmise, he was too familiar with the Ampère-Cauchy approach, I overlooked a simple fact that I only came across later in searching unsuccessfully for the provenance of the Theorem.

3.12.9 Lemma *Let f be differentiable in an interval (a, b) and suppose, for all $x, y \in (a, b)$ with $x \neq y$,*

$$f'\left(\frac{x+y}{2}\right) = \frac{f(x) - f(y)}{x - y}.$$

Then: f has derivatives of all orders in (a, b).

Proof. Obviously, we are not allowed to assume Theorem 3.12.6 as we wish to use this lemma to simplify the proof of that theorem.

Let $x, y \in (a, b)$ and let $h > 0$ be small enough so that $x \pm h, y \pm h \in (a, b)$. Note that

$$f'(x) = \frac{f(x+h) - f(x-h)}{2h}, \quad f'(y) = \frac{f(y+h) - f(y-h)}{2h}.$$

Thus

$$\begin{aligned}\frac{f'(x) - f'(y)}{x - y} &= \frac{f(x+h) - f(x-h) - f(y+h) + f(y-h)}{2h(x-y)} \\ &= \frac{1}{2h}\left[\frac{f(x+h) - f(y+h)}{x - y} - \frac{f(x-h) - f(y-h)}{x - y}\right]\end{aligned}$$

and

$$f''(y) = \lim_{x \to y} \frac{f'(x) - f'(y)}{x - y} = \frac{1}{2h}\left[f'(y+h) - f'(y-h)\right] \tag{3.83}$$

exists.

Thus f is twice differentiable on (a, b).

Letting $u, v \in (a, b)$, choose $y = (u+v)/2, h = (u-v)/2$ and note that (3.83) yields

$$f''\left(\frac{u+v}{2}\right) = \frac{f'(u) - f'(v)}{u - v},$$

i.e., f' satisfies the same conditions as f. Thus f''' exists, $f^{(4)}$ exists, etc. □

Now that we know f is twice differentiable, we can simplify the proof of Theorem 3.12.6. One way is to simplify the reduction of the Theorem to Lemma 3.12.8:

3.12.10 Exercise Assume f is twice differentiable on (a, b) and satisfies

$$f'(\lambda a + (1-\lambda)b) = \lambda f'(a) + (1-\lambda)f'(b) \tag{3.84}$$

for all $\lambda \in (0, 1)$. Noting that every $x \in (a, b)$ can be written $x = \lambda a + (1-\lambda)b$ for some $\lambda \in (0, 1)$, differentiate (3.84) with respect to λ to show that f'' is constant on (a, b).

And one can, in fact, bypass Lemma 3.12.8:

3.12.11 Exercise Let g satisfy

$$g\left(\frac{x+y}{2}\right) = \frac{g(x)+g(y)}{2}, \quad \text{for all } x, y \in (a, b), \tag{3.85}$$

and suppose g is continuously differentiable on (a, b). Let $x, y \in (a, b)$ and define a sequence,

$$x_0 = x, \quad x_{n+1} = \frac{x_n + y}{2}.$$

Show:
i. $g'(x_0) = g'(x_1) = g'(x_2) = \ldots$
ii. $\lim_{n\to\infty} g'(x_n) = g'(y)$
iii. $g'(x) = g'(y)$.
[Hint. Consider y to be constant and differentiate (3.85) with respect to x.]

3.12.12 Exercise Combine Lemma 3.12.9 and Exercise 3.12.11 to prove Theorem 3.12.6.

The literature on Theorem 3.12.6 is more extensive than one might expect. The early papers include the following:
Rudolf Rothe, "Zum Mittelwertsatze der Differentialrechnung", *Mathematische Zeitschrift* 9 (1921), pp. 300–325.
Tsuruichi Hayashi, title unknown, *Science Reports of the Tôhoku Imperial University*, Ser. 1, 13 (1925), pp. 385–??
Paul von Szász, "Über einen Mittelwertsatz", *Mathematische Zeitschrift* 25 (1926), pp. 116–120.
Rudolf Rothe, "Zum Mittelwertsatz und zur Tayorschen Formel", *Tôhoku Mathematical Journal* 29 (1928), pp. 145–157.

The paper most often cited today is
J. Aczél, "A mean value property of the derivative of quadratic polynomials—without mean values and derivatives", *Mathematics Magazine* 58 (1985), pp. 42–45.
Also commonly cited is
S. Haruki, "A property of quadratic polynomials", *The American Mathematical Monthly* 86 (1979), pp. 207–212.
The geometry of the result is studied in

Bettina Richmond and Tom Richmond, "How to recognize a parabola", *The American Mathematical Monthly* 116 (2009), pp. 910–922.

The final result I wish to consider in this subsection is from the following paper: Ganesh Prasad, "On the nature of θ in the Mean-Value Theorem of the Differential Calculus", *Bulletin of the American Mathematical Society* 36, (1930), pp. 289–291.

Prasad takes a different approach to the problem of producing $c \in (0, 1)$ satisfying

$$f(x+h) = f(x) + hf'(x+ch) \tag{3.86}$$

in accordance with the Mean Value Theorem. He assumes $x = a$ is fixed and considers c to be some function $\theta(h)$ of h. He assumes θ is *single-valued*, i.e., there is a unique value $c \in (0, 1)$ satisfying (3.86). He proves two results, one negative and one positive. The negative result depends on the existence of a pathological function and results in a function f for which θ is single-valued and continuous, but not differentiable. The positive result, which I intend to prove here, he states as follows:

3.12.13 Theorem *If $\theta(h)$ is single-valued, it is necessarily continuous for every value of h.*

Prasad's paper is not exactly a model of clarity and I will not quote from it, but will flesh out his argument and add a couple of unstated assumptions that, I believe, were implicit or overlooked. The first is to assume not only that f is continuous on some interval $[a, b]$ and differentiable on (a, b), but is in fact continuously differentiable on $[a, b]$. Second, I assume for each $h \in (0, b-a]$ there to be a unique $c \in [0, 1]$ satisfying

$$f(a+h) = f(a) + hf'(a+ch),$$

Since there is such a c in $(0, 1)$ satisfying this, the assumption rules out the possibility that either of the tangents to f at $x = a$ or $x = a+h$ has the same slope as the secant.

To begin the proof proper, we let θ denote the function supplying c for its argument h:

$$f(a+h) = f(a) + hf'\big(a + \theta(h) \cdot h\big).$$

Let θ fail to be continuous at some $\overline{h} \in (0, b-a]$. Then there is an $\epsilon > 0$ such that for all $\delta > 0$ there is an h with $|h - \overline{h}| < \delta$ and $\big|\theta(h) - \theta(\overline{h})\big| > \epsilon$. Set $\delta_0 = b - a$ and, for each n, set $\delta_{n+1} = \delta_n/2$. For each n, choose h_n such that $|h_n - \overline{h}| < \delta_n$ and $\big|\theta(h_n) - \theta(\overline{h})\big| > \epsilon$. Consider the sequence $\theta(h_0), \theta(h_1), \theta(h_2), \ldots$ in $[0, 1]$. By the Bolzano–Weierstrass Theorem, there is a convergent subsequence $\theta(h_{i_0}), \theta(h_{i_1}), \theta(h_{i_2}), \ldots$ with limit $\overline{\theta} \in [0, 1]$.

Now

$$\begin{aligned}\lim_{i\to\infty} \big(f(a+h_i) - f(a)\big) &= f(a+\overline{h}) - f(a), \text{ by continuity of } f\\ &= \overline{h} \cdot f'\big(a + \theta(\overline{h}) \cdot \overline{h}\big). \end{aligned} \tag{3.87}$$

But

$$\lim_{n\to\infty}\left(f(a+h_{i_n})-f(a)\right) = \lim_{n\to\infty} h_{i_n}\cdot f'\left(a+\theta(h_{i_n})\cdot h_{i_n}\right)$$
$$= \overline{h}\cdot f'(a+\overline{\theta}\cdot\overline{h}), \tag{3.88}$$

by the continuity of f'.

Combining (3.87) and (3.88),

$$f'(a+\overline{\theta}\cdot\overline{h}) = f'\left(a+\theta(\overline{h})\cdot\overline{h}\right).$$

But

$$\left|a+\overline{\theta}\cdot\overline{h}-\left(a+\theta(\overline{h})\cdot\overline{h}\right)\right| = \left|\overline{\theta}-\theta(\overline{h})\right|\cdot\left|\overline{h}\right| \geq \epsilon\cdot\left|\overline{h}\right| > 0. \tag{3.89}$$

By the assumption that the value of $\theta\in[0,1]$ such that

$$f'(a+\theta\overline{h}) = \frac{f(a+\overline{h})-f(a)}{\overline{h}}$$

is unique, we must have $a+\overline{\theta}\cdot\overline{h} = a+\theta(\overline{h})\cdot\overline{h}$, i.e., $\overline{\theta}=\theta(\overline{h})$, contrary to (3.89).

Thus $\theta(h)$ is continuous at $\overline{h}$.

3.12.3 Complex Considerations

It has not been my intention to discuss the validity or nonvalidity of the Mean Value Theorem in any context other than single-variable Calculus as presented in an introductory course. I have made the exception in briefly mentioning functions of several variables in discussing Lagrange, Ampère and Cauchy, as their contributions here relate directly to the single-variable case. In the present subsection I will break my rule again and discuss the mean value property for functions of a complex variable, as the results I wish to present relate directly to the questions discussed in the immediately preceding subsection.

The starting point of any discussion of the validity of the Mean Value Theorem for functions of a complex variable should be the following:

3.12.14 Example Let $f:\mathbb{C}\to\mathbb{C}$ be given by $f(z)=e^z$, let $a\in\mathbb{C}$ be any complex number, and let $b=a+2\pi i$. There is no $c\in\mathbb{C}$ such that

$$\frac{f(b)-f(a)}{b-a} = f'(c).$$

For,

$$\frac{f(b)-f(a)}{b-a}=\frac{f(a+2\pi i)-f(a)}{a+2\pi i-a}=\frac{e^{a+2\pi i}-e^a}{2\pi i}=\frac{e^a-e^a}{2\pi i}=0,$$

but $f'(c)=e^c\neq 0$ for any $c\in\mathbb{C}$.

This is fairly devastating. The exponential function is not some pathological counterexample deviously constructed to force the failure of the mean value property, but is one of the most basic functions of the theory, extremely well behaved with many useful properties. If, pertaining to our discussion in the preceding section, the mean value of the real exponential function $f(x)=e^x$ could readily be determined, then surely the determination would extend to the complex case. This is indeed what happens with the formulæ for finding where the mean value occurs for quadratic and cubic polynomials as presented in Examples 3.12.4 and 3.12.5:

3.12.15 Example i. Let $f(z)=Az^2+Bz+C$ be a quadratic polynomial with complex coefficients A, B, C. For any $a, b\in\mathbb{C}$, with $a\neq b$,

$$\frac{f(b)-f(a)}{b-a}=f'\left(\frac{a+b}{2}\right).$$

ii. Let $f(z)=Az^3+Bz^2+Cz+D$ be a cubic polynomial with complex coefficients A, B, C, D, and $A\neq 0$. For any $a, b\in\mathbb{C}$ with $a\neq b$, if we define $m=\dfrac{f(b)-f(a)}{b-a}$, then

$$m=f'(c)\ \text{ for }\ c=\frac{-B\pm\sqrt{B^2-3A(C-m)}}{3A}.$$

I leave it to the reader to verify that the simple algebraic proofs of the real forms of these statements remain valid in the complex domain. What is interesting here is what these results tell us. In the quadratic case, if we graph the points a, b, and c, we see that c lies at the mid-point of the line segment connecting a and b (see Fig. 3.23).

The situation is quite different in the cubic case.

3.12.16 Example Let $f(z)=z^3$, $a=i$, $b=1$, so that $m=i$ and

$$c=\frac{0\pm\sqrt{0-3(-i)}}{3}=\pm\frac{\sqrt{3i}}{3}=\pm\frac{\sqrt{6}+\sqrt{6}i}{6}.$$

Neither value of c lies on the segment connecting a and b (see Fig. 3.24).

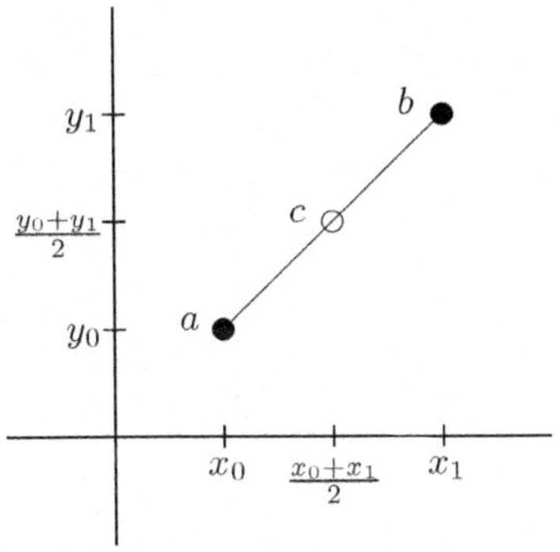

Fig. 3.23 c is a midpoint

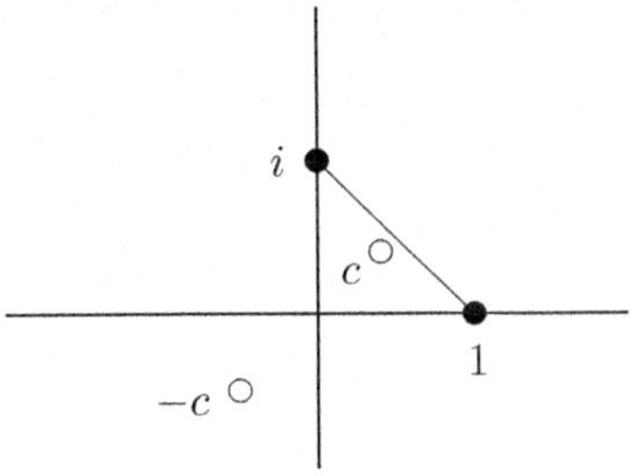

Fig. 3.24 Cubic "mean value"

3.12.17 Exercise i. Show directly there to be no real $\lambda \in [0, 1]$ satisfying

$$\lambda i + (1 - \lambda) \cdot 1 = \frac{\sqrt{6} + \sqrt{6}i}{6}. \tag{3.90}$$

ii. Does there exist a complex number $\lambda = \alpha + \beta i$, with α, β real that satisfies (3.90)? [Hint. i. Using the uniqueness of the expression of z in the form $x + yi$, for complex z and real x, y, assume (3.90) holds and solve for λ. Then show that (3.90) fails for this value of λ.]

It happens that, for any complex polynomial,

$$P(z) = A_n z^n + A_{n-1} z^{n-1} + \ldots + A_1 z + A_0,$$

of proper degree $n \geq 2$ (i.e., $A_n \neq 0$), and any complex numbers a, b with $a \neq b$, there are (counting multiplicities) $n - 1$ numbers $c_1, c_2, \ldots, c_{n-1}$ satisfying

$$P'(c_i) = \frac{P(b) - P(a)}{b - a},$$

but the c_i's are generally not on the line segment connecting a and b. Indeed, generalising Theorem 3.12.6, the quadratic, linear, and constant polynomials are the only analytic functions for which there is always such a c lying on the connecting segment.

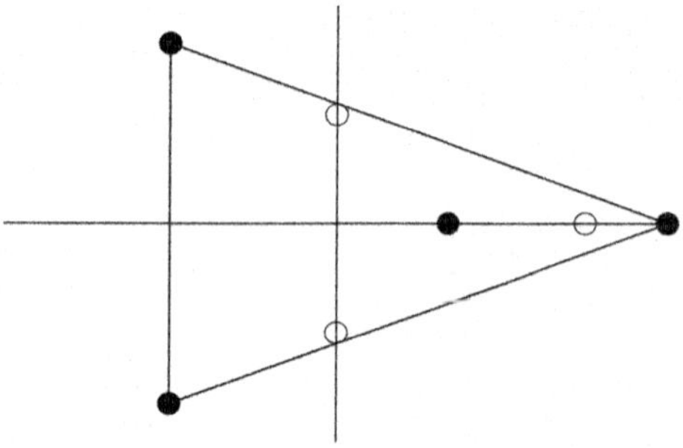

Fig. 3.25 "Mean values" in the 4th degree case

For higher degree polynomials, a weaker, but interesting mean value property can be proven.

To see what this weaker result is, we consider first another example:

3.12.18 Example Let $f(z) = z^4 - 3z^3 + 2z^2$ and consider

$$g(z) = \frac{f(z) - f(1)}{z - 1} = z^3 - 2z^2.$$

Now $g(3) = 27 - 18 = 9$.

The other solutions to $g(z) = 9$ are

$$z = \frac{-1 \pm \sqrt{-11}}{2}.$$

Solving $f'(z) = 4z^3 - 9z^2 + 4z = 9$ gives the solutions

$$z = \pm i, \frac{9}{4}.$$

Graphing the points $1, 3, \dfrac{-1 \pm \sqrt{-11}}{2}$ (black dots) as well as the zeroes of f' (white dots) we have Fig. 3.25.

The solutions to $f'(z) = 9$ all occur inside the convex hull of the points at which the difference quotients $g(z)$ equal 9.

The behaviour exhibited in this last example holds for all complex polynomials of degree at least 2.

3.12.19 Exercise Verify this for the polynomial f and points a, b of Example 3.12.16 above.

So there are two results that need to be proven in this subsection: first, that the polynomials of degree at most 2 are the only analytic functions for which the Mean Value Theorem holds in full generality; and, second, that for all polynomials of degree $n > 1$, the solutions c to

$$f'(c) = \frac{f(b) - f(a)}{b - a},$$

for any given a, b, all lie inside the convex hull of $\{b_1, b_2, \ldots, b_{n-1}, a\}$, where the b_j's are the solutions to

$$\frac{f(b_j) - f(a)}{b_j - a} = \frac{f(b) - f(a)}{b - a}.$$

Two additional goals could be set: Show how certain results familiarly proven in the real setting by appeal to the Mean Value Theorem can be proven in the complex case; and prove a complex version of the Mean Value Inequality for analytic functions. Strictly speaking, neither goal is germane to the discussion of the present book, but the first problem does arise in establishing the negative results for analytic functions that are not quadratic polynomials.

Of our two proper goals, the weak general result for all polynomials is older and makes more modest demands on our knowledge of Complex Analysis, so we shall start with this result. What we need to know here, aside from the Fundamental Theorem of Algebra, which is really an analytic result, are simple facts of complex algebra: Every complex number z can be written uniquely in the form $x + yi$ for real x, y called the *real* and *imaginary* parts of z, respectively. The number $z = x + yi$ has a *conjugate* $\overline{z} = x - yi$; the operation of conjugation $z \mapsto \overline{z}$ preserves addition and multiplication,

$$\overline{z_1 + z_2} = \overline{z_1} + \overline{z_2}, \quad \overline{z_1 \cdot z_2} = \overline{z_1} \cdot \overline{z_2};$$

and the multiplicative inverse of a non-zero complex number is given by

$$z^{-1} = \frac{\overline{z}}{|z|^2},$$

where $|z| = \sqrt{x^2 + y^2}$ is the *absolute value* or *norm* of z. And in the complex plane, the absolute value behaves like the absolute value operation on the real line.

The Fundamental Theorem of Algebra asserts that, if P is a complex polynomial of degree n, then there are complex numbers $a, a_0, a_1, \ldots, a_{n-1}$ such that

$$P(z) = a(z - a_0)(z - a_1) \cdots (z - a_{n-1}).$$

The numbers $a_0, a_1, \ldots, a_{n-1}$ are called the *roots* or *zeros* of P and are not necessarily distinct.

The Fundamental Theorem of Algebra had been around, but unproven, for some time when Carl Friedrich Gauss (1777–1855) finally proved the result in 1797, submitting it for his doctorate two years later. Gauss was so fond of the result that he gave three additional proofs of the theorem, and in a handwritten note on his copy

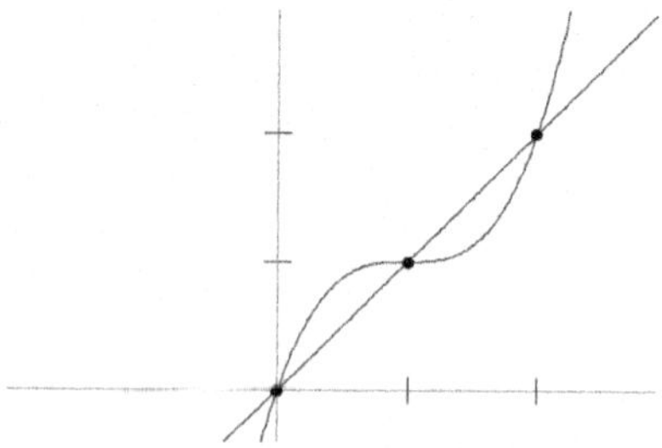

Fig. 3.26 Determining the (0, 1)-P-orbit

of his 1816 proof, he announced as an application the key result generalising the theorem we wish to prove.[378]

3.12.20 Definition Let P be a polynomial and $a \neq b$ be complex numbers. One can think of the intersection of the "line" with "slope" $m = \big(P(b) - P(a)\big)/(b - a)$ passing through $\langle a, P(a)\rangle$ with the graph of $w = P(z)$ as those points $\langle z, w\rangle$ simultaneously satisfying

$$\frac{w - P(a)}{z - a} = m \text{ and } P(z) = w.$$

The elements of the set B of z-coordinates (i.e., the abscissas) of these points of intersection we will call, for lack of a better term, the *(a, b)-P-orbit*.

It is probably best to explain this by referring to an example in the real case:

3.12.21 Example Let $P(x) = x^3 - 3x^2 + 3x$, $a = 0$, $b = 1$. Then $m = 1$ and the curve and line are as in Fig. 3.26. The line passing through $\langle 0, P(0)\rangle = \langle 0, 0\rangle$ and $\langle 1, P(1)\rangle = \langle 1, 1\rangle$ has slope 1 and intercepts the curve $y = P(x)$ in the points $\langle 0, 0\rangle$, $\langle 1, 1\rangle$, $\langle 2, 2\rangle$, whence the (0, 1)-P-orbit is $\{0, 1, 2\}$.

In the complex case, it is impossible to graph the curve $w = P(z)$ in complex 2-dimensional space, which is 4-dimensional real space. But we can think of a, b as determining a "line" passing through $\langle a, P(a)\rangle$ and $\langle b, P(b)\rangle$ with "slope" $m = \big(P(b) - P(a)\big)/(b - a)$ and look for other points of intersection, $\langle z, P(a) + m(z - a)\rangle = \langle z, P(z)\rangle$. Such points obviously satisfy $P(z) = P(a) + m(z - a)$, i.e., they are zeros of

$$Q(z) = P(z) - P(a) - m(z - a),$$

and, counting multiplicities, are n in number—by the Fundamental Theorem of Algebra. Observing that

$$Q'(z) = P'(z) - m,$$

[378] Karl Friedrich Gauss, *Werke* III, Königlichen Gesellschaft der Wissenschaften zu Göttingen, Göttingen, 1866, p. 112.

we see that the solutions c to

$$P'(c) = \frac{P(b) - P(a)}{b - a} = m \tag{3.91}$$

are just the zeros of Q'. Thus to establish the result we wish to prove, that the solutions to (3.91) lie in the convex hull of the (a, b)-P-orbit, it suffices to prove the following result:

3.12.22 Theorem (Gauss–Lucas Theorem) *Let $Q(z)$ be a complex polynomial of degree $n \geq 2$ with roots $a_0, a_1, \ldots, a_{n-1}$. The roots of $Q'(z)$ all lie within the convex hull of $\{a_0, a_1, \ldots, a_{n-1}\}$.*

Before proving this, we need a few simple facts about convex sets and convex hulls, beginning with their definitions.

3.12.23 Definitions A set of complex numbers $X \subseteq \mathbb{C}$ is *convex* if whenever $a, b \in X$, all numbers $\lambda a + (1 - \lambda)b$ for nonnegative real numbers $\lambda \in [0, 1]$ are also in X. The *convex hull* of a set $Y \subseteq \mathbb{C}$ is the intersection of all convex sets X containing Y.

3.12.24 Lemma *The convex hull of a set $X \subseteq \mathbb{C}$ is convex.*

This is trivial.

3.12.25 Lemma *Let X be convex, and $\lambda_0, \lambda_1, \ldots, \lambda_{n-1} \in [0, 1]$ be such that $\lambda_0 + \lambda_1 + \ldots + \lambda_{n-1} = 1$. Then for any $v_0, v_1, \ldots, v_{n-1} \in X$, $\lambda_0 v_0 + \lambda_1 v_1 + \ldots + \lambda_{n-1} v_{n-1} \in X$.*

Proof. By induction on n. The basis is trivial: If $n = 1$, $\lambda_0 = 1$ and $\lambda_0 v_0 = v_0 \in X$.

Suppose the statement is true for any set of k elements of X and let $v_0, v_1, \ldots, v_{k+1} \in X$, $\lambda_0 + \lambda_1 + \ldots + \lambda_{k+1} = 1$, with each $\lambda_j \in (0, 1)$.[379] Then

$$\begin{aligned} \lambda_0 v_0 + \lambda_1 v_1 + \ldots + \lambda_k v_k + \lambda_{k+1} v_{k+1} &= (1 - \lambda_{k+1}) \sum_{j=0}^{k} \frac{\lambda_j}{1 - \lambda_{k+1}} v_j + \lambda_{k+1} v_{k+1} \\ &= (1 - \lambda_{k+1}) v + \lambda_{k+1} v_{k+1}, \end{aligned}$$

where

$$v = \sum_{j=0}^{k} \frac{\lambda_j}{1 - \lambda_{k+1}} v_j \in X$$

by the induction hypothesis and the assumption that $v_{k+1} \in X$. But $(1 - \lambda_{k+1}) v + \lambda_{k+1} v_{k+1} \in X$ by convexity. □

[379] If any of the λ_j's is 0 or 1, there are fewer than $k + 1$ nonzero summands and $\sum \lambda_j v_j \in X$ by the induction hypothesis.

3.12.26 Corollary *The convex hull of a finite set* $\{v_0, v_1, \ldots, v_{n-1}\}$ *is the set of all numbers of the form* $\lambda_0 v_0 + \lambda_1 v_1 + \ldots + \lambda_{n-1} v_{n-1}$*, where, for each j,* $0 \le \lambda_j \le 1$*, and* $\lambda_0 + \lambda_1 + \ldots + \lambda_{n-1} = 1$.

Proof. By the Lemma, these numbers are in every convex set containing $\{v_0, v_1, \ldots, v_{n-1}\}$.

To see that the collection X of such numbers is convex, let $\mu \in [0, 1]$ and consider

$$v = \mu \sum_{j=0}^{n-1} \lambda_j v_j + (1-\mu) \sum_{j=0}^{n-1} \lambda'_j v_j = \sum_{j=0}^{n-1} \left(\mu\lambda_j + (1-\mu)\lambda'_j\right) v_j,$$

with each of $\mu, 1-\mu, \lambda_0, \ldots, \lambda_{n-1}, \lambda'_0, \ldots, \lambda'_{n-1} \in [0, 1]$ and $\sum_{j=0}^{n-1} \lambda_j = \sum_{j=0}^{n-1} \lambda'_j = 1$.

So, for each j,

$$\mu\lambda_j + (1-\mu)\lambda'_j \ge 0.$$

And

$$\sum_{j=0}^{n-1} \left(\mu\lambda_j + (1-\mu)\lambda'_j\right) = \mu \sum_{j=0}^{n-1} \lambda_j + (1-\mu) \sum_{j=0}^{n-1} \lambda'_j = \mu \cdot 1 + (1-\mu) \cdot 1 = 1,$$

whence, writing $\Lambda_j = \mu\lambda_j + (1-\mu)\lambda'_j$, we have each $\Lambda_j \in [0, 1]$, $\sum_{j=0}^{n-1} \Lambda_j = 1$, and

$v = \sum_{j=0}^{n-1} \Lambda_j v_j$ is in X and X is thus convex. □

We are now in position to prove Theorem 3.12.22, which has variously been attributed to Gauss and Félix Lucas.[380] Gauss's contribution appears in two nearly identically worded notes first published in his collected works[381] and is no more than an announcement of the result and a hint for the proof. The second of these notes reads in full:

> Let $\begin{matrix} a, b, c, \ldots, m, n \\ a', b', c', \ldots, m' \end{matrix}$ be the roots of the equation $\begin{matrix} fx = 0 \\ f'x = 0 \end{matrix}$, where $f'x = \frac{dfx}{dx}$, and where the corresponding points in the plane are denoted by the same letters, then, if one imagines at $a, b, c, \ldots, m, n$ equal repelling or attracting masses, which work in inverse proportion to distance, $a', b', c', \ldots, m'$ are balanced.

[380] For example, S. Saks and A. Zygmund, *Analytic Functions*, 3rd ed., Elsevier Publishing Company, Amsterdam, 1971, and PWN–Polish Scientific Publishers, Warsaw, 1971, p. 60 (Exercise 8), credit the result to Gauss, while Lars Ahlfors, *Complex Analysis*, 2nd ed., McGraw-Hill Book Company, New York, 1966, p. 29 (Theorem 1), refers to the result as Lucas's Theorem.

[381] Gauss, *Werke*, III (1866), p. 112, in connexion with his third proof of the Fundamental Theorem of Algebra, and in volume VIII (1900), p. 32.

Gauss's observation was rediscovered and more expansively explained in 1868 by Félix Lucas, and briefly restated by Lucas 20 years later. In 1900, in Gauss's collected works, Robert Fricke accompanied the above note by Gauss with a slightly expanded explanation:

> Gauss probably wrote the [...] note in the year 1846. To prove it, one orients the plane so that one of the solutions of $f'x = 0$ comes to rest at the origin. Then:
>
> $$\frac{1}{a} + \frac{1}{b} + \frac{1}{c} + \cdots + \frac{1}{n} = 0$$
>
> $$\frac{a_0}{aa_0} + \frac{b_0}{bb_0} + \frac{c_0}{cc_0} + \cdots + \frac{n_0}{nn_0} = 0,$$
>
> where a_0 to a, b_0 to b, etc. are complex conjugates. The mechanical meaning of the last equation yields Gauss's theorem.[382]

Fricke's remark still reads more like a hint than a proof, but, except for the unnecessary "reorientation" of the plane, it really offers the heart of the proof. If $P(z)$ is a polynomial of degree n, it can be written

$$P(z) = a(z - a_0)(z - a_1) \ldots (z - a_{n-1}), \tag{3.92}$$

where $a_0, a_1, \ldots, a_{n-1}$ are the n roots (counting multiplicities) of P. If we imagine b to be some point distinct from $a_0, a_1, \ldots, a_{n-1}$ and forces inversely proportional to $b - a_j$ attracting b to or repelling b from each a_j, then the sum of these forces will be proportional to

$$\sum_{j=0}^{n-1} \frac{1}{b - a_j},$$

which sum is known to be 0 if b is a root of P'. This means that, if $P'(b) = 0$, b is in equilibrium with respect to these forces, which means b must lie in the convex hull of $\{a_0, a_1, \ldots, a_{n-1}\}$. Turning this physical heuristic into a mathematical proof is fairly easy; in fact, I find it easier to follow the maths than the heuristics:

Proof of Theorem 3.12.22. If b is a root of both P and P', there is nothing to prove. So we assume $P'(b) = 0$ and $b \notin \{a_0, a_1, \ldots, a_{n-1}\}$.

By induction on $n \geq 1$ it is easily seen that

$$P'(z) = a(z - a_1)(z - a_2) \ldots (z - a_{n-1}) + a(z - a_0)(z - a_2) \ldots (z - a_{n-1}) + \ldots a(z - a_0)(z - a_1) \ldots (z - a_{n-2}),$$

whence

$$\frac{P'(z)}{P(z)} = \frac{1}{z - a_0} + \frac{1}{z - a_1} + \ldots + \frac{1}{z - a_{n-1}},$$

provided $z \notin \{a_0, a_1, \ldots, a_{n-1}\}$.

---

[382] *Ibid.*

Letting

$$t_j = \frac{1}{|b - a_j|^2}, \quad \lambda_j = \frac{t_j}{\sum_{k=0}^{n-1} t_k},$$

one has

$$0 = \sum_{j=0}^{n-1} \lambda_j(\overline{b - a_j}) = \sum_{j=0}^{n-1} \lambda_j \bar{b} - \sum_{j=0}^{n-1} \lambda_j \overline{a_j}.$$

Conjugating again, since λ_j's are non-negative real numbers,

$$\sum_{j=0}^{n-1} \lambda_j b = \sum_{j=0}^{n-1} \lambda_j a_j, \text{ i.e., } \left(\sum_{j=0}^{n-1} \lambda_j \right) b = \sum_{j=0}^{n-1} \lambda_j a_j.$$

But

$$\sum_{j=0}^{n-1} \lambda_j = \sum_{j=0}^{n-1} \frac{t_j}{\sum_{k=0}^{n-1} t_k} = \frac{1}{\sum_{k=0}^{n-1} t_k} \cdot \sum_{j=0}^{n-1} t_j = 1,$$

whence we see that

$$b = \sum_{j=0}^{n-1} \lambda_j a_j, \ \sum_{j=0}^{n-1} \lambda_j = 1,$$

explicitly representing b as an element of the convex hull of $\{a_0, a_1, \ldots, a_{n-1}\}$. □

The Gauss–Lucas Theorem bears a superficial resemblance to Rolle's Theorem and the comparison is often made. The grounds for its truth, however, are completely different and, unlike Rolle's actual theorem which separates the zeros of a polynomial by those of the derivative, it seemingly less helpfully merely places these latter zeros somewhere scattered *among* the former ones. It is, however, a first step in obtaining nontrivial bounds on the zeros of complex polynomials.[383]

Moving on to the negative result concerning the failure of the Mean Value Theorem for analytic functions other than quadratic, linear, and constant polynomials, we must first review a few more facts about complex numbers and functions.

Topologically, I assume the reader familiar with open sets, connected sets and the notation for the open disc of radius r centred at a:

$$D(a, r) = \{z \mid |z| < r\}.$$

In particular, a set $U \subseteq \mathbb{C}$ is open if, for each $a \in U$, there is a real $r > 0$ such that $D(a, r) \subseteq U$.

---

[383]Cf. Leopold Fejér, "Über die Wurzel vom kleinsten absoluten Betrage einer algebraischen Gleichung", *Mathematische Annalen* 65 (1908), pp. 413–423; and Paul Montel, "Sur quelques propriétés des différences divisées", *Journal de mathématiques pures et appliquées* (9) 16 (1937), pp. 219–231.

3.12.27 Definitions Let $U \subseteq \mathbb{C}$ be open and $f : U \to \mathbb{C}$. f is *differentiable at* $a \in U$ if the limit

$$f'(z) = \lim_{z \to a} \frac{f(z) - f(a)}{z - a}$$

exists. If f is differentiable at every $a \in U$, we say that f is *differentiable on* U and the function f' so defined is called the *derivative* of f. f is *continuously differentiable on* U if f is differentiable on U and f' is continuous. f is *analytic on* U if, for each $a \in U$, there is a real $r > 0$ such that f expands into a power series around a in $D(a, r)$: for all $z \in D(a, r)$,

$$f(z) = \sum_{k=0}^{\infty} a_k (z - a)^k.$$

In Complex Analysis, as in the Calculus, differentiability at an isolated point is rarely of interest and one concentrates on open sets where a basic result of Complex Analysis asserts the three concepts to be equivalent: if f is defined on an open set U, then

$$\begin{aligned} f \text{ is differentiable on } U &\Leftrightarrow f \text{ is continuously differentiable on } U \\ &\Leftrightarrow f \text{ is analytic on } U. \end{aligned}$$

That continuous differentiability implies differentiability is trivial, and the proof that analyticity implies continuous differentiability requires only that one develop the theory of power series enough to show that termwise differentiation of a power series yields the derivative—which in turn can be differentiated and must therefore be continuous. The converse implications are much deeper, depending on developing the theory of complex integration, which theory I do not intend to go into here. Some of the easier results of Complex Analysis, however, are obtained by splitting a complex function into its real and imaginary parts: one defines $u, v : \mathbb{R} \times \mathbb{R} \to \mathbb{R}$ for $z = x + yi$ by

$$f(z) = u(x, y) + iv(x, y).$$

Often useful in such applications are the Cauchy–Riemann Equations, named after Cauchy and Riemann:

3.12.28 Lemma (Cauchy–Riemann Equations) *Let f be differentiable on an open set U containing z_0. Write $f(z) = u(x, y) + iv(x, y)$ for $z = x + yi$, and u, v real functions. Then*

$$\frac{\partial u}{\partial x} = \frac{\partial v}{\partial y}, \quad \frac{\partial u}{\partial y} = -\frac{\partial v}{\partial x}.$$

Proof. Write $z_0 = x_0 + y_0 i$. By differentiability the derivative is independent of the direction z approaches z_0 from. In particular, letting $z \to z_0$ horizontally, we have

$$\lim_{x\to x_0}\frac{f(x+y_0i)-f(x_0+y_0i)}{x-x_0}=\lim_{z\to z_0}\frac{f(z)-f(z_0)}{z-z_0}.$$

Thus

$$\begin{aligned}f'(z_0)&=\lim_{x\to x_0}\frac{f(x+y_0i)-f(x_0+y_0i)}{x-x_0}\\&=\lim_{x\to x_0}\frac{u(x,y_0)+iv(x,y_0)-u(x_0,y_0)-iv(x_0,y_0)}{x-x_0}\\&=\lim_{x\to x_0}\frac{u(x,y_0)-u(x_0,y_0)}{x-x_0}+i\lim_{x\to x_0}\frac{v(x,y_0)-v(x_0,y_0)}{x-x_0}\\&=\frac{\partial u}{\partial x}(x_0,y_0)+i\frac{\partial v}{\partial x}(x_0,y_0).\end{aligned}\tag{3.93}$$

Likewise,

$$\begin{aligned}f'(z_0)&=\lim_{y\to y_0}\frac{f(x_0+yi)-f(x_0+y_0i)}{yi-y_0i}\\&=\frac{1}{i}\lim_{y\to y_0}\frac{u(x_0,y)-u(x_0,y_0)}{y-y_0}+\frac{i}{i}\lim_{y\to y_0}\frac{v(x_0,y)-v(x_0,y_0)}{y-y_0}\\&=\frac{1}{i}\frac{\partial u}{\partial y}(x_0,y_0)+\frac{\partial v}{\partial y}(x_0,y_0)\\&=\frac{\partial v}{\partial y}(x_0,y_0)-i\frac{\partial u}{\partial y}(x_0,y_0).\end{aligned}\tag{3.94}$$

Equating real and imaginary parts of (3.93) and (3.94) yields the equations. □

Almost as elementary is a converse: If u, v are continuously differentiable functions in an open set U in the real plane, then the function $f(z)=u(x,y)+iv(x,y)$ is continuously differentiable in U viewed as a subset of $\mathbb{C}$. This result can be proven early on in the Complex Analysis course. The corresponding result when continuity of the partials is not assumed is also true, but not elementary. We will only need Lemma 3.12.28 in what follows.

The simplest application of the Cauchy–Riemann equations is to show certain functions not to be differentiable. In fact, unlike the real case, it is easy to exhibit nonpathological nowhere differentiable functions:

3.12.29 Example The function $f(z)=\overline{z}$ is nowhere differentiable. For, writing $f(z)=u(x,y)+iv(x,y)$, we have

$$u(x,y)=x,\quad v(x,y)=-y,$$

and, writing u_x, u_y, v_x, v_y for the partial derivatives,

$$u_x=1,\quad u_y=0,\quad v_x=0,\quad v_y=-1.$$

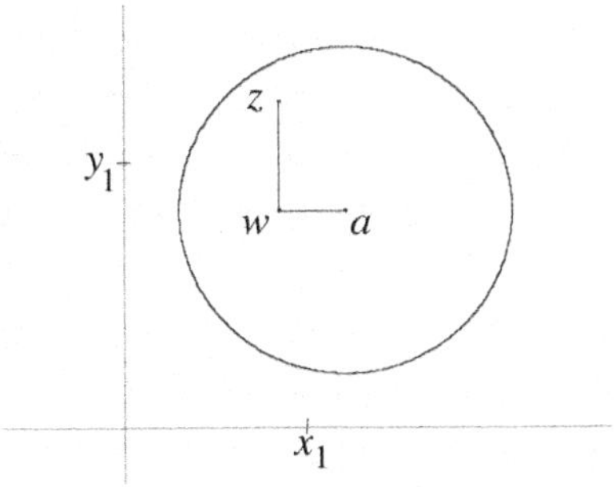

Fig. 3.27 Typical configuration of a, z, w

The second of the Cauchy–Riemann equations, $u_y = -v_x$, holds; but the first, $u_x = v_y$, does not.

3.12.30 Exercise Show $f(z) = |z|^2$ and $g(z) = |z|$ are not differentiable on any open set.

A more substantial application of the Cauchy–Riemann equations[384] is the important Constant Function Theorem.

3.12.31 Theorem (Constant Function Theorem for a Disc) *Let $a \in \mathbb{C}$, r a positive real number, and f a function differentiable on $D(a, r)$. If $f'(z) = 0$ for all $z \in D(a, r)$, then f is constant on $D(a, r)$.*

Proof. In the real case, the simplest proof was by appeal to the Mean Value Theorem:

$$f(x) - f(a) = f'(c)(x - a) = 0, \quad \text{whence } f(x) = f(a).$$

Unfortunately, as we have seen, the Mean Value Theorem does not hold in general in the complex case, and we need a different proof. We get this by applying the Mean Value Theorem for real functions to the real and imaginary parts of f.

Write $f(z) = u(x, y) + iv(x, y)$, $a = a_0 + b_0 i$, and let $z = x_0 + y_0 i \in D(a, r)$. The point $w = \langle x_0, b_0 \rangle$ also lies in $D(a, r)$, for

$$(x_0 - a_0)^2 + (b_0 - b_0)^2 = (x_0 - a_0)^2 \leq (x_0 - a_0)^2 + (y_0 - b_0)^2 < r.$$

(See Fig. 3.27.)

Now

$$f(z) - f(a) = f(z) - f(w) + f(w) - f(a). \tag{3.95}$$

Equating the real parts of (3.95) yields

$$u(x_0, y_0) - u(a_0, b_0) = u(x_0, y_0) - u(x_0, b_0) + u(x_0, b_0) - u(a_0, b_0)$$

[384] Or, rather, an application of the proof of the Cauchy–Riemann equations.

$$= \frac{\partial u}{\partial y}(x_0, y_1)(y_0 - b_0) + \frac{\partial u}{\partial x}(x_1, b_0)(x_0 - a_0) \qquad (3.96)$$

for some x_1 between x_0 and a_0 and some y_1 between y_0 and b_0—by the real Mean Value Theorem. But $f'(x_0 + y_1 i) = 0$, whence (3.94) yields

$$\frac{\partial u}{\partial y}(x_0, y_1) = 0;$$

and $f'(x_1 + b_0 i) = 0$ implies, via (3.93), that

$$\frac{\partial u}{\partial x}(x_1, b_0) = 0.$$

It follows that the right-hand side of (3.96) is 0, whence the real parts of $f(z)$ and $f(a)$ are the same.

Similarly, $f(z)$ and $f(a)$ have equal imaginary parts, whence $f(z) = f(a)$. As $z \in D(a, r)$ was arbitrary, this means f is constant. □

Alternate proof.[385] As before, write $f(z) = u(x, y) + iv(x, y)$ with u, v real. By (3.93) and (3.94), u_x, u_y, v_x, v_y are identically 0 if f' is.

Instead of looking at the horizontal and vertical segments connecting a to w and w to z, consider the diagonal line segment from a to z consisting of all points $a + t(z - a)$ where $0 \le t \le 1$. This line segment lies entirely in $D(a, r)$. (*Exercise.* Prove this.)

Let $g(t) = u\big(a_0 + t(x_0 - a_0), b_0 + t(y_0 - b_0)\big)$, apply Lagrange's Mean Value Theorem for functions $g : \mathbb{R} \to \mathbb{R} \times \mathbb{R}$ (formula 3.40 on page 263, above) to conclude

$$\begin{aligned} u(x_0, y_0) - u(a_0, b_0) &= g(1) - g(0) \\ &= \frac{dg}{dt}\big(a_0 + t_0(x_0 - a_0), b_0 + t_0(y_0 - b_0)\big)(1 - 0), \end{aligned}$$

for some $t_0 \in (0, 1)$. But

$$\begin{aligned} &\frac{dg}{dt}\big(a_0 + t_0(x_0 - a_0), b_0 + t_0(y_0 - b_0)\big) \\ &\quad = u_x\big(a_0 + t_0(x_0 - a_0), b_0 + t_0(y_0 - b_0)\big)(x_0 - a_0) \\ &\qquad + u_y\big(a_0 + t_0(x_0 - a_0), b_0 + t_0(y_0 - b_0)\big)(y_0 - b_0) \\ &\quad = 0 \cdot (x_0 - a_0) + 0 \cdot (y_0 - b_0) = 0. \end{aligned}$$

[385] The first proof is fairly traditional and I make no apologies for it. The second was pointed out to me by Robert Burckel, who also referred to another—interesting and more elementary—proof in: Darrell Desbrow, "On zero derivatives", *The American Mathematical Monthly* 103 (1996), pp. 410–411.

Thus $u(x_0, y_0) - u(a_0, b_0) = 0$.

Similarly, $v(x_0, y_0) - v(a_0, b_0) = 0$ and

$$f(z) = u(x_0, y_0) + iv(x_0, y_0) = u(a_0, b_0) + iv(a_0, b_0) = f(a).$$

□

In the Calculus one usually proves one's results for functions defined on intervals. This is generally sufficient because results for functions with a number of disjoint intervals as domain can usually be applied separately on the individual intervals. However, there are results, for example where unicity is invoked, that fail when the domain is not connected. When this happens, the result can only be stated for intervals. The complex analogue is to state results for functions with connected domains. In both theories, differentiation is assumed on open sets, albeit for different reasons.

There is a slight problem. While it is easy to show a set is disconnected when it is, it is not always easy to verify connectedness directly. What one often does is to use our knowledge that the image of a continuous curve is connected[386] and show the set in question to be *pathwise connected*.

Recall the definition of a connected set[387]:

3.12.32 Definitions A set T is *disconnected* by a pair of disjoint open sets U, V if there are nonempty sets $X, Y \subset T$ such that

$$X \subseteq U, \ Y \subseteq V, \ T = X \cup Y, \ U \cap V = \emptyset.$$

A set T that is not disconnected is called *connected*.

Identifying $\mathbb{C}$ with $\mathbb{R} \times \mathbb{R}$, Theorem 2.2.29 in Chap. 2 tells us that the image $\gamma(I)$ of any continuous function $f : I \to \mathbb{C}$, for any interval I, is connected. This suggests the following:

3.12.33 Definition A set $X \subseteq \mathbb{C}$ is *pathwise connected* if, for any $z, w \in X$, there are $\alpha, \beta \in \mathbb{R}$, with $\alpha < \beta$, and a continuous curve $\gamma : [\alpha, \beta] \to \mathbb{C}$ such that

i. $\gamma(\alpha) = z$

ii. $\gamma(\beta) = w$

iii. $\gamma([\alpha, \beta]) \subseteq X$.

3.12.34 Lemma *If $X \subseteq \mathbb{C}$ is pathwise connected, then X is connected.*

Proof. Suppose X were not connected. There would be disjoint open sets U, V and points $z, w \in X$ such that $z \in U$ and $w \in V$. But X is pathwise connected, so there is some curve $\gamma : [\alpha, \beta] \to X$ with $\alpha < \beta$ and $\gamma(\alpha) = z$, $\gamma(\beta) = w$. One easily sees that U, V disconnect the image $\gamma([\alpha, \beta])$, contrary to Theorem 2.2.29 in Chap. 2. □

---

386 Theorem 2.2.29 on page 69, above.

387 Definition 2.2.2, page 46, above.

Fig. 3.28 Dodging a

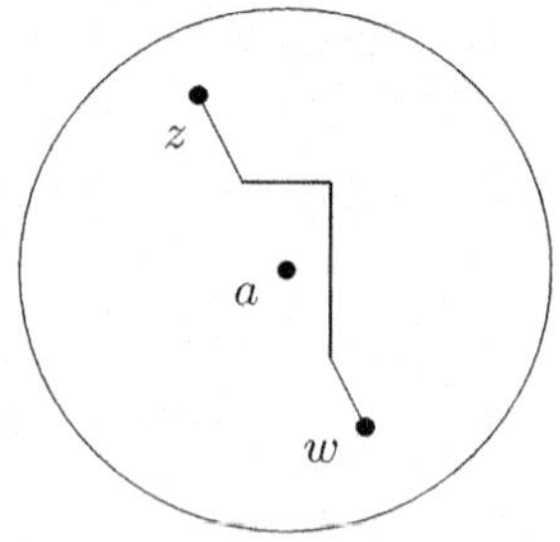

The converse also holds for open sets, but not for sets in general,[388] but we will not need this fact here.

The Lemma has a quick application:

3.12.35 Corollary *Let $a \in \mathbb{C}$, $r \in \mathbb{R}$, with $r > 0$.*
i. The open disc $D(a, r)$ is connected.
ii. The punctured disc $D' = D(a, r)\backslash\{a\} = \{z \in D(a, r) \mid z \neq a\}$ is connected.

Proof. i. To see this, one simply has to exhibit a curve lying entirely inside $D(a, r)$ connecting two given points z, w. We might as well use the obvious parametrisation of the line segment connecting z to w:

$$\gamma(t) = z + t(w - z), \quad t \in [0, 1].$$

γ is evidently continuous with $\gamma(0) = z$, $\gamma(1) = w$. It only remains to note that the image of γ lies entirely inside the disc:

$$\begin{aligned} |\gamma(t) - a| &= |z + t(w - z) - a| \\ &= |(1 - t)z + tw - a| \\ &= |(1 - t)z - (1 - t)a + tw - ta| \\ &\leq (1 - t)|z - a| + t|w - a| \\ &< (1 - t)r + tr = r. \end{aligned}$$

Thus $D(a, r)$ is pathwise connected and hence connected.

ii. If a does not lie on the segment connecting z to w, one can use it as one's γ connecting z to w as above. Otherwise one has to jog around a as in Fig. 3.28.

I leave the algebraic description of γ as an exercise to the reader. □

By part i of the Lemma, open discs are connected and we have thus proven the Constant Function Theorem for a special class of connected open sets. The result is easily lifted to all connected open sets:

3.12.36 Exercise i. Show by example that one can have $f'(z)$ identically 0 on an open set and yet f not be constant.

[388] A simple counterexample can be constructed from the graph of $y = \sin(1/x)$.

ii. Show that if U is a *connected* open set and $f'(z) = 0$ for all $z \in U$, then f is constant.
[Hint. ii. Pick $a \in U$ and show the sets

$$U_a = \{z \in U \,|\, f(z) = f(a)\}, \quad V_a = \{z \in U \,|\, f(z) \neq f(a)\}$$

separate U if f is not constant.]

In the real case, the most important and immediate application of the Constant Function Theorem is probably the uniqueness of the antiderivative up to an additive constant.

3.12.37 Corollary *Let f be defined on a connected open set U and let g, h be antiderivatives of f on U, i.e., for all $z \in U$, $g'(z) = h'(z) = f(z)$. Then g and h differ by a constant.*

Proof. Let $k(z) = g(z) - h(z)$. Then $k'(z) = g'(z) - h'(z) = f(z) - f(z) = 0$, whence Exercise 3.12.36 tells us k is constant. □

3.12.38 Exercise Let f be three times differentiable on a connected open set U.
i. If $f''(z) = 0$ for all $z \in U$, then f is linear or constant.
ii. If $f'''(z) = 0$ for all $z \in U$, then f is a polynomial of degree at most 2.

The Cauchy–Riemann equations also have consequences not having any real analogue:

3.12.39 Corollary *Let f be differentiable on a connected open set. If the real part of f, $u(x, y)$, or the imaginary part, $v(x, y)$, is constant, then f is constant.*

Proof. Assume, say, that u is constant. Let $z_0 = x_0 + y_0 i$ be an arbitrary element of U. By the constancy of u,

$$\frac{\partial u}{\partial x}(x_0, y_0) = \lim_{x \to x_0} \frac{u(x, y_0) - u(x_0, y_0)}{x - x_0} = \lim_{x \to x_0} \frac{0}{x - x_0} = 0.$$

Similarly,

$$\frac{\partial u}{\partial y}(x_0, y_0) = 0,$$

and the Cauchy–Riemann equations yield

$$\frac{\partial v}{\partial x}(x_0, y_0) = 0.$$

Thus, by (3.93), $f'(z_0) = 0$. But z_0 was arbitrary, whence the Constant Function Theorem tells us f is constant.

The proof for v constant is similar. □

With a little algebra, we can also derive the following Corollary.

3.12.40 Corollary *Let $f(z)$ be differentiable on a connected open set U and suppose $g(x, y) = u(x, y)^2 + v(x, y)^2$ is constant on U, where u, v are the real and imaginary parts of f: $f(z) = u(x, y) + iv(x, y)$. Then f is also constant on U.*

Proof. If $u(x, y)^2 + v(x, y)^2$ is identically 0, the result is trivial; so we assume $u^2 + v^2 = |f(z)|^2 = c \neq 0$.

Then $\overline{f}(z) = \dfrac{c}{f(z)}$ is differentiable on U and so are

$$(f - \overline{f})(z) = 2iv(x, y)$$
$$(f + \overline{f})(z) = 2u(x, y),$$

whence the real part of $f - \overline{f}$ and the imaginary part of $f + \overline{f}$ are zero. By Corollary 3.12.39, these functions are both constant, whence

$$2f(z) = f(z) - \overline{f}(z) + f(z) + \overline{f}(z)$$

is the sum of constant functions and is thus constant. □

3.12.41 Corollary *Let f be differentiable on a connected open set U and suppose $|f(z)|$ is constant on U. Then f is constant on U.*

Proof. Note that if $|f(z)|$ is constant, then so too is $|f(z)|^2$. □

We shall use Corollary 3.12.39 later. Corollaries 3.12.40 and 3.12.41 are cited here merely for amusement.

Lest we lose sight of our goal, it might be a good idea to recall it and start in on its proof. This will determine how much additional theory is needed. We begin with a couple of definitions.

3.12.42 Definitions Let f be differentiable on a connected open set U. We say that f has the *mean value property at* $a \in U$ if, for all $z \in U$ for which the segment $[a, z] = \{a + \lambda(z - a) \mid \lambda \in [0, 1]\}$ is contained in U, there is some $w \in [a, z]$ such that

$$f'(w) = \frac{f(z) - f(a)}{z - a}.$$

We say f has the *mean value property on* U if f has the mean value property at a for all $a \in U$.

3.12.43 Theorem *Let f be analytic on a connected open set U. If f has the mean value property on U, then f is a polynomial of degree at most 2.*

This is a relatively recent result. W.G. Dotson, Jr., proved this to be the case assuming f was a polynomial and conjectured the result to hold for all analytic functions in 1968.[389] His conjecture was verified the following year by Zalman Rubinstein,[390] who actually characterised those analytic functions which possessed the mean value property at a given point a. In 2011, Pietro Poggi-Corradini gave a short, more direct proof of Theorem 3.12.43. And Don Marshall is reported to have given another proof, but I have not seen it. The Poggi-Corradini proof makes fewer demands on one's knowledge of Complex Analysis, but it has not yet been published, so I start by following Rubinstein's proof. This approach will force us to establish results which, restricted to real functions, fill in some gaps in rigour in our earlier material.

3.12.44 Lemma *Let f be analytic on a connected open set U, f have the mean value property at $a \in U$, and suppose $f''(a) \neq 0$. Then, for some $\delta > 0$, $D(a, \delta) \subseteq U$ and f equals a polynomial of degree at most 2 on $D(a, \delta)$.*

Proof. By analyticity, f has continuous derivatives of all orders.[391] In particular, $f''(z)$ is continuous on U and there is some $\delta_0 > 0$ such that, for all $z \in U$,

$$|z - a| < \delta_0 \Rightarrow \left|f''(z) - f''(a)\right| < \frac{\left|f''(a)\right|}{2}.$$

But

$$\left|f''(z) - f''(a)\right| = \left|f''(a) - f''(z)\right| \geq \left|f''(a)\right| - \left|f''(z)\right|,$$

whence, for $z \in D(a, \delta_0)$,

$$\frac{\left|f''(a)\right|}{2} > \left|f''(a)\right| - \left|f''(z)\right|,$$

i.e.,

$$\left|f''(z)\right| > \left|f''(a)\right| - \frac{\left|f''(a)\right|}{2} = \frac{\left|f''(a)\right|}{2} > 0.$$

Thus $f''(z) \neq 0$ throughout $D(a, \delta_0)$.

We would like now to conclude from this that f' is one-to-one on $D(a, \delta_0)$, but, as the function $g(z) = e^z$ demonstrates, this is not possible. What is true, however, is that f' is one-to-one in some neighbourhood $D(a, \delta_1) \subseteq D(a, \delta_0)$. This is a nontrivial result of Complex Analysis:

3.12.45 Theorem *Let g be analytic on an open set $V \subseteq \mathbb{C}$ and suppose $g'(z) \neq 0$ for all $z \in V$. Then, for any $a \in V$, there is some $\delta > 0$ such that g is one-to-one on $D(a, \delta) \subseteq V$. Moreover, the function $h : g(D(a, \delta)) \to V$ inverse to g is differentiable at all points in its domain.*

---

[389] W.G. Dotson, Jr., "A note on complex polynomials having Rolle's property and the mean value property for derivatives", *Mathematics Magazine* 41 (1968), pp. 140–144.

[390] Zalman Rubinstein, "On analytic functions satisfying the mean value theorem and a conjecture of W.G. Dotson", *Mathematics Magazine* 42 (1969), pp. 256–259.

[391] This will in fact be proven later.

Properly speaking, the conclusion of this Theorem should also include the fact that the domain $g(D(a, \delta))$ of h is an open set, thus guaranteeing that the limit

$$\lim_{w \to w_0} \frac{h(w) - h(w_0)}{w - w_0}$$

is taken as w approaches w_0 from any direction or along any path and not merely as w approaches w_0 through values in $g(D(a, \delta))$. The openness of the range of any analytic function on any connected open set is a deeper result we will not establish here, and which we will not require in the sequel. Thus, here and below, we should read the differentiability of h at $g(z_0)$ as being analogous to the one-sided differentiability of a function at the endpoints of a closed interval.

We will discuss the proof of this later. For now we assume its truth and return to the proof of Lemma 3.12.44.

Applying Theorem 3.12.45 to $g = f'$, we know that f' is one-to-one on some neighbourhood $D(a, \delta_1) \subseteq D(a, \delta_0)$ of a. It therefore has an inverse, say, $h : f'(D(a, \delta_1)) \to D(a, \delta_1)$.

f has the mean value property at a on U, whence at a in $D(a, \delta_1)$. Thus, for any $z \in D(a, \delta_1)$,

$$f(z) = f(a) + (z - a)f'\big(a + \eta(z)(z - a)\big), \tag{3.97}$$

for some $\eta(z) \in (0, 1)$. This can be solved for η:

$$\frac{f(z) - f(a)}{z - a} = f'\big(a + \eta(z)(z - a)\big)$$
$$a + \eta(z)(z - a) = h\left(\frac{f(z) - f(a)}{z - a}\right),$$

whence

$$\eta(z) = \frac{h\left(\dfrac{f(z) - f(a)}{z - a}\right) - a}{z - a}.$$

Now η is a differentiable function on the punctured disc[392]

$$D' = D(a, \delta_1)\backslash\{a\} = \big\{z \in D(a, \delta_1) \,\big|\, z \neq a\big\}.$$

But $\eta(z) \in (0, 1)$, whence its imaginary part is the constant 0, whence Corollary 3.12.39 tells us η is a constant function on D': for some $c \in (0, 1)$, $\eta(z) = c$.

---

[392] Which is open: We do not need to know that the domain of h is open because we are interested in the differentiability of η, which holds throughout the open punctured disc.

Rewriting (3.97), we have

$$f(z) = f(a) + (z - a)f'\big(a + c(z - a)\big) \tag{3.98}$$

for all $z \in D'$. And simple computation shows this to hold for $z = a$ as well.

Because f is analytic, it can be expanded into a power series around a in some disc $D(a, \delta_2)$. Letting $\delta = \min\{\delta_1, \delta_2\}$, we have, for all $a \in D(a, \delta)$,

$$\begin{aligned} f(z) &= a_0 + a_1(z-a) + a_2(z-a)^2 + a_3(z-a)^3 \dots \\ f'(z) &= a_1 + 2a_2(z-a) + 3a_3(z-a)^2 + \dots \\ f'\big(a + c(z-a)\big) &= a_1 + 2a_2c(z-a) + 3a_3c^2(z-a)^2 + \dots \\ f(a) + (z-a)f'\big(a + c(z-a)\big) &= a_0 + a_1(z-a) + 2a_2c(z-a)^2 + \\ &\quad\; 3a_3c^2(z-a)^3 + \dots \end{aligned}$$

Equating coefficients[393] of the first and fourth of these, in accordance with (3.98), yields

$$nc^{n-1}a_n = a_n, \quad \text{for } n = 1, 2, 3, \dots.$$

For $n = 2$, this reads $2ca_2 = a_2$ and, since $a_2 = f''(a)/2 \neq 0$, we conclude $2c = 1$, i.e., $c = \frac{1}{2}$. And, for $n \geq 3$, since $n/2^{n-1} \neq 1$, we must have $a_n = 0$.

Thus, in $D(a, \delta)$, we have $f(z) = a_0 + a_1(z-a) + a_2(z-a)^2$, a quadratic polynomial. □

The next step in proving Theorem 3.12.43 is to point out that the set of zeros of f'' has no limit point in U. This is essentially a version of Lemma 3.7.5 for complex functions:

3.12.46 Theorem *Let g be analytic on a connected open set V and suppose the set of zeros of g, $Z = \{z \in V \mid g(z) = 0\}$ has a limit point in V. Then g is identically 0 on V.*

The proof given back on pages 319–320 for the real case made heavy use of the Mean Value Theorem and thus does not apply in the complex case. As with Theorem 3.12.45, we will discuss the proof of Theorem 3.12.46 later, applying it first to completing the proof of Theorem 3.12.43.

The bridge from Theorem 3.12.43 to Theorem 3.12.46 is the following topological lemma:

3.12.47 Lemma *Let V be a connected open set and $Z \subseteq V$ a set with no limit point in V. Then $V \backslash Z = \{z \in V \mid z \notin Z\}$ is a connected open set.*

Proof. First note that $V\backslash Z$ is open: For, if $a \in V\backslash Z$, one can find $\delta_a > 0$ small enough so that $D(a, \delta_a) \subseteq V$ and $D(a, \delta_a)$ contains no element of Z, i.e., $D(a, \delta_a) \subseteq V\backslash Z$.

---

[393] This, like the termwise differentiation of a power series, which we have used in the past, will be justified shortly.

Let W_1, W_2 be open sets disconnecting $V\backslash Z$. Letting

$$V_j = (V\backslash Z) \cap W_j,$$

we have $V\backslash Z$ exhibited as the union of the two disjoint, nonempty, open sets V_1, V_2.

Now, let $a \in Z$. Because a is not a limit point of Z, one can find $\delta_a > 0$ small enough so that $D(a, \delta_a) \subseteq V$ and $D(a, \delta_a) \cap Z = \{a\}$. Let D'_a be the punctured disc $D'_a = D(a, \delta_a)\backslash\{a\}$. D'_a is a connected open set by Corollary 3.12.35. Thus D'_a is entirely contained in one of V_1, V_2 (else V_1, V_2 each have nonempty intersections with $D'_a \subseteq V\backslash Z = V_1 \cup V_2$).

Let $A_j = \{a \in Z \mid D'_a \subseteq V_j\}$ and define $V_j^* = V_j \cup A_j$. The sets A_j are subsets of Z, which is disjoint from $V\backslash Z = V_1 \cup V_2$ and $A_1 \cap A_2 = \emptyset$ since $a \in A_1 \cap A_2$ would put $D'_a \subseteq V_1 \cap V_2$. Thus V_1, V_2, A_1, A_2 are pairwise disjoint, whence V_1^*, V_2^* are disjoint.

Also, $z \in V_j^*$ implies $z \in V_j$ or $z \in A_j$. In the first case, $D(z, \delta_z) \subseteq V_j \subseteq V_j^*$ for some $\delta_z > 0$ because V_j is open. In the latter case, z equals some $a \in A_j$, whence

$$D(a, \delta_a) = D'_a \cup \{a\} \subseteq V_j \cup A_j = V_j^*.$$

Thus $D(a, \delta_a) \subseteq V_j^*$ for all $a = z \in V_j^*$ and it follows that V_j^* is open.

But

$$\begin{aligned} V_1^* \cup V_2^* &= V_1 \cup A_1 \cup V_2 \cup A_2 = \left(V_1 \cup V_2\right) \cup \left(A_1 \cup A_2\right) \\ &= \left(V\backslash Z\right) \cup Z = V, \end{aligned}$$

and V is not connected, contrary to assumption. □

Proof of Theorem 3.12.43. Let f be analytic on a connected open set U and suppose f has the mean value property on U. f'' is also analytic on U and, provided f is not constant or linear, the set $Z = \{z \in U \mid f''(z) = 0\}$ has no limit point in U. The set $U\backslash Z$ is a connected open set.

By Lemma 3.12.44, for any $a \in U\backslash Z$, there is some $\delta_a > 0$ and a polynomial P_a of degree at most 2 such that f equals P_a on $D(a, \delta_a)$. Thus $f'''(z) = 0$ identically on $U\backslash Z$. By Exercise 3.12.38.ii, there is a polynomial P of degree at most 2 such that, for all $z \in U\backslash Z, f(z) = P(z)$.

But f and P are continuous on all of U, whence, for $a \in Z$,

$$f(a) = \lim_{z \to a} f(z) = \lim_{z \to a} P(z) = P(a),$$

i.e., f equals P on all of U □

The last step of the proof of Theorem 3.12.43 was handled differently by Rubinstein, who characterised those functions f analytic on a connected open set U which have the mean value property at a point $a \in U$ at which $f''(a) = 0$. It turns out that these are polynomials of the form

$$f(z) = A + B(z - a) + C(z - a)^k, \tag{3.99}$$

for k a nonnegative integer.

3.12.48 Exercise For a polynomial f of the form $A + B(z-a) + C(z-a)^k$, with k an integer greater than 2, for any $z \in \mathbb{C}$ find c on the line segment $[a, z]$ such that

$$f'(c) = \frac{f(z) - f(a)}{z - a}.$$

Show too that $f''(a) = 0$. Why must k be a positive integer > 2 for this last argument to work?

3.12.49 Exercise Assume Rubinstein's characterisation (3.99) of functions f analytic on a connected open set U and which have the mean value property at a point a at which $f''(a) = 0$. Suppose f has the mean value property at two distinct points $a, b \in U$. Show: f is a polynomial of degree at most 2. [Hint. Show that the coefficient C of (3.99) is 0 if f'' is 0 at both points.]

With this we have finished our discussion of Theorem 3.12.43 and have only to prove the all important lemmas that are Theorems 3.12.45 and 3.12.46.

I devote the rest of this subsection to giving as self-contained a presentation of proofs of these results as possible. This may be a waste of some readers' times. One who has not already had a course in Complex Analysis will find this heavy going, while the reader who is thoroughly familiar with the subject will find it superfluous. The presentation is probably only of value to readers like myself who are a bit rusty and need a review.[394]

There are several approaches one can take to proving Theorem 3.12.45. Without assuming the equivalence of differentiability, continuous differentiability, and analyticity, one can try to prove it directly assuming only continuous differentiability and the Cauchy–Riemann equations to derive first an analogue to Lemma 3.10.8:

3.12.50 Lemma *Let g be continuously differentiable on a connected open set U, $a \in U$, and z_n, z'_n two sequences, with $z_n \neq z'_n$, each converging to a. Then,*

$$\lim_{n\to\infty} \frac{g(z_n) - g(z'_n)}{z_n - z'_n} = g'(a).$$

The proof is a fairly straightforward computation,[395] which I shall forego here. Once this Lemma is established, the rest of the proof that g is one-to-one is simple: Let a fail to be a zero of g' and suppose g is not one-to-one in any neighbourhood of a. For each n, there are $z_n, z'_n \in D(a, 1/(n+1))$ with $z_n \neq z'_n$ such that $g(z_n) = g(z'_n)$. But

---

[394] It is over four decades since I learned the subject and over three since I taught it, so my own layer of rust is quite thick.

[395] One can find it in: Konrad Knopp (Lipman Bers, trans.), *Problem Book in the Theory of Functions I*, Dover Publications, Inc., New York, 1948, pp. 15, 66–67.

$$\frac{g(z_n) - g(z_n')}{z_n - z_n'} = 0 \text{ and } \frac{g(z_n) - g(z_n')}{z_n - z_n'} \to g'(a),$$

whence $g'(a) = 0$, a contradiction.

This proves that g is one-to-one in some neighbourhood $D(a, \delta)$ of a and thus has an inverse h on $g(D(a, \delta))$. Sadly, to prove the differentiability of h one must also prove that h is continuous, something we get for free in ordinary single-variable Calculus. To this end, I note that there is a very nice non-computational proof of Theorem 3.12.45 using only the ostensibly weaker assumption of continuous differentiability by appeal to a result proven independently in the 1930s by K. Noshiro[396] and Stefan E. Warschawski.[397]

3.12.51 Lemma (Noshiro–Warschawski Theorem) *Let G be continuously differentiable in a convex open set W and suppose* Re $G'(z)$, *the real part of $G'(z)$, is a positive real number for all $z \in W$. Then G is one-to-one on W and has a differentiable inverse H on $G(W)$.*

The proof is very clever and I cannot resist outlining it. One fixes $z_0 \in W$ and considers, for any $z \in W$, the line segment $\gamma(t) = z_0 + t(z - z_0)$, $t \in [0, 1]$, connecting z_0 to z. By the Chain Rule,

$$\frac{d}{dt}\frac{G(\gamma(t))}{z - z_0} = G'\big(z_0 + t(z - z_0)\big) = G'(\gamma(t)).$$

But

$$\int_0^1 G'(\gamma(t))dt = \left.\frac{G(\gamma(t))}{z - z_0}\right|_0^1 = \frac{G(z) - G(z_0)}{z - z_0}$$

and

$$\int_0^1 \text{Re}\, G'(\gamma(t))dt = \text{Re} \int_0^1 G'(\gamma(t))dt = \text{Re}\, \frac{G(z) - G(z_0)}{z - z_0}.$$

The Extreme Value Theorem yields a minimum value $m > 0$ of Re $G'(\gamma(t))$, whence

$$\text{Re}\, \frac{G(z) - G(z_0)}{z - z_0} \geq \int_0^1 m\, dt = m > 0.$$

Thus G is one-to-one and has an inverse H satisfying

$$\big|H(w) - H(w_0)\big| \leq \frac{1}{m}|w - w_0|,$$

---

396 K. Noshiro, "On the theory of Schlicht functions", *Journal of the Faculty of Science, Hokkaido Imperial University* 2 (1934–1935), pp. 129–155.

397 S. Warschawski, "On the higher derivatives at the boundary in conformal mapping", *Transactions of the American Mathematical Society* 38 (1935), pp. 310–340.

for $w_0 = G(z_0)$, $w = G(z)$, and H is continuous. The usual calculation of the derivative of the inverse function (which depends on the continuity of the inverse) carries through.

To conclude Theorem 3.12.45 from Lemma 3.12.51, start with $a \in V$ such that $g'(a) \neq 0$ and normalise g by defining $G(z) = g(z)/g'(a)$ and noting that $G'(a) = g'(a)/g'(a) = 1 > 0$. Choose $\delta > 0$ small enough so that, for all $z \in W = D(a, \delta)$, $|G'(z) - G'(a)| < 1/2$, i.e., $|G'(z) - 1| < 1/2$. Then $|\mathrm{Re}\, G'(z) - 1| < 1/2$, i.e.,

$$\frac{1}{2} < \mathrm{Re}\, G'(z) < \frac{3}{2},$$

and Lemma 3.12.51 applies and G has a differentiable inverse H on $G(W)$. But, for $w \in g'(a)W = \{g'(a)\zeta \mid \zeta \in W\}$,

$$\begin{aligned} w = g(z) &\Leftrightarrow \frac{w}{g'(a)} = \frac{g(z)}{g'(a)} \\ &\Leftrightarrow \frac{w}{g'(a)} = G(z) \\ &\Leftrightarrow H\left(\frac{w}{g'(a)}\right) = z, \end{aligned}$$

whence $h(w) = H\big(w/g'(a)\big)$ is inverse to g and it is differentiable with

$$h'(w) = H'\big(w/g'(a)\big) \cdot \frac{1}{g'(a)}.$$

One can also prove Theorem 3.12.45 by applying Theorem 3.12.46 and complex integration. Such a proof is perhaps the least elementary, but is moderately popular in textbooks.[398]

And one can base one's proof of Theorem 3.12.45 on analyticity. This seems most appropriate as i. we have used analyticity in the proof of Lemma 3.12.44, ii. we will need to use it in the proof of Theorem 3.12.46, and iii. it affords us an opportunity to discuss the basis for the theory of power series, a subject I have been cavalierly avoiding until now.[399]

---

[398] Two examples: Konrad Knopp (Frederick Bagemihl, trans.), *Theory of Functions; Part I*, Dover Publications, Inc., New York, 1945, pp. 135–136; and, Joseph Bak and Donald J. Newman, *Complex Analysis*, Springer-Verlag New York, Inc., New York, 1982, pp. 146–147. A related, but different, proof appears in: Serge Lang, *Complex Analysis*, Addison-Wesley Publishing Company, Reading (Mass.), 1977, pp. 159–160.

[399] There was no point in being rigorous in discussing Lagrange's use of Taylor series in Sect. 3.3 as his work with them was largely formal and intuitive, and decidedly nonrigorous. And it seemed harmless enough in Sect. 3.7 in discussing Lemma 3.7.5 as this was an aside. Now, however, we really should approach the subject rigorously.

There are works devoted to the history of infinite series to which the reader can refer for more information on the subject.[400] All we need to know here is that the first systematic study of the convergence of power series—as well as of series in general—was given in 1821 by Cauchy in his *Cours d'analyse*. Cauchy's treatment is a bit drawn out, but the basic facts about the radius of convergence as presented in modern textbooks are already to be found there. And, in the *Résumé des leçons* of 1823 one almost finds one of the standard proofs that one can differentiate a power series term-by-term.[401]

Cauchy's first contributions here were his *convergence criteria*: his axiomatic assumption of the convergence of Cauchy sequences and the derived tests of convergence—the *Ratio Test*[402] and the *Root Test*. It is the Root Test that is important here.

3.12.52 Definition Let $a_0, a_1, a_2, \ldots$ be a sequence of positive real numbers. If the sequence is bounded above, each delayed subsequence $a_n, a_{n+1}, a_{n+2}, \ldots$ has a least upper bound b_n. The sequence, $b_0, b_1, b_2, \ldots$ of these bounds is weakly decreasing,

$$m > n \Rightarrow b_m \leq b_n,$$

and thus has a limit $b = \lim_{n\to\infty} b_n \geq 0$. This limit is called the *limit supremum* of the original sequence $a_0, a_1, a_2, \ldots$ and is written

$$\overline{\lim_{n\to\infty}}\, a_n = \limsup_{n\to\infty} a_n = \lim_{n\to\infty} b_n = b.$$

3.12.53 Theorem (Root Test; Radius of Convergence) *Let $a, a_0, a_1, a_2, \ldots$ be complex numbers. Let $|a_0|, |a_1|, |a_2|, \ldots$ be bounded and $\lambda = \overline{\lim_{n\to\infty}} |a_n|^{1/n}$, and define $R = 1/\lambda$ (possibly $= \infty$). Consider the formal power series*

$$\sum_{n=0}^{\infty} a_n (z-a)^n. \tag{3.100}$$

---

[400] Two in my possession are Richard Reiff, *Geschichte der unendllichen Reihen*, Verlag der H. Laupp'schen Buchhandlung, Tübingen, 1899, and Giovanni Ferraro, *The Rise and Development of the Theory of Series Up to the Early 1820s*, Springer Science+Business, LLC, New York, 2008.

[401] I say "almost" because his definition of convergence is now seen to be equivalent to uniform convergence, whence his result that, if $f_0, f_1, f_2, \ldots$ converge to a function f and $f_0', f_1', f_2', \ldots$ converge to g then $f' = g$, is correct, while the similar claim, using our definition of functional convergence as pointwise convergence, is clearly not generally true.

[402] Some like to ascribe the Ratio Test to d'Alembert, who used the ratio of successive terms to analyse the convergence of the binomial series, but did not use this ratio to test for convergence, which he assumed as already given. (An English translation of d'Alembert's work can be found in Smoryński, *Treatise*, *op. cit.*) Gauss had used the Ratio Test in 1812 in rigorously establishing the convergence of his *hypergeometric series*, but this work attracted little attention at the time. It was thus Cauchy who brought the test to the attention of the mathematical public.

i. if $|z-a| < R$*, then the series* (3.100) *converges*
ii. if $|z-a| > R$*, then the series* (3.100) *diverges.*

The number R is called the *radius of convergence* of (3.100) because it says the series converges everywhere inside the open disc $D(a, R)$ of radius R centred at a and nowhere outside the closed disc $\overline{D}(a, R) = \{z \mid |z-a| \leq R\}$. It says nothing about the convergence or divergence of the series on the circle of radius R itself. If R is infinite, i.e., if $\lambda = 0$, this means the series converges for all complex numbers z.

Cauchy begins with a simple lemma.

3.12.54 Lemma *Let* $u_0, u_1, u_2, \ldots$ *be a sequence of nonnegative real numbers and suppose* $\overline{\lim_{n\to\infty}}\, u_n = \lambda$.

i. if $0 \leq \lambda < 1$*, then* $\sum_{n=0}^{\infty} u_n^n$ *converges;*
ii. if $1 < \lambda \leq \infty$*, then* $\sum_{n=0}^{\infty} u_n^n$ *diverges.*

Proof. λ is the limit of the descending sequence of least upper bounds,

$$b_n = \operatorname{lub}\{u_n, u_{n+1}, u_{n+2}, \ldots\}.$$

Thus, for any $\epsilon > 0$ there is a number n_0 so large that, for all $n > n_0$, $|b_n - \lambda| < \epsilon$.

i. Let $\lambda < 1$, choose ρ such that $\lambda < \rho < 1$, and set $\epsilon = \rho - \lambda$. For $n > n_0$, $|b_n - \lambda| < \epsilon$, whence $\lambda \leq b_n < \lambda + \epsilon = \rho$. But $u_n \leq b_n$, whence, for any $n > n_0$, $u_n \leq b_n < \rho$ and $u_n^n < \rho^n$. Let $m > n_0$ and note

$$s_m = \sum_{n=0}^{m} u_n^n = \sum_{n=0}^{n_0} u_n^n + \sum_{n=n_0+1}^{m} u_n^n < \sum_{n=0}^{n_0} u_n^n + \sum_{n=n_0+1}^{m} \rho^n.$$

But $\sum_{n=0}^{n_0} u_n^n$ is a fixed finite number and

$$\sum_{n=n_0+1}^{m} \rho^n < \sum_{n=n_0+1}^{\infty} \rho^n = \rho^{n_0+1} \sum_{k=0}^{\infty} \rho^k = \rho^{n_0+1} \cdot \frac{1}{1-\rho} < \frac{1}{1-\rho},$$

since $\rho < 1$ implies $\rho^{n_0+1} < 1$. Thus the strictly increasing sequence $s_0, s_1, s_2, \ldots$ of partial sums is bounded above and has a limit (namely the least upper bound of these sums).

ii. Let $\lambda > 1$, choose $\epsilon = \lambda - 1$, and note that for all large n,

$$\begin{aligned}|b_n - \lambda| < \epsilon \Rightarrow -\epsilon < b_n - \lambda < \epsilon \\ \Rightarrow \lambda - \epsilon < b_n \\ \Rightarrow 1 < b_n, \text{ by choice of } \epsilon.\end{aligned}$$

But b_n is the least upper bound of $u_n, u_{n+1}, u_{n+2}, \ldots$, whence there is some $m > n$ such that

$$1 + \frac{b_n - 1}{2} < u_m \le b_n.$$

And $1 < u_m$ implies $1 < u_m^m$, whence the general term of $\sum_{n=0}^{\infty} u_n^n$ does not tend to 0 and the series does not converge. □

Proof of Theorem 3.12.53. Let $a, a_0, a_1, a_2, \ldots$ be complex numbers satisfying the conditions of the Theorem and let $\lambda = \overline{\lim_{n\to\infty}} |a_n|^{1/n}$.

Let $u_n = |a_n|^{1/n} \cdot |z - a|$.

i. If $\lambda = 0$, $\overline{\lim_{n\to\infty}} u_n = 0$ for any z, whence $\sum_{n=0}^{\infty} u_n^n = \sum_{n=0}^{\infty} |a_n| \cdot |z - a|^n$ converges.

Likewise, if $0 < \lambda < 1$, and $|z - a| < R = 1/\lambda$,

$$\overline{\lim_{n\to\infty}} u_n = \overline{\lim_{n\to\infty}} |a_n|^{1/n} \cdot |z - a| = \lambda \cdot |z - a| < \lambda \cdot \frac{1}{\lambda} = 1,$$

and the lemma tells us $\sum_{n=0}^{\infty} |a_n| \cdot |z - a|^n$ converges.

Thus we have proven for $\lambda < 1$ the *absolute convergence* of the series $\sum a_n(z - a)^n$. Convergence follows easily enough: Let $\epsilon > 0$ be given and choose n_0 so large that for $m > n > n_0$, $\sum_{k=n}^{m} |a_k| \cdot |z - a|^k < \epsilon$ (by the Cauchy convergence of the series of absolute values) and observe that

$$\left| \sum_{k=n}^{m} a_k (z - a)^k \right| \le \sum_{k=n}^{m} |a_k| \cdot |z - a|^k < \epsilon.$$

Thus the power series is Cauchy convergent, whence convergent.[403]

[403] I am cheating here insofar as I have nowhere in this book proven the convergence of Cauchy sequences even in the real case, much less in the complex case. The convergence of such sequences in the complex case reduces quickly to that in the real case by splitting the sequence into its real and imaginary parts. In the real case, one can simply accept the convergence of Cauchy sequences as one's formulation of the Completeness Axiom as many do; or, one can derive it from one's preferred

ii. If $\lambda > 1$, then by the proof of the lemma, for any n, there is an $m > n$ such that $|a_m| \cdot |z-a|^m = |a_m(z-a)^m| > 1$, whence the general term does not tend to 0 and the series diverges. □

The existence of a radius of convergence is a matter of theoretical importance. The Root Test, as a test of convergence, is rendered practical by some means of calculating $|a_n|^{1/n}$ for various sequences $a_1, a_2, a_3, \ldots$ But even without such a calculation, conclusions can be drawn:

3.12.55 Exercise The two series,

$$\sum_{n=0}^{\infty} a_n(z-a)^n \text{ and } \sum_{n=1}^{\infty} a_n(z-a)^{n-1},$$

have the same radius of convergence.
[Hint. Show for any value of z that, if one series converges, then so does the other.]

A useful calculation of the limit supremum of the n-th roots of sequences of positive numbers is afforded by the following:

3.12.56 Lemma *i.* $\lim_{n\to\infty} n^{1/n} = 1$.
ii. For any positive real number r, $\lim_{n\to\infty} r^{1/n} = 1$.

Proof. i. Observe that $n^{1/n} = e^{(1/n)(\ln n)}$ and consider $f(x) = \frac{1}{x}\ln x$ for $x > 1$:

$$\frac{\ln x}{x} = \frac{2\ln\sqrt{x}}{x} < \frac{2\sqrt{x}}{x} = \frac{2}{\sqrt{x}} \to 0 \text{ as } x \to +\infty,$$

where the inequaltiy follows from the fact that, for $u > 1$, $\ln u < u$.[404] Thus

$$\lim_{n\to\infty} n^{1/n} = \lim_{x\to\infty} e^{(1/x)(\ln x)} = e^{\lim_{x\to\infty}(1/x)(\ln x)} = e^0 = 1.$$

If bringing logarithms and exponentiation to bear on the problem seems a bit much, there is a more elementary approach: Set $x_n = n^{1/n} - 1 > 0$ for $n \geq 2$. Observe

$$n = (1+x_n)^n = \sum_{j=0}^{n} \binom{n}{j} x_n^j \geq \binom{n}{2} x_n^2 = \frac{n(n-1)}{2} x_n^2.$$

(Footnote 403 continued)
formulation—in this book this is the Least Upper Bound Principle, which readily yields the result via the Bolzano–Weierstrass Theorem. I leave the details to the reader.

[404] This is obvious if one defines

$$\ln u = \int_1^u \frac{du}{u}$$

as is often done in the Calculus course.

Thus

$$n \cdot \frac{2}{n(n-1)} \geq x_n^2$$

and

$$0 < x_n^2 \leq \frac{2}{n-1},$$

i.e., $0 < x_n \leq \sqrt{2/(n-1)}$ and $x_n \to 0$ as $n \to \infty$, i.e., $n^{1/n} - 1 \to 0$, i.e., $n^{1/n} \to 1$.

ii. We can simply write

$$r^{1/n} = e^{(1/n)(\ln r)} \to e^{0(\ln r)} = e^0 = 1.$$

Again, one can avoid logarithms and exponentiation: For real $r \geq 1$ and $n > r$, one has $1 \leq r^{1/n} < n^{1/n}$, whence

$$1 = \lim_{n\to\infty} 1 \leq \lim_{n\to\infty} r^{1/n} \leq \lim_{n\to\infty} n^{1/n} = 1,$$

i.e., $\lim_{n\to\infty} r^{1/n} = 1$. If $0 < r < 1$, then $1/r > 1$ and

$$\lim_{n\to\infty} r^{1/n} = \lim_{n\to\infty} \frac{1}{(1/r)^{1/n}} = \frac{1}{\lim_{n\to\infty}(1/r)^{1/n}} = \frac{1}{1} = 1. \qquad \square$$

The immediate application of this lemma and Theorem 3.12.53 is the following:

3.12.57 Theorem *Let*

$$f(z) = \sum_{n=0}^{\infty} a_n (z-a)^n$$

have radius of convergence R (possibly infinite). The radius of convergence of its formal derivative,

$$g(z) = \sum_{n=0}^{\infty} n a_n (z-a)^{n-1},$$

is also R.

Proof. Observe

$$\overline{\lim_{n\to\infty}} \left|na_n\right|^{1/n} = \overline{\lim_{n\to\infty}} \left(|n|^{1/n} \cdot \left|a_n\right|^{1/n} \right).$$

It need not be the case in general that the limit supremum of a product of sequences will equal the product of the limit suprema of the individual sequences. (Let $b_n = 1 + (-1)^n$, $c_n = 1 + (-1)^{n+1}$.) What is clear is that, for nonnegative b_n, c_n,

$$\overline{\lim_{n\to\infty}} (b_n c_n) \leq \left(\overline{\lim_{n\to\infty}} b_n\right)\left(\overline{\lim_{n\to\infty}} c_n\right),$$

whence

$$\varlimsup_{n\to\infty} \left|na_n\right|^{1/n} \le \varlimsup_{n\to\infty} |n|^{1/n} \cdot \varlimsup_{n\to\infty} \left|a_n\right|^{1/n} = \varlimsup_{n\to\infty} \left|a_n\right|^{1/n}. \tag{3.101}$$

The reverse inequality is even easier: For any positive integer n,

$$\left|na_n\right|^{1/n} \ge \left|a_n\right|^{1/n},$$

whence

$$\varlimsup_{n\to\infty} \left|na_n\right|^{1/n} \ge \varlimsup_{n\to\infty} \left|a_n\right|^{1/n} \tag{3.102}$$

Thus (3.101) and (3.102) yield

$$\varlimsup_{n\to\infty} \left|na_n\right|^{1/n} = \varlimsup_{n\to\infty} \left|a_n\right|^{1/n},$$

and it follows that $g(z)(z-a) = \sum_{n=0}^{\infty} na_n(z-a)^n$ has the same radius of convergence as f, whence g has this radius of convergence. □

3.12.58 Exercise Show that the sequence $n^{1/n}$ is strictly decreasing by considering

$$(n+1)^n = \sum_{k=0}^{n} \binom{n}{k} n^k = 1 + n^2 + \sum_{k=2}^{n} \binom{n}{k} n^k, \text{ for } n > 2,$$

and showing:

i. $1 + n^2 < n^n$, if $n > 2$

ii. $\binom{n}{k} \le n^{n-k}$, for $n > 2$, $k \ge 2$.

iii. $(n+1)^n < n^{n+1}$.

$\left[\text{Hint. ii. Note that } \binom{n}{k} = \binom{n}{n-k} = \dfrac{n(n-1)\cdots(k+1)}{(n-k)!}.\right]$

The main result we are aiming at is the proof that one can differentiate a power series term-by-term. The classic proof since Cauchy is to show that the partial sums of the formal derivative g of a power series f converge uniformly in any proper subdisc $D(a,r) \subset D(a,R)$, where R is the radius of convergence of the formal derivative at a. One then appeals to the theorem from integration theory that the integrals of these partial sums, which happen to be the partial sums of the original series, converge uniformly, whence the series f is the integral/antiderivative of g. This requires one to develop the theory of complex integration and many authors prefer a more direct computational approach.

3.12.59 Theorem (Termwise Differentiation of Power Series) *Let the power series* $f(z) = \sum_{n=0}^{\infty} a_n(z-a)^n$ *have radius of convergence R (possibly infinite). Then f is differentiable in D(a, R) and*

$$f'(z) = \sum_{n=0}^{\infty} na_n(z-a)^{n-1}.$$

Moreover, f is uniformly differentiable in $D(a, \rho)$ *for any* $\rho < R$.

Proof. Replacing f by $F(z) = f(z+a)$ if necessary, we may assume for notational convenience that $a = 0$, and write

$$f(z) = \sum_{n=0}^{\infty} a_n z^n, \quad g(z) = \sum_{n=1}^{\infty} na_n z^{n-1}.$$

By Theorem 3.12.57, both functions have the same radius R of convergence. Let $0 < \rho < R$ and $0 < r < R - \rho$.[405] Let $z \in D(0, \rho)$. For $|w| < r$,

$$|z+w| \le |z| + |w| < \rho + r < R,$$

so $z + w \in D(0, R)$. Consider

$$\begin{aligned}
\frac{f(z+w)-f(z)}{w} - g(z) &= \sum_{n=0}^{\infty} \frac{a_n(z+w)^n - a_n z^n}{w} - \sum_{n=1}^{\infty} na_n z^{n-1} \\
&= \frac{a_0 \cdot 1 - a_0 \cdot 1}{w} + \frac{a_1(z+w) - a_1 z}{w} + \sum_{n=2}^{\infty} a_n \frac{(z+w)^n - z^n}{w} \\
&\quad - 1 \cdot a_1 z^0 - \sum_{n=2}^{\infty} na_n z^{n-1} \\
&= 0 + a_1 \frac{w}{w} + \sum_{n=2}^{\infty} a_n \frac{(z+w)^n - z^n}{w} - a_1 - \sum_{n=2}^{\infty} na_n z^{n-1} \\
&= \sum_{n=2}^{\infty} a_n \left[\frac{(z+w)^n - z^n}{w} - nz^{n-1} \right].
\end{aligned} \tag{3.103}$$

The term inside the square brackets in (3.103) equals

[405] For the sake of definiteness, one can take $r = (R - \rho)/2$. The point is that, given R, r depends only on ρ.

$$\frac{1}{w}\sum_{k=1}^{n}\binom{n}{k}w^k z^{n-k} - nz^{n-1} = \sum_{k=2}^{n}\binom{n}{k}w^{k-1}z^{n-k}.$$

But

$$\begin{aligned}\left|\sum_{k=2}^{n}\binom{n}{k}w^{k-1}z^{n-k}\right| &\le \sum_{k=2}^{n}\binom{n}{k}|w|^{k-1}\cdot|z|^{n-k}\\ &\le \sum_{k=2}^{n}\binom{n}{k}|w|\cdot r^{k-2}\cdot|z|^{n-k},\ \text{since } |w|<r\\ &\le \frac{|w|}{r^2}\sum_{k=2}^{n}\binom{n}{k}r^k|z|^{n-k}\\ &\le \frac{|w|}{r^2}(|z|+r)^n\\ &\le \frac{|w|}{r^2}(\rho+r)^n. \end{aligned} \tag{3.104}$$

By (3.103) and (3.104),

$$\left|\frac{f(z+w)-f(z)}{w} - g(z)\right| \le \frac{|w|}{r^2}\sum_{n=2}^{\infty}|a_n|\cdot(\rho+r)^n.$$

But $\rho + r < R$, whence $\sum_{n=2}^{\infty}|a_n|\cdot(\rho+r)^n$ has some finite value A and we have

$$\left|\frac{f(z+w)-f(z)}{w} - g(z)\right| \le \frac{|w|A}{r^2},$$

which is $< \epsilon$ provided $|w| < \delta = r^2\epsilon/A$.

It follows that

$$\lim_{w\to 0}\frac{f(z+w)-f(z)}{w} \text{ exists and equals } g(z)$$

for all $z \in D(0,\rho)$. Moreover, δ depends only on ϵ, r, ρ, i.e., on ϵ and ρ, for $z \in D(0,\rho)$. Thus f is uniformly differentiable in $D(0,\rho)$. □

This Theorem tells us immediately that if f is analytic on U then it is differentiable (hence, of course, also continuous) at every point $a \in U$, the derivative is obtained by termwise differentiation and is thus analytic itself in U (since it has the same radius of convergence at any point in U). Moreover, the Theorem tells us that f is uniformly differentiable in a small enough disc $D(a,\rho)$ around any point $a \in U$.

3.12.60 Remark I am somewhat awed by this proof, which was shown to me by Robert Burckel. It contains one small complication brought on by the need to establish uniform differentiability. This is the choice of $r < R - \rho$, which guarantees $\rho + r$ to be a uniform bound for $|z| + r$. If one wanted only to establish differentiability, it would have been sufficient to have chosen r small enough so that $D(z, r) \subseteq D(0, \rho)$, making ρ an upper bound on $|z| + r$. This choice, however, would yield uniform differentiability in $D(z, r)$ and not in $D(0, \rho)$ as r must be chosen smaller and smaller the closer z comes to the boundary of $D(0, \rho)$, making the factor $|w|/r^2$ larger and larger.

The other complication is the clever trick of bounding $|w|^{k-1}$ by $|w| \cdot r^{k-2}$ instead of by r^{k-1}. This by itself moves the proof outside the usual range of simple ϵ-δ arguments familiar from the first-year Calculus course.

The obvious application of this Theorem is the Maclaurinesque calculation of the coefficients of the power series:

3.12.61 Corollary *Let*

$$f(z) = \sum_{n=0}^{\infty} a_n (z-a)^n$$

have radius of convergence $R > 0$. Then f has derivatives of all orders in the disc $D(a, R)$, and for all n,

$$a_n = \frac{f^{(n)}(a)}{n!}.$$

I leave the easy (inductive) proof to the reader and turn my attention to the proofs of Theorems 3.12.45 and 3.12.46, which now follow fairly quickly.

Proof of Theorem 3.12.45. We must show that, if g is analytic on an open set V and $g'(a) \neq 0$, then g is one-to-one in some disc $D(a, \delta)$. To this end, let R be the radius of convergence of g and g' at $a \in V$ and let $\epsilon > 0$ be arbitrary. Let $\rho < R$ and by uniform differentiability find $\delta_0 > 0$ small enough so that for $z, w \in D(a, \rho)$

$$0 < |w - z| < \delta_0 \Rightarrow \left| \frac{g(z) - g(w)}{z - w} - g'(w) \right| < \epsilon.$$

Since g' is differentiable, it must be continuous, whence one can choose $\delta_1 > 0$ small enough so that

$$0 < |w - a| < \delta_1 \Rightarrow \left| g'(w) - g'(a) \right| < \epsilon.$$

If $\delta = \min\left\{\delta_0/2, \delta_1\right\}$, one has, for any $z, w \in D(a, \delta)$, if $z \neq w$,

$$0 < |w - a| < \delta \;\&\; 0 < |z - a| < \delta \Rightarrow 0 < |z - w| < \delta_0 \;\&\; 0 < |w - a| < \delta_1.$$

But then

$$\left|\frac{g(z)-g(w)}{z-w}-g'(a)\right| \le \left|\frac{g(z)-g(w)}{z-w}-g'(w)\right| + \left|g'(w)-g'(a)\right|$$
$$< \epsilon + \epsilon = 2\epsilon.$$

Now if $g(z) = g(w)$ for some $z, w \in D(a, \delta)$, with $z \neq w$, the term on the left would reduce to $\left|-g'(a)\right|$ and we would have $\left|g'(a)\right| = \left|-g'(a)\right| < 2\epsilon < \left|g'(a)\right|$, on choosing $\epsilon < \left|g'(a)\right|/2$. This yields a contradiction, whence $g(z) \neq g(w)$ for any distinct $z, w \in D(a, \delta)$, i.e., g is one-to-one in $D(a, \delta)$.

To see that the inverse h of g is continuous, choose $\epsilon < \left|g'(a)\right|/4$. For,

$$2\epsilon > \left|\frac{g(z)-g(w)}{z-w}-g'(a)\right| \ge \left|g'(a)\right| - \left|\frac{g(z)-g(w)}{z-w}\right|$$
$$\frac{2\left|g'(a)\right|}{4} > \left|g'(a)\right| - \left|\frac{g(z)-g(w)}{z-w}\right|$$
$$\left|\frac{g(z)-g(w)}{z-w}\right| > \frac{1}{2}\left|g'(a)\right| > 0.$$

Writing K for the positive constant $\left|g'(a)\right|/2$, this means

$$\left|g(z)-g(w)\right| > K|z-w|, \tag{3.105}$$

for all $z, w \in D(a, \delta)$ with $z \neq w$. But any $z', w' \in g(D(a, \delta))$ are of the forms $g(z) = z'$, $g(w) = w'$. In terms of z', w', (3.105) reads

$$|z'-w'| > K\left|h(z')-h(w')\right|.$$

Let $\epsilon' > 0$ be given, $\delta' = K\epsilon'$ and note that, for $z', w' \in g\big(D(a, \delta)\big)$,

$$0 < |z'-w'| < \delta' \Rightarrow K\left|h(z')-h(w')\right| < \delta'$$
$$\Rightarrow \left|h(z')-h(w')\right| < \frac{\delta'}{K} = \epsilon'.$$

Thus h is continuous on $g\big(D(a, \delta)\big)$.

Differentiability of h now follows by the usual argument:

$$\lim_{z'\to w'}\frac{h(z')-h(w')}{z'-w'} = \lim_{z'\to w'}\frac{h\big(g(z)\big)-h\big(g(w)\big)}{g(z)-g(w)}$$
$$= \lim_{z'\to w'}\frac{z-w}{g(z)-g(w)}$$

$$= \lim_{z \to w} \frac{z - w}{g(z) - g(w)}, \text{ by continuity of } h$$
$$= \frac{1}{g'(w)} = \frac{1}{g'\big(h(w')\big)}. \qquad \square$$

There is another proof relying more directly on the power series expansion of g. It proceeds by assuming $g'(a) \neq 0$, formally inverting the Taylor series expansion of g around a, and proving via direct computation that this inverse series has a positive radius of convergence. Fully carried out, the proof is more complicated than that given here,[406] but it does provide the extra information that $g\big(D(a, \delta)\big)$ is open for some $\delta > 0$.

Proof of Theorem 3.12.46. Let g be analytic on a connected open set V and suppose $Z = \big\{z \in V \,\big|\, g(z) = 0\big\}$ has a limit point $a \in V$. We must show g is identically 0 on V.

By choice of a, there is a sequence $z_0, z_1, z_2, \ldots$ of elements of Z such that $z_n \to a$. By the continuity of g,

$$g(a) = g(\lim_{n\to\infty} z_n) = \lim_{n\to\infty} g(z_n) = \lim_{n\to\infty} 0 = 0.$$

By the analyticity of g we can write

$$g(z) = \sum_{n=0}^{\infty} a_n(z - a)^n$$

on some disc $D(a, r) \subseteq V$. Now $a_0 = g(a) = 0$. Let $k > 0$ be the least integer k, if such exists, such that $a_k \neq 0$. Then

$$g(z) = \sum_{n=k}^{\infty} a_n(z - a)^n = (z - a)^k \sum_{n=k}^{\infty} a_n(z - a)^{n-k} = (z - a)^k h(z),$$

where h is analytic by k-fold application of Exercise 3.12.55.

Now, for each element z_n of the sequence of elements of Z converging on a, we have

$$0 = g(z_n) = (z_n - a)^k h(z_n),$$

whence $h(z_n) = 0$. As before, $h(a) = \lim_{n\to\infty} h(z_n) = 0$. But

$$h(z) = a_k + a_{k+1}(z - a) + a_{k+2}(z - a)^2 + \ldots$$
$$h(a) = a_k + 0 + 0 + \ldots = a_k.$$

Thus $a_k = 0$, contrary to choice of k. It follows that every a_n is 0 and

[406] Cf. Lang, *op. cit.*, pp. 67–71.

$$g(z) = \sum_{n=0}^{\infty} 0(z-a)^n = 0$$

everywhere on $D(a, r)$.

A simple connectedness argument shows g identically 0 on V: Let A be the set of limit points of Z in V. For each $a \in A$ choose r_a such that g is identically 0 on $D(a, r_a)$ and define

$$V_0 = \bigcup \{D(a, r_a) \mid a \in A\}$$
$$V_1 = \bigcup \{D(a, r) \mid D(a, r) \subseteq V \text{ contains at most one element of } Z\}.$$

Then $V_0 \cap V_1 = \emptyset$, V_0 and V_1 are open, and $V = V_0 \cup V_1$. The assumption that V is connected then entails that one of these sets is empty. Hence, if $A \neq \emptyset$, then $V_0 = V$ and g is identically 0. □

With this, we complete our discussion of complex functions and power series. It has been a somewhat lengthy digression from our discussion of the Mean Value Theorem in first-year Calculus, but I think the material will not be found to have been completely irrelevant: The Gauss–Lucas Theorem offers something of a complex version of the Mean Value Theorem for polynomials, Theorem 3.12.43 of Rubinstein nicely complements the negative results of Sect. 3.12.2, and the proofs of termwise differentiability of power series and of Theorem 3.12.46 apply equally well to real functions—and we have used these results, the former without proof, earlier in the book.

3.12.4 An Anticlimax

There is one last thing I want to say about the Mean Value Theorem before leaving the present chapter, and this concerns its proof in Nonstandard Analysis. Toward the end of the 19th century, infinitesimals were making a minor comeback in geometry as *nonarchimedean* geometries were devised to study how much could be proven without assuming the *Archimedean Axiom* that was so central to the old Greek method of exhaustion. Some mathematicians followed up with attempts at a nonarchimedean analysis, but they did not get very far. In his 1907/1908 lectures on elementary mathematics from an advanced standpoint, Klein had this to say on the matter:

> In the most recent mathematics, "actually" infinitely small quantities have come to the front again, but in entirely different connection, namely in the geometric investigations of Veronese[407] and also in Hilbert's *Grundlagen der Geometrie*.[408] The guiding thought of these

---

[407] Giuseppe Veronese, *Fundamenti di geometria*, Padua, 1891; a German translation *Grundzüge der Geometrie* appeared in 1894.

[408] David Hilbert, *Grundlagen der Geometrie*, B.G. Teubner, Leipzig, 1899. This book was immediately translated into French with some additions. The English translation by E.J. Townsend, *The*

investigations can be stated briefly as follows: A geometry is considered in which $x = a$ (a an ordinary real number) determines not only *one* point on the x-axis, but infinitely many points, whose abscissas differ by finite multiples of infinitely small quantities of different orders $\eta, \zeta, \ldots$ A point is thus determined only when one assigns

$$x = a + b\eta + c\zeta + \cdots,$$

where a, b, c are ordinary real numbers, and the $\eta, \zeta, \ldots$ actually infinitely small quantities of decreasing orders. Hilbert uses this guiding idea by subjecting these new quantities $\eta, \zeta, \ldots$ to such axiomatic assumptions as will make it evident that one can operate with them consistently. To this end it is of chief importance to determine appropriately the relation as to size between x and a second quantity $x_1 = a_1 + b_1\eta + c_1\zeta + \cdots$. The first assumption is that $x >$ or $< x_1$ if $a >$ or $< a_1$; but if $a = a_1$, the determination as to size rests with the second coefficient, so that $x \gtrless x_1$ according as $b \gtrless b_1$; and if, in addition, $b = b_1$, the decision lies with the c, etc. These assumptions will be clearer to you if you refrain from attempting to associate with the letters any sort of concrete representation.[409]

The coordinates and their ordering referred to by Klein have a simple description. One thinks of the quantities x, x_1 as ordinary sequences,

$$\begin{aligned} x &: \ (a, b, c, \ldots) \\ x_1 &: \ (a_1, b_1, c_1, \ldots) \end{aligned}$$

of real numbers, and orders them *lexicographically*: $x < x_1$ if the coordinate of x is less than that of x_1 at the first place they differ. Klein has not said how one is to add such sequences, but the componentwise operation would seem to be called for as one would expect

$$x + x_1 = (a + a_1) + (b + b_1)\eta + (c + c_1)\zeta + \cdots.$$

Multiplication would not have a clear geometric interpretation, but algebraically it would make sense if one used $1, \eta, \eta^2, \ldots$ in place of $1, \eta, \zeta, \ldots$ Then one would imagine the sequences as coefficients of a power series,

$$a + bx + cx^2 + dx^3 + \cdots$$

evaluated at an infinitesimal η. Multiplication would thus be convolution as described on page 241, above.

(Footnote 408 continued)
Foundations of Geometry, Open Court Publishing Company, LaSalle (Illinois), 1902, appeared shortly after. The book has gone through a number of editions with varying collections of appendices. The 7th edition (1939) included 5 papers by Hilbert on geometry, as well as edited versions of his most important papers on the foundations of mathematics. From the 8th edition on only the geometric papers were retained, but in the 10th edition (1968) supplementary appendices written by his former assistant Paul Bernays were added. The book was a major influence and is still in print.

[409] Felix Klein, *Elementary Mathematics from an Advanced Standpoint. Arithmetic. Algebra. Analysis.*, MacMillan & Co., London, 1932, p. 218. This is a translation by E. R. Hedrick and C.A. Noble of the 3rd edition of lecture notes of a course given by Klein in Göttingen in 1907/1908. The translation is still in print as one of two volumes by Dover Publishing Company in New York.

3.12.62 Example $\mathcal{R}_K = (\mathbb{R}^{\mathbb{N}}, +, *, <, \mathbf{0}, \mathbf{1})$ is a non-Archimedean extension of $\mathbb{R}$, where

$$\mathbf{0} = (0, 0, \ldots), \quad \mathbf{1} = (1, 0, 0, \ldots),$$

$a + b$ is defined componentwise, $*$ is convolution, $<$ is the lexicographical ordering, and $r \in \mathbb{R}$ is identified with $r \cdot \mathbf{1} = (r, 0, 0, \ldots)$. The elements

$$\eta = (0, 1, 0, 0, \ldots)$$
$$\eta^2 = (0, 0, 1, 0, \ldots)$$

etc. are all infinitesimal: $\eta < r$ for every positive real r (for: $(0, 1, 0, 0, \ldots)$ precedes $(r, 0, 0, \ldots)$ in "alphabetical" order; $(0, 0, 1, 0, \ldots)$ precedes $(0, 1, 0, \ldots)$; etc.).

Klein continues by explaining the meaning of "non-Archimedean" (or, "nonarchimedean"):

> Now it turns out that, after imposing upon these new quantities this rule, together with certain others, it is possible to operate with them as with finite numbers. One essential theorem, however, which holds in the system of ordinary real numbers, now loses its validity, namely the theorem: *Given two positive numbers e, a, it is always possible to find a finite integer n such that $n \cdot e > a$, no matter how small e is nor how large a may be.* In fact, it follows immediately from the above definition that an arbitrary finite multiple $n \cdot \eta$ of η is smaller than any positive finite number a, and it is precisely this property that characterizes the η as an infinitely small quantity. In the same way $n \cdot \zeta < \eta$, that is, ζ is an infinitely small quantity of higher order than η.[410]
> This number system is called non-Archimedean. The above theorem concerning finite numbers is called, namely, the *axiom of Archimedes*, because he emphasized it as an unprovable assumption, or as a fundamental one which did not need proof, in connection with the numbers which he used. The denial of this axiom characterizes the possibility of actually infinitely small quantities. The name *Archimedean axiom*, however, like most personal designations, is historically inexact. Euclid gave prominence to this axiom more than half a century before Archimedes; and it is said not to have been invented by Euclid, either, but, like so many of his theorems, to have been taken over from Eudoxus of Knidos. The study of non-Archimedean quantities, which have been used especially as coordinates in setting up a non-Archimedean geometry, aims at deeper knowledge of the nature of continuity and belongs to the large group of investigations concerning the logical dependence of different axioms of ordinary geometry and arithmetic. For this purpose, the method is always to set up artificial number systems for which only a part of the axioms hold, and to infer the logical independence of the remaining axioms from these.
> The question naturally arises whether, starting from such number systems, it would be possible to modify the traditional foundations of infinitesimal calculus, so as to include actually infinitely small quantities in a way that would satisfy modern demands as to rigor; in other words, to construct a non-Archimedean analysis. The first and chief problem of this analysis would be to prove the mean-value theorem
>
> $$f(x + h) - f(x) = h \cdot f'(x + \vartheta h)$$

[410]Is it Klein or the translators? In the previous quote $\eta, \zeta, \ldots$ were of "decreasing orders". Generally, an infinitesimal ζ is said to be a higher order infinitesimal than η if ζ/η is infinitesimal.

> from the assumed axioms. I will not say that progress in this direction is impossible, but it is true that none of the investigators who have busied themselves with actually infinitely small quantities have achieved anything positive.[411]

Klein's remarks had no great influence in the development of Nonstandard Analysis. Indeed, they were not cited when in 1958 Schmieden and Laugwitz published the first nonstandard proof[412] of the Mean Value Theorem, nor again in the 1960s when Robinson published his proof.[413] Nor, perhaps, should they have been. Since Weierstrass and Bonnet, the obvious proof of the Mean Value Theorem has been an algebraic reduction to the Extreme Value Theorem.[414] Once one has proven the Extreme Value Theorem, the standard and nonstandard proofs of the Mean Value Theorem are identical — which is why I label this section an anticlimax. The nonstandard proofs simply offer nothing new.[415] Klein has emphasised the Mean Value Theorem not, it seems, as an expert pinpointing where the next great challenge for nonarchimedean analysis lies — this was in the proofs of the Intermediate and Extreme Value Theorems —, but as a testimonial to the importance the Mean Value Theorem had assumed in the Calculus. A nonarchimedean analysis had no business being called analysis unless it could prove the Mean Value Theorem. The Calculus had thrived for two and a half centuries before Lagrange discovered the Mean Value Theorem, proving it under highly restrictive conditions. After a couple of decades Cauchy moved it to its current central position in the Calculus and within a few decades it was being called the Fundamental Theorem of the Differential Calculus, a result on par with what is now called the Fundamental Theorem of the Calculus, but which should perhaps be labelled the Fundamental Theorem of the Integral Calculus as it is for this half of the Calculus that it is fundamental.

And today we have gone from celebrating the importance of the Mean Value Theorem to having to consider reforms that would remove the Mean Value Theorem from the first year Calculus course!

---

411 Klein, *op. cit.*, pp. 218–219.

412 Curt Schmieden and Detlef Laugwitz, "Eine Erweiterung der Infinitesimalrechnung", *Mathematische Zeitschrift* 69 (1958), pp. 1–39; in particular, pp. 29–30.

413 Announced without proof in Abraham Robinson, "Non-standard analysis", *Nederl. Akad. Wetensch. Proc. Ser. A* 64, and *Indagationes Math.* 23 (1961), pp. 432 - 440. Reprinted in: H.J. Keisler, S. Körner, W.A.J. Luxemburg, and A.D. Young (eds.), *Selected Papers of Abraham Robinson, vol. 2, Nonstandard Analysis and Philosophy*, North-Holland Publishing Company, Amsterdam, 1979, pp. 3 - 11. Robinson published his first (excessively) detailed proof in his *Introduction to Model Theory and to the Metamathematics of Algebra*, North-Holland Publishing Company, Amsterdam, 1963, pp. 258–259, and gave a more pleasant version of the proof in his *Non-Standard Analysis*, North-Holland Publishing Company, Amsterdam, 1966, pp. 70–71.

414 Schmieden and Laugwitz actually proved the Mean Value Theorem by reduction to the Mean Value Theorem for Integrals; Robinson applied the Weierstrass–Bonnet method.

415 This subsection may also disappoint because I do not give the proof here. But it is simple: Take the nonstandard proof of the Extreme Value Theorem sketched on page 60, above, and follow up with the standard proofs of Rolle's Theorem and the Mean Value Theorem. Those wanting greater detail are referred to any exposition of Nonstandard Analysis or, indeed, to my write-up in Chap. III, Sect. 6, in Smoryński, *Formalism, op. cit.*

Chapter 4
Calculus Reform

4.1 The Great Debate

Designing the perfect Calculus course is an impossible goal and a thankless task. The Calculus is a broad and deep field; one must decide which topics need to be covered in the first year and how deeply one needs to go into each one chosen. It has been said that one only first masters the Calculus after teaching it the first time; one cannot expect all students to fully understand everything and one must decide on the level of understanding expected of the students. And, the Calculus is not a course aimed at mathematics majors, but is usually a service course for engineers and physicists; some of the topics and the order of presentation are dictated by the needs of the first-year Physics course.

One aspect of the Calculus that generally isn't taken into account in considerations of course design is the cultural and intellectual significance of the Calculus. I am in no position to say that it definitely has or has not been surpassed by the work on theoretical physics, but by the mid-20th century it was the supreme intellectual accomplishment of mankind. Mathematicians had taken on the problem of infinite processes with all their inherent difficulties and conquered them — removing paradoxes, developing routine algorithms for handling them, and even successfully forming abstract generalisations of great power and beauty. None of this is even hinted at in the standard textbooks.

I am not going to begin to attempt to discuss Calculus Reform in any reasonable depth. This is a book about the Mean Value Theorem and I intend to discuss only the hotly debated question of the rôle of this Theorem in the first year Calculus course. The literature abounds with pronouncements supported by fairly weak arguments against the Mean Value Theorem, and I would like to review and critique some of these arguments here. To this end, I list a number of publications on the role of the Mean Value Theorem in the first year Calculus course. For the purpose of this discussion I organise these into groups. The first group is as follows:

C. Smoryński, *MVT: A Most Valuable Theorem*,
DOI 10.1007/978-3-319-52956-1_4

1924

A.A. Bennett, "Discussions: The consequences of Rolle's Theorem", *The American Mathematical Monthly* 31 (1924), pp. 40 – 42.

1956

M.R. Spiegel, "Mean value theorems and Taylor series", *Mathematics Magazine* 29 (1956), pp. 263 – 266.

1958

C.L. Wang, "Classroom notes: Proof of the Mean Value Theorem", *The American Mathematical Monthly* 65 (1958), pp. 362 – 364.

1959

R.C. Yates, "The Law of the Mean", *The American Mathematical Monthly* 66 (1959), pp. 579 – 580.

1960

Roger Osborn, "Some geometric considerations related to the Mean Value Theorem", *Mathematics Magazine* 33 (1960), pp. 271 – 276.
Jacqueline P. Evans, "The Extended Law of the Mean by a translation-rotation of axes", *The American Mathematical Monthly* 67 (1960), pp. 580 – 581.
Louis C. Barrett and Richard A. Jacobson, "Extended Laws of the Mean", *The American Mathematical Monthly* 67 (1960), pp. 1005 – 1007.

1962

M.J. Poliferno, "A natural auxiliary function for the Mean Value Theorem", *The American Mathematical Monthly* 69 (1962), pp. 45 – 47.
L.C. Barrett, "Classroom notes: Methods of proving mean value theorems", *The American Mathematical Monthly* 69 (1962), pp. 50 – 52.

1989

Herb Silverman, "A simple auxiliary function for the Mean Value Theorem", *The College Mathematics Journal* 20 (1989), p. 323.

The papers of this first group do not question the rôle of the Mean Value Theorem in the Calculus, but accept it as a fundamental result of the theory. What they do is to decry the manner in which the result is presented by a reduction to Rolle's Theorem via an unmotivated choice of auxiliary function. I have discussed this issue already in Chap. 3,[1] where I cited the attitudes of several of these authors on the general expositional shortcomings of the standard presentation, as well as several of their approaches to motivating the choice of an auxiliary function. There is no pressing need to go into these matters in any depth here, but I would like to add a couple of short remarks.

Because Bennett's is the earliest of these papers and because he states the problem so nicely, I cannot resist quoting him. Following a formal statement of Rolle's Theorem, he states

[1] Pages 152–156, above.

Among the numerous important consequences of this theorem that have been enumerated there are two involving only first derivatives, that are given in most texts on the calculus. One due to Lagrange and given in all standard modern texts states that under suitable conditions

$$\frac{f(b)-f(a)}{b-a}=f'(\xi),\ \xi \textit{ being some point between } a \textit{ and } b.$$

The other, due to Cauchy, and somewhat less extensively quoted, states that

$$\frac{f(b)-f(a)}{g(b)-g(a)}=\frac{f'(\xi)}{g'(\xi)},\ \xi \textit{ as before.}$$

The former, on account of its extensive utility, is often called the "law of the mean," or "mean value theorem," the latter being styled, in contrast, the "second law of the mean," or "extended mean value theorem." These names have the advantage of avoiding reference to the discoverers or reputed discoverers, since the ascribing to historical personages is usually controversial, often misleading, of little educational advantage, and of no logical import. On the other hand, these particular terms are satisfactory, since many writers call the second of these extensions the "theorem of the mean" and prove it first.
Even the clearest texts do not hesitate in this connection to introduce apparently artificial functions which can be identified with the F of Rolle's theorem. These functions are justified by the fact that they serve to establish the desired theorem as a consequence of a known theorem, but are not psychologically motivated. The result is that the student is mystified, and the particular form of the function is difficult to keep in mind. There is no need of this artificiality, and in the case of Lagrange's theorem, there seems to be no excuse for it.[2]

I confess I haven't made the trip to a university library to read the rest of this paper to see how he removes the artificiality of the choice. We have seen a number of ways of removing this artificiality and one comes to a point where one feels the Law of Diminishing Returns kicking in. I haven't even looked up some of these papers. I can report that the papers of Wang and Evans discuss the reduction by rotation of the Mean Value Theorem and the Cauchy Mean Value Theorem, respectively, to Rolle's Theorem. And Poliferno's paper uses the distance function as discussed in Chap. 1 and carried out in Chap. 2.[3] The anthology edited by Apostol *et alia* cited in our earlier discussion includes the papers of Yates, Barrett and Jacobson, Evans, Spiegel, and Poliferno, thus yielding a convenient, representative selection of papers discussing the issue.[4]

The paper by Silverman is a departure in that he believes none of these approaches escapes artificiality. Referring to the auxiliary function g, he says

Despite our best efforts at making the function g...appear geometrically intuitive and therefore "natural," most students seem to think of g as artificial and the proof as magical...
To establish the existence of a tangent line parallel to the secant line joining $(a, f(a))$ to $(b, f(b))$, it certainly makes sense to construct an auxiliary function that differs from f by a linear function whose slope is $(f(b)-f(a))/(b-a)$. But why not choose the simplest

---

[2]Bennett, p. 40. In quoting from these papers I cite only the authors' names without the "*op. cit.*" I have copied the original spelling and punctuation as accurately as possible, but have omitted without mention those references to the literature given by numbers (e.g. "[2]").

[3]Pages 135–138, above.

[4]Apostol, *op. cit.*

> such linear function, the one passing through the origin? This is accomplished by replacing [the usual g] (2) with
> $$g(x) = f(x) - \frac{f(b) - f(a)}{b - a}x. \qquad (2')$$
> The simplicity of the auxiliary function in (2′) more than offsets the computation needed to verify that it satisfies (1′) [the equation $g(b) = g(a)$].

I reject Silverman's basic premise that the geometrical motivations for the standard auxiliary function are artificial. Is subtracting the line parallel to the secant and passing through the origin really any more "natural" than subtracting the secant line itself, i.e., is the new g more "natural" than the traditional choice? That said, Silverman's choice of g in (2′) has popped up before: cf. Remark 3.6.2[5] on Galois's use of g. But again, that proof, though extremely simple, is somewhat inspired, and, while the better students may find it delightful, it does have an accidental air about it.

The standard reductions may be a bit too computational (especially the approach via rotation), but they *are* natural and if one draws the pictures, the geometry of the situation does not depend on the secant line's being horizontal: the Mean Value Theorem is no different from Rolle's Theorem.

This last remark brings us to the next issue, the belief that the Mean Value Theorem and its proof themselves and not the usual classroom presentation of the latter are the problem. It is asserted that the Mean Value Theorem should therefore be replaced by something else. Publications touching on this issue form our next group of papers:

1960

Jean Dieudonné, *Foundations of Modern Analysis*, Academic Press, New York, 1960.

1967

L.W. Cohen, "On being mean to the Mean Value Theorem", *The American Mathematical Monthly* 74 (1967), pp. 581 – 582.
L. Bers, "On avoiding the Mean Value Theorem", *The American Mathematical Monthly* 74 (1967), p. 583.

1969

R.P. Boas, "Lhospital's rule without the Mean Value Theorem", *The American Mathematical Monthly* 76 (1969), pp. 1051 – 1053.

1981

R.P. Boas, "Who needs these mean-value theorems, anyway?", *Two-Year College Mathematics Journal* 12 (1981), pp. 178 – 181.

1986

Robert S. Smith, "Rolle over Lagrange—another shot at the Mean Value Theorem", *The College Mathematics Journal* 17, (1986), pp. 403 – 406.

[5] Page 295, above.

1996

Felipe Acker, "The missing link", *The Mathematical Intelligencer* 18, no. 3 (1996), pp. 4 – 9; reprinted in: Robin Wilson and Jeremy Gray, *Mathematical Conversations: Selections from The Mathematical Intelligencer*, Springer-Verlag, New York, Inc., 2001.

Dieudonné can probably be regarded as the grandfather of the movement to replace the Mean Value Theorem for his pronouncement in 1960 that the proper form of the Mean Value Theorem is as an inequality and not as an equality. For, the equation does not generalise to higher dimensions, and applications generally only need the estimates supplied by the inequality. Acker takes exception to this, proving a Mean Value Equation in higher dimensions. In the course of describing this, he quotes Dieudonné:

> Now let's turn to the **Mean Value Theorem** or, should I say, the Mean Value Equality: if
>
> $$f : [a, b] \to \mathbb{R}$$
>
> is continuous on $[a, b]$ and differentiable at each point of $]a, b[$, then there exists a point c in $]a, b[$ such that
>
> $$f'(c) = \frac{f(b) - f(a)}{b - a}.$$
>
> The trouble appears when we try to generalize this result to higher dimensions: the pretty and geometrical equality becomes an inequality. I think the best expression of what everybody seems to believe was given by Jean Dieudonné in his celebrated *Foundations of Modern Analysis*:
>
>> After the formal rules of Calculus have been derived (Sects. 8.1 to 8.4), the other sections of the chapter are various applications of what is probably the most useful theorem in Analysis, the mean value theorem, proved in Sect. 8.5. The reader will observe that the formulation of that theorem, which is of course given for vector-valued functions, differs in appearance from the classical mean value theorem (for real-valued functions), which one usually writes as an *equality* $f(b) - f(a) = f'(c)(b - a)$. The trouble with that classical formulation is that: 1°. **there is nothing similar to it as soon as f has vector values;** 2°. it completely conceals the fact that *nothing* is known on the number c, except that it lies between a and b, and for most purposes, all one needs to know is that $f'(c)$ is a number which lies between the g.l.b. and l.u.b.[6] of f' in the interval $[a, b]$ (and *not* the fact that it actually is a value of f'). In other words, **the real nature of the mean value theorem is exhibited by writing it as an *inequality*, and not as an equality.**
>
> Well, Dieudonné was wrong! The Mean Value Theorem does generalize to higher dimensions as an *equality*.[7]

I can neither agree nor disagree with Dieudonné's opinion. I prefer the traditional form of the Mean Value Theorem. Equations are psychologically more comforting than inequalities. From a geometrical perspective, the equational form for functions from $\mathbb{R}$ to $\mathbb{R}$ is the more obvious version and thus more basic. Historically, on the other hand, the first proofs initially established the inequality and derived the equality

---

[6] That is, the greatest lower bound and the least upper bound.

[7] Acker; p. 208 of the anthologised version.

by applying the Intermediate Value Theorem to the derivative. The inequality, but not the equality, generalises effortlessly to higher dimensions.[8] And, as Dieudonné says, applications generally only require the Mean Value Theorem in the form of an inequality. The real nature of the Mean Value Theorem would thus seem to depend more on the individual's background and perspective than on the Theorem itself. I tend to think in simple geometric pictures, like Swann's Fig. 4.1 on page 458, below, and the equational form is more obvious to me. Dieudonné was a prominent member of Bourbaki, an openly "secret" society of French mathematicians who worshipped abstraction and eschewed diagrams. He would consider the "real nature" to be whatever held in the most general setting, in this case the multidimensional version of the Mean Value Theorem, which he found to be the inequality.

What Dieudonné has not said is that the traditional form of the Mean Value Theorem has no place in the elementary course and it should be replaced by the inequality. Effectively, he said that, from a higher point of view, the real nature of the Theorem was exhibited by the inequality. A more direct assault on the Theorem was to come.

The paper of Cohen was a response to a comment Lipman Bers had made:

> With characteristic vigor, L. Bers announced in a recent conversation: "Who needs the mean value theorem! All we want as a start in elementary calculus is the proposition that if $f'(x) = 0$ for all x in $[a, b]$, then f is constant."[9]

Bers had asserted that the Mean Value Theorem could be replaced in the introductory Calculus course by the Constant Function Theorem. The title of Cohen's paper suggests that Cohen wished to defend the Theorem from a perceived onslaught, but other than asking the question in this title, he is silent on the matter. What he does after citing Bers is to give a proof of the Constant Function Theorem that makes no mention of the Mean Value Theorem, thus actually supporting Bers. His proof is a bisection argument very similar to the trisection argument used by Acker in his proof of the Mean Value Theorem.[10]

Bers's paper directly follows Cohen's in the journal. It begins with a short sketch of the proof of the Strictly Increasing Function Theorem by appeal to the Least Upper Bound Principle, similar to the proof given in Chap. 2, above.[11] This is followed by the reduction of the Increasing Function Theorem to the Strictly Increasing Function Theorem and the remark that the Constant Function Theorem follows. He finishes the list of results with a variant of the Mean Value Inequality:

> COROLLARY 2. *If* $f'(x) \leq K$ *for* $a \leq x \leq b$, *then* $f(b) \leq f(a) + K(b - a)$. (Apply Corollary 1 [i.e., the Increasing Function Theorem] to $K(b - a) - f(x)$.)[12]

He concludes the paper with a short expression of his opinion:

---

[8]Acker does generalise the equality, but only after some re-interpretation.

[9]Cohen, p. 581.

[10]Cf. Sect. 11 of the preceding chapter.

[11]He notes that this proof was hardly new and in a footnote asks if any readers know of a reference. None was given when the paper was reprinted in Apostol, *op. cit.*.

[12]Bers, p. 583.

> Using the intermediate value theorem and either Corollary 2 or the fundamental theorem of calculus, one obtains at once the mean value theorem for continuously differentiable functions. That's all one needs in calculus.
> The "full" mean value theorem, for differentiable but not continuously differentiable functions, is a curiosity. It may be discussed together with another curiosity, Darboux' theorem that every derivative obeys the intermediate value theorem.[13]

The most relevant thing for the present discussion about the first paper of Boas is the title, "Lhospital's rule without mean value theorems", as it hints at a desire to replace the Mean Value Theorem. His second paper, as indicated by the title, is more directly opposed to the Theorem. The first paper presents a proof of L'Hôpital's Rule that does not rely on the Mean Value Theorem, but it doesn't say much about the desirability of avoiding the Mean Value Theorem. The second of his papers in our list, on the other hand, contains no proofs or other additions to theory, but campaigns for replacing the Mean Value Theorem by the Mean Value Inequality, citing several points where the Inequality possesses an advantage over the equality, or, at least, is no less advantageous.

This first point is as follows:

> The first advantage of (2) [the inequality] over (1) [the equality] is that it avoids the perennial problem that we can't say where the point c is on (a, b). Many students are bothered by the indetermination. (They think that we *could* tell them where c is, if we only would. This belief is only reinforced by exercises that ask them to find c in special cases. Such exercises may be good for something else, but they don't help the understanding of the mean-value theorem.)[14]

I discount this argument for a simple reason. In the Calculus course one also considers problems involving extrema, and the Extreme Value Theorem is also an abstract existence theorem that does not state explicitly where its c is. In the course students are given lots of exercises to find c in special cases, but do these exercises help the understanding? If they do, does this suggest one assign more exercises in finding c in the case of the Mean Value Theorem? And, if they don't, should we consider dropping max-min problems from the Calculus course? My own view is that the Intermediate Value Theorem, the Extreme Value Theorem, and the Mean Value Theorem are gentle introductory abstract existence theorems and should all be discussed. In the first two cases one studies techniques for finding the promised values of c because the actual values of c are usually of some interest; one doesn't bother finding the value of c for the Mean Value Theorem although it can be done by the same numerical techniques, because the actual value is generally of no importance. It is a good example of concentrating on what is relevant. In the case of the Mean Value Theorem one generally only needs to know that *some* c exists and it doesn't matter what value it has because c is usually a means to an end (giving quick proofs of various results) and not an end in itself.

Note too that the Intermediate and Extreme Value Theorems do not need full proofs. A bit of a hand-waving bisection argument coupled with more efficient

---

[13] *Ibid.*.

[14] Boas, 1981, p. 179.

numerical methods giving experience at finding c are probably more convincing than a rigorous proof involving the completeness property of the reals in one of its many manifestations. The Mean Value Theorem, on the other hand, can be reduced quite easily to the Extreme Value Theorem and the student should be aware that *in principle* its c can also be found.

His second point is that the inequality is more intuitive than the equality and he cites speed as an example:

> Second, (2) is more intuitive than (1) if we think of x as time and $f(x)$ as the distance traveled up to time x. Then (1) says that at some instant you are moving at exactly your average speed. This seems not to be very intuitive. But (2) says that the average speed is between the minimum speed and the maximum speed...[15]

My response to this should be fairly obvious: Reread the Preface to see why I think the Mean Value Theorem is intuitively obvious. Is the Mean Value Inequality as geometrically obvious? Boas's favoured motivation works well for the inequality, not so well for the equality; with mine the situation is reversed. His argument should be that insofar as the Calculus course is a service course for the engineering students, his motivation is probably more suited to their needs. But not every decision about course content should be dictated by the needs of non-majors.[16]

Boas's third point is a refutation of an advantage the Mean Value Theorem is believed to have over the Mean Value Inequality, namely its usefulness for proofs of basic results. He cites the proof of the Strictly Increasing Function Theorem as an example and notes that, if $f'(x) > 0$ everywhere on $[a, b]$, then

$$f(b) - f(a) \geq (b - a) \min_{a \leq x \leq b} f'(x) > 0,$$

"and there is no reason (except for a century or two of tradition) for dragging in the nebulous point c", adding the remark:

> In any case, proving theorems ought not to be a principle aim (probably not even a proper aim) of a first course in calculus.[17]

First, I note that there is a reason for dragging in c: c lies in the interior of (a, b) where $f'(x)$ is assumed to exist. Boas's proof assumes f' to assume a minimum value on $[a, b]$, which is not guaranteed without the assumption that f' is continuous on $[a, b]$. One can get around this partially: If one only assumes f' exists and is continuous on (a, b), one can first restrict one's attention to the interval $[a+\delta, b-\delta]$ to establish that $f(b - \delta) > f(a + \delta)$ and then appeal to the continuity of f at a and b. We have already seen that the Increasing Function Theorem can be proven without appeal to the Mean Value Theorem, but that is by another, more complicated proof.

---

[15] *Ibid.*

[16] The introductory Physics course serves mathematics majors as well as engineers, but the course does not forego laboratory work just because maths majors don't need it for their purposes.

[17] Bers, P. 179.

And I disagree about proving theorems not being a proper aim in a first course in the Calculus. It is far easier to introduce proofs to Freshmen who don't know what to expect in a new school than to sophomores, juniors, and seniors who "know how things are done" in their school and react in subsequent courses with suspicion when one suddenly makes "unreasonable" demands on them. If the Calculus is once again to provide the "equivalent of mathematical maturity", as used to be touted among the prerequisites in the prefaces of advanced mathematics texts, it must be more than a methods course and students must be exposed to proofs and even expected to supply some of their own. This does not mean that they must be exposed to the proofs (like those given in Chap. 2 of the present book) of the Intermediate and Extreme Value Theorems via the Least Upper Bound Principle. But the basic ϵ-δ proofs should be given, perhaps initially as "How close does x have to be to a to make $f(x)$ within .1 of L? within .001 of L? within ϵ of L?" And, too, I think there is nothing wrong with presenting a proof, like that given for the Mean Value Theorem in Chap. 2, which is little more than the carrying out of a computation. And, what could one object to a short clever proof like the reduction of any of the variants of the Increasing Function Theorem to the Mean Value Theorem?

Boas's fourth point is that the standard application of the Mean Value Theorem to estimation invokes the Mean Value Theorem only to conclude the Mean Value Inequality and applies the latter to get the actual estimate. For this, one obviously needs only to invoke the Mean Value Inequality. He is absolutely correct on this. I think it is a psychological thing: one becomes so enamoured of the proof of a lemma that, instead of applying the lemma, one repeats the construction behind it. I've seen this also in my own area of specialisation, Mathematical Logic, where, for decades, authors would apply Gödel's diagonal construction each time they needed a self-referential sentence even though there was a standard existence lemma allowing them to bypass the construction.

Finally, Boas makes a fifth point that the equational form of the Mean Value Theorem "is no longer true for vector-valued functions, whereas an appropriate generalization of (2) is", and he cites Dieudonné on this.[18]

Having disposed of the Mean Value Theorem, he turns his sights on the Cauchy Mean Value Theorem, finding it more troubling for students than the Mean Value Theorem and saying that as far as he knew the only application of the Cauchy Mean Value Theorem in the elementary course was L'Hôpital's Rule, which he had previously proven without the use of any form of the Mean Value Theorem. His proof assumes continuous differentiability, in which case the proof given in Chap. 3 of the Urform of L'Hôpital's Rule (Theorem 3.1.22)[19] is far simpler.

Smith's paper suggests replacing the Mean Value Theorem by Rolle's Theorem:

> Apparently, instructors can accomplish much of what is needed in the calculus without the Mean Value Theorem. Boas has even demonstrated that Cauchy's Mean Value Theorem is superfluous. Certainly, Rolle's Theorem is equivalent to, less complicated than, and more intuitive than, either mean value theorem. In view of this evidence, perhaps an alternate approach in the calculus should be considered—depose the theorem of Lagrange (and that

---

[18]Boas, 1981, p. 181.

[19]Page 176, above.

of Cauchy, too) and enthrone the theorem of Rolle. Reinforce this new direction by proving Rolle's Theorem for your students. A variety of proofs are readily available in texts and the literature.[20]

One of the more interesting papers concerning the replacement of the Mean Value Theorem is Thomas Tucker's paper, which abandons the Mean Value Theorem entirely in favour of the Increasing Function Theorem and which is the lead item of our next group of papers.

1997

T.W. Tucker, "Rethinking rigor in calculus", *The American Mathematical Monthly* 104 (1997), pp. 231 – 240.
Howard Swann, "Commentary on rethinking rigor in calculus", *The American Mathematical Monthly* 104 (1997), pp. 241 – 245.
Lou Talman, "Re: MAA Book, 'A Century of Mathematics' ", online essay.

1999

Scott E. Brodie, "On 'Rethinking rigor in calculus...' or Why we don't do calculus on the rational numbers", *The College Mathematics Journal* 30 (1999), pp. 135 – 138.

2006

Xu Ji-hong, "An alternative approach about several theorems in Calculus", *Journal of Mathematical Research and Exposition* 26 (2006), pp. 63–66.
Yao Jing-sun, "The application of full cover and tagged partition in analysis", *College Mathematics* 22, no. 4 (2006), pp. 109 – 112.

2008

Yao Jing-sun, "Some new ways to prove Rolle's Theorem", *College Mathematics* 24, no. 4 (2008), pp. 131 – 133.

Tucker begins with criticism of the Mean Value Theorem and the call for a more intuitive replacement, one so intuitively true that it can serve as an axiom for the Calculus course:

1. INTRODUCTION. Mathematicians have been struggling with the theoretical foundations of the calculus ever since its inception. Bishop Berkeley's attack on Newton's "ghosts of departed quantities," Euler's claim that $1 - 1 + 1 - 1 \cdots = 1/2$, Cauchy's $\varepsilon - \delta$ definition of limit, all are part of the fascinating history of this struggle. Calculus instructors and textbooks face the same struggle, but the tack taken, although formal, is often not sensible or honest. Instead of an admission that Newton, Leibnitz, the Bernoullis, and Euler all managed quite well without any rigorous foundations, instead of the story how a rigorous calculus took mathematicians two hundred years to get right, the Mean Value Theorem is waved, like a cross in front of a vampire, to hold the difficulties at bay. The origin of the Mean Value Theorem in the structure of the real numbers is not addressed; that is much too difficult for a standard course. Maybe it is traced back to the Extreme Value Theorem, but the trail ends there. The result is that a technical existence theorem is introduced without proof and used to prove intuitively obvious statements, such as "if your speedometer reads zero, you are not going anywhere" (if $f' = 0$ on an interval, then f is constant on that interval). That's

[20] Smith, p. 403.

the sort of thing that gives mathematics a bad name: assuming the nonobvious to prove the obvious. And by the way, there is nothing obvious about the Mean Value Theorem without the hypothesis of continuity of the derivative. Cauchy himself was never able to prove it in that form.

I have serious reservations about the need for formal theorems and proofs in a standard calculus course. On the other hand, for those mathematicians who do feel that need, I have a suggestion for an alternative theoretical cornerstone to replace the Mean Value Theorem (MVT); I hope textbook authors adopt it. It is much easier to state, much more intuitively obvious, and much more powerful than most mathematicians realize. It is simply this:

The Increasing Function Theorem (IFT). *If $f' \geq 0$ on an interval, then f is increasing on that interval.*[21]

Following this call to arms, he devotes the rest of his paper to demonstrating how many of the applications of the Mean Value Theorem can be derived from the Increasing Function Theorem. He prefaces this development with a brief section in which he establishes the Increasing Function Theorem by a bisection argument, coming to Sect. 3 on "Immediate consequences of the IFT". These are the Decreasing Function Theorem, the Constant Function Theorem, the Strictly Increasing Function Theorem, the so-called Racetrack Principle, and the Mean Value Inequality. He ends the section with further support for his approach:

Theorem 1e is perhaps the most important, especially from a historical viewpoint. If the inequalities are rewritten:

$$m \leq \frac{f(x) - f(a)}{x - a} \leq M$$

we have the Mean Value Inequality. The Mean Value Theorem follows immediately if we know that f' is continuous and that the Intermediate Value Theorem holds. That is exactly what Cauchy did: he proved the Mean Value Inequality and assumed the continuity of f' and the Intermediate Value Theorem. His assumption of continuity should not be surprising since his proof of the Mean Value Inequality also assumes that the difference quotient $(f(x+h) - f(x))/h$ approaches $f'(x)$ uniformly as h approaches 0. Peter Lax has argued that, for the theoretical foundations of an introductory calculus course, one should always avoid pathology and assume uniform continuity and uniform convergence, just as Cauchy did. It is interesting to note that before Cauchy, Ampère saw the importance of the Mean Value Inequality and even used it as the defining property of the derivative. One could argue in a similar vein that the Mean Value Theorem should be the defining property of the derivative; Andrew Gleason has told me that a calculus book by Donald Richmond around 1960 did exactly that, but I have been unable to find the book.[22]

Following a few additional sections on developing the Calculus from the Increasing Function Theorem and its immediate consequences, he concludes with remarks emphasising the pædagogical nature of his proposal:

I sympathize with yearnings for an occasional foray into the theoretical structure of the calculus. I just ask that it be thoughtful and sensible. Use intuitive definitions. If a theorem is to be used without proof, like the Mean Value Theorem, keep it as simple and as "obvious" as possible. Don't use tricky proofs or deus-ex-machina auxiliary functions. Don't prove things in more generality than necessary; even analysts don't usually deal with the discontinuous derivatives allowed by the Mean Value Theorem.[23]

---

[21] Tucker, p. 231.

[22] *Ibid*, p. 234.

[23] *Ibid*, pp. 239–240.

The first response to Tucker was published in the same journal directly after his paper as a commentary on it. This is the slightly contentious, but highly enjoyable, rejoinder by Howard Swann, who pulls no punches:

> Professor Tucker's article joins the current deconstructive attack on traditional content and methods of teaching of calculus that seems to be part of the mission of the militant wing of the 'Calculus Reform Movement.' Here the primary targets are current textbooks' efforts to present the foundations of calculus and the frequent use of the mean value theorem.
> As the author remarks, the traditional presentation of the foundations of calculus is often poorly motivated and incomprehensible to most students. So in reforming the teaching of the calculus sequence, one should either omit the logical foundations or attempt to make them interesting and comprehensible. The author, who is one of the co-authors of the 'Harvard Calculus' text[24] where the first option is chosen and the concept of *mathematical* proof based on rigorous definitions is eliminated entirely, urges that we keep things as "intuitive..., simple and obvious as possible." Various demonstrations are our new "proofs;" I use the quotation marks to make the distinction. The author's favored replacement for the Mean Value Theorem (MVT), the Increasing Function Theorem (IFT), finds its intuitive justification in an *automotive* ('Racetrack') argument. Such automotive arguments are a new addition to our pantheon of "proofs." An automotive "proof" of the IFT is 'if the speedometer on a motor-car always reports a number greater than or equal to zero, then the car must be moving (weakly) forward.' The IFT is to be treated as an 'axiom,' yet the essential first foundational question for calculus is 'What is it that a speedometer is supposed to report?' Intuition falters here, for nature has yet to provide us with a speedometer.[25]

Putting the rhetoric aside, Swann scores some good points against Tucker. In defence of formal definitions and rigour, he cites Bertrand Russell,

> ...the implications so astonished Bertrand Russell that he pronounced:
>
> > ...all goes smoothly until we reach those studies in which the notion of infinity is employed—the infinitesimal calculus and the whole of higher mathematics. The solution of the difficulties which formerly surrounded the mathematically infinite is probably the greatest achievement of which our age has to boast.[26]
>
> Learning to understand and appreciate proofs is a gradual process; it surely is imperative to introduce the notion of mathematical proof in beginning multisemester calculus and keep it alive even though actual proofs are few. Such an introduction is essential for later mathematics courses, and students must be made aware that the assertions of mathematics can be *proved* to be true.[27]

There are two points here: we must teach students to appreciate the magnitude of the discovery and that one is not "assuming the nonobvious to prove the obvious",

[24]The "Harvard Calculus" is a book, *Calculus*, written by a dozen or so members of the Harvard faculty that go by the name "Calculus Consortium at Harvard", though the book is usually listed bibliographically as Deborah Hughes-Hallett, Andrew M. Gleason, *et alia*, *Calculus*, John Wiley and Sons, New York, 1994. It has gone through a number of editions, whence I conclude it to have had some level of success and not to be the unmitigated disaster one would expect from Swann's description.

[25]Swann, p. 241. Cf. also Lagrange's critique of Newton cited on p. 240, above.

[26]Bertrand Russell, *Mysticism and Logic*, W.W. Norton & Co., Inc., New York, 1929, p. 64.

[27]Swann, p. 242.

but showing the adequacy of the formal definition for the purpose of making precise the intuitive notions. He is explicit on this point on the next page when he sketches a more intuitive proof of the Mean Value Theorem, saying "So, if our definition of the derivative as the slope of a line that is *tangent* to the graph at $(c, f(c))$ is any good, the slope of this tangent line must be $f'(c)$". What is intuitively obvious, when interpreted precisely through some formalism, is not actually obvious; and learning how to establish formal versions of things that are intuitively obvious is an important part of mathematical education. The pathology that Tucker approvingly cites Lax as denigrating also plays its rôle in making obvious the nonobvious fact that the intuitively obvious is not always true. The Calculus has traditionally been the course introducing rigour to the mathematics majors, and one should think long and hard before eliminating rigour from the Calculus course. It may not be needed for all students taking an introductory course in the subject and introductory Calculus is a service course after all, but don't we do the mathematics majors a disservice if we attempt to postpone the maturation process to a later course?[28]

Another good point that Swann makes is that the Mean Value Theorem can be explained just as intuitively as, if not more so than the Increasing Function Theorem. He says,

> As for the mean value theorem, the author states, "And by the way, there is nothing obvious about the MVT without the hypothesis of continuity of the derivative." I believe that this is not true, for here is a pictorial "proof" of the MVT...[29]

He then sketches the Valerio-Cavalieri approach, the one I cited already in the Preface as useful for explaining the Theorem to the proverbial man-in-the-street. There is a slight difference in presentation: where I suggested sliding the secant up (say) without rotating it until it left the curve, he starts with a parallel to the secant outside the curve and slides it down until it meets the latter. Figure 4.1, below, shows the very pretty picture he uses to illustrate the procedure.

Swann next considers Tucker's "proof" of the Increasing Function Theorem, citing a line from Tucker: "For a proof, draw the obvious picture". According to Swann,

> The "obvious picture" encourages this assertion, but knowing that the art of converting a "proof" to a proof is one of the key skills our majors should learn, if we are giving a proof here, we must go further. Two *mathematical* proofs are immediately discovered; a proof by contradiction (four main cases) or a direct proof. The direct proof shows first that the result must be true if $m = 0$, and then uses the same 'deus-ex-machina' auxiliary function that annoys the author when it is employed to prove the MVT from Rolle's Theorem.[30]

The "proof" of the Increasing Function Theorem is in fact less intuitive than the "pictorial 'proof'" of the Mean Value Theorem sketched by Swann. Moreover,

---

[28] I taught both the Calculus and the undergraduate Real Analysis courses while at San José with Swann and I can report greater success in teaching the ϵ-δ technique to the freshmen in the first course than to the juniors and seniors in the latter. I found the students who had not been exposed to formal rigour in the earlier course to be actually resistant to the method in the latter.

[29] Swann, p. 243.

[30] *Ibid.*, p. 244.

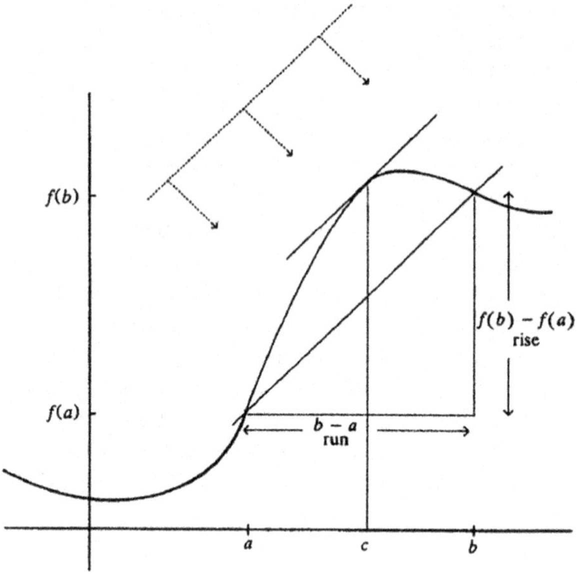

Fig. 4.1 Swann's proof of the MVT

> A one-line proof shows that the author's IFT follows from the MVT. The author's suggested proof for the MVT from the IFT requires, in addition to the usual assumptions for the MVT, that the function's derivatives be extendible to a continuous function on the closed interval, requires the extreme value theorem and the intermediate value theorem, and fails to establish that the sought-for value for c is strictly between points a and b. This is essential, for example, for showing that we can *repeat* the application of L'Hospital's rule a second time in evaluating a limit.[31]

To this I would add that Tucker's remark, "And by the way, there is nothing obvious about the MVT without the hypothesis of continuity of the derivative", is misleading. Look at Fig. 4.1 and tell me where the "hypothesis of continuity of the derivative" comes in. All that is needed is that the tangent exists at the point on the curve farthest from the secant line; it need not even be defined, much less continuous, nearby. Tucker is mistaking the intuitively obvious need for this hypothesis for a particular analytic proof for the intuitive obviousness of the hypothesis for the truth of the theorem, overlooking the centuries old geometric intuition.

Swann concludes with a few general comments:

> ...I do not find the main arguments of the paper to be persuasive. Those of us who, as the author says, "bemoan the absence of the Mean Value Theorem or the ε, δ definition of limit" regret that "it is time...to rethink the theory taught in standard calculus classes."...
> Whether or not the militants' 'final product" is 'better,' which is by no means established,[32] one thing is clear: books such as the "Harvard Calculus" are "enablers;' by legitimizing the

[31] *Ibid.*, p. 245.

[32] Swann here refers to a paper analysing the success and failure of using the "Harvard Calculus" at Oklahoma State: K. Johnson, "Harvard Calculus at Oklahoma State", *The American Mathematical Monthly* 102 (1995), pp. 794–797.

> abandonment of the concepts of mathematical proof, related rates, convergence of series, and so forth from the calculus sequence, other texts and teachers will feel free to follow. Mathematics is unique in its concern with rigorous foundations and proofs. Here its role as 'Queen and servant of the Sciences' is to offer the content of calculus as an anchor of certainty to aid the disciplines it serves. Should we not attempt to convey some sense of the remarkable way that the results of calculus can be proved to be true to those who will use it?[33]

Talman's response to Tucker and Swann was on an online forum (The Math Forum @ Drexel) in response to a collection of quotations compiled by Jerry Uhl from some volumes celebrating a century of mathematics in America.[34] The quotations dealt with pædagogical issues in general and during the discussion, which took place in the latter half of April of 1997, the subject of the Mean Value Theorem and the Tucker-Swann exchange on it was brought up. Talman's comments on the matter are most thoughtful, yet short and to the point:

> It seems to me that some important points have gotten lost in the debate over the Mean Value Theorem. Both Tucker and Swann, in their recent Monthly articles, seem to have missed this point, for example: The consensus over the past few decades seems to have been that proofs whose basis is the completeness of the reals are not appropriate fare for freshmen. This being the case, any attempt to give a "rigorous" proof of the MVT is doomed at the outset.
>
> . . .
>
> Another important point that many seem to me to have missed is this: In a freshman course, we ought to condition our choices of our basic principles on the ways in which our students will understand them and on the uses we will make of them, but not on our desire to have the strongest possible conclusions or on obedience to tradition. Neither of the latter two criteria seem to me to be pedagogically important.
>
> As to what students will understand, I know that I get more (at least an order of magnitude more) blank stares when I discuss the the MVT than I do when I discuss the Racetrack Principle.
>
> . . .
>
> If we can adjust other MVT arguments to rest instead on the Racetrack Principle *so that the adjusted arguments are no harder to follow than the originals*, then I see no good reason to insist upon the MVT. In fact, at the level of elementary calculus such adjustment is not difficult...
>
> . . .
>
> I suggest that none of elementary calculus requires the full strength of the MVT. The Racetrack Principle is easier to discuss with freshmen and supports all of the conclusions we want to draw. The cost of basing arguments on the RP is no higher than the cost of basing them on the MVT. (Advanced calculus, intermediate analysis, and beyond, are a different story.)

Talman's suggestion of the so-called Racetrack Principle as a suitable replacement for the Mean Value Theorem in an introductory course on the Calculus is supported anecdotally by the evidence of his own experience. Before accepting it, however, I would like to know two things. First, what was the presentation of the Mean Value

---

[33]Swann, p. 245. The inconsistency in the style of quotation marks is copied directly from the original.

[34]Peter L. Duren (ed.), *A Century of Mathematics in America*, 3 vols., American Mathematical Society, 1988–1989.

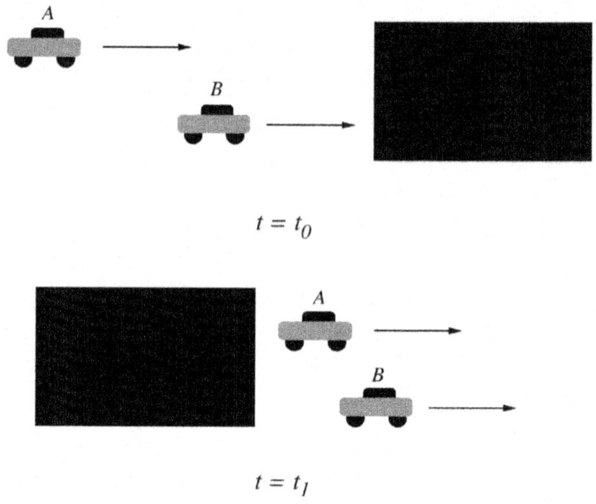

Fig. 4.2 Judging relative speeds

Theorem that drew blank stares? One can reduce it to Rolle's Theorem via an unexplained auxiliary function, one can motivate it *à la* Boas through an automotive analogy, or one can explain it geometrically. The three approaches could easily yield different numbers of blank stares.

Second, do the fewer blank stares under the Racetrack Principle indicate that the students understand or that they think they understand? As we saw in discussing Berkeley[35] and Lagrange,[36] or as Swann indicated when he said nature has no speedometer, not everyone is as clear on the concept of instantaneous velocity as Newton was. I would want to see the results of a simple experiment involving Fig. 4.2, below. Here two cars are each seen travelling in the same direction at constant speeds. Their positions are shown at times $t = t_0$ before they temporarily disappear behind a wall and at time t_1 after they emerge from behind the wall. Which car is moving faster? It is my understanding that college students often get the answer wrong when shown an animated version of this picture. I imagine the still images at the two instants make it easier to see that car A is travelling faster than car B, but if the students get this wrong, would their not displaying blank stares be an indication of understanding?

Brodie's paper references the Tucker-Swann exchange but is really a criticism of Tucker. he starts out summarising what he disagrees with and states his disagreement:

> In a recent "Point/Counterpoint" in the *American Mathematical Monthly*, it was suggested that the basic theorems on continuous functions and their derivatives (the Boundedness Theorem, the Extreme Value Theorem, the Intermediate Value Theorem, and, especially, the Mean Value Theorem) be omitted from the introductory calculus course. Reasons given were that "the origin of the Mean Value Theorem in the structure of the real numbers ... is too difficult for a standard course"; that these discussions are "the sort of thing that gives

[35]Page 113, above.

[36]Page 240, above.

> mathematics a bad name: assuming the nonobvious to prove the obvious"; that perhaps there is no "need for formal theorems and proofs in a standard calculus course"; and that, in any event, one shouldn't "prove things in more generality than is necessary; even analysts don't usually deal with the discontinuous derivatives allowed by the Mean Value Theorem."
> I demur. Without commenting on the pedagogical issues, I would like to point out that this program risks serious misdirection of the mathematical intuition of its students. In particular, I submit that the notion that these basic theorems are "obvious," save for obscure subtleties raised only by bizarre, pathological functions (which are scarcely encountered in practice) is incorrect.[37]

He then discusses the completeness of the real numbers and proceeds to demonstrate the failure in the rationals of many results of the Calculus: the Uniform Continuity Theorem, the Extreme Value Theorem, the Intermediate Value Theorem, the Mean Value Theorem —

> Even the watered-down "Increasing Function Theorem," profered [*sic*[38]] in [Tucker's paper] as a more sincere replacement for the Mean Value Theorem, fails over **Q**.[39]

His counterexamples are all constructed from polynomials, rational expressions, and square roots, simple functions encountered even in the simpler Business Calculus courses. To demonstrate the failure of the Mean Value Theorem in the rationals, for example, he cites the function

$$f(x) = 1 - \sqrt{4x^4 - 4x^2 + 1} \text{ on } [0, 1] \cap \mathbb{Q}.$$

4.1.1 Exercise *Show the following: For f as just defined,*

$$f(x) = \begin{cases} 2x^2, & 0 \le x < \frac{1}{\sqrt{2}} \\ 2 - 2x^2, & \frac{1}{\sqrt{2}} < x \le 1. \end{cases}$$

i. Conclude that f is differentiable at every rational $q \in [0, 1]$.
ii. Show that Rolle's Theorem fails for f.
iii. Is f continuously differentiable on $[0, 1] \cap \mathbb{Q}$?

Brodie concludes, offering his opinion:

> The great theorems of the calculus are not necessarily "obvious"—otherwise it would not have taken nearly 2,000 years of mathematical effort to discover them or their proofs. To hide from our students the persuasive arguments by which we have come to believe them is to do them a disservice.[40]

For a brief period in the 1960s, in response to the Soviet launch of Sputnik, ultra-rigorous textbooks on the Calculus appeared in the United States. Where I was a

---

[37] Brodie, p. 135.
[38] I am not sure if he means "preferred" or "proffered".
[39] Brodie, p. 137.
[40] *Ibid.*, p. 138.

student, George B. Thomas's wonderful textbook[41] was replaced by Johnson and Kiokemeister's book[42] which was almost more rigorous than the Advanced Calculus book then in use. I never cared for their text, probably for no good reason, but I do like Albert G. Fadell's pair of texts[43] which combine rigour and intuitive explanation seemingly effectively.[44] The days of such rigour in American Calculus are over. Proposals like those of Bers and Tucker, rightly or wrongly, are easily seen as a movement to no rigour — as evidently seen in the eyes of Swann and Brodie. Talman lies somewhere in between, wanting to give some of the theory but not that which is not needed. Fortunately for me, the issue at hand is not the design of the perfect Calculus course, but the more modest question of whether the Mean Value Theorem should be covered in a beginning course or whether it should be replaced by one of several supposedly more intuitive results.

Dieudonné wanted to replace the Mean Value Theorem by the Mean Value Inequality not because the latter was more intuitive and intelligible to students, but because the Inequality generalised more directly to higher dimensions and thus exhibited the "true nature" of the result. He accepted that some form of the Mean Value Theorem belonged in the Calculus course. It would seem from the works thus far cited that the opposition to teaching the Mean Value Theorem in beginning Calculus is a purely American phenomenon. Americans, as a rule, are provincial, but mathematicians less so than the average American. Yet the papers I've considered thus far all appeared in American journals. This feeling that the Mean Value Theorem as a matter of controversy is an American concern is reinforced by the papers of Xu and Yao. Their papers, which refer directly to those of Tucker and Swann, appear in Chinese counterparts to the pædagogically informed *The American Mathematical Monthly* and *The College Mathematical Journal*, and do not question the place of the Mean Value Theorem in the course at all. Instead, they make proposals for retaining the Theorem.

Xu begins with a statement of intent:

> There has been much discussion on the Increasing Function Theorem (IFT) and related results such as the Mean Value Theorem (MVT) [in] recent years in the Calculus reform in the USA…Our point of view is that the MVT needs to be retained, but its proof (and the proofs of other fundamental theorems in Calculus) can be modernized.[45]

---

[41] George B. Thomas, Jr., *Calculus*, 2nd edition, Addison-Wesley Publishing Company, Inc., Reading (Mass.), 1961.

[42] Johnson and Kiokemeister, *Calculus with Analytic Geometry*, 3rd edition, Allyn and Bacon, Inc., Boston, 1964.

[43] Albert G. Fadell, *Calculus with Analytic Geometry*, D. van Nostrand Company, Inc., Princeton, 1964, and *Vector Calculus and Differential Equations*, American Book·Van Nostrand·Reinhold, New York, 1968.

[44] "Seemingly": I already knew the advanced theory by the time I discovered Fadell's books, so I cannot base an opinion on my learning from them. Nor have I ever taught a course from them and cannot report how students took to them. I can only say that I like them.

[45] Xu, p. 63.

His approach is to prove the Heine–Borel Theorem by a bisection argument and apply it to obtain the rest.

Likewise Yao's 2008 paper offers several proofs of Rolle's Theorem by appeal to the Heine–Borel Theorem and *δ-fine tagged partitions*, which he first discusses in the 2006 paper. I won't define these special partitions here, but only say the reader has seen such already in the preceding chapter, Sect. 3.10.2 in discussing Mawhin's version of Flett's proof of the Peano-Gilbert result.

Summarising, the objections to the Mean Value Theorem seem to be these:

(1) the proof relies on an artificial auxiliary function and lacks motivation (e.g., Silverman, Smith);

(2) the proof relies on the completeness of the real numbers and does not belong in an introductory course (Tucker, Talman);

(3) the result is not intuitively obvious and does not make a good axiom like the Increasing Function Theorem or the so-called Racetrack Principle (Bers, Boas, Smith, Tucker, Talman);

(4) students wonder where c is (Dieudonné, Boas);

(5) it isn't needed in its full generality (Bers, Boas, Tucker);

(6) it doesn't generalise directly to higher dimensions (Dieudonné, Boas).

(1). Most of the writers who address the first objection do so to refute it. Rolle's Theorem is established quite easily from the Extreme Value Theorem and the vanishing of the derivative at a local extremum. Geometrically, if not analytically, the Mean Value Theorem is no different and it is natural to attempt an analytic reduction. The auxiliary function readily suggests itself. If this is not clear in the textbook, it is the fault of the exposition, not of the proof.

(2). As Brodie makes clear, the completeness of the real numbers must be explained. This does not mean one has to prove the Intermediate Value Theorem or the Extreme Value Theorem from the Least Upper Bound Principle or some other equivalent axiom as we did in Chap. 2. These results can be presented with the promise that proofs will be provided in a more advanced course. That the results depend on the completeness of the real line should be discussed, and that they require proof can be explained as further verification that the formal definition of continuity adequately captures the intuitive notion. This last remark may well be lost on the students, but it probably will be on their first exposure to such considerations regardless of the course. I believe that if you want students to understand such things in Advanced Calculus, you should plant the seed in introductory Calculus.

In any event, the proof of the Mean Value Theorem is a step removed from the completeness property and follows from the Extreme Value Theorem, which will be cited — usually without proof — in the introductory Calculus course. If the ultimate dependence of the Mean Value Theorem on the completeness of the reals is a reason for omitting it from the introductory Calculus course, it is equally an argument against presenting max-min problems and methods of solving equations like the secant method or the Newton-Raphson method.

(3). The Mean Value Theorem is intuitively obvious, as witnessed by the work of Valerio and Cavalieri, if one thinks geometrically. The velocity-motivated principles may be more intuitive to those more kinematically oriented, like the engineering and physics students, but are they as readily established as the Mean Value Theorem?

(4). The number c can be found, and by techniques discussed in the Calculus. Solving

$$f'(c) = \frac{f(b) - f(a)}{b - a}$$

differs from solving $f'(c) = 0$ only in the choice of a constant K in solving $f'(c) = K$. When exact solutions are desired, f is chosen so as to make the instance $f'(c) = 0$ easy to solve, but real-life problems often only admit of numerical solutions and the same general methods apply for other values of K. The difference between the c of the Mean Value Theorem and those of the Intermediate Value Theorem and the Extreme Value Theorem is that one generally doesn't need to know the value of c in the case of the Mean Value Theorem and the problem is thus not discussed in class.

(5). It may well be the case that most of the topics of the Calculus can profitably be discussed for the narrower class of continuously differentiable functions, but the Weierstrass-Bonnet proof of the Mean Value Theorem, which is the simplest proof of the result, holds under the weaker condition and there is no harm in mentioning that the proof requires neither the continuity of the derivative nor its existence at the endpoints of the interval in question. The only reason to emphasise that the extra generality is not needed is to justify one's decision to replace the Mean Value Theorem in its fundamental rôle by the Fundamental Theorem of the Calculus.

(6). Putting Acker's equational generalisation of the Mean Value Theorem to higher dimensions aside, my response to those who like to point out that the statement of the Mean Value Theorem has to be modified before it can be validated in higher dimensions is: so what? Is this the first time we've seen a one-dimensional phenomenon fail in higher dimensions? $\mathbb{R}^2, \mathbb{R}^3, \ldots$ are not linearly ordered. Do Dieudonné and Boas declare that monotone functions do not belong in the Calculus? The answer is no, because both men need the linear ordering of the reals to compare the norms of elements of $\mathbb{R}^2, \mathbb{R}^3, \ldots$ The absurdity of this objection is perhaps best brought out by applying the reasoning to physics: The first year physics course ought not to discuss Newtonian physics because the velocities of objects moving near the speed of light do not add up or because Newton's gravitational theory does not account for the orbit of Mercury.[46]

My conclusion is that there has been no compelling reason to look for a replacement for the Mean Value Theorem in the introductory course in the Calculus. Whether this means that the Mean Value Theorem must be taught or that the choice is merely a matter of personal preference has yet to be determined.

---

[46]Likewise, their argument against discussing quantum theory and general relativity up until a few years ago would have been: who needs it? Newton's theory is all that one needs in applications.

4.2 Beyond Polemics

There is no single standard course called "Calculus". *A* course in the Calculus can range anywhere from a mere methods course emphasising technique and application, to a theory course emphasising rigour and preparing the student for higher mathematics. Deciding where along this spectrum the course one wants to teach lies requires one to answer two important questions, raised in the titles of the following papers:

2011

Keith Stroyan, "Why do so many students take calculus?", *Notices of the AMS* 58 (2011), pp. 1122 – 1127.

2012

F. Quinn, "What should students get from calculus? (And how can we provide it?)", *Notices of the AMS* 59 (2012), pp. 839 – 841.

As Quinn points out, Stroyan actually addresses the second question. But the answer to that needs an answer to the first, and a first step in answering this question is to ask, "Which students take a Calculus course?" In the United States the answer to this question has changed over time. A bit of the history of the teaching of the Calculus in the United States can be found in

1989

George M. Rosenstein, Jr., "The best method. American Calculus textbooks in the nineteenth century", in: Peter Duren (ed.), *A Century of Mathematics in America, Part III*, American Mathematical Society, Providence, 1989.

Two things of particular interest to be learned from this are that, prior to the American Civil War, most colleges had fixed curricula which often included the Calculus and that the level of instruction in the Calculus was not very high:

> Although calculus was part of the curriculum in a number of American colleges during the first third of the nineteenth century, very little time was devoted to it. For example, at Harvard in 1830, sophomores studied trigonometry and its applications, topography and calculus. Furthermore, this third of a year was the only calculus they studied.[47]

Not much was covered and a textbook had to be accessible to students with little mathematical talent. The Civil War and its aftermath brought change:

> ...the elective system established itself when the demands of the public for a more practical education and the intellectual demand of the sciences for a larger piece of the curricular pie had to be met.
> Science, and mathematics with it, bloomed in the new land-grant colleges designed to encourage the study of agriculture and the mechanic arts, and authorized by Congress in the Morrill Act of 1862. It also flourished in the "Scientific Schools" formed at established colleges... Finally, science and mathematics benefited through the creation of universities, such as Cornell in 1869, and Johns Hopkins, in 1874, both named to honor their wealthy industrialist

[47] Rosenstein, p. 79.

> benefactors. In them, research and graduate education assumed a greater role than they had played in the colleges.
>
> ...
>
> Scientific education for more, but better motivated, students demanded more advanced mathematics texts.[48]

Specialisation in education and the needs of an increasingly industrialised nation brought more rigour into the Calculus textbooks. Calculus reform in the United States has been an ongoing practice since the early nineteenth century.

The cause of rigour got its biggest boost in the early 1960s after the Soviets launched Sputnik, racing ahead of the Americans in the Space Race. This reform probably went too far, ignoring the needs of most of the consumers of the Calculus course, and the course content has since been adjusted downward.

But who, now that the Calculus course is no longer universal, are the students taking the course today. Obviously there are the engineering students, physics and other physical science majors, and, of course, mathematics majors. My student days were at a large university located a few blocks away from a large medical school, whence there were a lot of pre-med students who were also required to take the course. And, indeed, there are still the occasional liberal arts students. The business majors generally take their own special course in Business Calculus and their needs are thus irrelevant in considering "the standard Calculus course".

The different types of students have different needs. The majority of engineering students probably don't need more than a methods course. A certain amount of intuitive explanation of concepts is probably necessary, but full mathematical rigour is not required for them. I suspect the same is true of the physics majors, and not even the mathematics majors need be exposed to a rigorous axiomatic development on their first exposure to the Calculus. Learning the Calculus is, however, a maturation process and mathematics majors should be exposed to some rigour even in the early Calculus sequence. So long as they take the same introductory course as the mathematics majors, the engineering and physics students will have to learn about rigour as well even though the benefits for them of such knowledge will not be as immediately evident.

I don't know what to say about the needs of the pre-med students. It was my understanding that pre-med students were forced to take the Calculus as a stamina requirement to test their self-discipline. If this is indeed the case, a watered-down course would not serve the purpose. For them a good, heavily rigorous approach would be called for. I am reminded here of the words of Robert Woodhouse,[49]

> ...mental discipline is all the good the generality of students derive from mathematics.

The importance of mathematics as a general training ground for the mind is a sentiment going back to Socrates and Plato, though it is a tradition often more honoured in the breach, as in mediæval universities where students memorised Euclid's proposi-

[48] *Ibid*, p. 86.

[49] Cited on p. 113, above.

tions without mastering the proofs or in modern Calculus courses in which no proofs are given.

Rigour, of course, is not the only feature of the Calculus bearing on stamina and discipline. Quinn speaks frankly on this:

> **Complex rules and accuracy.** It is a vital skill in science and engineering to be able to work accurately with complex rule-based systems. I feel that it is part of our job to develop this: calculus is certainly the best training ground in the current curriculum because the rules are realistically complicated, but are clear and concise and feedback is quick and accurate. This skill is also transferrable to many more domains than any specific content. But this is a skill that my students certainly don't have when they get here.
> Most high-school programs have de-emphasized rule skills in favor of "understanding" and working intuitively. If you can "see" the problem it should be easy. Calculator use has replaced a lot of rule-based work and attendant skills. AP calculus is a partial exception, but it is test-driven with greatly simplified rules used mechanically on short, routine problems.[50]

I would thus not accept any pre-med student into medical school unless he or she could evaluate

$$\int \sin^5 2\theta \cos^6 3\theta \, d\theta$$

by hand.

The needs of the liberal arts student who may opt to take the Calculus are different. Obviously, the mental discipline one is expected to derive from such a course is good for him or her. But the drudgery of numerical work and extensive practice of methods of integration are bound to be off-putting. I would think that exposure to theory and rigour, as well as exposure to — but not drill in — computation, is called for. They might actually take an interest in the evolution and refinement of notions such as continuity and tangency, as well as the extent to which mathematics is man-made and yet applies to the real world. And it wouldn't hurt for them to appreciate the Calculus as one of the supreme intellectual achievements of mankind, rivalled only by the accomplishments of theoretical physicists.

This last rhetorical flourish brings me to one point that I think has been overlooked by most of the reformers, except perhaps in the occasional derogatory reference to "tradition".[51] This is our duty to pass some of our cultural heritage on to our future mathematicians. The proposed replacements for the Mean-Value Theorem can be used to prove the Mean Value Theorem and its many consequences, but, especially since Weierstrass and Bonnet, the overall easiest rigorous development is that based on the Mean Value Theorem. And it is the Mean Value Theorem, not the Increasing Function Theorem or the so-called Racetrack Principle, that has been called — with justification — the Fundamental Theorem of the Differential Calculus. Moreover, half the motivation for the Calculus came from Geometry, not Physics. Tangents and areas are just as much a part of our mathematical cultural heritage as velocities and work. Rejecting the Mean Value Theorem because our supposed intuition of

---

[50]Quinn, p. 1 of a draft of his paper dated 20 October 2011 that I downloaded from the Internet. The published version omits the second sentence of the first paragraph (p. 839).

[51]Cf. the remarks by Boas and Talman quoted on pages 453 and 459, above.

instantaneous velocity does not apply to it as readily as it does to the Increasing Function Theorem or to the so-called Racetrack Principle can only be seriously contemplated by rejecting geometric intuition altogether. Presenting the Mean Value Theorem and attempting to make it plausible geometrically actually reminds us of this part of our heritage.

I suppose in taking it upon myself to discuss Calculus Reform at least insofar as it relates to the Mean Value Theorem, it is incumbent on me to make some specific recommendations even though my purpose is not to solve the problem but to provide some material relevant to its solution. For what it's worth, I suggest three courses — one designed primarily with engineering and physics students in mind, one designed for mathematics majors, and one for liberal arts students.

The splitting of the standard Calculus course into one for engineers and physicists and one for mathematicians might not be politically acceptable. When I once suggested it, the chairman of the Department objected that it would "track" students and a later change in major would result in a delay in graduation, something I would have expected anyway. In fact, the sort of division I suggest could effortlessly be accommodated in those schools already offering standard Calculus and Honours Calculus sequences. The former could be the course of choice for engineers and physicists and the latter a requirement for mathematics majors.

In the engineering Calculus course, the choice among the Increasing Function Theorem, the so-called Racetrack Principle, and the Mean Value Theorem as the Basic Fact relating differences and derivatives is arbitrary, a matter of personal taste: There is no compelling reason to choose any one of them over the other. My own preference is to opt for the Mean Value Theorem, motivated *à la* Valerio, Cavalieri, and Swann, and given a loose proof as in Chap. 2 on the theory that engineering students would respect the formula for the distance from a point to a line and regard the presentation as a calculation and not notice they were being subjected to a proof.

The finer points of our mathematical heritage are presumably not so important to these particular students and, given that *they* might have better intuition as regards velocity than geometry, one of the suggested replacements might be deemed more appropriate. Although Tucker feels all proofs are out of place in the first-year Calculus course, and would probably regard them as more out of place in a designated methods course as a course aimed at engineering students would be, he does go out of his way to prove the Increasing Function Theorem from the completeness of the real numbers, postulated via the Nested Interval Property. I think that, if one is going to use the Increasing Function Theorem and attempt to justify it, one can do better by applying the Least Upper Bound Principle, as this Principle is easily motivated: If X is a bounded nonempty set, the set of its upper bounds constitutes an interval, the left endpoint of which is the Least Upper Bound of X. This can be presented as an heuristic explanation of the truth of the Principle and why it is taken as an axiom, or as a proof of the Principle based on the obvious axiom that intervals have endpoints except when they are unbounded.

For the course aimed at mathematics majors, I would insist on establishing the Mean Value Theorem for cultural and pædagogical reasons. The Dieudonné–Boas complaint that students do not know where c is I take to be reason enough to prove it. Abstract existence theorems are commonplace in higher mathematics and this is a particularly simple example of such. The proof may not produce c explicitly or even approximate it numerically, but it does produce it graphically and one can announce that numerical methods can be applied to approximate c arbitrarily closely. In this way, it is a most gentle introduction to such existence proofs, unlike less intuitive such theorems the student will later encounter, like the existence of winning strategies in certain games for which such strategies have never been found. I would even go counter to Dieudonné and Boas and use the Mean Value Theorem instead of the Mean Value Inequality a few times just to demonstrate that one can use constants without knowing their exact values.

Finally, addressing the issue of a Calculus course for liberal arts students, I admit this might not be acceptable in that it would be a mathematics appreciation course: one Department chairman explained to me that the Department's attempt to set up such a course as a means of satisfying the general education's quantitative reasoning requirement was blocked by the faculty at large because the students should have to actually *do* mathematics for the course to qualify as such. It seems to me that most students would benefit more from learning how mathematics works and what it is good for than in being taught how to carry out all the computations. Probability and Statistics are areas where the understanding of basic principles is far more useful to the average man or woman in decision making than the ability to calculate the odds of various poker hands or, as we used to teach before the advent of powerful pocket calculators or simple software programs, to calculate standard deviations by hand — setting up data and frequency columns, subtracting a guessed value of the mean from the former, multiplying the deviations or the squares of the deviations by the corresponding entries of the frequency column, etc. With regards to the Calculus, I think the liberal arts student would gain from learning the evolution of central concepts, being exposed to the various methods applied through the ages to find tangents and areas, and of coming face to face with some of the paradoxes of the Calculus. And, of course, some proofs, presented as "explanations" would be good for them as well. They will derive no benefit being drilled in techniques of integration until they can evaluate

$$\int \sin^5 2\theta \cos^6 3\theta \, d\theta,$$

unless, of course, they later go to medical school and become outstanding surgeons...

4.3 Constructive Thoughts on the Subject

One issue not addressed in the above discussion is that of constructivity. There are, of course, the complaints that students wonder where the number c is that satisfies $f'(c) = \big(f(b)-f(a)\big)/(b-a)$, that the Mean Value Theorem is an abstract existence theorem providing no clue to the location of c. An abstract existence theorem presents a special challenge that a more direct existence proof algorithmically providing the solution doesn't. One can respond to this challenge in several different ways. One can declare the result unfit for an introductory course, as has been suggested for the Mean Value Theorem. One can seek a new constructive proof, perhaps under more restrictive conditions, as is the practice in Constructive Analysis. One can accept the result as true, justifying the search for the desired object, as is the case with various methods presented in the Calculus course for finding zeros of functions (the Newton-Raphson Method, bisection argument) and for finding extreme values of functions (solving $f'(x) = 0$, bisection argument). And, of course, one can demonstrate that there is a real problem by exhibiting examples where finding the elusive c is not so easy.

I should like to briefly consider the exhibition of the difficulty presented by the c of the Mean Value Theorem and the constructive response to this problem. I could justify this by noting that one suggested reform of the Calculus course is based on the constructive approach to Analysis, but the fact is that I simply find it interesting, which is not to say that I will refrain from discussing the indicated reform.

The underlying problem has, in fact, nothing to do with the Mean Value Theorem — or the Intermediate or Extreme Value Theorems. It has to do with the nature of the real numbers themselves. The integers and rational numbers are easily determined *exactly*; real numbers in general require *infinite precision*, which only exists *in principle* and not *in practice*. Brouwer used this to construct what are now often called *Brouwerian counterexamples* in his honour. The idea is to take a question we do not know the answer to and construct a real number from it which we cannot say is less than, equal to, or greater than 0. Using this number as a parameter, such counterexamples to the Intermediate, Extreme, and Mean Value Theorems can be constructed.

Brouwer begins with a question concerning the digits in the decimal expansion of π:

> Let d_ν be the ν-th cipher after the comma of the decimal expansion of π and $m = k_n$, if in the ongoing decimal expansion of π it happens for the n-th time at d_m, that the part $d_m d_{m+1} \ldots d_{m+9}$ of this decimal expansion forms a sequence 0123456789. Further, let $c_\nu = \left(-\frac{1}{2}\right)^{k_1}$, if $\nu \geq k_1$, otherwise $c_\nu = \left(-\frac{1}{2}\right)^{\nu}$, thus the infinite sequence $c_1, c_2, c_3, \ldots$ defines a real number r, for which neither $r = 0$, nor $r > 0$, nor $r < 0$ is valid.[52,53]

In this example, Brouwer asserts that the disjunction

$$r = 0 \text{ or } r > 0 \text{ or } r < 0$$

is not valid in the sense that we do not *know* which disjunct is true and thus cannot assert the disjunction. He is not saying that

[52]L.E.J. Brouwer, "Über die Bedeutung des Satzes vom ausgeschlossenen Dritten in der Mathematik, insbesondere in der Funktionenlehre", *Journal für die reine und angewandte Mathematik* 154, no. 1/2 (1923), pp. 1–7; here: p. 2.

[53]Note that k_1 will not be defined if 0123456789 never occurs in the decimal expansion of π. In this case $c_\nu = (-1/2)^\nu$ and $r = 0$, though we may not *know* this.

$$\neg(r = 0 \text{ or } r > 0 \text{ or } r < 0)$$

is true, i.e., that the disjunction is false, i.e., contradictory, as we may eventually come across the string 0123456789 of digits in the decimal expansion of π as we calculate more and more such digits. When that happens, of course, we can take another problem we do not know the answer to and construct a corresponding new counterexample. In short, we cannot tell from the sequence of elements as they are unfolding whether the limit will be positive, negative, or zero. Indeed, Brouwer more-or-less believed

$$\neg\forall r(r = 0 \text{ or } r > 0 \text{ or } r < 0). \qquad (4.1)$$

The apparent absurdity of (4.1) vanishes when one realises that the connectives $\neg, \&$, or, $\Rightarrow$ and quantifiers $\forall, \exists$ have different meanings to the constructivist. In classical mathematics one asserts a proposition P to declare P to be true. P or Q asserts that one of P and Q — we don't necessarily know which — is true. The constructivist asserts P to declare P to be known or proven. To him P or Q can only be asserted if one of P, Q has been proven. $\exists x P(x)$ means one can produce an x and a proof that $P(x)$ holds. $\forall x P(x)$ asserts one has a procedure which, given any x, will produce a proof that $P(x)$ holds. And negation asserts absurdity: $\neg P$ is provable just in case P leads to an absurdity. Thus (4.1) reads that it is absurd to think that we have a procedure which will decide for every real number whether it is 0, greater than 0, or less than 0. When one considers that real numbers are presented as, say, sequences of rationals and ask how we would set about deciding whether the limit is 0, > 0, or < 0, this assertion becomes eminently reasonable. If someone starts communicating a sequence of rational numbers to us, and we accept his guarantee that it will eventually converge and up till now he has presented us with a list of zeros, $0, 0, \ldots, 0$, how do we know whether the rest of the elements of the sequence will be 0's or whether they will suddenly switch and all become 1's or -1's?

In 1923, Brouwer was still producing only weak Brouwerian counterexamples, but they were interesting. Familiar fundamental results for which he supplied such counterexamples include trichotomy, the Extreme Value Theorem, and the Heine–Borel Theorem. By 1927, he had added the Intermediate Value Theorem to the list.[54]

To the average mathematician, Brouwerian counterexamples are not convincing. Any real number r presented as a convergent sequence of rational numbers converges to 0 or to a positive or to a negative number. This is true of Brouwer's r depending on the occurrence or non-occurrence of 0123456789 in the decimal expansion of π, whether or not we know it or where the string occurs. It either occurs or doesn't occur and, if it does occur, the string either begins at an even-numbered or an odd-numbered position in the expansion. If we knew which we could state where r sits relative to 0, but we don't know and we can't say. But one of the disjuncts is still true regardless of our ignorance.

---

[54]Cf. L.E.J. Brouwer, (D. van Dalen (ed.)) *Intuitionismus*, Bibliographisches Institut, Wissenschaftsverlag, Mannheim, 1992, p. 161.

Now, none of what I've said so far would be at all relevant if there weren't a next step. That next step is to reflect on the nature of constructive objects. Brouwer introduced a metaphysical element into his theory by postulating *free choice sequences*, a free choice sequence α being a sequence $a_0, a_1, a_2, \ldots$ of objects given by a succession of free choices rather than by a determinate law. The application of a function f to α, $f(\alpha) = \beta$, would be another sequence $b_0, b_1, b_2, \ldots$ The first element b_0 of β must be determined at some stage after only finitely many elements $a_0, a_1, \ldots, a_{n_0}$ are known. b_1 is determined after $a_0, a_1, \ldots, a_{n_1}$ are known. Etc. In short, finite initial segments of β are determined by finite initial segments of α. This is a continuity condition and every constructive function defined on all free-choice sequences turns out to be continuous. The same holds, albeit less obviously, for functions defined on all reals. In particular, the function

$$f(r) = \begin{cases} 0, & r = 0 \\ 1, & r > 0 \\ 2, & r < 0 \end{cases}$$

must be continuous if it is defined everywhere. It clearly is not continuous, whence it cannot be defined for all real numbers. But, if

$$\forall r (r = 0 \text{ or } r > 0 \text{ or } r < 0) \tag{4.2}$$

were constructively valid, the function f would be defined everywhere since for any real number r exactly one of the disjuncts would hold. Thus (4.2) is absurd (to use Brouwer's way of saying it is contradictory) and (4.1) is valid.

Using free choice sequences, the weak Brouwerian counterexamples can be turned into strong counterexamples — not specific instances that fail, but the absurdity of the universal declaration. For example, one does not produce a particular continuous function f which has no maximum on a particular closed interval $[a, b]$, but one shows that it is absurd to claim there is a uniform procedure that can be proven to find the maximum for any continuous function defined on the given interval.

This has evolved into an approach more acceptable to those not wedded to Brouwer's constructivist program: One starts with a theorem of the form

$$\forall \text{ continuous functions } f \text{ on } [a, b]\ \exists c \in [a, b] \text{ such that} \ldots$$

and shows that c cannot be produced uniformly from f by some "effective" procedure. One usually does this for a class of simple functions f_ε defined in terms of a parameter ε by showing that the function $\varepsilon \mapsto c$ is not continuous.

This discussion is perhaps a bit abstract and hard to follow on first reading, but it can be simply illustrated.

The idea behind showing that the intermediate value promised by the Intermediate Value Theorem cannot in general be supplied by a continuous function is fairly simple. Take a piecewise linear function that begins at $x = 0$, increases for a bit,

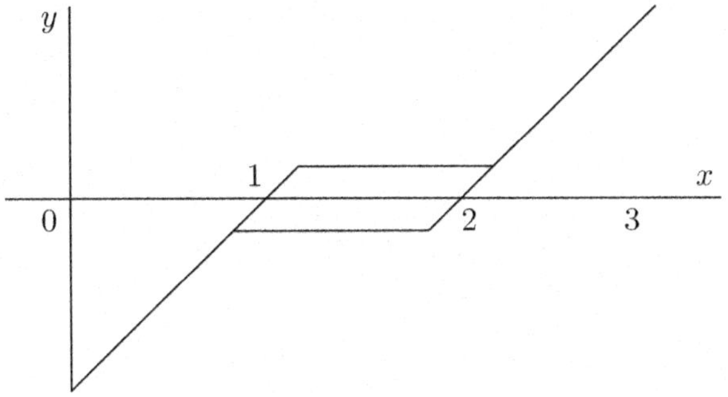

Fig. 4.3 Where is the intermediate value?

then somewhere near the x-axis turns horizontal for a while, and then increases for a bit, ending above the x-axis at $x = 3$ as in Fig. 4.3, above. There are three horizontal lines here, one representing those functions making the turn to the horizontal before hitting the x-axis, one for which the turn takes place at the axis, and one for which the turn takes place after crossing the axis. If ε represents the vertical displacement of the horizontal line from the x-axis, we see that the zero of the function, say f_ε, is near 2 for small values of $\varepsilon < 0$, near 1 for small values of $\varepsilon > 0$, and that every c between 1 and 2 is a zero for $\varepsilon = 0$. If g is a function satisfying $f_\varepsilon(g(\varepsilon)) = 0$, we see that g cannot be continuous at $\varepsilon = 0$.

There is some leeway in defining $f(\varepsilon, x) = f_\varepsilon(x)$. The following is one of many such definitions: Let, for $\varepsilon \in (-1, 1)$ and $x \in [0, 3]$,

$$f(\varepsilon, x) = \begin{cases} x - 1, & 0 \le x \le 1 + \varepsilon \\ \varepsilon, & 1 + \varepsilon < x < 2 + \varepsilon \\ x - 2, & 2 + \varepsilon \le x \le 3. \end{cases}$$

There is one subtle point here: $f(\varepsilon, x)$ is defined by undecidable cases and does not at first sight appear to be constructively defined. Representing x and ε as convergent sequences of rationals, the definition can be shown to result in a constructively defined function.[55] Once one accepts this, it is easy to see that, if ε_0 is Brouwer's number based on the digits of π, then f_{ε_0} affords a Brouwerian counterexample to the Intermediate Value Theorem. And, for general $\varepsilon \in (-1, 1)$, we have the following:

4.3.1 Exercise *i. Show that f is continuous on $(-1, 1) \times [0, 3]$.*
ii. Show that if $g : (-1, 1) \to [0, 3]$ satisfies $f(\varepsilon, g(x)) = 0$ then g is not continuous at $\varepsilon = 0$.

If we accept the constructive tenet that a constructive proof of an assertion $\forall x \exists y P(x, y)$ yields a constructive function g satisfying $\forall x P(x, g(x))$ and we accept

---

[55] I refer the reader to A.S. Troelstra and D. van Dalen, *Constructivism in Mathematics; An Introduction*, vol. I, North-Holland Publishing Company, Amsterdam, 1988, p. 261, for the theorem on which this claim is based.

that constructive functions of reals are continuous, we must conclude that the Intermediate Value Theorem lacks constructive validity.

I have left the work to the reader because our main interest here is not in the Intermediate Value Theorem, but in the Mean Value Theorem. For this reason I should like to skip lightly over the Extreme Value Theorem and give a direct strong counterexample to the continuous validity[56] of the Mean Value Theorem. However, Brouwer's student Arend Heyting[57] (1898–1980) gave such an elegant counterexample to the Extreme Value Theorem that I simply *must* give it here.

4.3.2 Exercise *For* ε *close to* 0 *define for, say,* $x \in [-2, 2]$

$$f_\varepsilon(x) = -3x^4 + 4\varepsilon x^3 + 6x^2 - 12\varepsilon x.$$

i. Show $f'(x) = -12(x+1)(x-1)(x-\varepsilon)$.
ii. Show that the least upper bound of f_ε *is* $3 + 8|\varepsilon|$.
iii. Show that the function $g(\varepsilon)$ *classically yielding the point where* f_ε *is maximum is not continuous.*

To handle the Mean Value Theorem, let $f(\varepsilon, t)$ be our counterexample to the Intermediate Value Theorem and define

$$F_\varepsilon(x) = F(\varepsilon, x) = \int_0^x f(\varepsilon, t)dt$$

for $\varepsilon \in (-1, 1)$, $x \in [0, 3]$. F is continuous as a function of ε and x.

We have

$$\begin{aligned}
F_\varepsilon(3) &= \int_0^{1+\varepsilon} (t-1)dt + \int_{1+\varepsilon}^{2+\varepsilon} \varepsilon dt + \int_{2+\varepsilon}^{3} (t-2)dt \\
&= \frac{t^2}{2} - t \Big|_0^{1+\varepsilon} + \varepsilon t \Big|_{1+\varepsilon}^{2+\varepsilon} + \frac{t^2}{2} - 2t \Big|_{2+\varepsilon}^{3} \\
&= \frac{(1+\varepsilon)^2}{2} - (1+\varepsilon) + \varepsilon(2+\varepsilon-1-\varepsilon) + \frac{9}{2} - 6 - \left(\frac{(2+\varepsilon)^2}{2} - 2(2+\varepsilon)\right) \\
&= \frac{1}{2} + \varepsilon + \frac{\varepsilon^2}{2} - 1 - \varepsilon + \varepsilon + \frac{9}{2} - 6 - \left(2 + 2\varepsilon + \frac{\varepsilon^2}{2} - 4 - 2\varepsilon\right) \\
&= -2 + \varepsilon + \frac{\varepsilon^2}{2} - \left(-2 + \frac{\varepsilon^2}{2}\right) \\
&= \varepsilon.
\end{aligned}$$

[56] By which is meant the continuity of the function witnessing the validity of an ∀∃ statement.

[57] Arend Heyting, *Intuitionism; An Introduction*, second revised edition, North-Holland Publishing Company, Amsterdam, 1966, p. 47.

Further, $F_\varepsilon(0) = 0$, whence

$$\frac{F_\varepsilon(3) - F_\varepsilon(0)}{3-0} = \frac{\varepsilon}{3}.$$

To solve

$$\frac{F_\varepsilon(3) - F_\varepsilon(0)}{3-0} = F'_\varepsilon(c_\varepsilon),$$

for some $c_\varepsilon \in (0, 3)$, we note that $F'_\varepsilon(x) = f_\varepsilon(x)$.

If $\varepsilon < 0$ there are three possibilities:

i. $0 \le c_\varepsilon \le 1+\varepsilon$. Then $f_\varepsilon(c_\varepsilon) = c_\varepsilon - 1 = \frac{\varepsilon}{3}$, and $c_\varepsilon = 1 + \frac{\varepsilon}{3}$, which, since $\varepsilon < 0$, is $> 1+\varepsilon$. Thus, this cannot be the case.

ii. $1+\varepsilon < c_\varepsilon < 2+\varepsilon$. Then $f_\varepsilon(c_\varepsilon) = \varepsilon = \frac{\varepsilon}{3}$ and this case cannot occur.

iii. $2+\varepsilon \le c_\varepsilon \le 3$. Then $f_\varepsilon(c_\varepsilon) = \varepsilon - 2 = \frac{\varepsilon}{3}$ and $c_\varepsilon = 2 + \frac{\varepsilon}{3} > 2+\varepsilon$.

Similarly, if $\varepsilon > 0$, one can show (*Exercise.*) $c_\varepsilon = 1 + \frac{\varepsilon}{3} < 1+\varepsilon$.

Now we see, if g is the function yielding c_ε, $\lim_{\varepsilon\to 0-} g(\varepsilon) = \lim_{\varepsilon\to 0-} 2+\varepsilon/3 = 2$, while $\lim_{\varepsilon\to 0+} g(\varepsilon) = \lim_{\varepsilon\to 0+} 1/3+\varepsilon = 1$ and g is not continuous at $\varepsilon = 0$.

As before we conclude that the Mean Value Theorem lacks constructive validity. Thus, despite my protestations against the opinions of Dieudonné and Boas on the elusive nature of the number c asserted to exist by the Mean Value Theorem, there is some sense in which they were correct. I don't know it is a sense they would find acceptable, and I am not convinced they would be pleased to read that the same holds true for the elusive c's of the Intermediate and Extreme Value Theorems, which I do not recall their having railed against.

Brouwer's constructive mathematics had been based on his view that mathematics is a system of mental constructions based on a primordial intuition of the passage of time, and he was led to reflect on the constructive process itself, thus proving things like (4.1) and the continuity of all real functions defined on an interval. These results diverge from traditional mathematics and not all constructivists accept the more metaphysical elements of Brouwer's theory. Errett Bishop (1928–1983), a leading American constructivist, took a more pragmatic approach, seeing the goal of constructive mathematics not to analyse mental constructions as such, but to ferret out the constructive content of traditional mathematics. His 1967 book on the *Foundations of Constructive Analysis*[58] performed this task surprisingly well. He explained

> The extent to which good constructive substitutes exist for the theorems of classical mathematics can be regarded as a demonstration that classical mathematics has a substantial underpinning of constructive truth.[59]

---

[58] McGraw-Hill Book Company, New York, 1967.

[59] *Ibid*, p. 9.

Bishop's method of finding this constructive underpinning is two-fold. First, one chooses formal definitions of one's concepts that can be dealt with successfully constructively, and, second, one makes explicit in the hypotheses of the theorems the conditions that must be met for the proofs to be carried out constructively. He explains his choices of definitions as follows:

> The task of making analysis constructive is guided by three basic principles. First, to make every concept affirmative. (Even the concept of inequality is affirmative.) Second, to avoid definitions that are not relevant. (The concept of a pointwise continuous function is not relevant. A continuous function is one that is uniformly continuous on compact intervals.) Third, to avoid pseudogenerality. (Separability hypotheses are freely employed...)[60]

Any concept can be defined in more than one way. Definitions that are equivalent classically, may not be so constructively. Inequality is a case in point. We can define the inequality of two numbers, $x \neq y$, either negatively as $\neg x = y$ or positively as the disjunction: $x < y$ or $y < x$. The negative definition has no constructive content and is difficult to work with constructively, while the positive one tells us where x and y stand with respect to one another: they are separated[61] and we even know which one is greater than the other.

The fundamental properties of continuous functions, such as the existence of the integral, are generally established by appeal to uniform continuity on closed, bounded subintervals. Moreover, most functions are as easily proven uniformly continuous on such intervals as they are proven continuous there. Bishop thus finds pointwise continuity irrelevant.

His eschewal of pseudogenerality he illustrates by reference to higher mathematics, topology in particular. He explains that he knows of no constructive topological space which has been proven to be non-separable. Thus he incorporates separability into his basic definitions.

Now, I do not want to go deeply into constructive mathematics. I just wish to describe his constructive versions of the Intermediate, Extreme, and Mean Value Theorems.

The Intermediate Value Theorem is simple enough that Bishop consigns it to the exercises:

> **11.** Let $f : [0, 1] \to \mathbf{R}$ be a continuous function, with $f(0) < 0$ and $f(1) > 0$. Show that for each $\epsilon > 0$ there exists x in $[0, 1]$ with $|f(x)| < \epsilon$.
> **12.** Construct a continuous function $f : [0, 1] \to \mathbf{R}$ with $f(0) < 0$ and $f(1) > 0$ such that there does *not* exist a point x in $[0, 1]$ with $f(x) = 0$.
> **13.** Let $f : [0, 1] \to \mathbf{R}$ be a continuous function, with $f(0) < 0$ and $f(1) > 0$, such that for arbitrary real numbers a and b with $0 \leq a < b \leq 1$ there exists x in $[a, b]$ with $f(x) \neq 0$. Show that there exists x in $[0, 1]$ with $f(x) = 0$.[62]

---

[60] *Ibid*, pp. ix–x.

[61] Brouwerian constructivists, known as *intuitionists*, say that x and y are *apart* and often write $x\#y$ in place of $x \neq y$ to emphasise the fact that they are using the positive notion.

[62] *Ibid*, p. 59.

Notice that he offers two positive versions (Exercises **11** and **13**) and one counterexample. The italicised "not" of Exercise **12** means that the reader is asked to construct a Brouwerian counterexample.

Bishop offers no clues on how to prove the positive results. They are not too difficult, however. To prove **11**, let $\epsilon > 0$ be given and choose n so large that, for all $x, y \in [0, 1]$,

$$|x - y| < \frac{1}{n} \Rightarrow |f(x) - f(y)| < \frac{\epsilon}{2},$$

which is possible because "continuous" is taken to mean "uniformly continuous". Let $x_i = i/n$, for $i = 0, 1, \ldots, n$. One would like to argue that, as one enumerates the values $f(x_0), f(x_1), \ldots, f(x_n)$ in turn, starting at $f(x_0) = f(0) < 0$ and ending at $f(x_n) = f(1) > 0$, either one hits upon the value 0 or there are successive pairs $f(x_i) < 0 < f(x_{i+1})$ and, since

$$\begin{aligned}
\left|f(x_i) - f(x_{i+1})\right| &= \left|f(x_i) - f\left(\frac{x_i + x_{i+1}}{2}\right) + f\left(\frac{x_i + x_{i+1}}{2}\right) - f(x_{i+1})\right| \\
&\le \left|f(x_i) - f\left(\frac{x_i + x_{i+1}}{2}\right)\right| + \left|f\left(\frac{x_i + x_{i+1}}{2}\right) - f(x_{i+1})\right| \\
&< \frac{\epsilon}{2} + \frac{\epsilon}{2} = \epsilon,
\end{aligned}$$

that, say, $|f(x_i)| < \epsilon$. For,

$$-f(x_i) < f(x_{i+1}) - f(x_i) < \epsilon \Rightarrow -\epsilon < f(x_i)$$

and, of course, $f(x_i) < 0 < \epsilon \Rightarrow f(x_i) < \epsilon$.

The problem, of course, is the failure of trichotomy in Constructive Analysis. However, one does have a good substitute for trichotomy: for all x, y, z,

$$x < y \Rightarrow x < z \text{ or } z < y. \tag{4.3}$$

The reason this holds, briefly, is that trichotomy fails when two numbers are so close to one another that we can't tell them apart. But if x and y are separated and z is given, either it is less than x whence less than y, greater than x, or indistinguishable from x in which case it is less than y.[63]

Now, for each i, one has

$$0 < f(x_i) \text{ or } f(x_i) < \epsilon \quad \text{and} \quad -\epsilon < f(x_i) \text{ or } f(x_i) < 0.$$

By the distributive law this means one of the following holds

[63] A real proof based on the definition of a real number as a Cauchy convergent sequence of rationals can be found in most texts on Constructive Analysis.

- $0 < f(x_i)$ & $-\epsilon < f(x_i)$, i.e., $0 < f(x_i)$
- $0 < f(x_i)$ & $f(x_i) < 0$, which is impossible
- $f(x_i) < \epsilon$ & $-\epsilon < f(x_i)$, i.e., $|f(x_i)| < \epsilon$
- $f(x_i) < \epsilon$ & $f(x_i) < 0$, i.e., $f(x_i) < 0$.

Thus, for each i,

$$f(x_i) < 0 \text{ or } f(x_i) > 0 \text{ or } |f(x_i)| < \epsilon,$$

i.e.,

$$\big(f(x_i) < 0 \text{ or } f(x_i) > 0\big) \text{ or } |f(x_i)| < \epsilon.$$

If we now go through the list $f(x_0), f(x_1), \ldots, f(x_n)$, we either find that $|f(x_i)| < \epsilon$ for some i, or we know that for all such i, $f(x_i) < 0$ or $f(x_i) > 0$. In the former case we are finished and in the latter we can carry out the argument that worked when trichotomy held.

Having already constructed a counterexample to the Intermediate Value Theorem, we can skip Exercise **12**. For Exercise **13**, find, for each subinterval $[x_i, x_{i+1}]$ an element y_i such that $f(y_i) \neq 0$. By the argument just given, we can find i and y_i such that $|f(y_i)| < \epsilon$. Call such an element a_0. Now repeat the argument on $[x_i, x_{i+1}]$ split into smaller pieces to find a_1 such that $|f(a_1)| < \epsilon/2$. Continuing in this manner, one constructs a Cauchy convergent sequence $a_0, a_1, a_2, \ldots$ for which $f(\lim_{i\to\infty} a_i) = \lim_{i\to\infty} f(a_i) = 0$. I leave the details as an exercise for the reader.

I suppose I should mention that Bishop adds two exercises to show that the result of Exercise **13** is not a pseudogenerality: the result applies to polynomials

$$P(x) = a_0 + a_1 x + \ldots + a_n x^n$$

for $n \geq 1$ provided $a_n \neq 0$.

The constructive treatment of the Extreme Value Theorem is a matter of greater subtlety. If ε is one of Brouwer's numbers based on the digits of π, so that one cannot tell whether it is $= 0$, > 0, or < 0, then the set $\{0, \varepsilon\}$ has neither a maximum nor a minimum. It does, however, have a least upper bound and a greatest lower bound. If ε is presented as the limit of a sequence $e_0, e_1, e_2, \ldots$ of rationals, the least upper bound is given by the limit of a sequence $a_0, a_1, a_2, \ldots$ where, for each n, $a_n = \max\{0, e_n\}$ and the greatest lower bound is given by $b_0, b_1, b_2, \ldots$ where, for each n, $b_n = \min\{0, e_n\}$.[64] More generally, it is shown in Constructive Analysis that a continuous (i.e., uniformly continuous) function on a closed, bounded interval $[a, b]$ with $a < b$ has a least upper bound B and a greatest lower bound A, and that for any $\epsilon > 0$ there are x, y such that $|f(x)| > B - \epsilon$ and $|f(y)| < A + \epsilon$. And

[64] The maximum and minimum exist in these cases because we can decide the ordering of rational numbers.

this is true even when maximum and minimum values cannot be found. I leave it to the reader to verify this himself or herself either by proving this or by consulting the literature.[65]

We are now finally coming to constructive versions of the Mean Value Theorem, or, perhaps, I should say constructive alternatives to the Mean Value Theorem. Following Bishop's desiderata for definitions, he ignores the "irrelevant" notion of pointwise differentiability and defines "differentiability" to be uniform differentiability:

Definition 13 Let f and g be continuous functions on a compact proper interval[66] I such that for each $\epsilon > 0$ there exists $\delta(\epsilon) > 0$ with

$$\left|f(y) - f(x) - g(x)(y - x)\right| \leq \epsilon|y - x| \quad \left(x, y \in I, |y - x| \leq \delta(\epsilon)\right)$$

Then f is said to be *differentiable* on I, g is called a *derivative* of f on I, and δ is called a *modulus of differentiability* for f on I. If f and g are continuous on the proper interval[67] J, then g is a *derivative* of f on J if it is a derivative of f on every compact proper subinterval I of J, and f is said to be *differentiable* on J.[68]

Bishop's definition of differentiability is essentially the Lagrange property (Definition 3.3.12 of page 258), with uniformity invoked and the continuity of both f and its derivative explicitly assumed.

Given that uniform differentiability is assumed, one might expect Bishop to give a proof of the Mean Value Theorem along the old Ampère–Cauchy lines. Instead, he opts for a constructive variant and proof of Rolle's Theorem along Lagrangian lines and a reduction of the Mean Value Theorem to it via a variant of the usual auxiliary function.

Lemma 6 *Let f be* differentiable on the interval $[a, b]$, and let $f(a) = f(b)$. Then for each $\epsilon > 0$ there exists x in $[a, b]$ with

(5.7) $$\left|f'(x)\right| \leq \epsilon$$

Proof Let ω be the modulus of continuity of f' on $[a, b]$ and δ the modulus of differentiability of f on $[a, b]$. Choose real numbers

$$a = x_0 < x_1 < \cdots < x_n = b$$

for which

$$x_{i+1} - x_i \leq \min\left\{\delta\left(\frac{\epsilon}{2}\right), \omega\left(\frac{\epsilon}{2}\right)\right\} \quad (0 \leq i \leq n - 1)$$

Then for $0 \leq i \leq n - 1$, we have

[65] Bishop, *op. cit.*, pp. 34–35 for the positive results, p. 59, Exercise **9** for the Brouwerian counterexample. Another good source is Troelstra and van Dalen, *op. cit.*, pp. 292–300 for the discussions of the Intermediate, Extreme, and Mean Value Theorems, the Extreme Value Theorem occupying pp. 294–295.

[66] A compact proper interval is one of the form $[a, b]$ with $a < b$.

[67] I.e., J is allowed to be open or half-open.

[68] Bishop, *op. cit.*, p. 40. Formula (5.1) has a typo: $f(x$ should read $f(x)$. Also, Bishop's typesetter seems to eschew periods.

$$f(x_{i+1}) - f(x_i) = f'(x_i)(x_{i+1} - x_i) + f(x_{i+1}) - f(x_i) - f'(x_i)(x_{i+1} - x_i)$$
$$\leq \left(f'(x_i) + \frac{\epsilon}{2}\right)(x_{i+1} - x_i) < \left(f'(x_i) + \epsilon\right)(x_{i+1} - x_i)$$

Therefore

$$0 = f(b) - f(a) = \sum_{i=0}^{n-1} f(x_{i+1}) - f(x_i) < \sum_{i=0}^{n-1} \left(f'(x_i) + \epsilon\right)(x_{i+1} - x_i)$$

By Proposition 7[69] it follows that $f'(x_i) > -\epsilon$ for at least one value of i, say for $i = j$. Similarly, $f'(x_i) < \epsilon$ for at least one value of i, say for $i = k$.
By the Corollary to Proposition 7[70] either (5.7) is valid or

(5.8) $$\left|f'(x_i)\right| > \frac{\epsilon}{2} \quad (0 \leq i \leq n - 1)$$

We may therefore assume (5.8). Since $\left|f'(x_{i+1}) - f'(x_i)\right| \leq \epsilon/2$, the quantities $f'(x_i)$ and $f'(x_{i+1})$ are either both positive or both negative. It follows that the quantities $f'(x_i)$ $(0 \leq i \leq n - 1)$ are either all positive or all negative. In the former case, $0 < f'(x_k) < \epsilon$, so that (5.7) holds with $x \equiv x_k$; and in the latter case, $-\epsilon < f'(x_j) < 0$, so that (5.7) holds with $x \equiv x_j$.
Rolle's theorem implies the *Law of the Mean*, which gives a basic estimate for the difference of two values of a differentiable function.
Theorem 7 *Let f be differentiable on the interval $[a, b]$. Then for each $\epsilon > 0$ there exists x in $[a, b]$ with*

(5.9) $$\left|f(b) - f(a) - f'(x)(b - a)\right| \leq \epsilon$$[71]

As I said, Bishop reduces his Theorem 7 to Rolle's Theorem via the usual trick. I leave the details to the reader as an exercise.

Classically, this formulation quickly yields the Mean Value Theorem for continuously differentiable functions. For, let $\epsilon_0, \epsilon_1, \epsilon_2, \ldots$ tend to 0 and choose $x_i \in [a, b]$ such that (5.9) holds for x_i and ϵ_i. By the Bolzano-Weierstrass Theorem, some subsequence $x_{i_0}, x_{i_1}, \ldots$ converges to a limit $x \in [a, b]$. But

$$\left|f(b) - f(a) - f'(x)(b - a)\right| = \lim_{n\to\infty} \left|f(b) - f(a) - f'(x_{i_n})(b - a)\right|$$
$$\leq \lim_{n\to\infty} \epsilon_{i_n} = 0.$$

Constructively, more work needs to be done before one can conclude

$$\frac{f(b) - f(a)}{b - a} = f'(c)$$

for some c. Bishop does not carry out such a task. The result is strong enough to obtain the Mean Value Inequality, which suffices for most applications.

---

[69]Proposition 7 asserted that, if $x_0 + x_1 + \ldots x_{n-1} > 0$, then some $x_i > 0$.

[70]I.e., by (4.3).

[71]*Ibid*, pp. 43–44. Troelstra and van Dalen, in the work cited in footnote 65, offer a slightly different proof of Rolle's Theorem and skip the Mean Value Theorem proceeding directly to the Higher Order Mean Value Theorem, which is also Bishop's next goal.

This brings my simplified account of Constructive Analysis to a close. Its relevance to the present chapter has been its bearing on the bemoaned elusiveness of c, offering some support for those who make the complaint and some refutation of their complaints at the same time. Constructive Analysis, however, offers more to the debate insofar as one proposal for a reform of the Calculus is based on the constructive experience:

1999

Mark Bridger and Gabriel Stolzenberg, "Uniform Calculus and the Law of Bounded Change", *The American Mathematical Monthly* 106 (1999), pp. 628 – 635.

Bridger and Stolzenberg lead off referring to the exchange between Tucker and Swann, announcing their programme as follows:

> ...Tucker and Swann work with *pointwise* continuity and differentiability, weak notions that make proving statements like the increasing function theorem more difficult. On closed finite intervals, uniform continuity and differentiability are as easy to verify, and using them as starting points permits a natural development of the calculus in which such difficulties do not arise.[72]

Bridger and Stolzenberg call their programme the *Uniform Calculus* and their paper offers a readable outline of important steps in the development in a manner of great simplicity and elegance. They opt for proving the Mean Value Inequality as a replacement for the Mean Value Theorem.

> **7 The Law of Bounded Change**
>
> THEOREM 7.1. *If f is uniformly differentiable and $A \leq f' \leq B$ on $[a, b]$, then $A(b-a) \leq f(b) - f(a) \leq B(b-a)$.*
>
> This is the law of bounded change. It says that bounds for the derivative are bounds for the difference quotient. Notice that the increasing function theorem is just the law of bounded change for $A = 0$ (and we don't care about B) and the law of bounded change is the increasing function theorem applied to the functions $Bx - f(x)$ and $f(x) - Ax$.
>
> PROOF. It suffices to prove that for all $\epsilon > 0$, the conclusion holds with A and B replaced by $A - \epsilon$ and $B + \epsilon$. The justification for this is the general truth that if $p < q + \epsilon$ for all $\epsilon > 0$, then $p \leq q$. That this holds for reals follows by rational approximation from the fact that it holds for the rationals.
>
> Since $F(u, v) \to f'(u)$[73] as $v \to u$, for each $\epsilon > 0$ there is a $\delta > 0$ such that $f'(u) - \epsilon < F(u, v) < f'(u) + \epsilon$ for $0 \leq v - u < \delta$. But $A \leq f'(u) \leq B$, so $f(v) - f(u) = F(u, v)(v - u)$ lies between $(A - \epsilon)(v - u)$ and $(B + \epsilon)(v - u)$.
>
> Hence, if we express $f(b) - f(a)$ as a telescoping sum of n differences $f(u_i) - f(u_{i-1})$, where $u_0 = a$ and each $u_i - u_{i-1} = (b - a)/n < \delta$, we have that $(A - \epsilon)(b - a) \leq f(b) - f(a) \leq (B + \epsilon)(b - a)$ □[74]

They follow this up with a number of corollaries. Corollary 7.2 is the Constant Function Theorem, given the proof, "This is just the law of bounded change with A and B equal to 0". Of particular interest is

---

[72]Bridger and Stolzenberg, p. 1 of an online version of their paper.

[73]$F(u, v)$ is the difference quotient $\big(f(u) - f(v)\big)/(u - v)$.

[74]Bridger and Stolzenberg, pp. 4–5 in the online version.

Corollary 7.7. *(Generalized Law of Bounded Change) If* $Ag' \leq f' \leq Bg'$ *on* $[a, b]$, *then* $A[g(b) - g(a)] \leq f(b) - f(a) \leq B[g(b) - g(a)]$.[75]

This is just the Cauchy Mean Value Inequality and they apply it to prove L'Hôpital's Rule.

They conclude their paper with the comment

> We believe that this development, which is in the constructivist manner of Errett Bishop and L.E.J. Brouwer, produces proofs that are shorter and more transparent than those encountered in classical treatments. The idea of working with uniform rather than pointwise notions is a hallmark of the constructivist tradition.[76]

This idea is also a hallmark of the early approach to the subject as we saw with Lagrange, Ampère, and Cauchy. Rigorous non-uniform proofs were slow in coming — the earliest non-uniform proof of the Mean Value Theorem I found, for example, was that in Weierstrass's 1861 lectures, coming after Bolzano's unpublished failed attempt of the 1830s. Undoubtedly Bridger and Stolzenberg have a point and their proposal should be taken seriously.

This proposal requires an exposition, and they promised such in a book, *A New Course in Analysis*. Parts of this book were written up — I found a couple of chapters online —, but the book itself was never completed. Stolzenberg's main interests went in another direction, thus leaving Bridger to write the following book alone:

2006

> Mark Bridger, *Real Analysis: A Constructive Approach*, John Wiley & Sons, New York, 2006.

This is the ultimate version of their programme of Uniform Calculus.

With their mention of the Tucker/Swann debate, Bridger and Stolzenberg give the impression that they are proposing the Uniform Calculus as an approach to first-year Calculus. Indeed, the proofs given in their paper could readily fit into such a course and in such a manner that should please Swann and be almost acceptable to Tucker. The presentation, for example, of the proof of the Fundamental Theorem, which depends on uniform continuity, suddenly becomes feasible. And the nicely motivated Ampère–Cauchy approach to establishing the Mean Value Inequality becomes accessible to students at that level.

However, Bridger's textbook, as the title doubly implies, is not a textbook in the Calculus, but instead is a textbook for the more advanced Real Analysis course — and, in fact, for a constructive version of this course. It establishes constructive versions of theorems, like the Law of Bounded Change, but not their classical counterparts like the Mean Value Theorem. I would think that this would play to the fears of those mathematicians who, following Hilbert, believe the goal of constructive mathematics is to throw overboard all their favourite results. Perhaps I am behind the times, but I recall a lecture *c.* 1970 given by Errett Bishop to a roomful of analysts. Their reception may have grown more hostile in my memory over the years, but I can say

---

[75] *Ibid*, p. 6.

[76] *Ibid*, p. 8.

with certainty that it was decidedly unenthusiastic. Of course, Bishop's insistence on attacking nonconstructive reasoning instead of taking a positive approach by saying that a great deal of classical mathematics can be made constructive or by emphasising the benefits of constructive proofs did nothing to convince the analysts that his work was meaningful. But I think one can also lay some blame for the coolness of their response on some residual hostility to the constructive enterprise that Hilbert had generated earlier in the 1920s in his supposed battle with Brouwer over the latter's intuitionistic mathematics. The battle was not the epic *Kampf um Dasein* portrayed by Hilbert, but it got a lot of press so to speak.[77] Anyway, constructive and classical mathematics can co-exist peacefully alongside one another. Indeed, they complement one another: classical proofs can be shorter and simpler to state, while constructive proofs, being more direct, may appear less magical and often give more information.

For the problem at hand — Calculus reform — Bridger's work is not the solution, but neither the more advanced theoretical nature nor the blatant constructivity of the work rules out its relevance to the Calculus Reform discussion. The basic idea of using uniform concepts to simplify the theory as it would be presented in the first-year course is a solid suggestion.

* * *

I am obviously not about to offer a final solution to the problem of Calculus Reform here. Indeed, my interest is narrower — the attacks on the Mean Value Theorem by the reformers. With respect to this, I should finish up by noting that uniform calculus *per se* has nothing against the Mean Value Theorem. Indeed, the original proofs of the Theorem made heavy use of uniformity. Bridger and Stolzenberg prefer the Law of Bounded Change to the Mean Value Theorem because they are constructivists: the Mean Value Theorem (like the Intermediate and Extreme Value Theorems) is simply not constructively valid. Thus they cannot use it. Of all the objections to the Mean Value Theorem considered in this chapter, this, for me, is the only one that has any weight to it — and that only for constructive mathematics.

---

[77] An unsensationalised, factual account of the Hilbert/Brouwer controversies can be found in Chap. 14 of D. van Dalen, *L.E.J. Brouwer—Topologist, Intuitionist, Philosopher: How Mathematics is Rooted in Life*, Springer, London, 2013. A more technical account of the foundational dispute presented paper-by-paper is C. Smoryński, "Hilbert's programme", *CWI Quarterly* vol. 1, no. 4 (1988), pp. 3–59; reprinted in: Eckart Menzler-Trott (Craig Smoryński and Edward Griffor, trans.), *Logic's Lost Genius: The Life of Gerhardt Gentzen*, American Mathematical Society, Providence (Rhode Island), 2007.

Appendix A
Short Bibliography

As bibliographic information is given only at the first mention of any source in the main text I list here those items referred to in later footnotes as "*op. cit.*" as a convenience for the reader.

References

1. André Marie Ampère, "Recherches sur quelques points de la théorie des fonctions dérivées qui conduisent à une nouvelle démonstration de la série de *Taylor*, et à l'expression finie des termes qu'on néglige lorsqu'on arrête cette série à un terme quelconque", *Journal de l'École Polytechnique* series 13, volume 6 (1806), pp. 148–181.
2. Kirsti Andersen, "Cavalieri's method of indivisibles", *Archive for History of Exact Sciences* 31 (1985), pp. 291–367.
3. Tom M. Apostol, Hubert E. Chrestenson, C. Stanley Ogilvy, Donald E. Richmond, and N. James Schoonmaker (eds.), *Selected Papers on Calculus*, The Mathematical Association of America, 1969.
4. Margaret E. Baron, *The Origins of the Infinitesimal Calculus*, Pergamon Press, Oxford, 1969.
5. June Barrow-Green, "From cascades to calculus: Rolle's theorem", in: Robson and Stedall, *op. cit.*
6. George Berkeley, *The Analyst; or, a Discourse Addressed to an Infidel Mathematician*, London, 1734.
7. Errett Bishop, *Foundations of Constructive Analysis*, McGraw-Hill Book Company, New York, 1967.
8. Umberto Bottazzini (Warren van Egmond (trans.)), *The* Higher Calculus*: A History of Real and Complex Analysis from Euler to Weierstrass*, Springer-Verlag, New York, 1986.
9. Carl B. Boyer, *History of Analytic Geometry*, The Scholars Bookshelf, Princeton Junction (NJ), 1988.
10. Robert E. Bradley and C. Edward Sandifer (eds. and trans.), *Cauchy's* Cours d'Analyse*; An Annotated Translation*, Springer Science + Business Media, LLC, New York, 2009.
11. Florian Cajori, *A History of the Conceptions of Limits and Fluxions in Great Britain from Newton to Woodhouse*, The Open Court Publishing Company, Chicago and London, 1919.
12. Dennis M. Cates, *Cauchy's Calcul Infinitesimal; A Complete English Translation*, Fairview Academic Press, Walnut Creek (California), 2012.
13. Augustin Louis Cauchy, *Cours d'analyse de l'École Royale Polytechnique; Ire Partie. Analyse algébrique*, de Bure, Paris, 1821.

C. Smoryński, *MVT: A Most Valuable Theorem*,
DOI 10.1007/978-3-319-52956-1

14. Augustin Louis Cauchy, *Résumé des leçons données a l'École Royale Polytechnique sur le calcul infinitesimal*, de Bure, Paris, 1823.
15. J.M. Child (ed. and trans.), *The Early Mathematical Manuscripts of Leibniz*, The Open Court Publishing Company, Chicago and London, 1920.
16. Marshall Clagett, ed., *The Science of Mechanics in the Middle Ages*, The University of Wisconsin Press, Madison, 1959.
17. Gaston Darboux, "Mémoire sur les fonctions discontinues", *Annales scientifiques de l'École Normale Supérieure*, 2nd series, vol. 4 (1875), pp. 57–112.
18. Bibhutibhushan Datta and Avadesh Narayan Singh (Kripa Shankar Shukla, reviser), "Use of Calculus in Hindu mathematics", *Indian Journal of History of Science* 19, No. 2 (1984), pp. 95–109.
19. René Descartes (David Eugene Smith and Marcia L. Latham, trans.), *The Geometry of René Descartes*, Dover Publications, Inc., New York, 1954.
20. P.P. Divakaran, "Notes on *Yuktibhāṣā*: the birth of calculus", to appear.
21. Joseph Edward, *Differential Calculus with Applications and Numerous Examples: An Elementary Treatise*, Macmillan and Co., London, 1886.
22. C.H. Edwards, Jr., *The Historical Development of the Calculus*, Springer-Verlag, New York, 1979.
23. Bernard R. Gelbaum and John M.H. Olmsted, *Counterexamples in Analysis*, Holden-Day, Inc., San Francisco, 1964.
24. Judith V. Grabiner, "The changing concept of change: the derivative from Fermat to Weierstrass", *Mathematics Magazine* 56 (1983), pp. 195–206.
25. Judith V. Grabiner, *The Origins of Cauchy's Rigorous Calculus*, The MIT Press, Cambridge (Mass.), 1981.
26. Johann Philipp Grüson (trans.), *Lagrange's Theorie der analytischen Funktionen*, F.T. Lagarde, Berlin, 1798.
27. Hans Hahn, "Die Krise der Anschauung", in: *Krise und Neuaufbau in den exakten Wissenschaften*, F. Deuticke, Leipzig and Vienna, 1933.
28. Thomas Little Heath, *The Thirteen Books of Euclid's Elements*, 3 volumes, Cambridge University Press, Cambridge, 1908.
29. T.L. Heath, *The Works of Archimedes Edited in Modern Notation with Introductory Chapters by T.L. Heath with a Supplement The Method of Archimedes Recently Discovered by Heiberg*, Dover Publications, Inc., New York, no date given.
30. Eduard Heine, "Die Elemente der Functionenlehre", *Journal für die reine und angewandte Mathematik* 74 (1872), pp. 172–188.
31. Marquis de L'Hôpital, *Analyse de infiniment petits*, Paris, 1768.
32. Vojtěch Jarník, "Bolzano's 'Functionenlehre' ", in: Vojtěch Jarník, *Bolzano and the Foundations of Mathematical Analysis*, Society of Czechoslovak Mathematicians and Physicists, Prague, 1981.
33. Hubert C. Kennedy, *Peano, Life and Works of Giuseppe Peano*, D. Reidel Publishing Company, Dordrecht, 1980.
34. Hubert C. Kennedy (ed. and trans.), *Selected Works of Giuseppe Peano*, George Allen & Unwin Ltd, London, 1973.
35. Serge Lang, *Complex Analysis*, Addison-Wesley Publishing Company, Reading (Mass.), 1977.
36. H. Laurent, *Traité d'analyse*, Gauthier-Villars, Paris, 1885
37. Jean Mawhin, "Some contributions of Peano to analysis in the light of the work of Belgian mathematicians", in: Fulvia Skof (ed.), *Giuseppe Peano between Mathematics and Logic*, Springer-Verlag Italia, Milan, 2011.
38. Augustus de Morgan, *The Differential and Integral Calculus*, Society for the Diffusion of Useful Knowledge, London, 1836–1842.
39. Isaac Newton (John Colson ed. and trans.), *The Method of Fluxions and Infinite Series with its Application to the Geometry of CURVE-LINES*, 1736.
40. K. Ramasubramanian and M.D Srinivas, "Development of Calculus in India", in: C.S. Seshadri (ed.), *Studies in the History of Indian Mathematics*, Hindustan Book Agency, New Delhi, 2010.

41. Eleanor Robson and Jacqueline Stedall (eds.), *The Oxford Handbook of the History of Mathematics*, Oxford University Press, Oxford, 2009.
42. Steve Russ (ed.), *The Mathematical Works of Bernard Bolzano*, Oxford University Press, Oxford, 2004.
43. Bertrand Russell, *Principles of Mathematics*, 2nd ed., W.W. Norton & Company, Inc., New York, no date given.
44. Bimala Prasada Sidhānta Sarasvatī (trans.) and Danavir Goswami (ed.), *The Sūrya-siddhānta and Siddhānta-śiroma?i*, Gopsons Papers Ltd., Noida, 2007.
45. P.C. Sengupta, "Infinitesimal calculus in Indian mathematics — its origins and development", *Journal of the Department of Letters (Calcutta University)* 22 (1932), pp. 1–17.
46. David Eugene Smith (ed.), *A Source Book in Mathematics*, 1929; reprinted: Dover Publications, Inc., New York, 1959.
47. Craig Smoryński, *Adventures in Formalism*, College Publications, London, 2012.
48. Craig Smoryński, *A Treatise on the Binomial Theorem*, College Publications, London, 2012.
49. Murray R. Spiegel, "Mean value theorems and Taylor series", in: Apostol *et al.*, *op. cit.*
50. Jacqueline Stedall (ed.), *Mathematics Emerging: A Sourcebook 1540 – 1900*, Oxford University Press, Oxford, 2008.
51. John Stewart, *Sir Isaac Newton's Two Treatises of the Quadrature of Curves, and Analysis by Equations of an infinite Number of Terms, explained*, London, 1745.
52. Dirk Struik (ed.), *A Source Book in Mathematics, 1200 – 1800*, Harvard University Press, Cambridge (Mass.), 1969.
53. A.S. Troelstra and D. van Dalen, *Constructivism in Mathematics; An Introduction*, vol. I, North-Holland Publishing Company, Amsterdam, 1988
54. Klaus Volkert, *Geschichte der Analysis*, Bibliographisches Institut & F.A. Brockhaus AG, Zürich, 1988.
55. Karl Weierstrass and Hermann Amandus Schwarz, *Differential Rechnung, nach einer Vorlesung des Herrn Professor Weierstrass in Sommersemester 1861*, Hdschr. Koll. N 37 (Humboldt Universität zu Berlin).
56. Lancelot Wilkinson and Pandit Bápú Deva Śāstrí, *Hindu Astronomy II. The Siddhánta Śiromani*, Calcutta, 1861.

Index

C. Smoryński, *MVT: A Most Valuable Theorem*,
DOI 10.1007/978-3-319-52956-1

Y

Z

CPSIA information can be obtained
at www.ICGtesting.com
Printed in the USA
LVOW13s1617130417
530744LV00006B/77/P

9 783319 52955